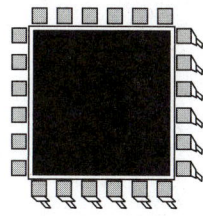

Introduction to Embedded Microcomputer Systems:

Motorola 6811 and 6812 Simulation

andcc	8-bit logical and to RegCC	lbne	long branch if result is nonzero
bgnd	enter background debug mode	lbpl	long branch if result is positive
call	subroutine in expanded memory	lbra	long branch always
dbeq	decrement and branch if result=0	lbrn	long branch never
dbne	decrement and branch if result≠0	lbvc	long branch if overflow clear
ediv	RegY=(Y:D)/RegX, unsigned divide	lbvs	long branch if overflow set
edivs	RegY=(Y:D)/RegX, signed divide	leas	16-bit load effective addr to SP
emacs	16 by 16 signed mult, 32-bit add	leax	16-bit load effective addr to X
emaxd	16-bit unsigned maximum in RegD	leay	16-bit load effective addr to Y
emaxm	16-bit unsigned maximum in memory	maxa	8-bit unsigned maximum in RegA
emind	16-bit unsigned minimum in RegD	maxm	8-bit unsigned maximum in memory
eminm	16-bit unsigned minimum in memory	mem	determine the membership grade
emul	RegY:D=RegY*RegD unsigned mult	mina	8-bit unsigned minimum in RegA
emuls	RegY:D=RegY*RegD signed mult	minm	8-bit unsigned minimum in memory
etbl	16-bit look up and interpolation	movb	8-bit move memory to memory
exg	exchange register contents	movw	16-bit move memory to memory
ibeq	increment and branch if result=0	orcc	8-bit logical or to RegCC
ibne	increment and branch if result≠0	pshc	push 8-bit RegCC onto stack
idivs	16-bit by 16-bit signed divide	pshd	push 16-bit RegD onto stack
lbcc	long branch if carry clear	pulc	pop 8 bits off stack into RegCC
lbcs	long branch if carry set	puld	pop 16 bits off stack into RegD
lbeq	long branch if result is zero	rev	Fuzzy logic rule evaluation
lbge	long branch if signed ≥	revw	weighted Fuzzy rule evaluation
lbgt	long branch if signed >	rtc	return sub in expanded memory
lbhi	long branch if unsigned >	sex	sign extend 8-bit to 16-bit reg
lbhs	long branch if unsigned ≥	tbeq	test and branch if result=0
lble	long branch if signed ≤	tbl	8-bit look up and interpolation
lblo	long branch if unsigned <	tbne	test and branch if result≠0
lbls	long branch if unsigned ≤	tfr	transfer register to register
lblt	long branch if signed <	trap	illegal instruction interrupt
lbmi	long branch if result is negative	wav	weighted Fuzzy logic average

Motorola 6812 assembly instructions (in addition to the 6811)

example	addressing mode	Effective Address
ldaa #u	immediate	EA is 8-bit address (0 to 255)
ldaa u	direct	EA is 8-bit address (0 to 255)
ldaa U	extended	EA is a 16-bit address
ldaa m,r	8-bit index	EA=r+m (0 to 255)

Motorola 6811 addressing modes

example	addressing mode	Effective Address
ldaa m,r	5-bit index	EA=r+m (-16 to 15)
ldaa v,+r	pre-increment	r=r+v, EA=r (1 to 8)
ldaa v,-r	pre-decrement	r=r-v, EA=r (1 to 8)
ldaa v,r+	post-increment	EA=r, r=r+v (1 to 8)
ldaa v,r-	post-decrement	EA=r, r=r-v (1 to 8)
ldaa A,r	Reg A offset	EA=r+A, zero padded
ldaa B,r	Reg B offset	EA=r+B, zero padded
ldaa D,r	Reg D offset	EA=r+D
ldaa q,r	9-bit index	EA=r+q (-256 to 255)
ldaa W,r	16-bit index	EA=r+W (-32768 to 65535)
ldaa [D,r]	D indirect	EA={r+D}
ldaa [W,r]	indirect	EA={r+W} (-32768 to 65535)

Motorola 6812 addressing modes (in addition to the 6811)

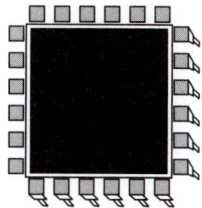

Introduction to Embedded Microcomputer Systems:

Motorola 6811 and 6812 Simulation

Jonathan W. Valvano

University of Texas at Austin

THOMSON

BROOKS/COLE

Australia • Canada • Mexico • Singapore • Spain •
United Kingdom • United States

THOMSON

BROOKS/COLE

Publisher: *Bill Stenquist*
Editorial Coordinator: *Valerie Boyajian*
Technology Project Manager: *Burke Taft*
Marketing Manager: *Tom Ziolkowski*
Marketing Assistant: *Darcie Pool*
Project Manager, Editorial-Production: *Mary Vezilich*
Print/Media Buyer: *Vena Dyer*

Permissions Editor: *Karyn Morrison*
Production Service: *RPK Editorial*
Cover Designer: *Laurie Albrecht*
Cover Image: *Todd Daman*
Compositor: *Carlisle Communications, Ltd.*
Printer: *Phoenix Color Corporation*

Printed in the United States of America
1 2 3 4 5 6 7 06 05 04 03 02

For more information about our products, contact us at:
Thomson Learning Academic Resource Center
1-800-423-0563

For permission to use material from this text,
contact us by: **Phone:** 1-800-730-2214
Fax: 1-800-730-2215
Web: http://www.thomsonrights.com

Motorola 6811 and 6812 Instruction Sets by permission of Motorola.

Library of Congress Control Number: 2002107445

ISBN 0-534-39177-x

Brooks/Cole–Thomson Learning
511 Forest Lodge Road
Pacific Grove, CA 93950
USA

Asia
Thomson Learning
5 Shenton Way #01-01
UIC Building
Singapore 068808

Australia
Nelson Thomson Learning
102 Dodds Street
South Melbourne, Victoria 3205
Australia

Canada
Nelson Thomson Learning
1120 Birchmount Road
Toronto, Ontario M1K 5G4
Canada

Europe/Middle East/Africa
Thomson Learning
High Holborn House
50/51 Bedford Row
London WC1R 4LR
United Kingdom

Latin America
Thomson Learning
Seneca, 53
Colonia Polanco
11560 Mexico D.F.
Mexico

Spain
Paraninfo Thomson Learning
Calle/Magallanes, 25
28015 Madrid, Spain

Contents

4

Assembly Language Programming   142

5

I/O Programming   176

6

Microcomputer Interfacing   215

10
Elementary Data Structures   378

11
Interrupt Synchronization   422

Appendix 1. Embedded System Development Using TExaS   481

Appendix 2. Glossary of Terms   505

Appendix 3. Solutions Manual   519

Index   549

Preface

Embedded computer systems are electronic systems that include a microcomputer to perform a specific dedicated application. The computer is hidden inside these products. Embedded systems are ubiquitous. Every week millions of tiny computer chips come pouring out of factories like Motorola, Intel, Philips, and Mitsubishi, finding their way into our everyday products. Our global economy, our production of food, our transportation systems, our military defense, our communication systems, and even our quality of life depend on the efficiency and effectiveness of these embedded systems. Engineers play a major role in all phases of this effort: planning, design, analysis, manufacturing, and marketing. This book provides an introduction to embedded systems, including both hardware interfacing and software fundamentals.

Although real applications require actual physical devices to perform desired operations, the approach taken in this book is to develop hardware and software using simulation. The simulator that accompanies this book is called *Test EXecute And Simulate* (**TExaS**). This simulator, like all good applications, has an easy learning curve. Although a professional engineer could use **TExaS** to develop actual embedded systems, its intended use is as an educational tool. It provides a self-contained approach to writing and testing microcomputer hardware and software. It is unique from other simulators in two aspects. If enabled, the simulator shows activity internal to the chip like the read/write address/data bus, the instruction register and effective address register. In this way, you can use **TExaS** to learn how a computer works. In particular, you can experience the architecture by observing activity inside the microcomputer. On the other end of the spectrum, you have the ability to connect external hardware devices like switches, LEDs, LCDs, keyboards, serial port devices, motors, and analog circuits. You can use logic probes, voltmeters, oscilloscopes and logic analyzers to observe the external hardware. The simulator supports most of the I/O port functions of the Motorola 6811 and 6812 microcomputers, like interrupts, serial port, input capture, output compare, key wakeup, STRA, timer overflow, real-time interrupt, and the ADC. With these features, you can use **TExaS** to learn microcomputer programming and interfacing. You can develop software either in assembly language using the **TExaS** assembler, or program in C using a cross-compiler. The freeware ICC11 compiler is included on the CD that accompanies this book. It can be found in the ICC11 folder as part of the **TExaS** installation. **TExaS** can import object code from most cross-compilers for the 6811 or 6812. So, combined with a commercial-grade cross-compiler, you will have a powerful hardware/software development system.

P.1 Objectives of the Book

The overall objective of this book is to present basic computer architecture, teach assembly language programming, and present an introduction to interfacing. This book develops these topics around the **TExaS** simulator. Although **TExaS** can be used to develop hardware/software embedded systems on the Motorola 6805, 6808, 6811 or 6812, this book focuses on just the 6811 and 6812. The book describes both the general processes and the specific details involved in microcomputer simulation. In particular, detailed case studies are used to illustrate fundamental concepts, and laboratory assignments are provided. The specific objectives of this book include the understanding of

- the basic procedures involved in hardware/software simulation,
- how information is represented on the computer,
- the basic arithmetic and logical operations performed by the computer,
- the fundamental architecture of the 6811 and 6812 microcomputers,
- the input/output operations,
- assembly language programming, considering both function and style,
- simple hardware interfaces, including switches, keyboards, LEDs, LCDs, DC motors, DACs, ADCs, and serial ports,
- debugging techniques breakpoints, ScanPoints, profiles, monitors, voltmeters, oscilloscopes, and logic analyzers,
- program structures with a comparison between assembly and C,
- modular programming,
- elementary data structures,
- interrupt programming.

This book does not discuss in detail every 6811/6812 instruction, but rather presents some of the instructions and uses them to discuss the general issues of representation of information, computer architecture, developing embedded system software. In contrast, the Motorola programming reference guides do give details of each assembly instruction. In a similar manner, the Motorola microcomputer technical reference manuals explain all the I/O port functions. In other words, you will use this book along with the manuals from Motorola. Two short manuals are included as appendices and many of the Motorola manuals are available as pdf documents on the CD. It might be better to order physical documents from Motorola's literature center, or to download the latest version from the Motorola web site.

P.2 Prerequisites

This book is intended for an introductory laboratory course in microcomputer programming and/or microcomputer interfacing. It is assumed the student has some knowledge of C programming, and digital logic and analog circuits. In fact, simple C programs are used throughout the book to assist the development of assembly language programs. Fortunately, located on the CD that accompanies this book, there is a reference called *Developing Embedded Software in C using ICC11/ICC12* written as an HTML document. This reference is a complete C programming book, written specifically for embedded software on the 6811 and/or 6812. For more advanced information about microcomputer interfacing and programming, see *Embedded Microcomputer Systems: Real Time Interfacing* by Jonathan W. Valvano, published by Brooks/Cole, copyright © 2000.

P.3 Special Features

This book incorporates a number of special features specifically designed for the beginning engineer. An effective educational approach is to learn by doing. The first *action component* of the book is the use of *checkpoints,* which can be found throughout the book. A checkpoint is a short question meant as an immediate feedback mechanism for the reader to evaluate his or her level of comprehension. Checkpoints should be performed while reading the chapter. Answers to checkpoints are given in Appendix 3. The second action component of the book is the *tutorial.* They are included at the end of each chapter. The purpose of the tutorial is to immediately reinforce the specific topics of that chapter. Each tutorial includes a sequence of actions (specific things for the reader to do) and a list of questions. Tutorials are meant to be performed without supervision, and should be performed after reading the chapter, but before attempting the labs or homework. Answers to the tutorial questions are also included in Appendix 3. The most important action components of the book are the *laboratory assignments* included at the end of the chapters. Each laboratory solution can be built and tested using the **TExaS** simulator. Only by performing the laboratory assignments can the reader truly assimilate the hardware and software concepts introduced in this book. Laboratories are meant to be performed under the supervision of an instructor, and involve the classic engineering processes of design, construction, debugging, and evaluation. *Homework problems* can also be found at the end of each chapter. These problems are less detailed, and are intended to evaluate the reader's understanding of specific topics introduced in the chapter.

Sections labeled with a * are advanced topics and can be skipped without loss of continuity. Chapter 9 contains a set of software style guidelines.

P.4 How to Teach a Course Based on This Book

The first step in the design of any course is to create a list of educational objectives. This book along with the two HTML documents on the CD could be used to teach introductory microcomputer programming and/or microcomputer interfacing. Specific educational objectives that are supported in this book are microcomputer architecture, number systems, assembly language programming, I/O device interfacing, subroutines, local variables, data structures, and interrupts. Chapters 1 through 8 could be used in this introductory assembly language programming class. This book can also be used for a more advanced course in microcomputer interfacing, using Chapters 6 through 11. In this more advanced class, assignments could be developed in assembly or in C. An *Instructor's Solutions Manual* (ISBN 0534-39178-8) is available free to adopters of this text.

The next important decision to make is the organization of the student laboratory. The importance of practical "hands on" experience is critical in the educational process. Unfortunately, space, staff, and money constraints force all of us to compromise, doing the best we can. On the other hand, the role of simulation is becoming increasingly important as the race for technological superiority is run with shorter and shorter design cycle times. Consequently, it is important to expose our students to the all phases of engineering design including problem specification, conceptualization, simulation, construction, and analysis. Universities that adopt this book will be allowed to download, rewrite, print out, and distribute the laboratory assignments presented in this book. Universities that adopt this book will also be allowed to install **TExaS** on university network computers.

The first laboratory configuration is based entirely on material included with this book, and involves no extra costs. Each book comes with a CD that allows the student to install the **TExaS** application on a single computer. Students, for the most part, work off campus and come to a TA station for help or lab grading. In this configuration you can either develop software in assembly using the **TExaS** assembler or develop C programs using the demo version of ICC11. The simulator itself becomes the platform on which the lab assignments are developed and tested. For examples of this ICC11/**TExaS** combination run some of the demonstration examples in the ICC11 subdirectory of the **TExaS** application. Some laboratories are intended to be developed in assembly language using the **TExaS** simulator, while others could be performed in either assembly or C.

The second configuration performs simulation of C programs. In this laboratory setup, a standard PC-compatible laboratory room can be used. **TExaS** and a professional cross-compiler (like ImageCraft ICC11 or ICC12) are installed, and the students perform laboratory assignments in this central facility. Other than the professional cross-compiler, no additional setup costs are required. The advantage of this approach is that professional-style C software can be developed with only a modest expense. Laboratories can be adapted to be used with C instead of assembly language.

A third laboratory configuration combines simulation with some real microcomputer experiments. Labs can be first simulated, then run on a real microcomputer. Alternatively, some labs can be simulated, while other labs are developed on a real microcomputer. Students can work off campus on the simulation aspects of the labs. In this configuration, you can either develop software in assembly using the **TExaS** assembler or develop C programs using a cross-compiler. Object files generated by **TExaS** can be programmed into a real microcomputer, and debugged using standard debugging systems. This is the more expensive than the other two configurations because actual microcomputer hardware and debugging systems are required.

P.5   What's on the CD?

1. The `Readme.exe` is a 15-minute introductory tutorial about developing assembly language programs on **TExaS**. This animation does not need to be copied to your hard drive; you can simply watch the movie by double-clicking the `Readme.exe` that is on the CD itself.
2. The `Readme2.exe` is a 10-minute introductory tutorial about developing C language programs on **TExaS**. Again, this animation does not need to be copied to your hard drive to be viewed.
3. **TExaS** is a complete editor, assembler, and simulator for the 6805, 6808, 6811 and 6812 microcomputers. It simulates external hardware, I/O ports, interrupts, memory, and program execution. It is intended as a learning tool for embedded systems. This software is not freeware, but the purchase of the book entitles the owner to install one copy of the program. The **TExaS** directory contains the installer, which you must execute before using the application. Once installed **TExaS** creates these five subdirectories
   - ■ `MC6811`   6811 assembly examples
   - ■ `MC6812`   6812 assembly examples
   - ■ `ICC11`   6811 C examples using the freeware compiler

- ■ ICC11A   6811 C examples using the professional ICC11 compiler
- ■ ICC12   6812 C examples using the professional ICC12 compiler

The subdirectory ICC11 also contains the ImageCraft Freeware compiler. You can run the existing ICC11A and ICC12 examples, but to edit and recompile you will need the commercial C compiler. For **TExaS** upgrades look at: **www.ece. utexas.edu/~valvano**.

4. The **PDF** directory contains many data sheets in Adobe's pdf format. This information does not need to be copied to your hard drive; you can simply read the data sheets from the CD itself. In particular there are data sheets for microcomputers, digital logic, memory chips, op amps, ADCs, DACs, timer chips and interface chips.

5. The **assembly** directory contains a short HTML document describing how to program in assembly for embedded systems using the **TExaS** application. This document does not need to be copied to your hard drive; you can simply read the HTML document from the CD itself. The **TExaS** application itself also contains a lot of information about assembly language development as part of its on-line help.

6. The **embed** directory contains an HTML document describing how to program in C for embedded systems using ImageCraft's ICC11/ICC12. This document does not need to be copied to your hard drive; you can simply read the HTML document from the CD itself.

7. The **tutorial** directory on the CD contains files needed to perform the tutorials in this book. To perform the tutorials, you will copy either the Mc6811T or Mc6812T subdirectory to a hard drive. The movies in this directory can be viewed directly from the CD itself.

P.6 Acknowledgments

Many shared experiences contributed to the development of this book. First I would like to acknowledge the many excellent teaching assistants I have had the pleasure of working with. Some of these hardworking, underpaid warriors include Pankaj Bishnoi, Rajeev Sethia, Adson da Rocha, Bao Hua, Raj Randeri, Santosh Jodh, Naresh Bhavaraju, Ashutosh Kulkarni, Bryan Stiles, V. Krishnamurthy, Paul Johnson, Craig Kochis, Sean Askew, George Panayi, Jeehyun Kim, Vikram Godbole, Andres Zambrano, Ann Meyer, Hyunjin Shin, and Icaro Santos. Ann Meyer developed most of the code for the HD44780 LCD simulation. My teaching assistants have contributed greatly to the contents of this book and particularly to its laboratory assignments. In the similar manner, my students have recharged my energy each semester with their enthusiasm, dedication, and quest for knowledge.

Secondly, I appreciate the patience and expertise of my fellow faculty members here at the University of Texas at Austin. From a personal perspective, Dr. John Pearce provided much needed encouragement and support throughout my career. In addition, Dr. John Cogdell, Dr. Charles Roth, and Dr. Francis Bostick supported my efforts to develop undergraduate laboratory classes in our department. Dr. Robert Flake provided the mathematical foundation used in the simulated motor. I continue to appreciate the encouragement and support of Dr. G. Jack Lipovski. An outside observer might conclude that Dr. Jack and I enjoy taking opposite sides of every issue. In actuality, this friendly competition causes us to organize our otherwise erratic thoughts, and in the process everything we do is the better for it.

Also, I wish to thank the following reviewers: Shuvra S. Bhattacharyya, University of Maryland at College Park; Brian T. Davis, Michigan Technological University; Roger Kieckhafer, Michigan Technological University; Krist Petersen, New Mexico State University; and David J. Waldo, Oklahoma Christian University.

Lastly, I appreciate the valuable lessons of character and commitment taught to me by my parents and grandparents. I recall how hard my parents and grandparents worked to make the world a better place for the next generation. My Dad taught me the lesson that extreme effort yields a quality product. My Mom showed me how love melts barriers. Most significantly, I acknowledge the love, patience, and support of my wife, Barbara, and my children, Ben, Dan, and Liz. In particular, Ben helped with the Web site and the animations.

Good luck!

Jonathan W. Valvano

Introduction to Embedded Microcomputer Systems

Chapter 1 objectives are to

❑ introduce embedded microcomputer systems,
❑ introduce software development,
❑ introduce C language development,
❑ outline the basic steps in developing microcomputer systems.

Learning new techniques by doing them is an effective approach. But the dilemma in learning a laboratory-based topic like microcomputer programming is the tremendous volume of details to be learned before microcomputer hardware and software systems can be designed. The approach taken in this book is to learn by doing. **Test EXecute And Simulate** (**TExaS**) is an educational product designed for the purpose of studying computer organization, microcomputer programming, and microcomputer interfacing. Basically, it simulates a complete microcomputer hardware/software development system. **TExaS** includes an editor to write programs, an assembler to convert the programs into machine code, a microcomputer simulator to run the programs, and a debugger to test the system. So when you are ready to get started, read Appendix 1. Rather than introduce the voluminous details in an encyclopedic fashion, the book is organized by basic concepts, and the details are introduced as they are needed. We will start with simple systems and progressively add complexity.

1.1 Overview

An *embedded computer system* includes a microcomputer with mechanical, chemical, and electrical devices attached to it, programmed for a specific dedicated purpose, and packaged as a complete system. Any electrical, mechanical, or chemical system that involves inputs, decisions, calculations, analyses, and outputs is a candidate for implementation as an embedded system. Electrical, mechanical, and chemical sensors collect information. Electronic interfaces convert the sensor signals into a form acceptable for the microcomputer. For example, a *tachometer* is a sensor that measures the revolutions per second of a rotating shaft. Microcomputer software performs the necessary decisions, calculations, and analyses. Additional interface electronics convert the microcomputer outputs into the necessary form. *Actuators* can be used to create mechanical or chemical outputs. For example, an electrical motor converts electrical power into mechanical power. The embedded computer systems in this book will contain a Motorola 6811 or 6812, which will be programmed to perform a specific dedicated application.

Microcomputers like the 6811 or 6812 include a processor, memory, and input/output devices. Software for embedded systems typically solves only a limited range of problems. The microcomputer is embedded or hidden inside the device. *Read Only Memory,* or ROM, is a type of memory where the information is programmed or burned into the device, and the data are still stored even if power is removed and reapplied. In an embedded system, the software is usually programmed into ROM and therefore fixed. Even so, *software maintenance* (e.g., verifying that the system is operating properly, updating the software, eliminating bugs, adding features, extending features to new applications, and modifying end user configurations) is still extremely important. The role of simulation is becoming increasingly important in today's marketplace as we race to build better and better machines with shorter and shorter design cycles. In this book, the hardware/software systems will be developed using the **TExaS** simulator.

A typical automobile now contains an average of ten microcontrollers. In fact, upscale homes may contain as many as 150 microcontrollers and the average consumer now interacts with microcontrollers up to 300 times a day. Embedded microcomputers impact virtually all aspects of daily life:

■ consumer electronics,
■ communication systems,
■ automotive systems,
■ military hardware,
■ business applications,
■ medical devices.

Table 1.1 presents typical embedded microcomputer applications and the function performed by the embedded microcomputer. Each microcomputer accepts inputs, performs calculations, and generates outputs.

In contrast, a general-purpose computer system typically has a keyboard, disk and graphics display and can be programmed for a wide variety of purposes. Typical general-purpose applications include word processing, electronic mail, business accounting, scientific computing, and database systems. The user of a general-purpose computer does have access to the software that controls the machine. In other words, the user decides which operating system to run and which applications to launch. Because the general-purpose computer has a removable disk or network interface, new programs can easily be added to the system. The most common type of general-purpose computer is the personal computer, e.g., the Pentium-based IBM-PC-compatible computer. Computers more powerful than the personal computer can be grouped in the workstation category, ranging from $10,000 to $50,000 range. Supercomputers cost above $50,000. These computers often employ multiple processors and have much more memory than the typical personal computer. The workstations and supercomputers are used for handling large amounts of information (business applications) or performing large calculations (scientific research.) This book will not specifically cover the general-purpose computer, although many of the basic principles of embedded computers do apply to all types of systems.

Checkpoint 1.1: *There is a microcomputer embedded in a digital watch. List three operations the software must perform.*

Table 1.1
Embedded system
applications

Application	function performed by the microcomputer
consumer electronics	
washing machine	controls the water and spin cycles
exercise equipment	measures speed, distance, calories, heart rate
remote controls	accept key touches and send infrared pulses
clocks and watches	maintain the time, alarm, and display
games and toys	entertain the user, accept joystick input, display video output
audio/video electronics	interact with the operator and enhance performance
set-back thermostats	adjust day/night thresholds saving energy
camera, camcorder	record and organize images
television, VCR, cable box	accept inputs and process audio/visual signals
communication systems	
answering machines	play outgoing message, save and organize messages
telephones	interactive switching, and information retrieval
fax machines	send and receive images
radios	send and receive audio, reject noise
cellular phones, pagers	accept key pad input, output sound, and enable communication
automotive systems	
automatic braking	optimize stopping on slippery surfaces
noise cancellation	improve sound quality
locks	enable keyless entry, alarm systems
electronic ignitions	control sparks and fuel injectors
power windows and seats	remember preferred settings for each driver
cruise control	maintain constant speed
collision avoidance	reduce accidents
climate control	improve comfort
emission control	reduce pollution
instrumentation	collect and provide necessary information
military hardware systems	
smart weapons	recognize friendly targets
missile guidance	direct ordnance at the desired target
global positioning	determine where you are on the planet
surveillance	collect information about enemy activities
business applications	
cash registers	accept inputs and manages money
vending machines	collect money and dispense product
ATM machines	provide both security and convenience
traffic controllers	sense car positions and control traffic lights
robots	accept input from sensors, control the motors
bar code readers and writers	inventory control and shipping
automatic sprinklers	control the wetness of the soil
elevator controllers	maximize traffic, minimize waiting time
security systems	detect or prevent illegal activities or fire
lighting and heating	maximize comfort and minimize cost
medical devices	
monitors	measure important functions
drug delivery systems	administer proper doses
cancer treatments	control doses of radiation, drugs, or heat
pacemakers	help the heart beat regularly
prosthetic devices	increase mobility for the handicapped
dialysis machines	perform functions normally done by the kidney

1.2 Attitude

Good engineers employ well-defined design processes when developing complex systems. When we work within a structured framework, it is easier to prove our system works (verification) and to modify our system in the future (maintenance). As our software systems become more complex, it becomes increasingly important to employ well-defined software design processes. Throughout this book, a very detailed set of software development rules will be presented. This book focuses on real-time embedded systems written in assembly language, but most of the comments should apply to other situations as well. At first, it may seem radical to impose such a rigid structure upon software. We might wonder if creativity will be sacrificed in the process. True creativity is more about good solutions to important problems and not about being sloppy and inconsistent. Because software maintenance is a critical task, the time spent organizing, documenting, and testing during the initial development stages will reap huge dividends throughout the life of the software project.

Observation: *The easiest way to debug is to write software without any bugs.*

We define *clients* as programmers who will use our software. A client develops software that will call our functions. We define *coworkers* as programmers who will debug and upgrade our software. Coworkers, possibly ourselves, develop, test, and modify our software.

Writing quality software has a lot to do with attitude. We should be embarrassed to ask our coworkers to make changes to our poorly written software. Since so much software development effort involves maintenance, we should create software modules that are easy to change. In other words, we should expect that each piece of our code will be read in the future by another engineer whose job it will be to make changes to our code. We might be tempted to quit a software project once the system is running, but the short time we might thereby save by not organizing, documenting, and testing will be lost many times over in the future when it is time to update the code.

As project managers, we must reward good behavior and punish bad behavior. A company, in an effort to improve the quality of their software products, implemented the following policies: The employees in the customer relations department receive a bonus for every software bug that they can identify. These bugs are reported to the software developers, who in turn receive a bonus for every bug they fix.

Checkpoint 1.2: *Why did the above policy fail horribly?*

We should demand of ourselves that we deliver bug-free software to our clients. Again, we should be embarrassed when our clients report bugs in our code. We should be mortified when other programmers find bugs in our code. There are a few steps we can take to facilitate this important aspect of software design.

Test it now. When we find a bug, fix it immediately. The longer we put off fixing a mistake the more complicated the system becomes, making it harder to find. Remember that bugs do not go away on their own, but we can make the system so complex that the bugs will manifest themselves in a mysterious and obscure fashion. For the same reason, we should completely test each module individually, before combining them into a larger system. We should not add new features before we are convinced the existing system is bug-free. In this way, we start with a working system, add features, then debug this system until it is working again. This incremental approach makes it easier to track progress. It allows us to undo bad decisions, because we can always revert back to a previous working system. Adding new features before the old ones are debugged is very risky. With this sloppy approach, we could easily reach the project deadline with 100% of the features implemented, but have a system that doesn't run. In addition, once a bug is introduced, the longer we wait to remove it, the harder it will be to correct. This is particularly true when the bugs interact with each other. Conversely, with the incremental approach, when the project schedule slips, we can deliver a working system at the deadline that supports some of the features.

Maintenance Tip: *Go from working system to working system.*

Plan for testing. Module testing methodology should be considered at the start of a project. In particular, testing should be included as part of the software design. Our testing and the client's usage go hand in hand. In particular, how we test the software module will help the client understand the context and limitations of how our software is to be used. On the other hand, a clear understanding of how the client wishes to use our software is critical for both the software design and its testing.

Maintenance Tip: *It is better to have some parts of the system that run with 100% reliability than to have the entire system with bugs.*

Get help. Use whatever features are available for organization and debugging. Pay attention to warnings, because they often point to misunderstandings about data or functions. Misunderstanding of assumptions can cause bugs when the software is upgraded, or reused in a different context than originally conceived. Remember that computer time is a lot cheaper than programmer time.

Maintenance Tip: *It is better to have a software system that runs slow than one that does not run at all.*

In the early days of microcomputer systems, software size could be measured in hundreds of lines of source code or thousands of bytes of object code. These early systems, due to their small size, were inherently simple. The explosion of hardware technology (both in speed and size) has lead to a similar increase in the size of software systems. The only hope for success in a large software system will be to break it into simple modules. In most cases, the complexity of the problem itself can not be avoided (e.g., there is just no simple way to get to the moon). Nevertheless, a complex system can be created out of simple components. A real creative effort is required to orchestrate simple building blocks into larger modules, which themselves are grouped. But we must first use our creativity to break a complex problem into simple components, rather than developing complex solutions to simple problems.

Observation: *There are two ways of constructing a software design: one way is to make it so simple that there are* obviously *no deficiencies and the other way is make it so complicated that there are no* obvious *deficiencies. C.A.R. Hoare, "The Emperor's Old Clothes," CACM Feb. 1981.*

1.3   Basic Components of an Embedded System

As shown in Figure 1.1, the term "embedded microcomputer system" refers to a device that contains one or more microcomputers inside. In this context, the word "embedded" means "hidden inside so one can't see it." A *computer* is an electronic device with a processor, memory, and I/O ports. The *processor* executes software, which performs specific predefined operations. The processor includes registers (which are high speed memory), an arithmetic logic unit or ALU (to execute math functions), a bus interface unit (which communicates with memory and I/O), and a control unit (for making decisions). *Memory* is a high-speed storage medium for software and data. *Software* consists of a sequence of commands that are usually executed in order. In an embedded system, we use Read Only Memory (ROM) for storing the software and fixed constant data, and Random Access Memory (RAM) for storing temporary information. The information in the ROM is *nonvolatile,* meaning the contents are not lost when power is removed. A *port* is a physical connection between the computer and its outside world. It allows information to enter and exit the system. Information enters via the input ports and exits via the output ports. Other names used to describe ports are I/O ports, I/O devices, interfaces, or sometimes just devices. The I/O devices are a crucial part of an embedded system because they provide necessary functionality. Examples include parallel ports, serial ports, timer, and analog to digital convertor (ADC). The software together with the I/O ports and associated interface circuits give an embedded computer system its distinctive characteristics. A *device driver* is a set of software functions that facilitate the use of an I/O port.

Figure 1.1
An embedded system includes a microcontroller interfaced to external devices.

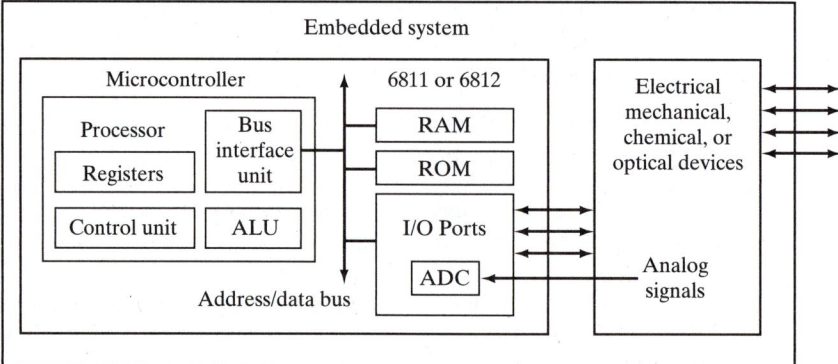

Checkpoint 1.3: *What is an embedded system?*

The term *microcomputer* means a small computer. Small in this context describes its size, not its computing power, so a microcomputer can refer to a very wide range of products from the very simple (e.g., the MC68HC705J1A is a 20-pin chip with 1240 bytes of ROM, 64 bytes RAM, and 14 I/O pins) to the most powerful Pentium. One typically restricts the term "embedded" to refer to systems that do not look and behave like a typical computer. Most embedded systems do not have a keyboard, a graphics display, or secondary storage (disk). There are two ways to develop embedded systems. The first technique uses a *microcontroller,* which is a microcomputer system that incorporates the processor, RAM, ROM and I/O devices into a single package. The 6811 and 6812 devices used in this book are microcontrollers. These devices are suitable for low-cost, low-performance systems. On the other hand, one can develop a high-performance embedded system around the PC architecture. These systems are first developed on a standard PC, and then the software and hardware are migrated to a stand-alone embedded-PC platform.

> ***Checkpoint 1.4:*** *What is a microcomputer?*

The external devices attached to the microcontroller allow the system to interact with its environment. An *interface* is defined as the hardware and software that combine to allow the computer to communicate with the external hardware. We must learn how to interface a wide range of inputs and outputs that can exist in either digital or analog form. This book provides an introduction to microcomputer programming, while *Embedded Microcomputer Systems: Real Time Interfacing* by Jonathan W. Valvano, also published by Brooks/Cole, provides detailed information about hardware interfacing and the design of embedded systems. In general, we can classify I/O interfaces into four categories;

> *parallel.* Binary data are available simultaneously on groups of lines
> *serial.* Binary data are available one bit at a time on a single line
> *analog.* Data are encoded as a variable voltage
> *time.* Data are encoded as a period, frequency, pulse width, or phase shift.

The other general concept involved in most of these systems is that they run in *real-time*. In a real-time computer system, we can put an upper bound on the time required to perform the input–calculation–output sequence. A real-time system can guarantee a worst case upper bound on the response time between when the new input information becomes available and when that information is processed. Another real-time requirement that exists in many embedded systems is the execution of periodic tasks. A periodic task is a task that must be performed at equal time intervals. A real-time system can put a small and bounded limit on the time error between when a task should be run and when it is actually run. Because of the real-time nature of these systems, microcomputers have a rich set of features to handle all aspects of time.

> ***Checkpoint 1.5:*** *An input device allows information to be entered into the computer. List some of the input devices available on a general-purpose computer.*

> ***Checkpoint 1.6:*** *An output device allows information to exit the computer. List some of the output devices available on a general-purpose computer.*

1.4   Flowcharts and Structured Programming

Next, we introduce the flowchart syntax that will be used throughout the book. Programs themselves are written in a linear or one-dimensional fashion. Writing programs this way is a natural process, because the computer itself usually executes the program in a top-to-bottom sequential fashion. This one-dimensional format is fine for simple programs, but conditional branching and function calls may create complex behaviors that are not easily observed in a linear fashion. Flowcharts are one way to describe software in a two-dimensional format, specifically providing convenient mechanisms to visualize conditional branching and function calls. You will not have to draw a flowchart for every program you write. However, flowcharts are very useful in the initial design stage of a software system to define complex algorithms. Flowcharts can also be used in the final documentation stage of a project, once the system is operational, in order to assist in its use or modification.

> **Observation: TExaS** is one of the few software development systems that allow you to add flowcharts directly into your software as part of its documentation.

Figure 1.2 illustrates the flowchart syntax, showing both the flowcharts and corresponding C program. The oval shapes define entry and exit points. The main *entry point* is the starting point of the software. Each function, or subroutine, also has an entry point. The *exit point* returns the flow of control back to the place from which the function was called. When the software runs continuously, as is typically the case in an embedded system, there will be no main exit point. Rectangles specify *process* blocks. In a high-level flowchart, a process block might involve many operations, but in a low-level flowchart, the exact operation is defined in the rectangle. Parallelograms define *input/output* operations. Some flowchart artists use rectangles for both processes and

Figure 1.2
Example flowchart showing some common flowchart symbols.

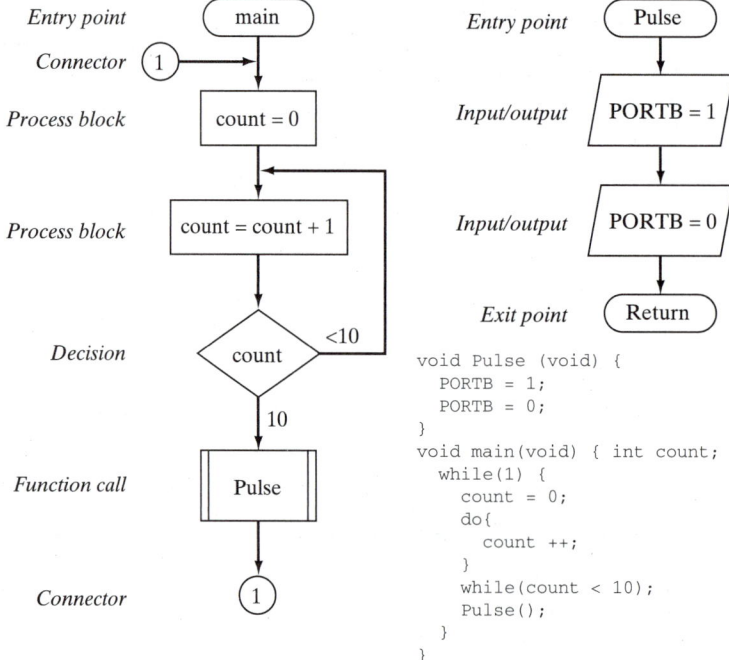

```
void Pulse (void) {
    PORTB = 1;
    PORTB = 0;
}
void main(void) { int count;
    while(1) {
        count = 0;
        do{
            count ++;
        }
        while(count < 10);
        Pulse();
    }
}
```

input/output. Since input/output operations are an important part of embedded systems, we will use the parallelogram format, which will make it easier to identify input/output in our flowcharts. Diamond-shaped objects define branch points or *decision* blocks. Rectangles with double lines on the side specify a call to a *predefined function*. In this book, functions, subroutines and procedures are all terms that refer to a well-defined section of code that performs a specific operation. Functions usually return a result parameter, while procedures usually do not. "Functions" and "procedures" are terms used when describing a high-level language, while "subroutines" are terms often used when describing assembly language. When a function (or subroutine or procedure) is called, the software execution path jumps to the function, the specific operation is performed, and the execution path returns to the point immediately after the function call. Circles are used as *connectors*.

> **Common error.** In general, it is bad programming style to develop software whose flowcharts require a lot of connectors.

> **Checkpoint 1.7:** Using a flowchart, describe the control algorithm that a toaster must use to cook toast. Assume the inputs are toast temperature in F, and desired temperature in F. The output is heat (on/off).

There are an almost infinite number of operations one can perform on a computer, and the key to developing great products is to select the correct ones. Just as for hiking through the woods, we need to develop guidelines (like maps and trails) to keep us from getting lost. One of the fundamental issues when developing software, especially in assembly language, is to maintain a consistent structure. One such framework is called **structured programming.** Most high-level languages (with the exception of `goto`) force the programmer to write structured programs. Structured programs are built from three basic building blocks: the **sequence,** the **conditional,** and the **while-loop.** At the lowest level, the process block contains simple and well-defined commands, like the process blocks shown in Figure 1.2. I/O functions are also low-level building blocks. Structured programming involves combining existing blocks into more complex structures, as shown in Figure 1.3.

Figure 1.3
Flowchart showing the basic building blocks of structured programming.

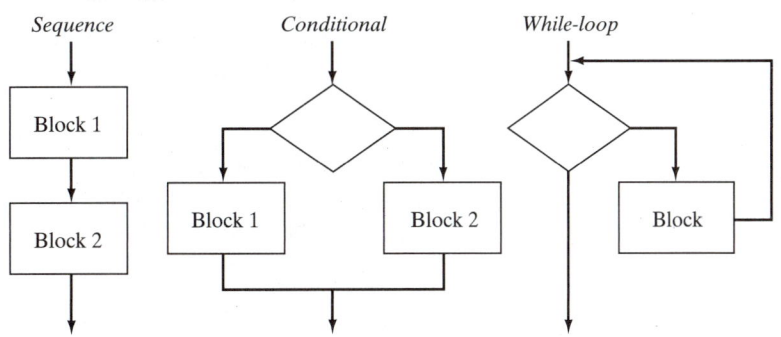

1.5   Product Development Cycle

In this section, we will introduce the product development process in general. The basic approach is introduced here, and the details of these concepts will be presented throughout the remaining chapters of the book. As we learn software/hardware development tools and techniques, we can place them into the framework presented in this section. As illustrated in Figure 1.4, the development of a product follows an analysis–design–implementation–testing cycle. For complex systems with long life-spans, we traverse multiple times around the development cycle. For simple systems, a one-time pass may suffice.

Figure 1.4
Software development cycle.

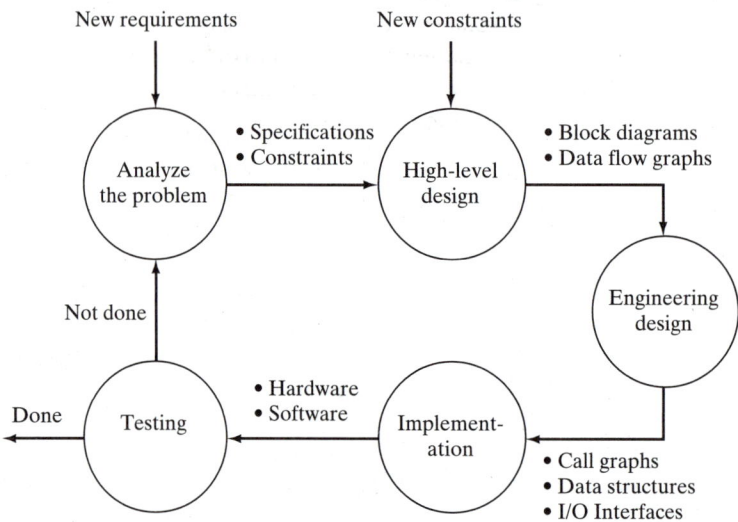

During the **analysis phase,** we discover the requirements and constraints for our proposed system. We can hire consultants and interview potential customers in order to gather this critical information. A *requirement* is a specific parameter that the system must satisfy. We begin by rewriting the system requirements, which are usually written in general form, into a list of detailed *specifications*. In general, specifications are detailed parameters describing how the system should work. For example, a requirement may state that the system should fit into a pocket, whereas a specification would give the exact size and weight of the device. For example, suppose we wish to build a thermometer. During the analysis phase, we would determine obvious specifications such as range, resolution, accuracy, and speed. There may be less obvious requirements to satisfy, such as weight, size, battery life, product life, ease of calibration, display readability, and reliability. A *constraint* is a limitation within which the system must operate. The system may be constrained to such factors as compatibility with other products, use of specific electronic and mechanical parts as other devices, interfaces with other instruments and test equipment, and development schedule.

Checkpoint 1.8: What's the difference between a requirement and a specification?

During the **high-level design** phase, we build a conceptual model of the hardware/software system. It is in this model that we exploit as much abstraction as appropriate. The project is broken into modules or subcomponents. Modular design will be presented in Chapter 9. During this phase, we estimate the cost, schedule, and expected performance of the system. At this point we can decide if the project has a high enough potential for profit. A *data flow graph* is a block diagram of the system, showing the flow of information. Arrows point from source to destination. The rectangles represent hardware components and the ovals are software modules. We use data flow graphs in the high-level design, because they describe the overall operation of the system while hiding the details of how it works. A data flow graph for a simple thermometer is shown in Figure 1.5. The sensor converts temperature in an electrical resistance. The amplifier converts resistance into the 0 to +5V voltage range required by the ADC. The ADC converts analog voltage into a digital sample. The `ADC routines`, using the ADC and timer hardware, collect samples and calculate voltages. The `calculation` software uses a table data structure to convert voltage to temperature. Voltage and temperature data are represented as fixed-point numbers within the computer. The temperature data is passed to the `LCD routines` creating ASCII strings, which will be sent to the *liquid crystal display* (LCD) module. The user will be able to select the Fahrenheit or Centigrade scale using a switch.

Figure 1.5
A data flow graph showing how the temperature signal passes through a simple thermometer.

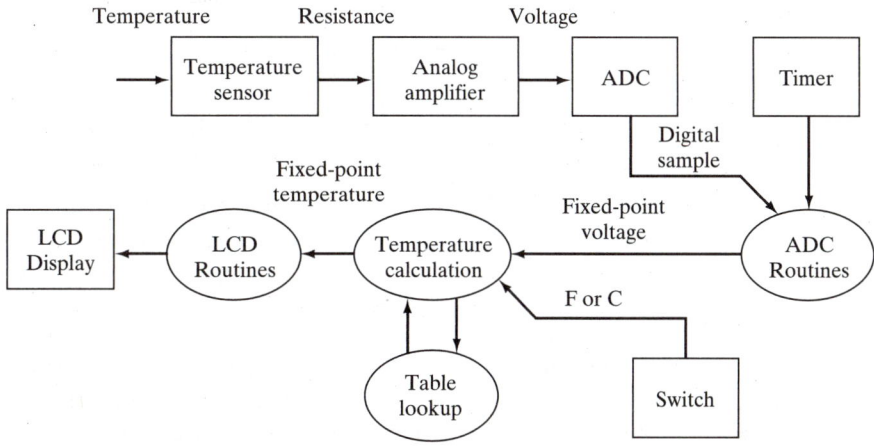

The next phase is **engineering design.** We begin by constructing a preliminary design. This system includes the overall top-down hierarchical structure, the basic I/O signals, shared data structures and overall software scheme. At this stage there should be a simple and direct correlation between the hardware/software systems and the conceptual model developed in the high-level design. Next, we finish the top-down hierarchical structure, and built mock-ups of the mechanical parts (connectors, chassis, cables, etc.) and user software interface. Sophisticated 3-D CAD systems can create realistic images of our system. Detailed hardware designs must include mechanical drawings. It is a good idea to have a second source, which is an alternative supplier that can sell our parts if the first source can't deliver on time. *Call-graphs* are a graphical way to define how the software/hardware modules interconnect. *Data structures,* which will be presented in Chapter 10, include both the organization of information and mechanisms to access the data. A call graph for a simple thermometer is shown in Figure 1.6. Again, rectangles represent hardware components and ovals show software modules. An arrow points from the calling routine to the module it calls. The I/O ports are organized into groups and placed at the bottom of the graph. A high-level call graph, like the one shown in Figure 1.6, shows only the high-level hardware/software modules. A detailed call-graph would include each software function and I/O port. Normally, hardware is passive and the software initiates hardware/software communication, but as we will learn in Chapter 11, it is possible for the hardware to interrupt the software and cause certain software modules to be run. In this system, the timer hardware will cause the `ADC` software to collect a sample. The `main program` gets the next sample from the `ADC` software, converts it to temperature, and displays the result by calling the `LCD` interface software.

Figure 1.6
A call flow graph for a simple thermometer.

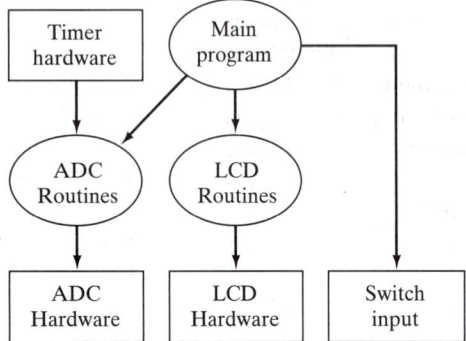

Observation: If module A calls module B, and B returns data, then a data flow graph will show an arrow from B to A, but a call graph will show an arrow from A to B.

The next phase is **implementation.** An advantage of a top-down design is that implementation of subcomponents can occur concurrently. During the initial iterations of the development cycle, it is quite efficient to implement the hardware/software using simulation. One major advantage of simulation is that it is usually quicker to implement an initial product on a simulator versus constructing a physical device out of actual components. Rapid prototyping, important in the early stages of product development, allows for more loops around the analysis–design–implementation–testing cycle, which in turn lead to a more sophisticated product. Chapter 4 includes the details of assembly language programming. I/O devices are described in Chapter 5. Interfacing will be presented in Chapter 6. Chapter 7 describes the debugging process. Chapter 8 presents more sophisticated programming techniques. The solution to the ADC example shown in Figures 1.5 and 1.6 will be developed as a laboratory assignment in Chapter 11.

Recent software and hardware technological developments have made significant impacts on the software development for embedded microcomputers. The simplest approach is to use a cross-assembler or cross-compiler to convert source code into the machine code for the target system. The machine code can then be loaded into the target machine. Debugging embedded systems with this simple approach is very difficult for two reasons. First, the embedded system lacks the usual keyboard and display that assist us when we debug regular software. Second, the nature of embedded systems involves the complex and real-time interaction between the hardware and software. These real-time interactions make it impossible to test software with the usual single-stepping and print statements.

The next technological advancement that has greatly affected the manner in which embedded systems are developed is simulation. Because of the high cost and long times required to create hardware prototypes, many preliminary feasibility designs are now performed using hardware/software simulations. A simulator is a software application that models the behavior of the hardware/software system. If both the external hardware and software program are simulated together, even although the simulated time is slower than the clock on the wall, the real-time hardware/software interactions can be studied.

During the **testing** phase, we evalutate the performance of our system. First, we debug the system and validate basic functions. Next, we use careful measurements to optimize performance such as static efficiency (memory requirements), dynamic efficiency (execution speed), accuracy (difference between truth and measured), and stability (consistent operation). Debugging techniques will be presented in Chapter 7.

Maintenance is the process of correcting mistakes, adding new features, optimizing for execution speed or program size, porting to new computers or operating systems, and reconfiguring the system to solve a similar problem. No system is static. Customers may change or add requirements or constraints. To be profitable, we probably will wish to tailor each system to the individual needs of each customer. Maintenance is not really a separate phase, but rather involves additional loops around the development cycle.

1.6   Quality Programming

Embedded system development is similar to other engineering tasks. We can choose to follow well-defined procedures during the development and evaluation phases, or we can meander in a haphazard way and produce code that is hard to test and harder to change. The ultimate goal of the system is to satisfy stated objectives such as accuracy, stability, and input/output relationships. Nevertheless it is appropriate to separately evaluate the individual components of the system. Therefore in this section, we will evaluate the quality of our software. There are two categories of performance criteria with which we evaluate the "goodness" of our software. Quantitative criteria include dynamic efficiency (speed of execution), static efficiency (memory requirements), and accuracy of the results. Qualitative criteria center around ease of software maintenance. Another qualitative way to evaluate software is ease of understanding. If your software is easy to understand then it will be:

> easy to debug (fix mistakes)
> easy to verify (prove correctness)
> easy to maintain (add features)

> **Common error:** *Programmers who sacrifice clarity in favor of execution speed often develop software that runs fast, but doesn't work and can't be changed.*

Golden Rule of Software Development
Write software for others as you wish they would write for you.

**1.6.1
Quantitative
Performance
Measurements**

In order to evaluate our software quality, we need performance measures. The simplest approaches to this issue are quantitative measurements. *Dynamic efficiency* is a measure of how fast the program executes. It is measured in seconds or CPU cycles. *Static efficiency* is the number of memory bytes required. Since most embedded computer systems have both RAM and ROM, we specify memory requirement in global variables, stack space, fixed constants and program object code. The global variables plus maximum stack size must be less than the available RAM. Similarly, the fixed constants plus program size must be less than the ROM size. We can judge our software system according to whether or not it satisfies given constraints, like software development costs, memory available, and time-table.

**1.6.2
Qualitative
Performance
Measurements**

Qualitative performance measurements include those parameters to which we can not assign a direct numerical value. Often in life the most important questions are the easiest to ask, but the hardest to answer. Such is the case with software quality. We therefore ask the following qualitative questions. Can we prove our software works? Is our software easy to understand? Is our software easy to change? Since there is no single approach to writing the best software, we can only hope to present some techniques that you may wish to integrate into your own software style. In fact, this book devotes considerable effort to the important issue of developing quality software. In particular, we will study self-documented code, abstraction, modularity, and layered software. These issues indeed have a profound effect on the bottom-line financial success of our projects. Although the issues are quite real, because there is often not an immediate and direct relationship between software quality and profit, we may be mistakenly tempted to dismiss its importance.

To get a benchmark on how good a programmer you are, take the following two challenges. In the first test, find a major piece of software that you have written over 12 months ago, then see if you can still understand it enough to make minor changes in its behavior. The second test is to exchange with a peer a major piece of software that you have both recently written (but not written together), and in the same manner, see if you can make minor changes to each other's software.

> **Observation:** *You can tell if you are a good programmer if 1) you can understand your own code 12 months later, and 2) others can make changes to your code.*

1.7 Tutorial 1. Getting Started

Tutorials in this book represent short activities for you to do on your own. Each chapter has a tutorial that allows you to have a hands-on experience to support the basic concepts introduced in that chapter. An **action** defines a specific task that you should perform. The answers to the **questions** can be found at the end of the book. The objective of this first tutorial is to provide an overview embedded system development in general and of the **TExaS** simulator in particular. When you are ready to use the **TExaS** simulator to develop your own programs, first perform this tutorial, then read Appendix 1.

Action: Watch the *getting started* movie.
First time users of **TExaS** should watch the `ReadMe.exe` animation located on the CD. This movie introduces the major components of the application. It takes about 15 minutes and provides a narrated overview of the **TExaS** application. You need not install **TExaS** or copy any files from the CD to your hard drive; just double-click the `ReadMe.exe` icon.

Question 1.1. The branch instruction causes what four operations to execute over and over?

Question 1.2. What external components are connected to PORTA, PORTB, and PORTC?

Question 1.3. What does the red arrow in the listing file signify?

Question 1.4. What are the special features of the **TExaS** editor help in writing assembly programs?

Question 1.5. What microcomputer is being simulated?

Question 1.6. What does the backdump command allow?

Question 1.7. Does the **TExaS** application simulate both hardware and software or just software?

1.8   Homework Assignments

Homework 1.1: A microcomputer is embedded in a vending machine. List three operations the software must perform.

Homework 1.2: What is a port?

Homework 1.3: What does nonvolatile mean?

Homework 1.4: What do the acronyms RAM ROM I/O ALU ADC mean?

Homework 1.5: What is the difference between a microcomputer and a microcontroller?

Homework 1.6: Using a flowchart, describe the control algorithm that a thermostat must use to maintain constant temperature. Assume the inputs are current temperature in F, the desired temperature in F, and an AC/off/heat three-way switch. The outputs are AC (on/off) and heat (on/off).

Homework 1.7: Using a flowchart, describe the cruise control algorithm that a car must use to maintain constant speed. Assume the inputs are current speed in mph, brake (on/off), and a cruise on/off momentary button. The output is accelerator position (0 to 100%). The desired speed is the current speed at the time the cruise control is activated. Touching the brake turns off the system.

Homework 1.8: Draw a flowchart of the following C program. PORTB is an output. This is an incremental controller that maintains the motor at a constant speed of 100.

```c
void main(void) { unsigned char power,speed;
  power = 0;
  ADC_Init();   /* turn on ADC power */
  while(1){
    PORTB = power; /* output to actuator */
    speed = ADC_Read();
    if(speed < 100){ /* too slow */
        if(power < 255) power++;
    }
    else{  /* too fast */
        if(power > 1) power--;
    }
  }
}
```

Homework 1.9: Write C code for the flowchart shown in Figure 1.7. PORTB is an output connected to a stepper motor. PORTA is an input connected to a toggle switch.

Figure 1.7
Flowchart showing a stepper motor controller, used for Homework 1.9.

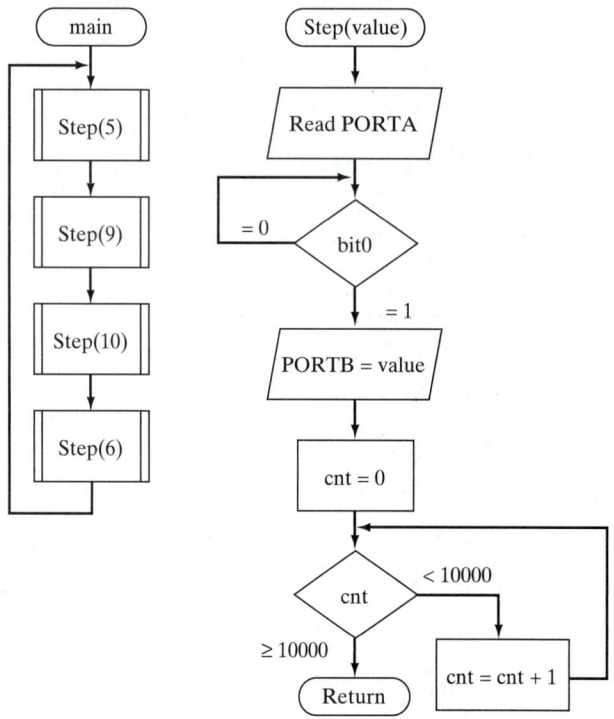

Homework 1.10: Draw a data flow graph of the thermostat algorithm developed in Homework 1.6.

6 **Homework 1.11:** Draw a data flow graph of the cruise control algorithm developed in Homework 1.7.

Homework 1.12: Draw a flowchart of this C program using just the three basic building blocks of structured programming. In particular, first draw the flowchart in the regular way, then show the groupings that define each basic block.

```
int data[100],sum;
void calc(void) {int i; sum=0;
  for(i=0;i<100;i++)
    sum=sum+data[i];
}
```

Homework 1.13: Draw a flowchart of this C program using just the three basic building blocks of structured programming. In particular, first draw the flowchart in the regular way, then show the groupings that define each basic block.

```c
int decide(int in){int out;
  switch(in){
    case 0:  out=1; break;
    case 1:  out=2; break;
    default: out=3;
  }
  return out;
}
```

Homework 1.14: Draw a flowchart of this C program using just the three basic building blocks of structured programming. In particular, first draw the flowchart in the regular way, then show the groupings that define each basic block. Hint: look at Homework 1.13.

```c
int decide(int in){
  if(in==0) return 1;
  if(in==1) return 2;
  return 3;
}
```

2 Information

Chapter 2 objectives are to

❏ introduce the concept of how numbers are stored on the computer,
❏ discuss how characters are represented,
❏ define terms like precision, basis, hexadecimal, big and little endian,
❏ review arithmetic and logic operations,
❏ explain the usage of condition code bits,
❏ develop mechanisms to convert between character strings and binary numbers,
❏ introduce fixed-point and floating point numbers.

Numbers are stored on the computer in binary form. In other words, information is encoded as a sequence of 1's and 0's. *Precision* is the number of distinct or different values. We express precision in alternatives, decimal digits, bytes, or binary bits. We use the expression 4½ decimal digits to mean about 20,000 alternatives and the expression 4¾ decimal digits to mean more than 20,000 alternatives but less than 100,000 alternatives. The ½ decimal point means twice the number of alternatives or one additional binary bit. Let the operation [[x]] be the greatest integer of x. E.g., [[2.1]] is rounded up to 3. Table 2.1 illustrates various representations of precision. *Alternatives* is defined as the total number of possibilities. For example, an 8-bit number scheme can represent 256 different numbers. An 8-bit *digital to analog converter* (DAC) can generate 256 different analog outputs. An 8-bit *analog to digital converter* (ADC) can measure 256 different analog inputs.

Table 2.1
Relationships between various representations of precision.

binary bits	bytes	alternatives	decimal digits
8	1	256	2½
10		1024	3
12		4096	3¾
16	2	65536	4¾
20		1,048,576	6
24	3	16,777,216	7½
30		1,073,741,824	9
32	4	4,294,967,296	9¾
n	[[n/8]]	2^n	$\log_{10}(2^n)$

Observation: A good rule of thumb to remember is $2^{10 \cdot n} \approx 10^{3 \cdot n}$

Checkpoint 2.1: How many binary bits is 3½ decimal digits?

Checkpoint 2.2: About how many decimal digits is 64 binary bits? You can answer this without a calculator, just using the "rule of thumb."

For large numbers we use abbreviations, as shown in Table 2.2. For example, 16K means 16 • 1024, which equals 16384. Computer engineers use the same symbols as other scientists, but with slightly different values. The scientific meaning of 2 kilovolts is 2000 volts, but 2 kilobytes is 2048 bytes.

Table 2.2
Common abbreviations for large numbers.

abbreviation	pronunciation	Computer Engineering Value	Scientific Value
k	"kay"	2^{10} 1024	10^3
M	"meg"	2^{20} 1,048,576	10^6
G	"gig"	2^{30} 1,073,741,824	10^9
T	"tera"	2^{40} 1,099,511,627,776	10^{12}
P	"peta"	2^{50} 1,125,899,906,843,624	10^{15}
E	"exa"	2^{60} 1,152,921,504,606,846,976	10^{18}

Checkpoint 2.3: How big is 2 Tbytes?

2.1 Hexadecimal Representation

The hexadecimal number system uses base 16 as opposed to our regular decimal number system, which uses base 10. Hexadecimal is a convenient mechanism for us humans to represent binary information, because it is extremely simple for us to convert back and forth between binary and hexadecimal. Hexadecimal number system is often abbreviated as "hex." A *nibble* is defined as 4 binary bits. Each value of the 4-bit nibble is mapped into a unique hex digit, as shown in Table 2.3. In mathematics, a subscript of 2 means binary, but in assembly language we will use the prefix % to signify binary numbers.

Table 2.3
Definition of hexadecimal representation.

Hex Digit	Decimal Value	Binary Value
0	0	%0000
1	1	%0001
2	2	%0010
3	3	%0011
4	4	%0100
5	5	%0101
6	6	%0110
7	7	%0111
8	8	%1000
9	9	%1001
A or a	10	%1010
B or b	11	%1011
C or c	12	%1100
D or d	13	%1101
E or e	14	%1110
F or f	15	%1111

Computer programming environments use a wide variety of symbolic notations to specify the numbers in various bases. Table 2.4 illustrates various formats for number constants.

Table 2.4
Comparison of various formats.

environment	binary format	hexadecimal format	decimal format
Motorola assembly language	%01111010	$7A	122
Intel and TI assembly language	01111010B	7AH	122
C language		0x7A	122

As illustrated in Figure 2.1, to convert from binary to hexadecimal we can:

1) divide the binary number into right justified nibbles,
2) convert each nibble into its corresponding hexadecimal digit.

Figure 2.1
Example of conversion from binary to hexadecimal.

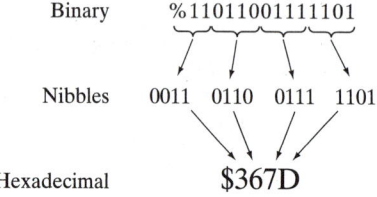

Binary	%11011001111101
Nibbles	0011 0110 0111 1101
Hexadecimal	$367D

Checkpoint 2.4: Convert the binary number %01000101 to hexadecimal.

Checkpoint 2.5: Convert the binary number %110010101011 to hexadecimal.

As illustrated in Figure 2.2, to convert from hexadecimal to binary we can:

1) convert each hexadecimal digit into its corresponding 4-bit binary nibble,
2) combine the nibbles into a single binary number.

Figure 2.2
Example of conversion from hexadecimal to binary.

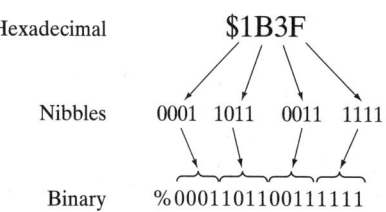

Hexadecimal	$1B3F
Nibbles	0001 1011 0011 1111
Binary	%0001101100111111

Checkpoint 2.6: Convert the hex number $40 to binary.

Checkpoint 2.7: Convert the hex number $63F to binary.

Checkpoint 2.8: How many binary bits does it take to represent $123456?

2.2   Boolean Information

A boolean number has two states. The two values represent logical true and false. The positive logic representation defines true as a 1 or high, and false as a 0 or low. In C programming, a false is represented by a zero, and a true as any nonzero value. If you were controlling a motor, light, heater or air conditioner, the boolean could mean on or off. Figure 2.3 shows the simulation using **TExaS** of a simple switch that has two states, and a LED that can be on or off. PB0 is a digital output of the microcomputer, which can be either high or low. The output of the LED driver is low or HiZ (shown as **z** in Figure 2.3.) In communication systems, we represent the information as a sequence of booleans: mark or space. For black or white graphic displays we use booleans to specify the state of each pixel. The most efficient storage of booleans on a computer is to map each boolean into one memory bit. In this way, we can pack 8 booleans into each byte. If we have just one boolean to store in memory, out of convenience we allocate an entire byte for it. A common positive logic definition is

> False is defined as all zeros, and
> True is defined as any nonzero value.

Figure 2.3
External to the microcomputer, boolean information is encoded as voltage (0 or 5 V), position of a switch (off, on), and the presence of light (dark, light).

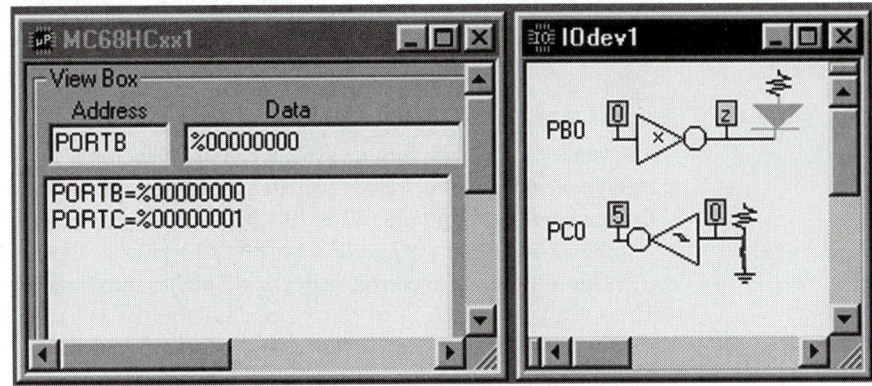

Checkpoint 2.9: Give an example of a switch that is not binary.

2.3   8-bit Numbers

2.3.1
8-bit Unsigned
Numbers

A byte contains 8 bits as shown in Figure 2.4, where each bit $b_7, \ldots, b_0$ is binary and has the value 1 or 0. We specify b_7 as the *most significant bit* or MSB, and b_0 as the least significant bit or LSB.

Figure 2.4
8-bit binary format.

b7	b6	b5	b4	b3	b2	b1	b0

If a byte is used to represent an unsigned number, then the value of the number is

$$N = 128 \cdot b_7 + 64 \cdot b_6 + 32 \cdot b_5 + 16 \cdot b_4 + 8 \cdot b_3 + 4 \cdot b_2 + 2 \cdot b_1 + b_0$$

Notice that the significance of bit n is 2^n. There are 256 different unsigned 8-bit numbers. The smallest unsigned 8-bit number is 0 and the largest is 255. For example, %00001010 is 8+2 or 10. Other examples are shown in Table 2.5.

Table 2.5
Example of conversions from unsigned 8-bit binary to hexadecimal and to decimal.

binary	hex	Calculation	decimal
%00000000	$00		0
%01000001	$41	64+1	65
%00010110	$16	16+4+2	22
%10000111	$87	128+4+2+1	135
%11111111	$FF	128+64+32+16+8+4+2+1	255

Checkpoint 2.10: *Convert the binary number %01101010 to unsigned decimal.*

Checkpoint 2.11: *Convert the hex number $45 to unsigned decimal.*

The *basis* of a number system is a subset from which linear combinations of the basis elements can be used to construct the entire set. The basis represents the "places" in a "place-value" system. For positive integers, the basis is the infinite set {1, 10, 100, . . .}, and the "values" can range from 0 to 9. Each positive integer has a unique set of values such that the dot-product of the value-vector times the basis-vector yields that number. For example, 2345 is (. . . , 2,3,4,5) • (. . . , 1000,100,10,1), which is 2*1000+3*100+4*10+5. For the unsigned 8-bit number system, the basis is

$$\{1, 2, 4, 8, 16, 32, 64, 128\}$$

The values of a binary number system can only be 0 or 1. Even so, each 8-bit unsigned integer has a unique set of values such that the dot-product of the values times the basis yields that number. For example, 69 is (0,1,0,0,0,1,0,1) • (128,64,32,16,8,4,2,1), which equals 0*128+1*64+0*32+0*16+0*8+1*4+0*2+1*1.

One way for us to convert a decimal number into binary is to use the basis elements. The overall approach is to start with the largest basis element and work towards the smallest. One by one, we ask ourselves whether or not we need that basis element to create our number. If we do, then we set the corresponding bit in our binary result and subtract the basis element from our number. If we do not need it, then we clear the corresponding bit in our binary result. We will work through the algorithm with the example of converting 100 to 8-bit binary. (See Table 2.6) We start with the largest basis element (in this case 128) and ask whether or not we need to include it to make 100. Since our number is less than 128, we do not need it, so bit 7 is zero. We go the next largest basis element, 64 and ask, "do we need it?" We do need 64 to generate our 100, so bit 6 is one and we subtract 100 minus 64 to get 36. Next, we go the next basis element, 32 and ask, "do we need it?" Again, we do need 32 to generate our 36, so bit 5 is one and we subtract 36 minus 32 to get 4. Continuing along, we do not need basis elements 16 or 8, but we do need basis element 4. Once we subtract the 4, are working result is zero, so basis elements 2 and 1 are not needed. Putting it together, we get %01100100 (which means 64+32+4).

Observation: *If the least significant binary bit is zero, then the number is even.*

Observation: *If the right-most n bits (least significant) are zero, then the number is divisible by 2^n.*

Observation: *Bit 7 of an 8-bit number determines whether its value is greater than or equal to 128.*

Table 2.6

Example of conversion from decimal to unsigned 8-bit binary to hexadecimal.

Number	Basis	Need it?	bit	Operation
100	128	no	bit 7=0	none
100	64	yes	bit 6=1	subtract 100-64
36	32	yes	bit 5=1	subtract 36-32
4	16	no	bit 4=0	none
4	8	no	bit 3=0	none
4	4	yes	bit 2=1	subtract 4-4
0	2	no	bit 1=0	none
0	1	no	bit 0=0	none

Checkpoint 2.12: *Give the representations of the decimal 45 in 8-bit binary and hexadecimal.*

Checkpoint 2.13: *Give the representations of the decimal 200 in 8-bit binary and hexadecimal.*

2.3.2
8-bit Signed
Numbers

One of the first schemes to represent signed numbers was called *one's complement*. It was called one's complement because to negate a number, you complement (logical not) each bit. For example, 25 equals 00011001 in binary, and -25 is 11100110. An 8-bit one's complement number can vary from -127 to $+127$. The most significant bit is a sign bit, which is 1 if the number is negative. The difficulty with this format is that there are two zeros: $+0$ is 00000000, and -0 is 11111111. Another problem is that one's complement numbers do not have basis elements. These limitations led to the use of two's complement.

The *two's complement* number system is the most common approach used to define signed numbers. It was called two's complement because to negate a number, you complement each bit (like one's complement), then add 1. For example, 25 equals 00011001 in binary, and -25 is 11100111. If a byte is used to represent a signed two's complement number then the value of the number is

$$N = -128 \cdot b_7 + 64 \cdot b_6 + 32 \cdot b_5 + 16 \cdot b_4 + 8 \cdot b_3 + 4 \cdot b_2 + 2 \cdot b_1 + b_0$$

Observation: *One usually means two's complement when one refers to signed integers.*

There are 256 different signed 8-bit numbers. The smallest signed 8-bit number is -128 and the largest is 127. For example, %10000010 equals $-128+2$ or -126. Other examples are shown in Table 2.7.

Table 2.7
Example conversions
from signed 8-bit binary
to hexadecimal and to
decimal.

binary	hex	Calculation	decimal
%00000000	$00		0
%01000001	$41	64+1	65
%00010110	$16	16+4+2	22
%10000111	$87	$-128+4+2+1$	-121
%11111111	$FF	$-128+64+32+16+8+4+2+1$	-1

Checkpoint 2.14: *Convert the signed binary number %11101010 to signed decimal.*

Checkpoint 2.15: *Are the signed and unsigned decimal representations of the 8-bit hex number $45 the same or different?*

For the signed 8-bit number system the basis is

$$\{1, 2, 4, 8, 16, 32, 64, -128\}$$

Observation: *The most significant bit in a two's complement signed number will specify the sign.*

Notice that the same binary pattern of %11111111 could represent either 255 or -1. It is very important for the software developer to keep track of the number format. The computer can not determine whether the 8-bit number is signed or unsigned. You, as the programmer, will determine whether the number is signed or unsigned by the specific assembly instructions you select to operate on the number. Some operations like addition, subtraction, and shift left (multiply by 2) use the same hardware (instructions) for both unsigned and signed operations. On the other hand, multiply, divide, and shift right (divide by 2) require separate hardware (instruction) for unsigned and signed operations. For example, the 6811 multiply instruction, `mul`, operates only on unsigned values. So if you use the `mul` instruction, you are implementing unsigned arithmetic. The Motorola 6812 has both unsigned, `mul`, and signed, `smul`, multiply instructions. So if you use the `smul` instruction, you are implementing signed arithmetic.

Similar to the unsigned algorithm, we can use the basis to convert a decimal number into signed binary. We will work through the algorithm with the example of converting -100 to 8-bit binary, as shown in Table 2.8. We start with the largest basis element (in this case -128) and decide whether we need to include it to make -100? The answer is yes (without -128, we would be unable to add the other basis elements together to get any negative result), so we set bit 7 and subtract the basis element from our value. Our new value equals -100 minus -128, which is 28. We go to the next largest basis element, 64 and ask, "do we need it?" We do not need 64 to generate our 28, so bit 6 is zero. Next we go to the next basis element, 32 and ask, "do we need it?" We do not need 32 to generate our 28, so bit 5 is zero. Now we need the basis element 16, so we set bit 4, and subtract 16 from our number 28 ($28-16=12$). Continuing, we need basis elements 8 and 4 but not 2 and 1. Putting it together we get %10011100 (which means $-128+16+8+4$).

Table 2.8

Example conversion from decimal to signed 8-bit binary.

Number	Basis	Need it	bit	Operation
-100	-128	yes	bit 7=1	subtract $-100 - -128$
28	64	no	bit 6=0	none
28	32	no	bit 5=0	none
28	16	yes	bit 4=1	subtract $28-16$
12	8	yes	bit 3=1	subtract $12-8$
4	4	yes	bit 2=1	subtract $4-4$
0	2	no	bit 1=0	none
0	1	no	bit 0=0	none

Observation: *To take the negative of a two's complement signed number we first complement (flip) all the bits, then add 1.*

A second way to convert negative numbers into binary is to first convert them into unsigned binary, then do a two's complement negate. For example, we earlier found that $+100$ is %01100100. The two's complement negate is a two step process. First we do a logic complement (flip all bits) to get %10011011. Then add one to the result to get %10011100.

A third way to convert negative numbers into binary is to first add 256 to the number, then convert the unsigned result to binary using the unsigned method. For example, to find -100, we add 256 plus -100 to get 156. Then we convert 156 to binary resulting in %10011100. This method works because in 8-bit binary math adding 256 to a number does not change the its value (e.g., $256-100$ has the same value as -100).

Checkpoint 2.16: *Give the representations of −45 in 8-bit binary and hexadecimal.*

Checkpoint 2.17: *Why can't you represent the number 200 using 8-bit signed binary?*

Sign-magnitude representation dedicates one bit as the sign, leaving the remaining bits to specify the magnitude of the number. If b_7 is 1 then the number is negative, otherwise the number is positive.

$$N = -1^{b_7} \cdot (64 \cdot b_6 + 32 \cdot b_5 + 16 \cdot b_4 + 8 \cdot b_3 + 4 \cdot b_2 + 2 \cdot b_1 + b_0)$$

Unfortunately, there is no basis set for the sign-magnitude number system. For example, %10000010 equals $-1 \cdot 2$ or -2. Other examples are shown in Table 2.9.

Table 2.9

Example of conversions from sign-magnitude 8-bit binary to hexadecimal and to decimal.

binary	hex	Calculation	decimal
%00000000	$00		0
%01000001	$41	64+1	65
%00010110	$16	16+4+2	22
%10000111	$87	$-1 \cdot (4+2+1)$	-7
%11111111	$FF	$-1 \cdot (64+32+16+8+4+2+1)$	-127

One problem with sign-magnitude is that there are two representations of the number 0: "00000000" and "10000000". But, the biggest advantage of two's complement signed numbers over sign-magnitude is that the same addition and subtraction hardware (e.g., the `adda`, `suba` instructions) can be used for both signed and unsigned numbers. We also can use the same hardware for shift left or multiply by 2 (e.g., `asla` is the same instruction as `lsla`). Although the hardware for these three operations works for both signed and unsigned numbers, the overflow error) conditions are distinct. The C bit in the condition code register (CCR) signifies unsigned overflow, and the V bit in the CCR means a signed overflow has occurred. Unfortunately, we must use separate signed and unsigned operations for multiply, divide, and shift right or divide by 2.

> **Common Error:** *An error will occur if you use signed operations on unsigned numbers, or use unsigned operations on signed numbers.*

> **Maintenance Tip:** *To improve the clarity of our software, always specify the format of your data (signed versus unsigned) when defining or accessing the data.*

**2.3.3
Character
Information**

We can use bytes to represent characters with the **American Standard Code for Information Interchange** (ASCII) code. Standard ASCII is actually only 7 bits, but is stored using 8-bit bytes with the most significant byte equal to 0. For example, the capital 'V' is defined by the 8-bit binary pattern %01010110. Table 2.10 shows the ASCII code for some of the commonly-used nonprinting characters.

Table 2.10

Common special characters and their ASCII representations.

Abbr.	ASCII character	binary	hexadecimal	decimal
BS	delete or backspace	%00001000	$08	8
HT	tab	%00001001	$09	9
CR	enter or return	%00001101	$0D	13
ESC	escape or esc	%00011011	$1B	27
SP	space	%00100000	$20	32

Some computer systems use the 8th bit of the ASCII code to define additional characters such as graphics and letters in other alphabets. The 7-bit ASCII code definitions are given in the Table 2.11. For example, the letter 'V' is in the $50 column and the 6 row. Putting the two together yields hexadecimal $56.

Table 2.11
Standard 7-bit ASCII.

BITS 4 to 6

BITS 0 to 3	0	1	2	3	4	5	6	7
0	NUL	DLE	SP	0	@	P	`	p
1	SOH	DC1	:	1	A	Q	a	q
2	STX	DC2	!	2	B	R	b	r
3	ETX	DC3	#	3	C	S	c	s
4	EOT	DC4	$	4	D	T	d	t
5	ENQ	NAK	%	5	E	U	e	u
6	ACK	SYN	&	6	F	V	f	v
7	BEL	ETB	'	7	G	W	g	w
8	BS	CAN	(	8	H	X	h	x
9	HT	EM	)	9	I	Y	i	y
A	LF	SUB	*	:	J	Z	j	z
B	VT	ESC	+	;	K	[	k	{
C	FF	FS	,	<	L	\	l	;
D	CR	GS	-	=	M	]	m	}
E	SO	RS	.	>	N	^	n	~
F	S1	US	/	?	O	_	o	DEL

Checkpoint 2.18: How is the character 0 represented in ASCII?

Standard ASCII code used only 7 bits and can only represent 127 different characters. In order to define even more characters, *extended ASCII* can be used. 8-bit extended ASCII uses all 8 bits of the character and can represent 128 more characters. When even more characters are required ASCII can be extended to 16 bits per character. The codes for 0 to 127 match standard ASCII. On the IBM-PC, combinations of Alt-Ctrl and the special function keys can be used to generate extended ASCII. The displayed character depends on the font, but examples of extended ASCII include à, å, ½, ╟, and π.

One way to encode a character string is to use null-termination. In this way, the characters of the string are stored one right after the other, and the end of the string is signified by the NUL character (0). For example, the string "Valvano" is encoded as the following 8 bytes $56, $61, $6C, $76, $61, $6E, $6F, $00.

Checkpoint 2.19: How is "Hello World" encoded as a null-terminated ASCII string?

2.3.4
8-bit Binary Coded Decimal (BCD)

When communicating with humans (input or output), computers need to store information in an easy-to-read decimal format. One such format is *binary coded decimal* or BCD. The 8-bit BCD format contains two decimal digits, and each decimal digit is encoded in four-bit binary. For example, the number 72 is stored as $72 or %01110010. We can represent numbers from 0 to 99 using 8-bit BCD.

Checkpoint 2.20: What binary values are used to store the number 25 in 8-bit BCD format?

2.4 Extended Precision Numbers

2.4.1
16-bit Unsigned Numbers

A *word* or *double byte* contains 16 bits, where each bit $b_{15}, \ldots, b_0$ is binary and has the value 1 or 0, as shown in Figure 2.5.

Figure 2.5
16-bit binary format.

b15	b14	b13	b12	b11	b10	b9	b8	b7	b6	b5	b4	b3	b2	b1	b0

If a word is used to represent an unsigned number, then the value of the number is

$$N = 32768 \cdot b_{15} + 16384 \cdot b_{14} + 8192 \cdot b_{13} + 4096 \cdot b_{12}$$
$$+ 2048 \cdot b_{11} + 1024 \cdot b_{10} + 512 \cdot b_9 + 256 \cdot b_8$$
$$+ 128 \cdot b_7 + 64 \cdot b_6 + 32 \cdot b_5 + 16 \cdot b_4 + 8 \cdot b_3 + 4 \cdot b_2 + 2 \cdot b_1 + b_0$$

There are 65536 different unsigned 16-bit numbers. The smallest unsigned 16-bit number is 0 and the largest is 65535. For example, %0010000110000100 or $2184 is 8192 + 256 + 128 + 4 or 8580. Other examples are shown in Table 2.12.

Table 2.12
Example conversions from unsigned 16-bit binary to hexadecimal and to decimal.

binary	hex	Calculation	decimal
%0000000000000000	$0000		0
%0000010000000001	$0401	1024+1	1025
%0000110010100000	$0CA0	2048+1024+128+32	3232
%1000111000000010	$8E02	32768+2048+1024+512+2	36354
%1111111111111111	$FFFF	32768+16384+8192+4096+2048+1024 +512+256+128+64+32+16+8+4+2+1	65535

Checkpoint 2.21: Convert the 16-bit binary number %0010000001101010 to unsigned decimal.

Checkpoint 2.22: Convert the 16-bit hex number $1234 to unsigned decimal.

For the unsigned 16-bit number system the basis is

$$\{1, 2, 4, 8, 16, 32, 64, 128, 256, 512, 1024, 2048, 4096, 8192, 16384, 32768\}$$

Checkpoint 2.23: Convert the unsigned decimal number 1234 to 16-bit hexadecimal.

Checkpoint 2.24: Convert the unsigned decimal number 10000 to 16-bit binary.

2.4.2
16-bit Signed Numbers

There are also 65536 different signed 16-bit numbers. The smallest two's complement signed 16-bit number is −32768 and the largest is 32767. For example, %1101000000000100 or $D004 is −32768+16384+4096+4 or −12284. Other examples are shown in Table 2.13.

Table 2.13
Example of conversions from signed 16-bit binary to hexadecimal and to decimal.

binary	hex	Calculation	decimal
%0000000000000000	$0000		0
%0000010000000001	$0401	1024+1	1025
%0000110010100000	$0CA0	2048+1024+128+32	3232
%1000010000000010	$8402	−32768+1024+2	−31742
%1111111111111111	$FFFF	−32768+16384+8192+4096+2048+1024 +512+256+128+64+32+16+8+4+2+1	−1

If a word is used to represent a signed two's complement number, then the value of the number is

$$N = -32768 \cdot b_{15} + 16384 \cdot b_{14} + 8192 \cdot b_{13} + 4096 \cdot b_{12}$$
$$+ 2048 \cdot b_{11} + 1024 \cdot b_{10} + 512 \cdot b_9 + 256 \cdot b_8$$
$$+ 128 \cdot b_7 + 64 \cdot b_6 + 32 \cdot b_5 + 16 \cdot b_4 + 8 \cdot b_3 + 4 \cdot b_2 + 2 \cdot b_1 + b_0$$

Checkpoint 2.25: Convert the 16-bit hex number $1234 to signed decimal.

Checkpoint 2.26: Convert the 16-bit hex number $ABCD to signed decimal.

For the signed 16-bit number system the basis is

$$\{1, 2, 4, 8, 16, 32, 64, 128, 256, 512, 1024, 2048, 4096, 8192, 16384, -32768\}$$

Common Error: An error will occur if you use 16-bit operations on 8-bit numbers, or use 8-bit operations on 16-bit numbers.

Maintenance Tip: To improve the clarity of your software, always specify the precision of your data when defining or accessing the data.

Checkpoint 2.27: Convert the signed decimal number 1234 to 16-bit hexadecimal.

Checkpoint 2.28: Convert the signed decimal number −10000 to 16-bit binary.

2.4.3 Big and Little Endian

When we store 16-bit data into memory it requires two bytes. Since the memory systems on most computers are byte addressable (a unique address for each byte), there are two possible ways to store in memory the two bytes that constitute the 16-bit data. Motorola microcomputers implement the *big endian* approach that stores the most significant part first. Intel microcomputers implement the *little endian* approach that stores the least significant part first. The PowerPC is *biendian*, because it can be configured to efficiently handle both big and little endian. Figure 2.6 shows two ways to store the 16-bit number 1000 ($03E8) at locations $50-$51.

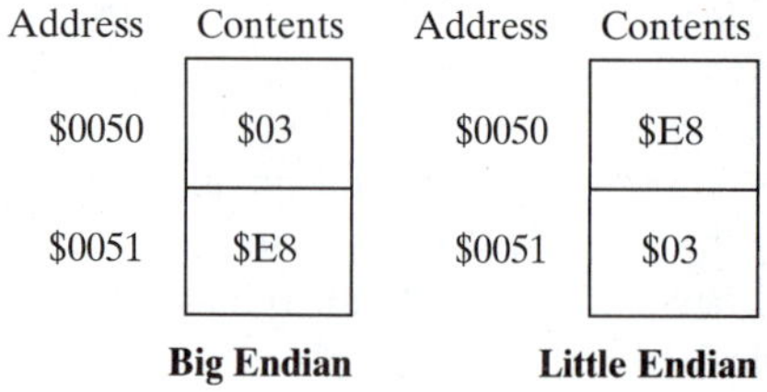

Figure 2.6
Example of big and little endian formats of a 16-bit number.

Address	Contents
$0050	$03
$0051	$E8

Big Endian

Address	Contents
$0050	$E8
$0051	$03

Little Endian

We also can use either the big and little endian approach when storing 32-bit numbers into memory that is byte (8-bit) addressable. Figure 2.7 shows the big and little endian formats that could be used to store the 32-bit number $12345678 at locations $50-$53.

Figure 2.7
Example of big and little endian formats of a 32-bit number.

Address	Contents	Address	Contents
$0050	$12	$0050	$78
$0051	$34	$0051	$56
$0052	$56	$0052	$34
$0053	$78	$0053	$12

Big Endian   Little Endian

In the above two examples we normally would not pick out individual bytes (e.g., the $12), but rather capture the entire multiple byte data as one nondivisible piece of information. On the other hand, if each byte in a multiple byte data structure is individually addressable, then both the big and little endian schemes store the data in first to last sequence. For example, if we wish to store the 4 ASCII characters '6811', which is $36383131 at locations $50-$53, then the ASCII '6'=$36 comes first in both big and little endian schemes, as illustrated in Figure 2.8.

Figure 2.8
Character strings are stored in the same for both big and little endian formats.

Address	Contents
$0050	$36
$0051	$38
$0052	$31
$0053	$31

Big and Little Endian

The terms "big and little endian" come from Jonathan Swift's satire *Gulliver's Travels*. In Swift's book, a Big Endian refers to a person who cracks their egg on the big end. The Lilliputians were Little Endians because they insisted that the only proper way is to break an egg on the little end. The Lilliputians considered the Big Endians as inferiors. The Big and Little Endians fought a long and senseless war over the best way to crack an egg.

Common Error: An error will occur when data is stored in Big Endian by one computer and read in Little Endian format on another.

2.4.4
Arbitrary Length Integers

Consider an unsigned number with n bits, where each bit $b_{n-1}, \ldots, b_0$ is binary and has the value 1 or 0, as shown in Figure 2.9.

Figure 2.9
Arbitrary length binary format.

bn – 1	bn – 2		b1	b0

If an n-bit binary number is used to represent an unsigned number, then the value of the number is

$$N = 2^{n-1} \bullet b_{n-1} + 2^{n-2} \bullet b_{n-2} + \ldots + 2 \bullet b_1 + b_0 = \sum_{i=0}^{n-1} 2^i \bullet b_i$$

There are 2^n different unsigned n-bit numbers. The smallest unsigned n-bit number is 0 and the largest is $2^n - 1$. For the unsigned n-bit number system, the basis is

$$\{1, 2, 4, \ldots, 2^{n-2}, 2^{n-1}\}$$

If an n-bit binary number is used to represent a signed two's complement number then the value of the number is

$$N = -2^{n-1} \bullet b_{n-1} + 2^{n-2} \bullet b_{n-2} + \ldots + 2 \bullet b_1 + b_0 = -2^{n-1} \bullet b_{n-1} + \sum_{i=0}^{n-2} 2^i \bullet b_i$$

There are also 2^n different signed n-bit numbers. The smallest signed n-bit number is -2^{n-1} and the largest is $2^{n-1} - 1$. For the signed n-bit number system, the basis is

$$\{1, 2, 4, \ldots, 2^{n-2}, -2^{n-1}\}$$

Maintenance Tip: When programming in C, we will use data types `char` `short` and `long` when we wish to explicitly specify the precision as 8-bit, 16-bit or 32-bit. We will use the `int` data type only when we don't care about precision, and we wish the compiler to choose the most efficient way to perform the operation.

Observation: When programming in assembly, we will always explicitly specify the precision of our numbers and calculations.

2.4.5
Binary Coded
Decimal (BCD)

The *binary coded decimal* or BCD format is convenient for storing data that has just been input or is just about to be output. Each byte or a BCD number contains two decimal digits, and each decimal digit is encoded in four-bit binary. For example, the number 1,234,567 is stored in four bytes as $01234567. If `m` is the number of bytes, then the numbers from 0 to 100^m-1 can be stored.

Checkpoint 2.29: What hexadecimal values are used to store the number 3456 in 16-bit BCD format?

2.5 Programming Numbers in Assembly Language

In the next few chapters, we will study software development in detail, but for now we introduce some assembly language terminology in order to understand some of the arithmetic and logical operations explained in this chapter. The following notation will be used in describing the action of specific assembly language instructions.

```
w     is a signed 8-bit -128 to +127 or unsigned 8-bit 0 to 255
n     is a signed 8-bit -128 to +127
u     is an unsigned 8-bit 0 to 255
W     is a signed 16-bit -32787 to +32767 or unsigned 16-bit 0 to 65535
N     is a signed 16-bit -32787 to +32767
U     is an unsigned 16-bit 0 to 65535
=[addr]  specifies an 8-bit read from addr
={addr}  specifies a 16-bit read from addr using "big endian"
[addr]=  specifies an 8-bit write to addr
{addr}=  specifies a 16-bit write to addr using "big endian"
```

Here is a review of the instructions introduced in the first chapter. Notice that the 6811/6812 can use either an 8-bit (direct) or a 16-bit (extended) format for specifying memory addresses.

```
ldaa   #w    RegA=w      load accumulator A, immediate addressing
ldaa   u     RegA=[u]    load accumulator A, direct addressing
ldaa   U     RegA=[U]    load accumulator A, extended addressing
staa   u     [u]=RegA    store accumulator A, direct addressing
staa   U     [u]=RegA    store accumulator A, extended addressing
bra    U     PC=U        branch, PC relative addressing
```

The `ldaa` instruction brings 8-bit data from memory or an I/O port into Register A, as illustrated in Figure 2.10.

Figure 2.10
The `ldaa Data` instruction loads data from RAM into Register A.

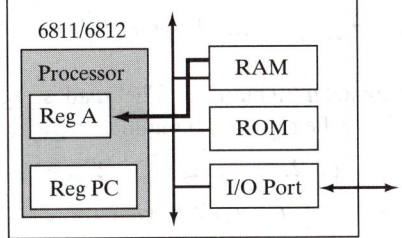

The `staa` instruction sends 8-bit data from Register A to memory or an I/O port, as illustrated in Figure 2.11.

Figure 2.11
The `staa PORTB` instruction stores data from Register A out to Port B.

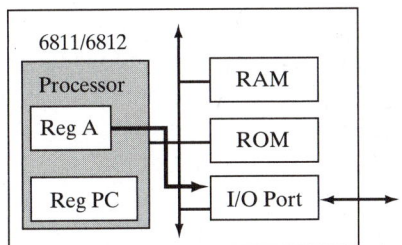

Checkpoint 2.30: Write assembly code that copies the data from memory location 10 to memory location 20.

Checkpoint 2.31: Write assembly code that writes the binary %11000111 to Port B.

2.6　Logical Operations

Software uses logical operations to combine information, to extract information and to test information. A **unary** operation produces its result given a single input parameter. For example, negate, increment, and decrement are unary operations. A **binary** operation produces a single result given two inputs. The **logical and (&)** operation yields a true result if both input parameters are true. The **logical or (|)** operation yields a true result if either input parameter is true. The **exclusive or (^)** operation yields a true result if exactly one input parameter is true. The logical operators are summarized in Table 2.14 and shown as digital gates in Figure 2.12. CMOS refers to complementary metal oxide semiconductor The "HC" in 74HC08 stands for high-speed CMOS. Most microcomputers (all the 6811 and 6812 microcomputers) are made with high-speed CMOS.

Table 2.14
Logical operations.

A	B	A&B	A|B	A^B
0	0	0	0	0
0	1	0	1	1
1	0	0	1	1
1	1	1	1	0

Figure 2.12
Logical operations can be implemented with discrete digital gates.

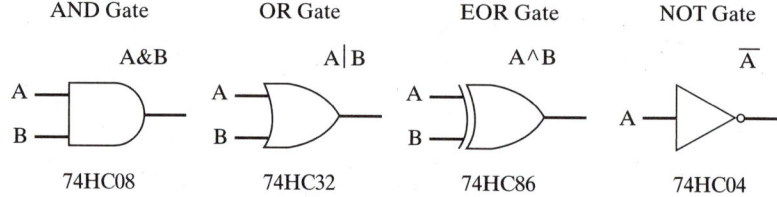

Checkpoint 2.32: *Using just the gates shown in Figure 2.12, design an equals circuit, such that the output is 1 if and only if input A equals input B. There will be two input signals and one output signal.*

The 6811 and 6812 perform logical operations in a bit-wise fashion on two 8-bit parameters yielding an 8-bit result. For example, the calculation r=m&n means each bit is calculated separately, $r_7 = m_7 \& n_7$, $r_6 = m_6 \& n_6, \ldots, r_0 = m_0 \& n_0$.

In discrete digital logic, the complement operation is called a NOT gate, as shown in Figure 2.12. The complement function is defined in Table 2.15. Again, the 6811 and 6812 perform the complement in a bitwise fashion. For example, the calculation r=~n means each bit is calculated separately, $r_7 = {\sim}n_7$, $r_6 = {\sim}n_6, \ldots, r_0 = {\sim}n_0$.

Table 2.15
Logical complement.

A	~A
0	1
1	0

As shown in the following list of 6811/6812 instructions, the computer usually performs logical operations using a register like Register A. These instructions are very important for embedded systems.

```
anda   #w    RegA=RegA&w       and accummulator A
anda   u     RegA=RegA&[u]     and accummulator A
anda   U     RegA=RegA&[U]     and accummulator A
oraa   #w    RegA=RegA|w       or accummulator A
oraa   u     RegA=RegA|[u]     or accummulator A
oraa   U     RegA=RegA|[U]     or accummulator A
eora   #w    RegA=RegA^w       exclusive or accummulator A
eora   u     RegA=RegA^[u]     exclusive or accummulator A
eora   U     RegA=RegA^[U]     exclusive or accummulator A
coma         RegA=~RegA        complement accummulator A
```

The following C code uses the **and** operation to extract, or *mask,* individual bits from an 8-bit value. Figure 2.13 shows two switch interfaces. In particular, PORTA bit 0 contains a signal that is high or low depending on the position of the switch. In both circuits, the signal is high ($\mathbf{q}=1$, which is $+5V$) if the switch is on, and low ($\mathbf{q}=0$, which is 0V) if the switch is open. Some switches bounce, which means there will be multiple open/closed cycles when the switch is changed. The interface with the capacitor can be used to remove these multiple transitions. The simple circuit can be used if the switch doesn't bounce or if the bouncing doesn't matter. When the computer reads PORTA it gets all 8 bits of the input port. On the other hand, the expression PORTA&0x01 will be zero, if and only if bit 0 of PORTA is zero. The following C code will set a variable to true (nonzero) if the switch is pressed.

```
Pressed = PORTA&0x01; /* true if Port C bit 0 is high */
```

Figure 2.13

Interface of a switch to a microcomputer input.

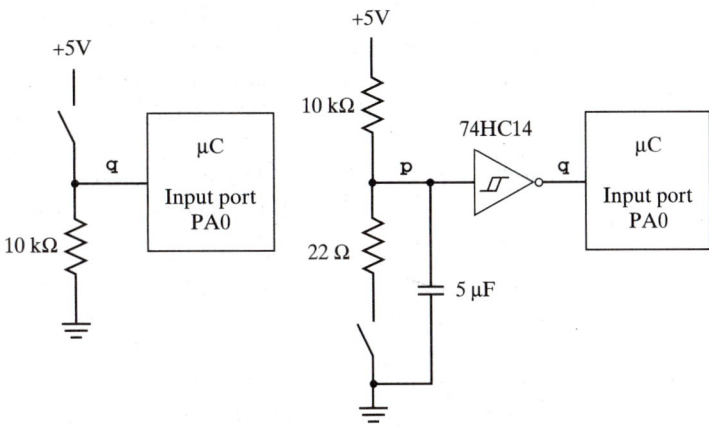

The following 6811/6812 assembly code uses the anda instruction to perform the same operation.

```
ldaa    PORTA    read input Port A
anda    #$01     clear all bits except bit 0
staa    Pressed  true iff Port A bit 0 is high
```

To illustrate how the above program works, let $a_7\ a_6\ a_5\ a_4\ a_3\ a_2\ a_1\ a_0$ be the values of the 8 individual bits in PORTA. The ldaa instruction brings these values into register A. The anda instruction clears all bits except bit 0, and the staa instruction stores the result into the variable called Pressed.

a_7	a_6	a_5	a_4	a_3	a_2	a_1	a_0	value of PORTA
0	0	0	0	0	0	0	1	$01 constant
0	0	0	0	0	0	0	a_0	result of the anda instruction

The **and** operation can be used to clear bits. The following C code turns off the LED connected to PB3 by clearing bit 3 of PORTB, while the other seven bits remain constant. Notice that 0xF7 is %11110111 in binary.

```
PORTB &= 0xF7; /* clear just bit 3, PB3 becomes 0 */
```

The following 6811/6812 assembly code uses the anda instruction to perform the same operation.

```
ldaa    PORTB    read output Port B
anda    #$F7     clear just bit 3, other 7 bits left unchanged
staa    PORTB    update the actual output port
```

To illustrate how the above program works, let b_7 b_6 b_5 b_4 b_3 b_2 b_1 b_0 be the values of the original 8 bits in PORTB. The `ldaa` instruction brings these values into register A. The `anda` instruction clears bit 3, and the `staa` instruction stores the result back to PORTB.

b_7	b_6	b_5	b_4	b_3	b_2	b_1	b_0	value of PORTB
1	1	1	1	0	1	1	1	$F7 constant
b_7	b_6	b_5	b_4	0	b_2	b_1	b_0	result of the anda instruction

The **complement** operation can be used to make this last function easier to read. The following C code also clears bit 3 of PORTB. Notice that ~0x08 is 0xF7. Since this complement is executed at compile-time rather than at run-time, the compiler will generate identical machine code for the two C implementations.

```
PORTB &= ~0x08; /* clear bit 3, PB3 becomes 0 */
```

The following C code uses the **or** operation to set bits 4 and 5 of the register DDRC. The other six bits of DDRC remain constant. *Friendly* software modifies just the bits that need to be modified.

```
DDRC |= 0x30; /* set bits 4 and 5, making PC4 and PC5 outputs */
```

The following 6811/6812 assembly code uses the `oraa` instruction to perform the same operation.

```
ldaa   DDRC    read previous value of DDRC
oraa   #$30    set bits 4 and 5, other 6 bits left unchanged
staa   DDRC    update the actual direction register
```

To illustrate how the above program works, let c_7 c_6 c_5 c_4 c_3 c_2 c_1 c_0 be the values of the original 8 bits in DDRC. The `ldaa` instruction brings these values into register A. The `oraa` instruction sets bits 4 and 5, and the `staa` instruction stores the result back to DDRC.

c_7	c_6	c_5	c_4	c_3	c_2	c_1	c_0	value of DDRC
0	0	1	1	0	0	0	0	$30 constant
c_7	c_6	1	1	c_3	c_2	c_1	c_0	result of the oraa instruction

Maintenance Tip: *When interacting with just some of the bits of an I/O register it is better to modify just the bits of interest, leaving the other bits unchanged. In this way, the action of one piece of software does not undo the action of another piece.*

The **exclusive or** operation can be used to toggle bits. E.g., refer to the microcomputer system presented in Figure A1.7. The following C code toggles the LED connected to PB3 by inverting bit 3 of PORTB, while the other seven bits remain constant. Notice that 0x08 is %00001000 in binary.

```
PORTB   ^= 0x08; /* toggle PB3 from 0 to 1 or from 1 to 0 */
```

The following 6811/6812 assembly code uses the `eora` instruction to perform the same operation.

```
ldaa    PORTB    read output Port B
eora    #$08     toggle just bit 3, other 7 bits left unchanged
staa    PORTB    update the actual output port
```

To illustrate how the above program works, let b_7 b_6 b_5 b_4 b_3 b_2 b_1 b_0 be the values of the original 8 bits in PORTB. The `ldaa` instruction brings these values into register A. The `eora` instruction toggles bit 3, and the staa instruction stores the result back to PORTB.

| b₇ | b₆ | b₅ | b₄ | b₃ | b₂ | b₁ | b₀ | value of PORTB |

$$b_7 \quad b_6 \quad b_5 \quad b_4 \quad b_3 \quad b_2 \quad b_1 \quad b_0 \quad \text{value of PORTB}$$
$$\underline{0 \quad 0 \quad 0 \quad 0 \quad 1 \quad 0 \quad 0 \quad 0} \quad \text{\$08 \quad constant}$$
$$b_7 \quad b_6 \quad b_5 \quad b_4 \quad \sim b_3 \quad b_2 \quad b_1 \quad b_0 \quad \text{result of the eora \quad instruction}$$

The next example uses the **exclusive or** operation to toggle two output bits. The infinite loop program will cause out of phase squarewaves on Port C bit 1 and 0. The other six bits of Port C remain constant.

```
DDRC |= 0x03;                       /* make PC1 PC0 output */
PORTC = (PORTC&0xFC)|0x01;   /* PC1=0, PC0=1 */
while(1){
   PORTC ^= 0x03;                   /* toggle bits 1 and 0 */
}
```

Program 2.1 uses assembly instructions to perform the same operation. The period of the squarewave is determined by the speed of the microcomputer. Figure 2.14 shows the simulated waveforms running on a 2MHz 6811. This program could have been optimized by removing the `ldaa PORTC` instruction at loop.

Program 2.1
6811/6812 assembly program that generates two square waves.

```
main ldaa   DDRC    read previous value of DDRC
     oraa   #$03    set bits 1 and 0
     staa   DDRC    make PC1, PC0 outputs, leaving other bits as is
     ldaa   PORTC   read previous value of PORTC
     anda   #$FC    clear bits 1,0
     oraa   #$01    set bit 0
     staa   PORTC   make PC1=0, PC0=1, leaving other bits as is
loop ldaa   PORTC   read previous value of PORTC
     eora   #$03    toggle bits 1,0
     staa   PORTC   change PC1 PC0, leaving other bits as is
     bra    loop
```

Figure 2.14
Scope window showing the execution of Program 2.1.

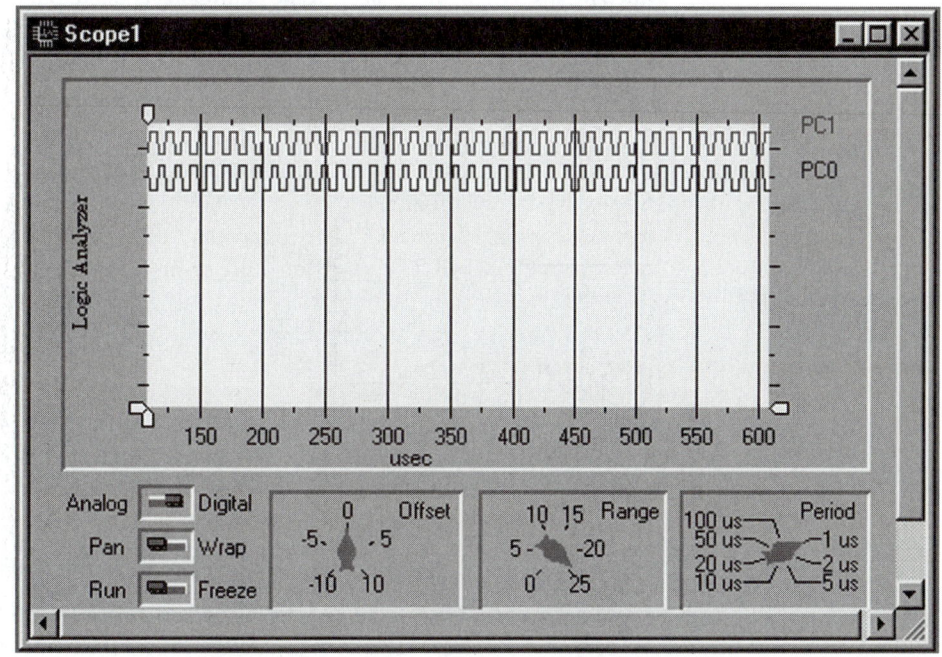

Other convenient logical operators are summarized in Table 2.16 and shown as digital gates in Figure 2.15. The **NAND** operation is defined by an AND followed by a NOT. Similarly, the **NOR** operation is an OR followed by a NOT, and the **exclusive NOR** operation is an EOR followed by a NOT. Notice also that the exclusive NOR implements the equals operation.

Table 2.16
Logical operations.

A	B	NAND	NOR	exclusive NOR
0	0	1	1	1
0	1	1	0	0
1	0	1	0	0
1	1	0	0	1

Figure 2.15
Other logical operations can also be implemented with discrete digital gates.

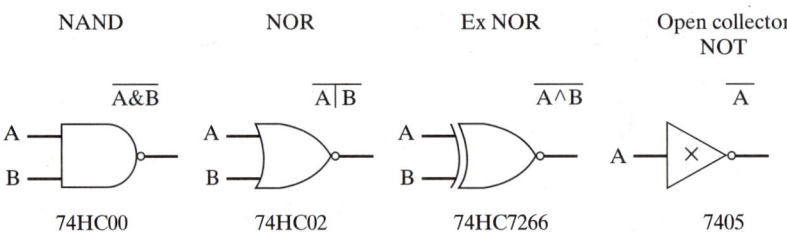

The output of an open collector gate, drawn with the 'x', has two states: low (0V) and HiZ (floating.) In general, we can use an **open collector NOT** gate to control the current to a device, such as a relay, an LED, a solenoid, a small motor and a small light. We used the open collector NOT gate in the LED interface shown in Figure 2.16 to control the current to our diode. When input to the 7405 is high (**p**=1, which means +5V), the output is low (**q**=0, which means 0V). In this state, a 10 mA current is applied to the diode, and it lights up. But when the input is low (**p**=0, which means +0V), the output floats (**q**=HiZ, which is neither high or low). This floating output state causes the LED current to be zero, and the diode is dark. The **TExaS** simulator signifies this floating state with a **z**, as seen in Figure 2.3.

Figure 2.16
LED interface.

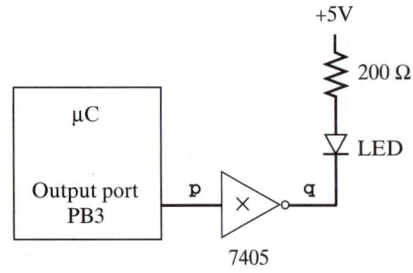

Checkpoint 2.33: *Write assembly code that clears bit 1 of Port B.*

Checkpoint 2.34: *Write assembly code that sets bit 7 of Port B.*

Checkpoint 2.35: *Assume Register A contains an ASCII code 0 to 9. Write assembly code that converts the ASCII code into the corresponding decimal number.*

While we're introducing digital circuits, we need to consider two more type of digital devices that are are essential for microcomputer design. First, the D flip-flop, shown in Figure 2.17, can be used to store information. It is a basic building block of memory on the computer. To save information, we first place the digital value we wish to remember on the **D** input, then give a rising edge to the **clock** input. After the rising edge of the **clock**, the value is available at the Q output, and the D input is free to change. This operation is also defined in Table 2.17. The 74HC374 is an 8-bit D flip-flop, such that all 8 bits are stored on the rising edge of a single clock. The 74HC374 is similar in structure and operation to a register, which is high speed memory inside the processor.

Figure 2.17
A 1-bit and an 8-bit D flip-flop.

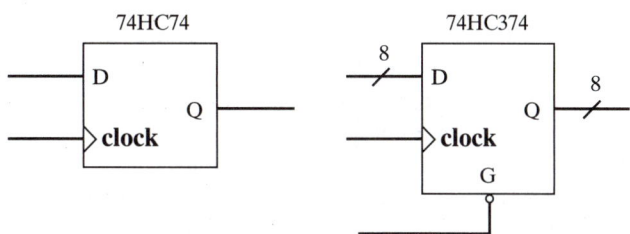

Table 2.17
D flip-flop operation. Q_{old} is the value of the D input at the time of the last rising edge (↑) of the clock.

D	clock	Q
0	0	Q_{old}
0	1	Q_{old}
1	0	Q_{old}
1	1	Q_{old}
0	↑	0
1	↑	1

Second, the tristate driver, shown in Figure 2.18, can be used to dynamically control signals within the computer. The tristate driver is an essential component from which computers are built. To activate the driver, we make its gate (G) low. When the driver is active, its output (Y) equals its input (A). To deactivate the driver, we make its G high. When the driver is deactived, its output Y floats independent of A. We saw this floating state with the open collector logic, and it is also called HiZ or high impedence. The HiZ output means the output is neither driven high or low. The operation of a tristate driver is defined in Table 2.18. The 74HC244 is an 8-bit tristate driver, such that all 8 bits are active or deactive controlled by a single control line. The 74HC374 8-bit D flip-flop includes tristate drivers on its outputs. Normally, we can't connect to digital outputs together. The tristate driver provides a way to connect multiple outputs to the same signal, as long as at most one of the gates is active at a time.

Figure 2.18
A 1-bit and an 8-bit tristate driver.

Table 2.18
Tristate driver operation.
HiZ is the floating state,
such that the output is
not high or low.

A	G	Y
0	0	0
0	1	HiZ
1	0	1
1	1	HiZ

2.7 Shift Operations

In C programming the shift is a binary operation. In other words, the $<<$ and $>>$ operators take two inputs and yield one output, e.g., $r=m>>n$. But at the machine level (i.e., assembly programming), the shift operators are actually unary operations, e.g., $r=m>>1$. The assembly instructions used for shifting will shift one bit at a time. If you want to shift multiple times, you will have to execute the instruction multiple times. The logical shift right (LSR) is the equivalent to an unsigned divide by 2, as shown in Figure 2.19. A zero is shifted into the most significant position, and the carry flag will hold the bit shifted out.

Figure 2.19
8-bit logical shift right.

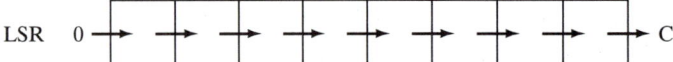

Consider the top row of 8 D flip-flops of Figure 2.20 as a register containing an 8-bit value. The LSR function can be implemented in hardware as a two-step process. The first step, which occurs on the falling edge of **shift** (rising edge of **copy**), is to make a copy of the 8 bits into the lower row of D flip-flips. Then, on the rising edge of the **shift** signal, the new shifted value is clocked back in the top row.

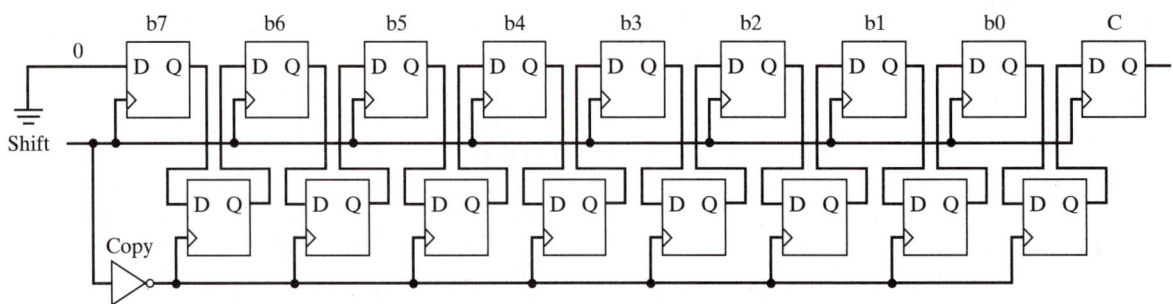

Figure 2.20
8-bit logical shift right hardware.

The arithmetic shift right (ASR) is the equivalent to a signed divide by 2, as shown in Figure 2.21. Notice that the sign bit is preserved and the carry flag will hold the bit shifted out.

Figure 2.21
8-bit arithmetic shift right.

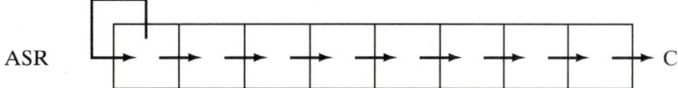

Checkpoint 2.36: Use D flip-flops like Figure 2.20 to build an 8-bit ASR function.

The same shift left operation works for both unsigned and signed multiply by 2, as shown in Figure 2.22. In other words, the arithmetic shift left (ASL) is identical to the logical shift left (LSL). A zero is shifted into the least significant position, and the carry bit will contain the bit that was shifted out.

Figure 2.22
8-bit shift left.

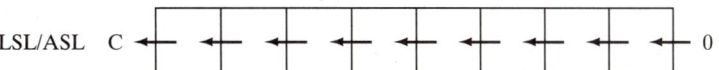

The **roll** operation can be used to create multiple-byte shift functions. Roll right and roll left are shown in Figure 2.23. In each case, the carry is shifted into the 8-bit byte, and the carry bit will contain the bit that was shifted out.

Figure 2.23
8-bit roll right and 8-bit roll left.

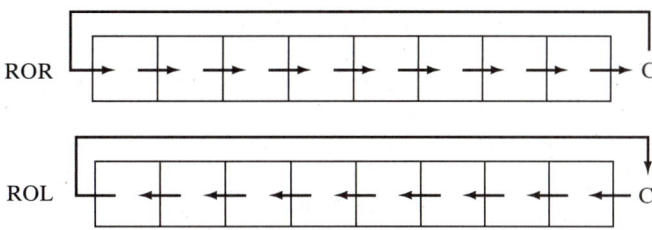

The simplest way to perform a shift operation on the microcomputer is to use a register like Register A. These instructions are a few of the shift functions available on the 6811/6812. `asla` and `lsla` instructions have identical machine codes. The two assembly language names allow the programmer to write clearer code. The following instructions use inherent addressing which means no memory address is required.

```
asla    RegA=RegA*2      arithmetic shift left RegA
lsla    RegA=RegA*2      logical shift left RegA
asra    RegA=RegA/2      arithmetic shift right RegA
lsra    RegA=RegA/2      logical shift right RegA
rola    roll left RegA
rora    roll right RegA
```

Maintenance Tip: *Use the* `asla` *instruction when manipulating signed numbers, and use the* `lsla` *instruction when shifting unsigned numbers.*

The following C code uses the **shift** and **or** operations to combine two parts into one number. `High` and `Low` are unsigned 4-bit components, which will be combined into a single unsigned 8-bit `Result`. We will assume both `High` and `Low` are bounded within the range of 0 to 15. The expression `High<<4` will perform four logical shift lefts.

```
Result = (High<<4)|Low;
```

The assembly code for this operation is

```
ldaa    High      read value of High
lsla              shift into position
lsla
lsla
lsla
oraa    Low       combine the two parts together
staa    Result    save answer
```

To illustrate how the above program works, let $0\ 0\ 0\ 0\ h_3\ h_2\ h_1\ h_0$ be the value of `High`, and let $0\ 0\ 0\ 0\ l_3\ l_2\ l_1\ l_0$ be the value of `Low`. The `ldaa` instruction brings `High` into register A. The four `lsla` instructions move the `High` into bit positions 4-7, the `oraa` instruction combines `High` and `Low` and the `staa` instruction stores the combination into `Result`.

0	0	0	0	h_3	h_2	h_1	h_0	value of `High`
0	0	0	h_3	h_2	h_1	h_0	0	after first `lsla`
0	0	h_3	h_2	h_1	h_0	0	0	after second `lsla`
0	h_3	h_2	h_1	h_0	0	0	0	after third `lsla`
h_3	h_2	h_1	h_0	0	0	0	0	after last `lsla`
0	0	0	0	l_3	l_2	l_1	l_0	value of `Low`
h_3	h_2	h_1	h_0	l_3	l_2	l_1	l_0	result of the `oraa` instruction

The C code `Result=Low&0x0F; Result|=(High&0x0F)<<4;` first clears the bits 7,6,5,4 of `High` and `Low`. The assembly code for this new operation, which combines the bits 3,2,1,0 of `High` and `Low` regardless of the high nibbles of the two inputs is

```
ldaa    Low
anda    #$0F
staa    Result
ldaa    High      read value of High
anda    #$0F
lsla              shift into position
lsla
lsla
lsla
oraa    Result    combine the two parts together
staa    Result    save answer
```

2.8 Arithmetic Operations

When software executes arithmetic instructions, the operations are performed by digital hardware inside the processor. Even though the design of such logic is complex, we will present a brief introduction, in order to provide a little insight as to how the computer performs arithmetic. It is important to remember that arithmetic operations (addition, subtraction, multiplication, and division) have constraints when performed with finite precision on a microcomputer. An overflow error occurs when the result of an arithmetic operation can not fit into the finite precision of the result. For example, when two 8-bit numbers are added, the sum may not fit back into the 8-bit result. Previously, we stated that the same digital hardware (instructions) could be used to add and subtract unsigned and signed numbers. Unfortunately, we will have to design separate overflow detection for signed and unsigned addition and subtraction.

Checkpoint 2.37: How many bits does it take to store the result of two unsigned 8-bit numbers added together?

Checkpoint 2.38: How many bits does it take to store the result of two signed 8-bit numbers added together?

Checkpoint 2.39: How many bits does it take to store the result of two unsigned 8-bit numbers multiplied together?

Checkpoint 2.40: How many bits does it take to store the result of two signed 8-bit numbers multiplied together?

It is common for computers to perform arithmetic operations using an *accumulator*, like Register A, which is a high-speed storage register inside the processor with which arithmetic and logic operations can be performed. The following instructions are a few of the arithmetic functions available on the 6811/6812, which fetch data from memory and add/subtract it from Register A. With immediate mode (#w) the 8-bit constant is located in the instruction itself. With direct mode (u) the 8-bit data is fetched from memory location u (u is an address from 0 to 255). With extended mode (U) the 8-bit data is fetched from the 16-bit memory location U. Recall that direct/extended mode affects the size of the address, not the size of the data. Each of these instructions use 8-bit data. The size of the data will be determined by the size of the register into which the operation will be performed. We will discuss multiplication and division in the next chapter.

```
adda  #w   RegA=RegA+w     add to register A
adda  u    RegA=RegA+[u]   add to register A
adda  U    RegA=RegA+[U]   add to register A
suba  #w   RegA=RegA-w     subtract from register A
suba  u    RegA=RegA-[u]   subtract from register A
suba  U    RegA=RegA-[U]   subtract from register A
```

All microcomputers have a *condition code register* (CC or CCR) that specifies the status of the most recent operation. In this section, we will introduce the four condition code bits common to most microcomputers, shown in Table 2.19. If the two inputs to an addition or subtraction operation are considered as unsigned, then the C bit (carry) will be set if the result does not fit. In other words, after an unsigned addition, the C bit is set if the answer is wrong. If the two inputs to an addition or subtraction operation are considered as signed, then the V bit (overflow) will be set if the result does not fit. In other words, after a signed addition, the V bit is set if the answer is wrong.

Table 2.19
Condition code bits contain the status of the previous arithmetic or logical operation.

bit	name	meaning after addition or subtraction
N	negative	result is negative
Z	zero	result is zero
V	overflow	signed overflow
C	carry	unsigned overflow

We begin the design of an adder circuit with a simple subcircuit called a binary full adder, as shown in Figure 2.24. There are two binary data inputs A, B and a carry input, C_{in}. There is one data output S_{out}, and one carry output, C_{out}. As shown in Table 2.20, C_{in}, A, and B are three independent binary inputs each having a significance or 0 or 1. These three inputs are added together (the sum could be 0, 1, 2, or 3) and the result is encoded in the two-bit binary result with C_{out} as the most significant bit and S_{out} as the least significant bit. C_{out} is true if the sum is 2 or 3, and S_{out} is true if the sum is 1 or 3.

Figure 2.24
A binary full adder.

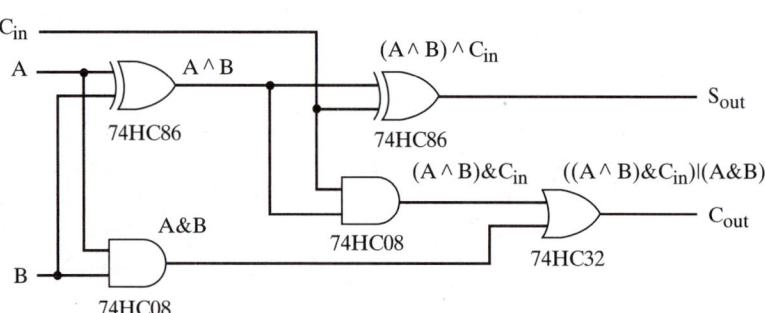

Table 2.20
Input/output response of a binary full adder.

A	B	C_{in}	$A+B+C_{in}$	C_{out}	S_{out}
0	0	0	0	0	0
0	0	1	1	0	1
0	1	0	1	0	1
0	1	1	2	1	0
1	0	0	1	0	1
1	0	1	2	1	0
1	1	0	2	1	0
1	1	1	3	1	1

We can build an 8-bit adder by concatenating eight binary full adders together, as shown in Figure 2.25. The carry into the 8-bit adder is zero, and the carry out will be saved in the carry bit of the CCR.

Figure 2.25
We make an 8-bit adder using eight binary full adders.

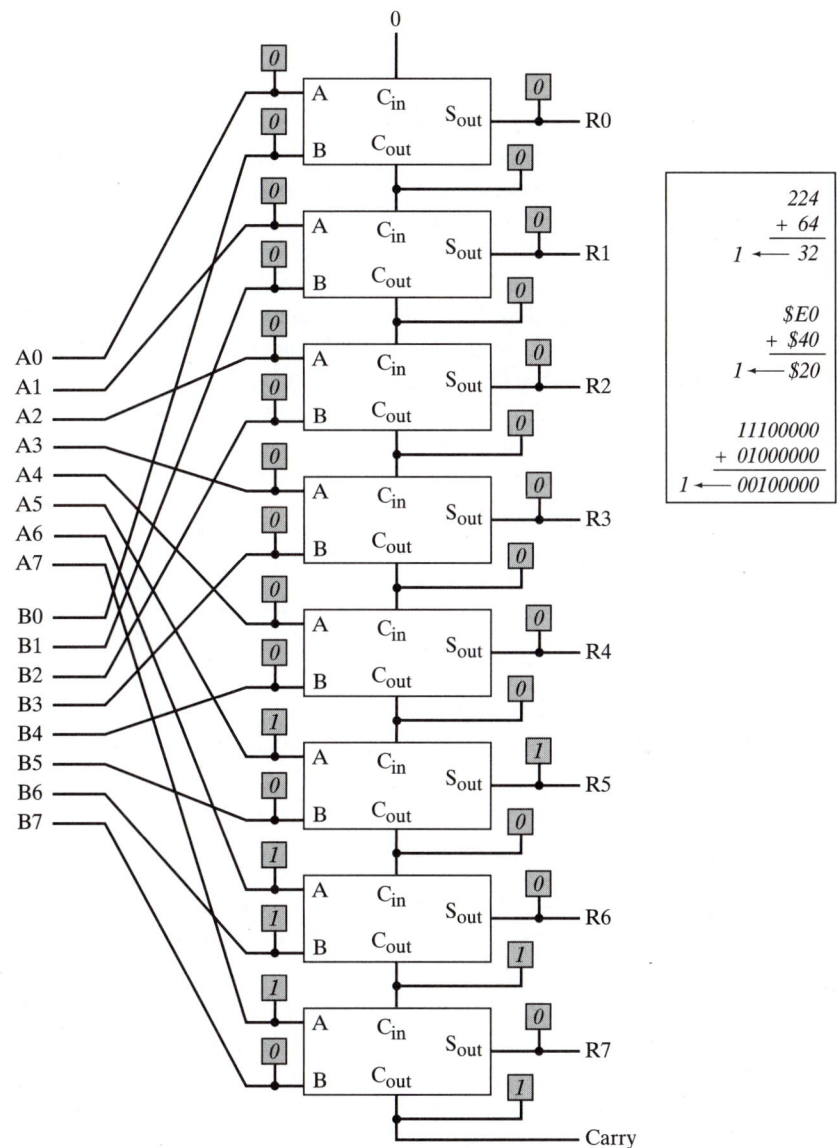

We are now ready to understand the sequence of events required to execute the instruction `adda #64`. First the current value of Register A and the constant 64 are connected to the two 8-bit inputs of the 8-bit binary adder. Next, the result of the addition is stored in back in Register A, and the condition code bits are appropriately set. Assume Register A is initially 224. The shaded boxes and *italicized* values in Figure 2.25 show this particular case where the A input to the adder is 224 (11100000_2) and the B input equals 64 (01000000_2). For bits 0, 1, 2, 3, and 4, the A and B inputs are 0, so the carry out and result bits will also be 0. Since A5 is 1 and B5 equals 0, the result R5 will also be 1, and the carry from 5 to 6 will be 0. For bit 6, both A6 and B6 are 1, so the result R6 is zero, but the carry between bits 6 and 7 is 1. For bit 7, the carry in and A7 are 1, so the result R7 is 0, and the carry out is 1. The carry out of bit 7 will represent the unsigned overflow for the entire 8-bit addition.

For an 8-bit unsigned number, there are only 256 possible values, which are 0 to 255. We can think of the numbers as positions along a circle, like a clock. There is a discontinuity at the 0|255 interface; everywhere else adjacent numbers differ by ±1. If we add two unsigned numbers, we start at the position of the first number and move in a clockwise direction the number of steps equal to the second number. As shown in Figure 2.26, if 96 + 64 is performed in 8-bit unsigned precision, the correct result of 160 is obtained. In this case, the carry bit will be 0 signifying the answer is correct. On the other hand, if 224+64 is performed in 8-bit unsigned precision, the incorrect result of 32 is obtained. In this case, the carry bit will be 1, signifying the answer is wrong.

Figure 2.26
Number wheel showing
96+64 and 224+64.

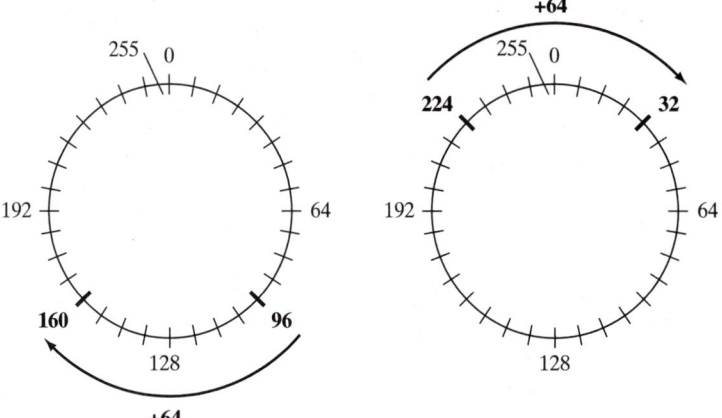

To calculate the negative of a two's complement number, we complement all of the bits and add 1. For example, the 8-bit binary representation for −100 is 10011100. The complement of this binary value is 01100011. When we add 1 to 01100011, we get the binary 01100100, which is the proper representation for 100. Using this fact, we can build an 8-bit subtractor (R = A−B) by first negating B, then using eight binary full adders to add A plus −B, as shown in Figure 2.27. The carry into the 8-bit adder is one, and the carry out is inverted and saved in the carry bit of the CCR.

Figure 2.27
We make an 8-bit subtractor using eight binary full adders.

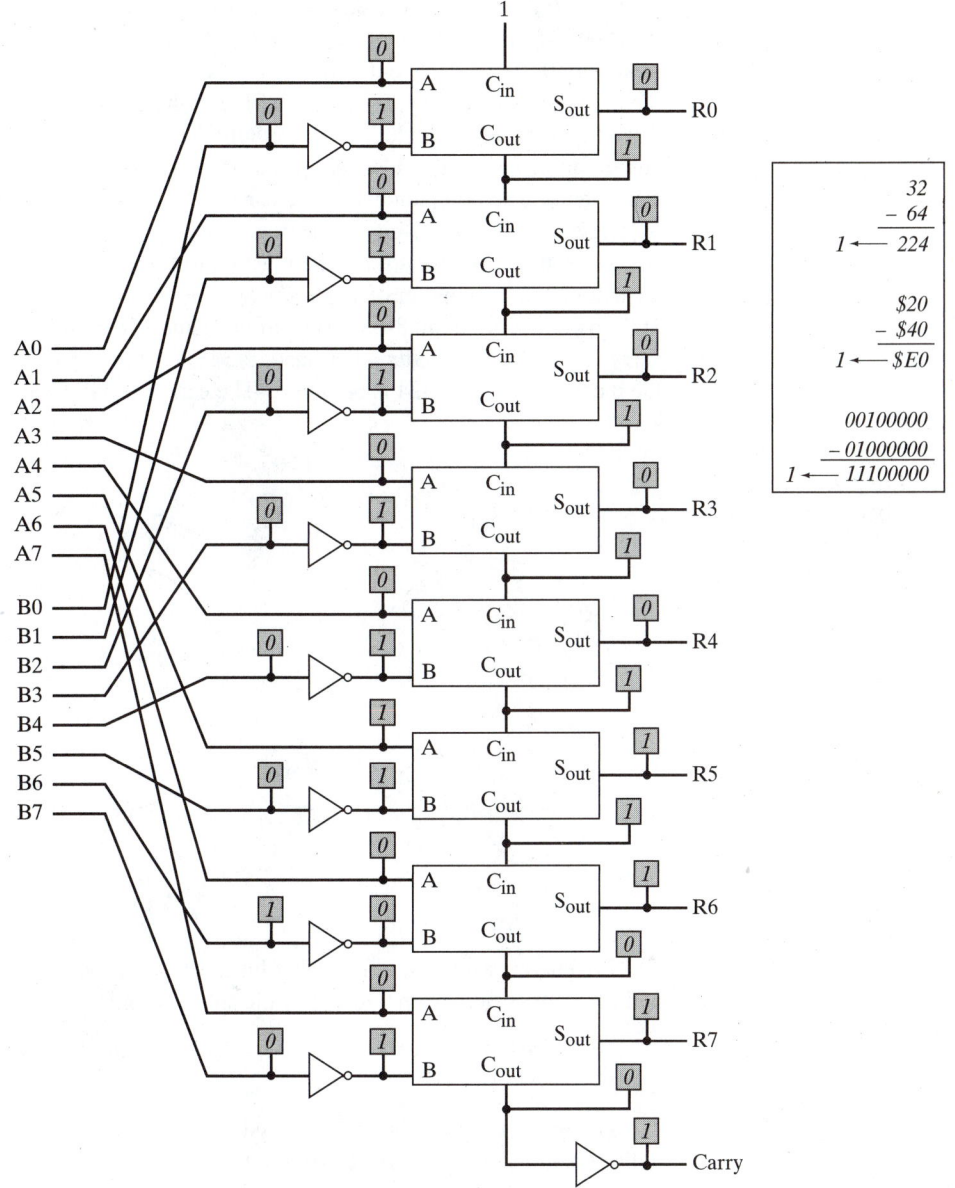

Again, let's consider the sequence of events required to execute the instruction `suba #64`. First the current value of Register A and the constant 64 are connected to the two 8-bit inputs of the 8-bit binary subtractor. The eight NOT gates perform the complement of B. The eight binary adders perform the addition of A and the complement of B. Notice that the carry into the 8-bit adder is 1 as needed to implement the subtraction A−B. Lastly, the result of the subtraction is stored in back in Register A, and the condition code bits are appropriately set. Assume Register A is initially 32. The shaded boxes and *italicized*

values in Figure 2.27 show this particular case where the A input to the subtractor is 32 (00100000_2) and the B input equals 64 (01000000_2). For bits 0, 1, 2, 3, and 4, the A and B inputs are 0, so the carry out will be 1 and result bits will be 0. The full adder in bit 5 sees three ones, so the result R5 will also be 1, and the carry from 5 to 6 will be 1. For bit 6, the full adder sees just a single 1, so the result R6 is 1, and there is no carry from bit 6 to bit 7. Again for bit 7, the full adder sees just a single 1, so the result R7 is 1, and there is no carry out of bit 7. The complement of the carry out of bit 7 will be the represent the unsigned overflow for the entire 8-bit subtraction.

For subtraction, we start at the position of the first number a move in a counterclockwise direction the number of steps equal to the second number. As shown in Figure 2.28, if $160 - 64$ is performed in 8-bit unsigned precision, the correct result of 96 is obtained. (The carry bit will be 0.) On the other hand, if $32 - 64$ is performed in 8-bit unsigned precision, the incorrect result of 224 is obtained. (The carry bit will be 1.)

Figure 2.28
Number wheel showing
$160 - 64$ and $32 - 64$.

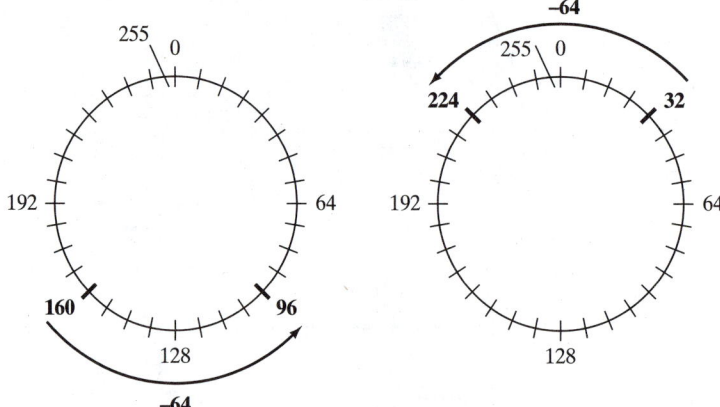

In general, we see that the carry bit is set when we cross over from 255 to 0 while adding or cross over from 0 to 255 while subtracting.

Observation: The carry bit, C, is set after an unsigned addition or subtraction when the result is incorrect.

For an 8-bit signed number, the possible values range from -128 to 127. Again there is a discontinuity, but this time it exists at the $-128|127$ interface. Everywhere else adjacent numbers differ by ±1. The meanings of the numbers with bit 7=1 are different from meanings of unsigned numbers, but we add and subtract signed numbers on the number wheel in a similar way (e.g., addition of a positive number moves clockwise.) Therefore, we can use the same hardware (Figures 2.25 and 2.27) to add and subtract two's complement signed numbers. The only difference is that the carry out generated by the circuits do not represent an error when adding or subtracting two's complement signed numbers. Instead a new bit, called overflow or **V**, will be calculated to signify errors when operating on signed numbers. Adding a negative number is the same as subtracting a positive number; hence, this operation would cause a counterclockwise motion. As shown in Figure 2.29, if $-32 + 64$ is performed, the correct result of 32 is obtained. In this case, the overflow bit will be 0 signifying the answer is correct. On the other hand, if $96 + 64$ is performed, the incorrect result of -96 is obtained. In this case the overflow bit will be 1, signifying the answer is wrong.

Figure 2.29
Number wheel showing
$-32+64$ and $96+64$.

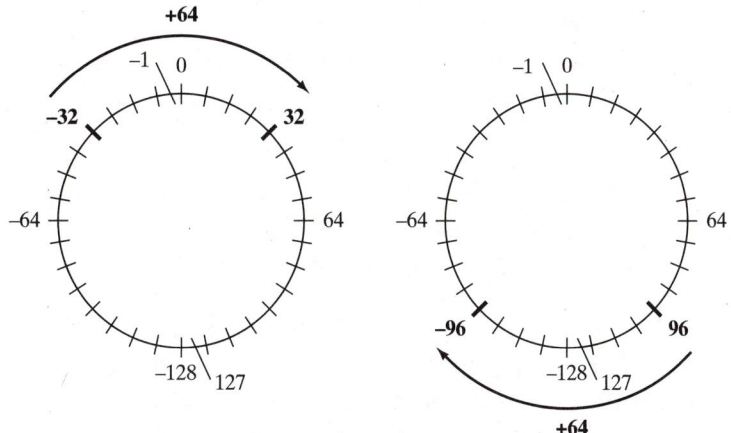

For subtracting signed numbers, we again move in a counterclockwise direction. Subtracting a negative number is the same as adding a positive number, hence this operation would cause a clockwise motion. As shown in Figure 2.30, if $32-64$ is performed, the correct result of -32 is obtained. (The overflow bit will be 0.) On the other hand, if $-96-64$ is performed, the incorrect result of 96 is obtained. (The overflow bit will be 1.)

Figure 2.30
Number wheel showing
$32-64$ and $96-64$.

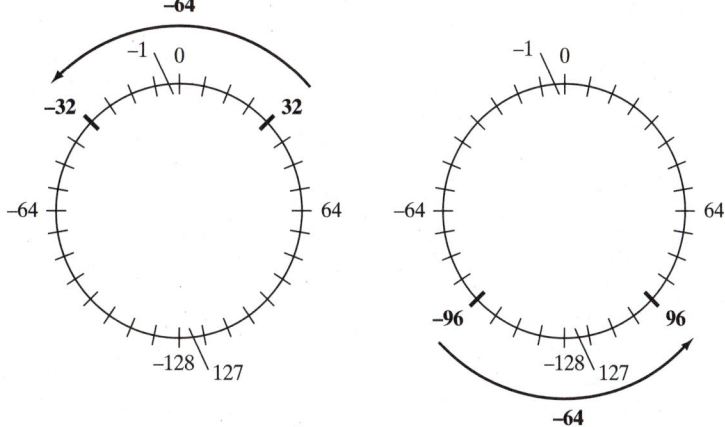

In general, we see that the overflow bit, **V**, is set when we cross over from 127 to -128 while adding or cross over from -128 to 127 while subtracting.

Observation: The overflow bit, V, is set after a signed addition or subtraction when the result is incorrect.

Another way to determine the overflow bit after an addition is to consider the carry out of bit 6. The V bit will be set if there is a carry out of bit 6 (into bit 7) but no carry out of bit 7 (into the C bit). It is also set if there is no carry out of bit 6 but there is a carry out of bit 7. Let A7,A6,A5,A4,A3,A2,A1,A0 and B7,B6,B5,B4,B3,B2,B1,B0 be the individual binary bits of the two 8-bit numbers that are to be added, and let R7,R6,R5,R4,R3,R2,R1,R0 be individual binary bits of the 8-bit sum, as implemented in Figure 2.25. The N bit is set if the unsigned result is above 127 or if the signed result is negative.

$$N = R7$$

The Z bit is set if the result is zero. The Z bit will be clear if any of the result bits is one.

$$Z = \overline{R7} \ \& \ \overline{R6} \ \& \ \overline{R5} \ \& \ \overline{R4} \ \& \ \overline{R3} \ \& \ \overline{R2} \ \& \ \overline{R1} \ \& \ \overline{R0}$$

If the V bit is set after a signed addition, then the result is incorrect because a signed over-flow occurred. The first term of the following equation is true if you add two negative num-bers together and get a positive result. The second term is true if you add two positive numbers together and get a negative result.

$$V = A7 \ \& \ B7 \ \& \ \overline{R7} \ + \ \overline{A7} \ \& \ \overline{B7} \ \& \ R7$$

If the C bit is set after an unsigned addition, then the result is incorrect because an unsigned overflow occurred. The first term of the following equation is true if you add two numbers both above 127. The second term is true if the A input is above 127, but the result is less than or equal to 127. The third term is true if the B input is above 127, but the result is less than or equal to 127.

$$C = A7 \ \& \ B7 \ + \ A7 \ \& \ \overline{R7} \ + \ B7 \ \& \ \overline{R7}$$

Checkpoint 2.41: *Assume Register A is initially* -100. *After executing the instruction* adda #64 *what is the value in Register A, and the NZVC bits?*

Checkpoint 2.42: *Assume Register A is initially* -100. *After executing the instruction* adda #-64 *what is the value in Register A, and the NZVC bits?*

In a similar manner, let the result R be the result of the subtraction A−B, as implemented in Figure 2.27. The N and Z bits are the same as with addition.

$$N = R7$$

$$Z = \overline{R7} \ \& \ \overline{R6} \ \& \ \overline{R5} \ \& \ \overline{R4} \ \& \ \overline{R3} \ \& \ \overline{R2} \ \& \ \overline{R1} \ \& \ \overline{R0}$$

If the V bit is set after a signed subtraction (R=A−B), then the result is incorrect because a signed overflow occurred. The first term of the following equation is true if you subtract a negative number minus a positive number together and get a positive result (a negative number minus a positive number should still be negative). The second term is true if you subtract a positive number minus a negative number together and get a negative result (a positive number minus a negative number should still be positive).

$$V = A7 \ \& \ \overline{B7} \ \& \ \overline{B7} \ + \ \overline{A7} \ \& \ B7 \ \& \ R7$$

If the C bit is set after an unsigned subtraction (R=A−B), then the result is incorrect be-cause an unsigned overflow occurred. The first term of the following equation is true if you subtracted a number greater than 127 (B>127) from a number less than 128 (A<128). The second term is true if the B input is above 127, but the result is greater than 127. The third term is true if the A input is less than or equal to 127, but the result is greater than 127.

$$C = \overline{A7} \ \& \ B7 \ + \ B7 \ \& \ R7 \ + \ \overline{A7} \ \& \ R7$$

Checkpoint 2.43: *Assume Register A is initially 200. After executing the instruction* suba #64 *what is the value in Register A, and the NZVC bits?*

Checkpoint 2.44: *Assume Register A is initially 200. After executing the instruction* suba #-64 *what is the value in Register A, and the NZVC bits?*

Common Error: *Ignoring overflow (signed or unsigned) can result in significant errors.*

Observation: *Microcomputers have two sets of conditional branch instructions (if statements) that make program decisions based on either the C or V bit.*

Some instructions operate only on signed numbers, and others work only for unsigned numbers. An error will occur if you use unsigned instructions after operating on signed numbers, and vice versa. There are some applications where arithmetic errors are not possible. For example if we had two 8-bit unsigned numbers that we knew were in the range of 0 to 100, then no overflow is possible when they are added together.

Typically the numbers we are processing are either signed or unsigned (but not both), so we need only consider the corresponding C or V bit (but not both the C and V bits at the same time). In other words, if the two numbers are unsigned, then we look at the C bit and ignore the V bit. Conversely, if the two numbers are signed, then we look at the V bit and ignore the C bit. There are two appropriate mechanisms to deal with the potential for arithmetic errors when adding and subtracting. The first mechanism, used by most compilers, is called *promotion.* Promotion involves increasing the precision of the input numbers, and performing the operation at that higher precision. An error can still occur if the result is stored back into the smaller precision. Fortunately, the program has the ability to test the intermediate result to see if it will fit into the smaller precision. To promote an unsigned number we add zeros to the left side. In a previous example, we added the unsigned 8-bit 224 to 64, and got the wrong result of 32. With promotion we first convert the two 8-bit numbers to 16 bits, then add

decimal	8-bit	16-bit
224	1110,0000	0000,0000,1110,0000
+ 64	+0100,0000	+0000,0000,0100,0000
288	0010,0000	0000,0001,0010,0000

We can check the 16-bit intermediate result (e.g., 288) to see if the answer will fit back into the 8-bit result. In Figure 2.31, A and B are 8-bit unsigned inputs, A16, B16, and R16 are 16-bit intermediate values, and R is an 8-bit unsigned output. The oval symbol represents the entry and exit points, the rectangle is used for calculations, and the diamond shows a decision.

Figure 2.31
Flowcharts showing how to use promotion to detect and correct unsigned arithmetic errors.

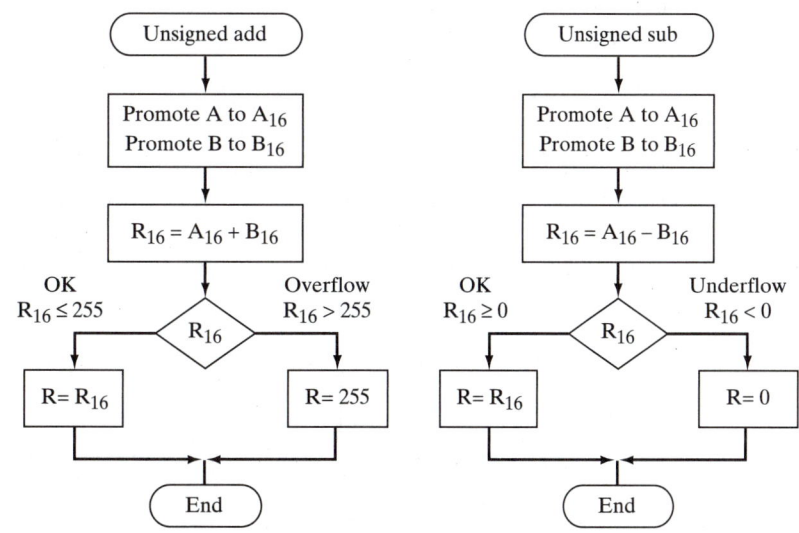

The C code in Program 2.2 adds and subtracts two 8-bit values, using promotion to detect for errors.

Program 2.2
Using promotion to detect and compensate for unsigned overflow errors.

```
unsigned char A,B,R;
void add(void) {  unsigned short result;
  result = A+B;     /* promote and perform 16-bit addition */
  if(result>255){   /* check for overflow */
    result = 255;   /* yes, overflow occurred */
  }
  R = result;       /* demote back to 8 bits */
}
void sub(void) {  short result;
  result = A-B;     /* promote and perform 16-bit subtraction */
  if(result<0) {    /* check for underflow */
    result = 0;     /* yes, underflow occurred */
  }
  R = result;       /* demote back to 8 bits */
}
```

To promote a signed number, we duplicate the sign bit as we add binary digits to the left side. Earlier, we performed the 8-bit signed operation $-96 - 64$ and got a signed overflow. With promotion we first convert the two numbers to 16 bits, then subtract

decimal	8-bit	16-bit
-96	1010,0000	1111,1111,1010,0000
-64	-0100,0000	-0000,0000,0100,0000
-160	0110,0000	1111,1111,0110,0000

We can check the 16-bit intermediate result (e.g., -160) to see if the answer will fit back into the 8-bit result. In Figure 2.32, A and B are 8-bit signed inputs, A16, B16, and R16 are 16-bit signed intermediate values, and R is an 8-bit signed output.

Figure 2.32
Flowcharts showing how to use promotion to detect and correct signed arithmetic errors.

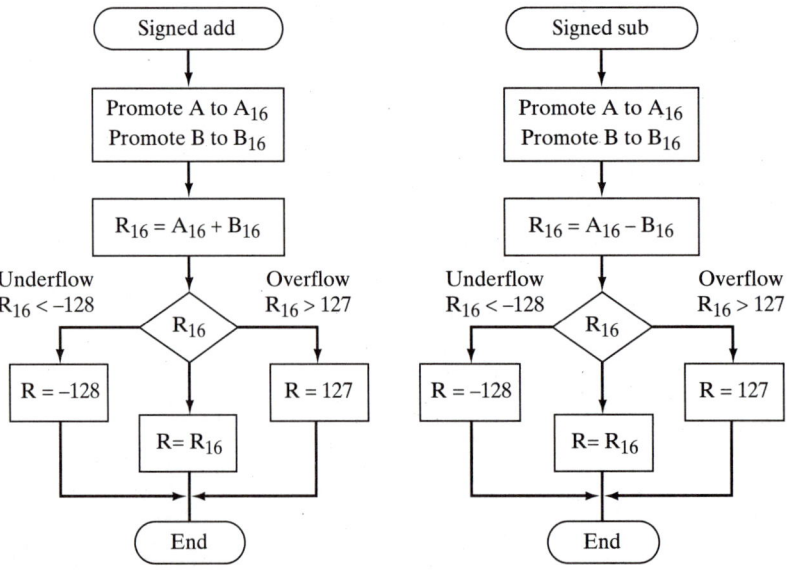

The C code in Program 2.3 adds and subtracts two 8-bit signed numbers. The compiler will automatically promote A and B to signed 16-bit values before the addition.

Program 2.3
Using promotion to detect and compensate for signed overflow errors.

```
char A,B,R;
void add(void) {  short result;
   result = A+B;        /* promote and perform 16-bit addition */
   if(result>127) {     /* check for overflow */
     result = 127;      /* yes, overflow occurred */
   }
   if(result<-128) {    /* check for underflow */
     result = -128;     /* yes, underflow occurred */
   }
   R = result;          /* demote back to 8 bits */
}
void sub(void) {  short result;
   result = A-B;        /* promote and perform 16-bit subtraction
*/
   if(result>127) {     /* check for overflow */
     result = 127;      /* yes, overflow occurred */
   }
   if(result<-128) {    /* check for underflow */
     result = -128;     /* yes, underflow occurred */
   }
   R = result;          /* demote back to 8 bits */
}
```

Observation: *When performing calculations on 8-bit numbers, most C compilers for the 6811 and 6812 will first promote to 16 bits, perform the operations using 16-bit operations, then demote the result back to 8 bits.*

Common Error: *Even though most C compilers automatically promote to a higher precision during the intermediate calculations, they do not check for overflow when demoting the result back to the original format.*

We will put off implementing the functions of Programs 2.2 and 2.3 in assembly language until Chapter 8 (Programs 8.29 and 8.30), after the necessary registers and instructions have been introduced. On the other hand, with just a couple of additional instructions, we can use another approach to detect and correct overflow errors occurring during addition and subtraction. The following instructions are a few of the conditional branch instructions available on the 6811/6812. In each of these cases, the instruction will test one of the condition code bits and branch (change the PC) if the condition exists. If the condition is false, then the program does not branch, and the computer will continue execution with the instruction immediately following the conditional branch. `location` can be any nearby[1] label within our program.

```
bcc location   branch on carry clear,    jump to location if C=0
bcs location   branch on carry set,      jump to location if C=1
bvc location   branch on overflow clear, jump to location if V=0
bvs location   branch on overflow set,   jump to location if V=1
bpl location   branch on plus,           jump to location if V=0
bmi location   branch on minus,          jump to location if N=1
beq location   branch on equal,          jump to location if Z=0
bne location   branch on not equal,      jump to location if Z=1
```

---

[1] The branch location must be within 127 bytes of the current location.

The other mechanism for handling addition and subtraction errors is called *ceiling and floor* It is analogous to movements inside a room. If we try to move up (add a positive number or subtract a negative number) the ceiling will prevent us from exceeding the bounds of the room. Similarly, if we try to move down (subtract a positive number or add a negative number) the floor will prevent us from going too low. The ceiling and floor prevent us from leaving the room. For our 8-bit addition and subtraction, we will prevent the 0 to 255 and 255 to 0 crossovers for unsigned operations and -128 to $+127$ and $+127$ to -128 crossovers for signed operations. These operations are described by the flowcharts in Figure 2.33. If the carry bit is set after an unsigned addition, the result is adjusted to the largest possible unsigned number (ceiling). If the carry bit is set after an unsigned subtraction, the result is adjusted to the smallest possible unsigned number (floor).

Figure 2.33
Flowcharts showing how to use overflow bits to detect and correct unsigned arithmetic errors.

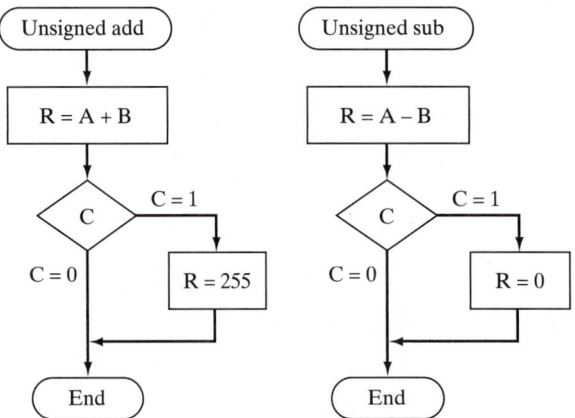

There are no mechanisms in C to access the condition code bits of the processor. So, implementation of this approach must be performed in assembly language. The pseudo-op `ds` stands for **define storage**. The operand field specifies the number of bytes, and we use it to create global variables. Assume `A8` `B8` and `R8` are three 8-bit (1-byte) global variables defined in RAM.

```
A8   ds   1   Input
B8   ds   1   Input
R8   ds   1   Output
```

The following assembly language adds two unsigned 8-bit numbers, using the algorithm presented in Figure 2.33.

```
      ldaa   A8   get first input
      adda   B8   A8+B8
      bcc    OK1  if C=0, then no error, so skip to the end
      ldaa   #255 overflow
OK1   staa   R8
```

The following assembly language subtracts two unsigned 8-bit numbers.

```
      ldaa   A8     get first parameter
      suba   B8     A8-B8
      bcc    OK2    if C=0, then no error, so skip to the end
      ldaa   #0     underflow
OK2   staa   R8
```

Signed addition and subtraction are described by the flowcharts in Figure 2.34. If the overflow bit is set after a signed operation the result is adjusted to the largest (ceiling) or smallest (floor) possible signed number depending on whether it was a −128 to 127 cross over (N=0) or 127 to −128 cross over (N=1). Notice that after a signed overflow, bit 7 of the result is always wrong because there was a cross over.

Figure 2.34
Flowcharts showing how to use overflow bits to detect and correct signed arithmetic errors.

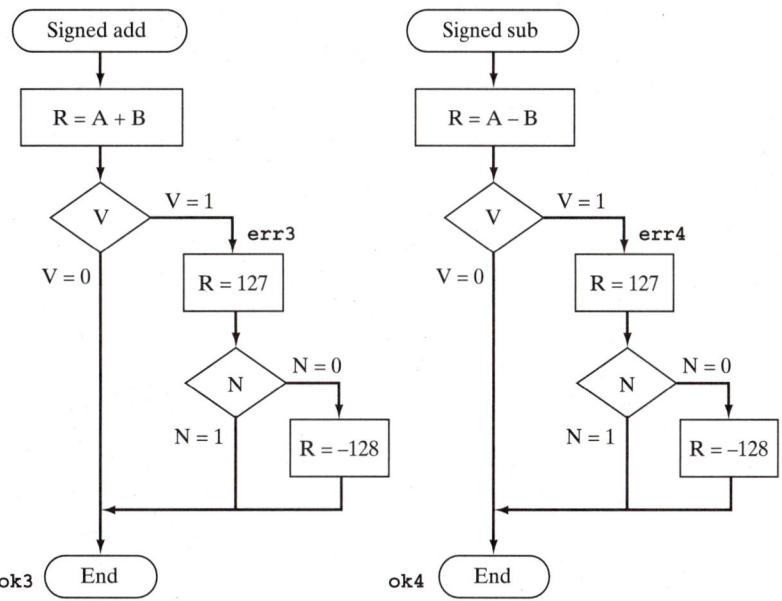

The following assembly language adds two signed 8-bit numbers, using the algorithm presented in Figure 2.34.

```
      ldaa A8     get first input
      adda B8     A8+B8
      bvc  ok3    if V=0, then no error, so skip to the end
err3  ldaa #127   assume it was overflow
      bmi  ok3    if V=1 and N=1, it was overflow, so skip to ok3
      ldaa #-128  if V=1 and N=0, it was underflow
ok3   staa R8
```

The following assembly language subtracts two signed 8-bit numbers.

```
      ldaa   A8     get first parameter
      suba   B8     A8-B8
      bvc    ok4    if V=0, then no error, so skip to the end
err4  ldaa   #127   assume overflow
      bmi    ok4    if V=1 and N=1, it was overflow, so skip to ok4
      ldaa   #-128  if V=1 and N=0, it was underflow
ok4   staa   R8
```

The algorithms for multiply and divide are included in this embedded systems book for two reasons. First, as programmers, we need to know how the computer works, so we can understand the strengths and limitations of our computer. Second, many embedded micro-computers have limited ability for performing mathematical operations. For example, the only multiplication the 6811 processor can perform is 8-bit by 8-bit unsigned multiplica-tion. Although the 6812 has signed and unsigned 16-bit multiplication, sometimes we need more precision than is supported by the processor. In these situations, we must use a more powerful microcomputer (if speed is important) or develop software algorithms for ex-tended precision arithmetic (if speed is not important).

We can perform unsigned multiplication using a combination of shift and addition op-erations. Let A and B be two unsigned 8-bit numbers, and $R = A \cdot B$. Simple calculations of $0 \cdot 0 = 0$ and $255 \cdot 255 = 65025$ illustrate the fact that the multiplication of two 8-bit num-bers will fit into a 16-bit product. To develop the algorithm used by the multiplication hard-ware, define one of the multiplicands in its basis representation.

$$B = 128 \cdot b_7 + 64 \cdot b_6 + 32 \cdot b_5 + 16 \cdot b_4 + 8 \cdot b_3 + 4 \cdot b_2 + 2 \cdot b_1 + b_0$$

Next, we distribute multiplication over addition

$$R = A \cdot 128 \cdot b_7 + A \cdot 64 \cdot b_6 + A \cdot 32 \cdot b_5 + A \cdot 16 \cdot b_4 + A \cdot 8 \cdot b_3 + A \cdot 4 \cdot b_2 + A \cdot 2 \cdot b_1 + A \cdot b_0$$

We can simplify the equation leaving only one-bit shifts

$$R = 2 \cdot (2 \cdot (2 \cdot (2 \cdot (2 \cdot (2 \cdot (2 \cdot A \cdot b_7 + A \cdot b_6) + A \cdot b_5) + A \cdot b_4) + A \cdot b_3) + A \cdot b_2) + A \cdot b_1) + A \cdot b_0$$

The multiplication by a power of 2 is a logical shift left, and the multiplication by a binary bit (0 or 1) is an add or no-add conditional. This equation motivates the following multi-plication algorithm. The multiplication function will be implemented as digital hardware in the processor, and available as an assembly language instruction, but here it is shown as a C function in Program 2.4. In particular, this exact function is available as the assembly in-struction `mul`, which will be presented in the next chapter. For an 8-bit multiply, we will use 16-bit shifts and additions, yielding a 16-bit product. Since the product, R, is a 16-bit unsigned number, there can be no overflow error in this 8 by 8 into 16-bit multiply.

Program 2.4
Unsigned 8-bit times 8-bit multiplication yielding a 16-bit product.

```
unsigned short mul(unsigned char A, unsigned char B) {
unsigned short R = 0;    /*  result, R=A*B    */
int n;
  for(n=0; n<8;  n++){
    R = R<<1;    /* shift left */
    if(B&0x80){        /* should we add? */
       R = R+A;           /* A is promoted first, then added */
    }
    B = B<<1;    /* move next bit into bit 7 position */
  }
  return R;
}
```

To better understand multiplication, we will hand execute Program 2.4 for A=100, B=10. In particular, Table 2.21 shows the values of n, B, and R after the if-statement, but before the B=B<<1.

Table 2.21
Example multiplication
of 10 by 100.

n	B	R	comments
0	10 = $0A	0	B7 is zero, no addition
1	20 = $14	0	B6 is zero, no addition
2	40 = $28	0	B5 is zero, no addition
3	80 = $50	0	B4 is zero, no addition
4	160 = $A0	100	B3 is one, add 100 to R
5	64 = $40	200	B2 is zero, no addition
6	128 = $80	500	B1 is one, add 100 to R
7	0 = $00	1000	B0 is zero, no addition

For signed multiplication we will first check the sign of each input, perform an unsigned multiplication on the absolute values of the inputs, then negate the product if necessary. Simple calculations of $-128 \bullet -128 = 16384$, $-128 \bullet 127 = -16256$, and $127 \bullet 127 = 16129$ illustrate the fact that the multiplication of two 8-bit signed numbers will always fit into a signed 16-bit product. The algorithm for signed multiplication is presented as a C function in Program 2.5.

Program 2.5
Signed 8-bit times 8-bit
multiplication yielding a
16-bit product.

```
short smul(char A, char B)  {
int sign = 0;     /* 0 means positive */
  if(A < 0){
    sign++;       /* A is negative */
    A = -A;       /* absolute value */
  }
  if(B < 0){
    sign--;       /* B is negative */
    B = -B;       /* absolute value */
  }
  if(sign)
    return -mul(A,B);  /* product is negative */
  else
    return mul(A,B);   /* product is positive */
}
```

We can perform unsigned division using a combination of shift and subtract operations. Let N be the unsigned 16-bit dividend and M be the unsigned 8-bit divisor. The 8-bit quotient is Q=N/M, and the 8-bit remainder is R=N%M. Assuming the remainder is less than the divisor, there is a unique solution (Q,R) such that N equals M*Q+R. In C, the division (/) and modulo (%) are separate operators, but in digital hardware and assembly language one operation produces both results. The overflow can occur two ways in a 16-bit by 8-bit division. A divide by zero causes an error, and an overflow error occurs if the quotient does not fit into 8 bits.

Binary long division is very similar to decimal long division, the way you learned to divide in elementary school. We line the divisor up under the dividend shifted as far to the left as possible so that the line-up (shifted) divisor is less than the dividend. In decimal division, we have to determine the decimal digit 0 to 9 of the quotient and multiply the divisor by that digit. In binary division, the binary digits are 0 or 1, so no multiplication step is required. We just subtract the shifted divisor from the dividend and record a 1 in that place-value for the quotient. We repeat the operation on the result of the subtraction. Figure 2.35 shows the two shift/subtract operations required to divide 1004 by 100, yielding a quotient of 10 and a remainder of 4. The divisor is subtracted twice yielding the two binary bits in the quotient of 10.

Figure 2.35
Binary long division example showing 1004 divided by 100.

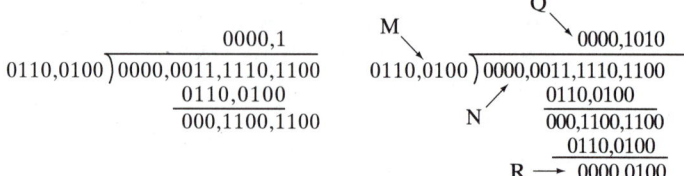

Program 2.6, written in C, uses global variables to input/output parameters. The dividend, N, will be modified during execution and become the remainder. Again, division on an actual computer is usually performed in hardware by the processor, and the C program is presented here only to illustrate the division algorithm. The available assembly instructions for division will be presented in the next chapter. A 64-bit by 32-bit assembly language version of this algorithm can be found in the file math.rtf, which is included as one of the **TExaS** examples, presented in Chapter 8.

Program 2.6   Unsigned 16-bit by 8-bit division yielding an 8-bit quotient and an 8-bit remainder.

```
unsigned short N;    /*  dividend, becomes the remainder */
unsigned char M;     /*  divisor */
unsigned char Q;     /*  quotient */
int error;           /*  -1 is divide by zero, 0 is OK, 1 is overflow */
void div(void) { unsigned short M16; int i;
  if(M)
    error = 0;
  else{
    error = -1;
    return;  /* divide by zero */
  }
  Q = 0;          /* actually this step can be omitted */
  M16 = M<<8;   /* M16 is divisor left-justified in 16 bits */
  for(i=0; i<8; i++){
    M16 = M16>>1;   /* logical shift right */
    Q = Q<<1;       /* logical shift left */
    if(N > M16){    /* should we subtract? */
      Q = Q|0x01;   /* yes, set bit in the quotient */
      N -= M16;     /* reduce dividend, transform into remainder */
    }               /* table values collected at this point */
  }
  if(N > M)         /* N is now the remainder */
    error = 1;      /* overflow if it is bigger than the divisor */
}
```

Hand execution, or *desk checking,* is a convenient way to test a software algorithm. Table 2.22 shows the values of i, N, M16, and Q within the for-loop after the if-statement. Assume initially N equals 1004, M equals 100. The statement M16=M<<8; will initialize M16 to 25600.

Table 2.22
Example division of 1004 divided by 100.

i	N	M16	Q	comments
0	1004	12800	0	Q7 is zero
1	1004	6400	0	Q6 is zero
2	1004	3200	0	Q5 is zero
3	1004	1600	0	Q4 is zero
4	204	800	1	Q3 is one, add bit to Q, subtract from N
5	204	400	2	Q2 is zero
6	4	200	5	Q1 is one, add bit to Q, subtract from N
7	4	100	10	Q0 is zero

For signed division, we force the remainder to be the same sign as the dividend. The absolute value of the remainder will be less than the divisor, and N equals M*Q+R.

Checkpoint 2.45: *Why do we have to specify a rule for the sign of the remainder?*

2.9 Conversions

We will illustrate the conversion between ASCII strings and binary numbers by developing high level functions in C. The purpose of introducing C programs is to introduce the conversion process. Later we will learn how to perform these operations in assembly language. In the first example, let Data be a fixed length string of three ASCII characters. Each entry of Data is an ASCII character 0 to 9. Let Data [0] be the ASCII code for the hundreds digit, Data[1] be the tens digit and Data[2] be the ones digit. Let n be an unsigned 16-bit integer. We will also need an index, i. From Table 2.11, we see the decimal digits 0 to 9 are encoded in ASCII as 0x30 to 0x39. So, to convert a single ASCII digit to binary, we simply subtract 0x30. To convert his string of 3 decimal digits into binary we can simply calculate

```
n = 100*(Data[0]-0x30) + 10*(Data[1]-0x30) + (Data[2]-0x30);
```

This 3-digit ASCII string could also be calculated as

```
n = (Data[2]-0x30) + 10*((Data[1]-0x30) + 10*(Data[0]-0x30));
```

If Data were a string of 5 decimal digits we could put the above function into a loop

```
n = 0;
for (i=0; i<5; i++) {
  n = 10*n + (Data[I]-0x30);
}
```

If Data were a variable length string of ASCII characters terminated with a null character (0), we could convert it to binary using a while loop, as shown in Program 2.7.

Program 2.7
Unsigned ASCII string to
decimal conversion.

```
/*  Convert ASCII string to unsigned 16-bit decimal */
unsigned short Str2UDec(unsigned char Data[]) {
  unsigned short n = 0;   /* the number */
  unsigned int i = 0;     /* index into Data */
  while (Data[i] != 0){
    n = 10*n + (Data[i]-0x30);
    i++;
  }
  return n;
}
```

The example, shown in Program 2.8, also converts an ASCII string into a 16-bit unsigned
decimal number. The program uses pointer syntax to access the string, and processes only
the ASCII digits '0' to '9'. Other ASCII characters in the string are ignored.

Program 2.8
Unsigned ASCII to
decimal conversion.

```
unsigned short Str2UDec(unsigned char *sPt) {
unsigned short n = 0;     /* the number */
unsigned char character;
  while(*sPt){                /* accepts until null */
    character = (*sPt++);   /* fetch next character */
    if((character >= '0') && (character <= '9')) {
      n = 10*n+(character-0x30);  /*  might overflow */
    }
  }
  return n;
}
```

The example, shown in Program 2.9, uses an I/O device capable of sending and receiving
ASCII characters. For 6811/6812 systems this device could be a PC computer connected to
the serial port of the 6811/6812 microcomputer. For a simulation of this configuration see
TUT2, TUT4 or SCI. The function `InChar()` returns an ASCII character from the I/O device. The function `OutChar()` sends an ASCII character to the I/O device. The function
`InUDec()` will accept characters from the device until a carriage return (the `Enter` key)
is typed. Only the numbers are echoed.

Program 2.9
Input an unsigned
decimal number.

```
#define CR 0x0D
/* InUDec accepts ASCII input in unsigned decimal format
    and converts to a 16-bit unsigned number up to 65535
    If n>65535, it will truncate without reporting the error */
unsigned short InUDec(void){ unsigned short n=0;
unsigned char character;
  while((character=InChar()) != CR){  /* accepts until <enter>
*/
    if((character > '0') && (character <= '9')){
      n = 10*n+(character-0x30); /* overflows if above 65535 */
      OutChar(character); /* echo this character */
    }
  }
  return n;
}
```

If the ASCII characters were to contain optional "+" and "−" signs, we could look for the presence of the sign character in the first position, as shown in Program 2.10. If there is a minus sign then set a flag. Next use our unsigned conversion routine to process the rest of the ASCII characters and generate the unsigned number, `n`. If the flag was previously set, we can negate the value `n`. The `length` is used to handle the situation where the operator uses backspace to erase the entire input.

Program 2.10
Input a signed decimal number.

```
/*  InSDec accepts ASCII input in signed decimal  format
        and converts to a signed 16-bit number
        with an absolute value up to 32767
    If you enter a number above 32767 or below -32767,
        it will truncate without reporting the error
    Backspace will remove last digit typed */
#define BS 0x08
short InSDec(void){
short n=0, sign=1; /* sign flag 1=positive 0=negative */
unsigned int length=0;
unsigned char character;
  while((character=InChar()) != CR){ /* accepts until <enter> */
    if (!length){ /* + or - only valid as first char */
      if (character == '-'){
        sign = -1;
        length++;
        OutChar('-'); /* if - inputted, sign is negative */
      }
      else if (character == '+'){
        length++;
        OutChar('+'); /* if + inputted, sign is positive */
      }
    }
    if((character >= '0') && (character <= '9')){
      n = n*10+character-'0'; /* overflows if above 32767 */
      length++;
      OutChar(character);
    }
/* If the input is a backspace, then n is changed and a backspace
   is outputted to the screen. If erases a minus, then sign is
   reset to positive */
    else if ((character==BS) && length){
      n = n/10;
      length--;
      if (!length)
          sign = 1;
      OutChar(BS);
    }
  }
  return sign*n;
}
```

To convert an unsigned integer into a fixed length string of ASCII characters, we could use the integer divide. Assume n is an unsigned integer less than or equal to 999:

```
Data[0] = n/100 + 0x30;
n = n%100;            /* n is now between 0 and 99 */
Data[1] = n/10 + 0x30;
n = n%10;             /* n is now between 0 and 9 */
Data[2] = n + 0x30;
```

To convert an unsigned integer into a variable length string of ASCII characters, we convert the digits in reverse order, then switch them. This conversion technique is shown in Program 2.11.

Program 2.11
Unsigned decimal to ASCII string conversion.

```
unsigned char Data[10];
void UDec2Str(unsigned short n){ unsigned int size,i;
unsigned char temp;
  size = 0; /* the total number of ASCII characters */
  do{
    Data[size] = (n%10) + 0x30; /* Start with the one's digit */
    n = n/10; /* go to next digit */
    size++;
  }
  while(n != 0);
  for(i=0; i< size/2; i++ ){
    temp = Data[i]; /* Reverse order */
    Data[i] = Data[size-1-i]; /* exchange, swap */
    Data[size-1-i] = temp;
  }
  Data[size] = 0; /* Null terminated ASCII string */
}
```

The example shown in Program 2.12 performs the same unsigned conversion, but uses **recursion**. The function OutCh() sends one ASCII character to the output device. A recursive function calls itself. There are two possibilities. If the number is between 0 and 9, the function does not call itself. The code n+'0' will convert it to ASCII. If the number is between 10 and 99, the n/10 will calculate the tens digit and the n%10 will calculate the one digit.

Program 2.12
Print unsigned 16-bit decimal to an output device.

```
/* Variable format 1 to 5 digits with no space before or after
   This function uses recursion to convert a decimal number of
      unspecified length as an ASCII string */
void OutUDec(unsigned short n){
  if(n >= 10){
    OutUDec(n/10);
    n=n%10;
  }
  OutChar(n+'0'); /* n is between 0 and 9 */
}
```

To convert a signed number into ASCII, we can first test the sign of the input. If it is negative we issue the first ASCII character as the minus sign ($-$), negate the input so that it is now positive, and use our unsigned conversion routine to generate the rest of the ASCII characters. This conversion technique is shown in Program 2.13.

Program 2.13
Print signed 16-bit decimal to an output device.

```
void OutSDec(short n) {
  if(n<0){
    n = -n;          /* take absolute value */
    OutChar('-' ); /* the number is negative */
  }
  OutUDec(n);          /* print the absolute value */
}
```

Maintenance Tip: Even though the machine will process data in binary, we can specify numbers in many formats (e.g., binary, decimal, hexadecimal, etc.). When writing software, use the format that makes your software easiest to understand. There is no one format that is best for all situations.

2.10   Fixed-Point Numbers

We will use fixed-point numbers when we wish to express values in our software that have noninteger values. A **fixed-point number** contains two parts. The first part is a **variable integer**, called **I**. This integer may be signed or unsigned. An unsigned fixed-point number is one that has an unsigned variable integer. A signed fixed-point number is one that has a signed variable integer. The **precision** of a number is the total number of distinguishable values that can be represented. The precision of a fixed-point number is determined by the number of bits used to store the variable integer. On the 6811 or 6812, we typically use 8 bits or 16 bits. Extended precision can be implemented, but the execution speed will be slower because the calculations will have to be performed using software algorithms rather than with hardware instructions. This integer part is saved in memory and is manipulated by software. These manipulations include but are not limited to add, subtract, multiply, divide, convert to BCD, convert from BCD. The second part of a fixed-point number is a **fixed constant**, called Δ. This value is fixed, and can not be changed. The fixed constant is not stored in memory. Usually we specify the value of this fixed constant using software comments to explain our fixed-point algorithm. The value of the fixed-point number is defined as the product of the two parts:

$$\text{fixed-point number} = \mathbf{I} \cdot \Delta$$

The **resolution** of a number is the smallest difference that can be represented. In the case of fixed-point numbers, the resolution is equal to the fixed constant (Δ). Sometimes we express the resolution of the number as its units. For example, a decimal fixed-point number with a resolution of 0.001 volts is really the same thing as an integer with units of mV. Sometimes it may be convenient to use **decimal fixed-point** With decimal fixed-point the fixed constant is a power of 10.

$$\text{decimal fixed-point number} = \mathbf{I} \cdot 10^m \text{ for some constant integer } m$$

Again, the integer **m** is fixed and is not stored in memory. Decimal fixed-point will be easy to display, while **binary fixed-point** will be easier to use when performing mathematical calculations. With binary fixed-point the fixed constant is a power of 2.

$$\text{binary fixed-point number} = \mathbf{I} \cdot 2^m \text{ for some constant integer } m$$

> **Observation:** *If the range of numbers is known and small, then the numbers can be represented in a fixed-point format.*

> **Checkpoint 2.46:** *Give an approximation of π using the decimal fixed-point ($\Delta = 0.001$) format.*

> **Checkpoint 2.47:** *Give an approximation of π using the binary fixed-point ($\Delta = 2^{-8}$) format.*

In the first example, we will develop the equations that a 6811/6812 would need to implement a digital voltmeter. The 6811 and 6812 have a built-in analog to digital converter (ADC) that can be used to transform an analog signal into digital form. The 8-bit ADC analog input range is 0 to +5 V, and the ADC digital output varies 0 to 255 respectively. Let V_{in} be the analog voltage in volts and N be the digital ADC output; then the equation that relates the analog to digital conversion is

$$V_{in} = 5*N/255 = 0.019607843*N.$$

From this equation, we can see that the smallest change in voltage that the ADC can detect is about 20 mV. In other words, the analog voltage must increase or decrease by 20 mV for the digital output of the ADC to change by at least one bit. It would be inappropriate to save the voltage as an integer, because the only integers in this range are 0,1,2,3,4,5. Since the 6811 and 6812 do not support floating point, the voltage data will be saved in fixed-point format. Decimal fixed-point is chosen because the voltage data for this voltmeter will be displayed. A fixed-point resolution of $\Delta = 0.01$ V is chosen because it is slightly smaller (better) than the ADC resolution. Table 2.23 shows the performance of the system. The table shows us that we need to store the variable part of the fixed-point number in a 16-bit variable.

Table 2.23
Performance data of a microcomputer-based voltmeter.

V_{in} (V) Analog input	N ADC digital output	I (10 mV) variable part of the fixed-point data
0.00	0	0
0.02	1	2
1.00	51	100
2.50	128	250
5.00	255	500

The software performs the following fixed-point calculation to convert N into I.

$$I = (100*N + 25)/51$$

It is very important to carefully consider the order of operations when performing multiple integer calculations. There are two mistakes that can happen. The first error is **overflow**, and it is easy to detect. Overflow occurs when the result of a calculation exceeds the range of the number system. The two solutions of the overflow problem, promotion and ceiling/floor, were discussed earlier. The other error is called **drop-out**. Drop-out occurs after a right shift or a divide, and the consequence is that an intermediate result loses its ability to represent all of the values. To avoid drop-out, it is very important to divide last when performing multiple integer calculations. If you divided first, e.g., I=100*(N/51), then the values of I would be only 0, 100, or 200. The addition of "25" has the effect of rounding to the closest integer. The value 25 is selected because it is about one-half of the denominator. For example, the calculation (100*N)/51=1 for N=1, whereas the "(100*1+25)/51" calculation yields the better answer of 2. The display algorithm is given as Program 2.14.

Program 2.14
Print unsigned 16-bit decimal fixed-point number to an output device.

```
void OutFDec(unsigned short n){ /* fixed constant is 0.01 */
   OutUDec(n/100);         /* digits to the left of the decimal */
   OutChar('.');           /* decimal point */
   OutUDec((n%100)/10);    /* tenths digit */
   OutUDec(n%10);          /* hundredths digit */
   OutChar('V');           /* units */
}
```

When adding or subtracting two fixed-point numbers with the same Δ, we simply add or subtract their integer parts. First, let x, y, z be three fixed-point numbers with the same Δ. Let $x = I \cdot \Delta$, $y = J \cdot \Delta$, and $z = K \cdot \Delta$. To perform $z = x + y$, we simply calculate $K = I + J$. Similarly, to perform $z = x - y$, we simply calculate $K = I - J$. When adding or subtracting fixed-point numbers with different fixed parts, then we must first convert two the inputs to the format of the result before adding or subtracting. This is where binary fixed-point is more convenient, because the conversion process involves shifting rather than multiplication/division.

In this next example, let x, y, z be three binary fixed-point numbers with the different Δs. In particular, we define x to be $\mathbf{I} \cdot 2^{-5}$, y to be $\mathbf{J} \cdot 2^{-2}$, and z to be $\mathbf{K} \cdot 2^{-3}$. To convert x to the format of z, we divide I by 4 (right shift twice). To convert y, to the format of z, we multiply J by 2 (left shift once). To perform $z = x + y$, we calculate

$$K = (I >> 2) + (J << 1)$$

For the general case, we define x to be $\mathbf{I} \cdot 2^{n}$, y to be $\mathbf{J} \cdot 2^{m}$, and z to be $\mathbf{K} \cdot 2^{p}$. To perform any general operation, we derive the fixed-point calculation by starting with desired result. For addition, we have $z = x + y$. Next, we substitute the definitions of each fixed-point parameter

$$\mathbf{K} \cdot 2^{p} = \mathbf{I} \cdot 2^{n} + \mathbf{J} \cdot 2^{m}$$

Lastly, we solve for the integer part of the result

$$\mathbf{K} = \mathbf{I} \cdot 2^{n-p} + \mathbf{J} \cdot 2^{m-p}$$

For multiplication, we have $z = x \cdot y$. Again, we substitute the definitions of each fixed-point parameter

$$\mathbf{K} \cdot 2^{p} = \mathbf{I} \cdot 2^{n} \cdot \mathbf{J} \cdot 2^{m}$$

Lastly, we solve for the integer part of the result

$$\mathbf{K} = \mathbf{I} \cdot \mathbf{J} \cdot 2^{n+m-p}$$

For division, we have $z=x/y$. Again, we substitute the definitions of each fixed-point parameter

$$\mathbf{K} \bullet 2^p = \mathbf{I} \bullet 2^n / \mathbf{J} \bullet 2^m$$

Lastly, we solve for the integer part of the result

$$\mathbf{K} = \mathbf{I}/\mathbf{J} \bullet 2^{n-m-p}$$

Again, it is very important to carefully consider the order of operations when performing multiple integer calculations. We must worry about overflow and drop out. In particular, in the division example, if $(n-m-p)$ is positive then the left shift $(\mathbf{I} \bullet 2^{n-m-p})$ should be performed before the divide $(/\mathbf{J})$.

We can use these fixed-point algorithms to perform complex operations using the integer functions of our 6811/6812. For example, consider the following digital filter calculation:

$$y = x - 0.0532672 \bullet x_1 + x_2 + 0.0506038 \bullet y_1 - 0.9025 \bullet y_2$$

In this case, the variables y, y_1, y_2, x, x_1, and x_2 are all integers, but the constants will be expressed in binary fixed-point format. The value -0.0532672 will be approximated by $-14 \bullet 2^{-8}$. The value 0.0506038 will be approximated by $13 \bullet 2^{-8}$. Lastly, the value -0.9025 will be approximated by $-231 \bullet 2^{-8}$. The fixed-point implementation of this digital filter is

$$y = x + x_2 + (-14 \bullet x_1 + 13 \bullet y_1 - 231 \bullet y_2) >> 8$$

Common Error: *Lazy or incompetent programmers use floating-point in many situations where fixed-point would be preferable.*

Observation: *As the fixed constant is made smaller, the accuracy of the fixed-point representation is improved, but the variable integer part also increases. Unfortunately, larger integers will require more bits for storage and calculations.*

Checkpoint 2.48: *Using a fixed constant of 2^{-8}, rewrite the digital equation $F = 1.8 \bullet C + 32$ in binary fixed-point format.*

Checkpoint 2.49: *Using a fixed constant of 10^{-3}, rewrite the digital filter $y = x - 0.0532672 \bullet x_1 + x_2 + 0.0506038 \bullet y_1 - 0.9025 \bullet y_2$ in decimal fixed-point format.*

Checkpoint 2.50: *Assume resistors R_1, R_2, R_3 are 16-bit unsigned binary fixed-point numbers with a fixed constant of 2^{-4}. Write an equation to calculate $R_3 = R_1 \| R_2$ (parallel combination.)*

2.11   *Floating-Point Numbers

Although we will not consider floating-point implementations on the 6811/6812 in this introductory text it is appropriate to know the definition of floating point. Floating point is similar in format to fixed point, except the exponent is allowed to change. The **mantissa** is similar to the integer part of a fixed point number. Consequently, both the exponent and the mantissa will be stored. Just like with fixed-point numbers we will use binary exponents for internal calculations, and decimal exponents when interfacing with humans. This number system is called floating-point because as the exponent varies the binary point or decimal point moves.

Observation: If the range of numbers is unknown or large, then the numbers must be represented in a floating-point format.

Observation: Floating-point implementations on computers like the 6811/6812 that do not have hardware support are extremely long and very slow. So, if you really need floating point, a computer with hardware support is highly desirable.

On the PC, the floating-point functions are implemented in hardware using a math co-processor. All floating-point literals and constant invocations deposit a floating-point number on a floating-point stack. On the floating-point stack there is only one floating-point format, 80-bit temporary real. In the regular computer memory, there are three floating-point formats.

The floating-point format, **f**, for the **short real** data type is shown in Figure 2.36.

Bit 31	Mantissa sign, **s** $= 0$ for positive, **s**$=1$ for negative
Bits 30:23	8-bit biased binary exponent $0 \leq \mathbf{e} \leq 255$
Bits 22:0	24-bit mantissa, **m**, expressed as a binary fraction, a binary 1 as the most significant bit is implied.

$$\mathbf{m} = 1.\mathbf{m}_1\,\mathbf{m}_2\,\mathbf{m}_3\ldots\mathbf{m}_{23}$$

Figure 2.36
32-bit short real floating-point format.

The value of a short real floating-point number is

$$\mathbf{f} = (-1)^{\mathbf{s}} \cdot 2^{\mathbf{e}-127} \cdot \mathbf{m}$$

The range of values that can be represented in the short real format is about $\pm 10^{-38}$ to $\pm 10^{+38}$. The 24-bit mantissa yields a precision of about seven decimal digits. The floating-point value is zero if both **e** and **m** are zero. Because of the sign bit, there are two zeros, positive and negative, which behave the same during calculations. To illustrate floating point, we will calculate the short real representation of the number 10. To find the binary representation of a floating-point number, first extract the sign.

$$10 = (-1)^0 \cdot 10$$

Step 2, you multiply or divide by two until you get a number greater than or equal to 1, but less than 2.

$$10 = (-1)^0 \cdot 2^3 \cdot 1.25$$

Step 3, the exponent **e** is equal to the number of divide by twos plus 127.

$$10 = (-1)^0 \cdot 2^{130-127} \cdot 1.25$$

Step 4, separate the 1 from the mantissa. Recall that the 1 will not be stored.

$$10 = (-1)^0 \cdot 2^{130-127} \cdot (1 + 0.25)$$

Step 5, express the mantissa as a binary fixed-point number with a fixed constant of 2^{-23}.

$$10 = (-1)^0 \cdot 2^{130-127} \cdot (1 + 2097152 \cdot 2^{-23})$$

Step 6, convert the exponent and mantissa components to hexadecimal.

$$10 = (-1)^0 \cdot 2^{\$82-127} \cdot (1 + \$200000 \cdot 2^{-23})$$

Step 7, extract **s**, **e**, **m** terms, convert hexadecimal to binary

$$10 = (0,\$82,\$200000) = (0,10000010,01000000000000000000000)$$

Sometimes this conversion does not yield an exact representation, as in the case of 0.1. In particular, the floating-point representation of 0.1 is only an approximation.

Step 1	$0.1 = (-1)^0 \cdot 0.1$
Step 2	$0.1 = (-1)^0 \cdot 2^{-4} \cdot 1.6$
Step 3	$0.1 = (-1)^0 \cdot 2^{123-127} \cdot 1.6$
Step 4	$0.1 = (-1)^0 \cdot 2^{123-127} \cdot (1+0.6)$
Step 5	$0.1 \approx (-1)^0 \cdot 2^{123-127} \cdot (1+5033165 \cdot 2^{-23})$
Step 6	$0.1 \approx (-1)^0 \cdot 2^{\$7B-127} \cdot (1+\$4CCCCD \cdot 2^{-23})$
Step 7	$0.1 \approx (0,\$7B,\$4CCCCD) = (0,01111011,10011001100110011001101)$

The following example shows the steps in finding the floating-point approximation for π.

Step 1	$\pi = (-1)^0 \cdot \pi$
Step 2	$\pi \approx (-1)^0 \cdot 2^1 \cdot 1.570796327$
Step 3	$\pi \approx (-1)^0 \cdot 2^{128-127} \cdot 1.570796327$
Step 4	$\pi \approx (-1)^0 \cdot 2^{128-127} \cdot (1+0.570796327)$
Step 5	$\pi \approx (-1)^0 \cdot 2^{128-127} \cdot (1+4788187 \cdot 2^{-23})$
Step 6	$\pi \approx (-1)^0 \cdot 2^{\$80-127} \cdot (1+\$490FDB \cdot 2^{-23})$
Step 7	$\pi \approx (0,\$80,\$490FDB) = (0,10000000,10010010000111111011011)$

There are some special cases for floating-point numbers. When **e** is 255, the number is considered as plus or minus infinity, which probably resulted from an overflow during calculation. When **e** is 0, the number is considered **denormalized**. The value of the mantissa of a denormalized number is less than 1. A denormalized short number has the value

$$\mathbf{f} = (-1)^s \cdot 2^{-126} \cdot \mathbf{m}$$

where

$$\mathbf{m} = 0.\mathbf{m}_1\mathbf{m}_2\mathbf{m}_3\ldots\mathbf{m}_{23}.$$

Observation: The floating-point zero is stored in denormalized format.

The floating-point format for the **long real** data type is shown in Figure 2.37.

Bit 63	Mantissa sign, $\mathbf{s} = 0$ for positive, $\mathbf{s} = 1$ for negative
Bits 62:52	11-bit biased binary exponent $0 \leq \mathbf{e} \leq 2047$
Bits 51:0	52-bit mantissa, **m**, expressed as a binary fraction, a binary 1 as the most significant bit is implied. $\mathbf{m} = 1.\mathbf{m}_1\mathbf{m}_2\mathbf{m}_3\ldots\mathbf{m}_{52}$

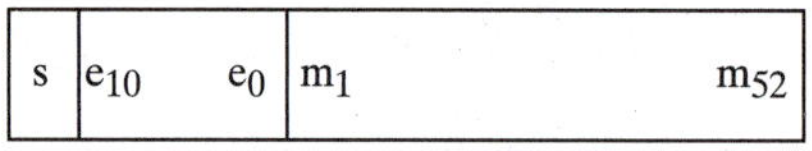

Figure 2.37
64-bit long real floating-point format.

The value of a long real floating-point number is

$$\mathbf{g} = (-1)^s \cdot 2^{e-1023} \cdot \mathbf{m}$$

The range of values that can be represented in the long real format is about $\pm 10^{-308}$ to $\pm 10^{+308}$. The 53-bit mantissa yields a precision of about 15 decimal digits. There are two zeros, positive and negative, which behave the same during calculations.

The floating-point format, **t**, for the **temporary real** data type is shown in Figure 2.38.

Bit 79   Mantissa sign, **s** = 0 for positive, **s** = 1 for negative

Bits 78:64   15-bit biased binary exponent $0 \leq \mathbf{e} \leq 32767$

Bits 63:0   entire 64-bit mantissa, **m**, expressed as a binary fraction.

$$\mathbf{m} = \mathbf{m_0.m_1m_2m_3}...\mathbf{m_{63}}$$

Figure 2.38

80-bit temporary real floating-point format.

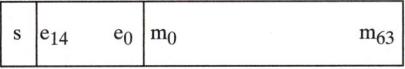

The value of a temporary real floating-point number is

$$\mathbf{f} = (-1)^s \cdot 2^{e-16383} \cdot \mathbf{m}$$

The range of values that can be represented in the temporary real format is $\pm 10^{-4932}$ to $\pm 10^{+4932}$. The 64 bit mantissa yields a precision of about 19 decimal digits. There are two zeros, positive and negative, which behave the same during calculations. Notice, that the most significant mantissa bit, $\mathbf{m_0}$, is explicitly stored. Normalized float numbers have $\mathbf{m_0}$ equal to 1. When $\mathbf{m_0}$ equals 0, the number is **unnormalized**. Mantissas for unnormalized and denormalized numbers are both less than 1. The difference is that the unnormalized exponent can be any value, whereas the exponent of a denormalized number is the smallest possible value.

When two floating-point numbers are added or subtracted, the smaller one is first unnormalized. The mantissa of the smaller number is shifted right and its exponent is incremented until the two numbers have the same exponent. Then the mantissas are added or subtracted. Lastly, the result is normalized. To illustrate the floating-point addition, consider the case of $10 + 0.1$. First, we show the original numbers in floating-point format. The mantissa is shown in binary format.

```
 10.0 = (-1)^0 • 2^3  • 1.0100000000000000000000
+ 0.1 = (-1)^0 • 2^-4 • 1.1001100110011001100110
```

Every time the exponent is incremented the manitissa is shifted to the right. Notice that 7 binary digits are lost. The 0.1 number is unnormalized, but now the two numbers have the same exponent. Often the result of the addition or subtraction will need to be normalized.

```
 10.0 = (-1)^0 • 2^3 • 1.0100000000000000000000
+ 0.1 = (-1)^0 • 2^3 • 0.0000001100110011001100110011001 1001101
 10.1 = (-1)^0 • 2^3 • 1.0100001100110011001100110011001
```

When two floating-point numbers are multiplied, their mantissas are multiplied and their exponents are added. When dividing two floating-point numbers, their mantissas are divided and their exponents are subtracted. After multiplication and division, the result is normalized. To illustrate the floating-point multiplication, consider the case of $10*0.1$. Let m_1, m_2 be the values of the two mantissas. Since the range is $1 \leq m_1, m_2 < 2$, their product will vary from $1 \leq m_1 * m_2 < 4$.

```
 10.0 = (-1)^0 • 2^3  • 1.0100000000000000000000
* 0.1 = (-1)^0 • 2^-4 • 1.1001100110011001100110 1
 1.0  = (-1)^0 • 2^-1 • 10.000000000000000000000000
```

The result needs to be normalized.

$$1.0 = (-1)^0 \cdot 2^0 \cdot 1.0000000000000000000000000$$

Checkpoint 2.51: Why can't you use your calculator to find the s, e, and m terms of a number in temporary real format?

Roundoff is the error that occurs as a result of an arithmetic operation. For example, the multiplication of two 64-bit mantissas yields a 128-bit product. The final result is normalized into a normalized floating-point number with a 64-bit mantissa. Roundoff is the error caused by discarding the least significant bits of the product. Roundoff during addition and subtraction can occur in two places. First, an error can result when the smaller number is shifted right. Second, when two n-bit numbers are added the result is $n + 1$ bits, so an error can occur as the $n + 1$ sum is squeezed back into an n-bit result. **Truncation** is the error that when a number is converted from one format to another. For example, when a temporary real number is converted to a short real format, 40 bits are lost as the 64-bit mantissa is truncated to fit into the 24-bit mantissa. Recall, the number 0.1 could not be exactly represented as a short real floating-point number. This is an example of truncation as the true fraction was truncated to fit into the finite number of bits available.

Observation: Computers use binary floating point because it is faster to shift than it is to multiply/divide by 10.

2.12   Tutorial 2. Arithmetic and Logical Operations

The purpose of the second tutorial is to study numbers, logical operations, and arithmetic operations. Computers internally store numbers in binary, but use decimal or hexadecimal when interacting with humans. Some computers implement floating point, but in this tutorial the formats will be restricted to signed and unsigned integers. In the first tutorial, you observed the registers in the **ViewBox**. *Registers* are high-speed storage elements inside the processor. The 6811/6812 has registers named CCR, A, B, X, Y, SP, and PC. Registers A and B generally contain data and are used for arithmetic and logical operations. Registers A and B concatenated together can also be referred to as Register D. Registers X and Y generally contain addresses and are used for pointers. The condition code register (CCR) contains the N Z V and C bits. The N bit is set if the most significant bit of the result is true. The Z bit is set if the result is zero. In general, the C bit is set on an unsigned overflow, and the V bit is set on a signed overflow.

The **TExaS ViewBox** is part of the microcomputer window, and is used to display and/or modify information, see Figures A1.13 and A1.14. To change the value of a parameter in the **ViewBox**: click on the entry, type the new value in the **Data** field, then hit enter. The available **ViewBox** formats are listed in Table 2.24. True condition code bits are displayed as upper case letters and false bits are shown as lower case letters. For example sXhInzvc means S = 0, X = 1, H = 0, I = 1, N = 0, Z = 0, V = 0, and C = 0.

Table 2.24
Available formats for displaying information in the TExaS **ViewBox**.

format	descriptions	examples
h	8-bit unsigned hexadecimal	$00 $12 $FF
d	8-bit unsigned decimal	0 18 255
b	8-bit unsigned binary	%00000000 %00010010 %11111111
H	16-bit unsigned hexadecimal	$0000 $1234 $FFFF
D	16-bit unsigned decimal	0 4660 65535
B	16-bit unsigned binary	%0001001000110100
−h or +h	8-bit signed hexadecimal	−$80 +$12 +$7F
−d or +d	8-bit signed decimal	−128 +18 +127
−b or +b	8-bit signed binary	−%10000000 +%00010010 +%01111111
−H or +H	16-bit signed hexadecimal	−$8000 +$1234 +$7FFF
−D or +D	16-bit signed decimal	−32768 +4660 +32767
−B or +B	16-bit signed binary	−%0001001000110100
b3	3-bit binary (least significant bits)	%000 %111
b4	4-bit binary (least significant bits)	%0000 %1111
cc	8-bit binary showing bits in the CCR	sXhInzvc
c or C	ASCII character	'A' 'x' '0'
s or S	NULL or EOT terminated ASCII string	"Hello World"
v	value of the address itself, unsigned decimal	2048
V	value of the address itself, unsigned hex	$0800
+v or −v	value of the address itself, signed decimal	−32768 +0 +32767
+V or −V	value of the address itself, signed hex	−$8000 +0 +$7FFF

Action. Execute **TExaS** and open the `Chap2.rtf` and `Chap2.uc` files. Bring the **ViewBox** to the front. We will begin talking about unsigned numbers, so we will use the "d" format to observe values in Register A, and use the "D" format for Register X.

Question 2.1. Register A contains an 8-bit integer. Its precision is 8 bits. As an unsigned number, its range of values is 0 to 255. What happens when you try to set Register A to 256?

Question 2.2. What happens when you try to set Register A to −1?

Question 2.3. Register X contains a 16-bit integer. Its precision is 16 bits. As an unsigned number, its range of values is 0 to 65535. What happens when you try to set Register X to 65536?

Question 2.4. What happens when you try to set Register X to −1?

Action. Next, we will study signed numbers. Change the format of Register A to the "+d", and change the format of Register X to the "+D". To change the format of a parameter in the **ViewBox**: click on the **ViewBox** entry, type the new format in the **Format** field, then hit enter.

Question 2.5. As a signed number, the range of values in Register A is −128 to +127. What happens when you try to set Register A to +128?

Question 2.6. What happens when you try to set Register A to −129?

Question 2.7. As a signed number, the range of values in Register X is −32768 to +32767. What happens when you try to set Register X to 32768?

Question 2.8. What happens when you try to set Register X to −32769?

Question 2.9. Use the help system of **TExaS** to look up the instruction `ldaa` and answer the question, "Even though the `ldaa` instruction does not perform any arithmetic or logical operations, does it modify the condition code bits, N, Z, V, and C?" Within the **TExaS** application execute **Help->HelpTopics**, double click **6811 assembly language**, double-click **6811 memory access instructions**, click `ldaa`.

Action. Assemble the `Chap2.rtf` program, and bring the `TheList.rtf` `TheLog.rtf` and `Chap2.uc` windows to the front. Notice that the instructions `lsla` and `asla` have the same object code.

Question 2.10. First perform the following logical operations by hand, and record what you think the result will be in 8-bit unsigned hexadecimal. In addition, record your expectation for the N and Z bits. Within **TExaS**, change the format of Register A to unsigned hexadecimal "h". Reset and single-step the program through part 1. Correct your answers by recording the proper values of Register A and the CCR bits N and Z. The logical operations clear the V bit. The complement instruction is the only one that sets the C bit, while the other logical operations affect only the N, Z, and V bits.

```
$0F&$85
$0F|$85
$0F^$85
~$0F
```

Question 2.11. First perform the following unsigned arithmetic operations by hand, and record what you think the result will be in 8-bit unsigned decimal format. In addition, record your expectation for the N, Z, and C bits. Although the processor will set the V bit during the calculation, we will ignore it when operating on unsigned integers. Within **TExaS**, change the format of Register A to unsigned decimal "d". Single-step the program through part 2. Correct your answers by recording the proper values of Register A and the CCR bits N, Z, and C.

```
155>>1
50<<1
96+64
224+64
160-64
32-64
```

Question 2.12. First perform the following signed arithmetic operations by hand, and record what you think the result will be in 8-bit signed decimal. In addition, record your expectation for the N, Z, and V bits. Although the processor will set the C bit during the calculation, we will ignore it when operating on signed integers. Within **TExaS**, change the format of Register A to signed decimal "+d". Single-step the program through part 3. Correct your answers by recording the proper values of Register A and the CCR bits N, Z, and V.

```
-101>>1
-50<<1
-32+64
96+64
32-64
-96-64
```

2.13 Homework Assignments

Homework 2.1: How many binary bits is 2¾ decimal digits?

Homework 2.2: About how many decimal digits is 14 binary bits?

Homework 2.3: Complete the following table so that there is an equal value for each row. Assume the decimal values are unsigned. The first row illustrates the process.

binary	hexadecimal	decimal
%01101001	$69	105
	$45	
		45
%10001111		
	$E4	
		99
%111001001101		
	$2B9	
		1000

Homework 2.4: Complete the following table so that there is an equal value for each row. Assume the decimal values are unsigned. The first row illustrates the process.

binary	hexadecimal	decimal
%10101101	$AD	173
	$78	
		123
%11111		
	$1234	
		36
%1000100001111101		
	$2456	
		54321

Homework 2.5: Complete the following table so that there is an equal value for each row. Assume each value is 8 bits and the decimal numbers are signed. The first row illustrates the process.

binary	hexadecimal	decimal
%01011110	$5E	94
	$A2	
		−47
%11000011		
	$D1	
		75
%00101011		
	$B7	
		−100

Homework 2.6: Complete the following table so that there is an equal value for each row. Assume each value is 8 bits and the decimal numbers are signed. The first row illustrates the process.

binary	hexadecimal	decimal
%11111110	$FE	−2
	$BD	
		−88
%00111011		
	$94	
		52
%11100000		
	$11	
		−126

Homework 2.7: How many binary bits does it take to represent 12,345,678? How many bytes?

Homework 2.8: How many binary bits does it take to represent 9,876,543,210? How many bytes?

Homework 2.9: In C, a `char` is 8 bits, a `short` is 16-bits, and a `long` is 32 bits. Assuming each is signed, give the range of each type of number.

Homework 2.10: In C, an `unsigned char` is 8 bits, an `unsigned short` is 16-bits, and an `unsigned long` is 32 bits. Assuming each is unsigned, give the range of each type of number.

Homework 2.11: What is the difference between the character 0 and the number 0?

Homework 2.12: How is "3.14159" encoded as a null-terminated ASCII string?

Homework 2.13: Complete the following table so that there is an equal value for each row. Assume the values are unsigned BCD. The first row illustrates the process.

binary	hexadecimal	BCD decimal
%01111000	$78	78
	$69	
		45
%10000111		
	$94	
		99
%00100110		
	$29	
		52

Homework 2.14: Using just the NAND gates, design an equals circuit, such that the output is 1 if and only if input A equals input B. There will be two input signals and one output signal.

Homework 2.15: Using just the NOR gates, design an equals circuit, such that the output is 1 if and only if input A equals input B. There will be two input signals and one output signal.

Homework 2.16: Design a digital circuit that takes the output of the 8-bit adder shown in Figure 2.25, and implements the Z, N, and V bits. The figure already includes the C bit (carry).

Homework 2.17: Design a digital circuit that takes the output of the 8-bit subtractor shown in Figure 2.27, and implements the Z, N, and V bits. The figure already includes the C bit (carry).

Homework 2.18: Write assembly software that copies 32 bits of data from addresses $0010-$0013 to addresses $0020-$0023.

Homework 2.19: Write assembly software that clears PORTA bits 3 and 4 and sets PORTB bits 7 and 6.

Homework 2.20: Write assembly software that makes PORTC bits 1, 3, 5, and 7 outputs and the rest inputs.

Homework 2.21: High and Low are unsigned 8-bit components, which need to be combined into a single unsigned 16-bit Result. We will assume both High and Low are bounded within the range of 0 to 255. The expression High<<8 will perform eight logical shift lefts. Write assembly software to implement Result = (High<<8)|Low;

Homework 2.22: Use D flip-flops like Figure 2.20 to build an 8-bit ASL function.

Homework 2.23: Use D flip-flops like Figure 2.20 to build an 8-bit ROR function.

Homework 2.24: Use D flip-flops like Figure 2.20 to build an 8-bit ROL function.

Homework 2.25: Let A and B be two 8-bit inputs to an 8-bit binary adder. Fill in the table showing R=A+B and the four CCR bits after each addition. The first row illustrates the process.

A	B	R	NZVC
10	100	110	0000
$40	$A3		
$C3	$6F		
100	−50		
150	180		
−5		0	
41	−50		
120	136		
20			0101

Homework 2.26: Let A and B be two 8-bit inputs to an 8-bit binary subtractor. Fill in the table showing R = A − B and the four CCR bits after each subtraction. The first row illustrates the process.

A	B	R	NZVC
100	10	90	0000
55	$93		
$DD	$9F		
−50	−70		
200	180		
12		0	
41	−50		
255	136		
	87		~~0101~~ 0100 *error*

Homework 2.27: Let A be an 8-bit input. Fill in the table showing the promotion to 16-bit unsigned and 16-bit signed. Give all answers in 16-bit hexadecimal. The first row illustrates the process.

A	Unsigned 16-bit	Signed 16-bit
$80	$0080	$FF80
$55		
$DD		
$00		
$FF		
$45		
$90		
$27		
$A4		

Homework 2.28: Let A be an 8-bit input. Fill in the table showing the promotion to 16-bit unsigned and 16-bit signed. Give all answers in 16-bit hexadecimal. The first row illustrates the process.

A	Unsigned 16-bit	Signed 16-bit
$70	$0070	$0070
$A5		
$1D		
$90		
$DF		
$85		
$52		
$B6		
$4B		

Homework 2.29: Modify the C code in Program 2.9 so that it detects an input greater than 65535. If the input does overflow, return with a value of 65535. Hint: consider extending the precision of n to 32 bits by defining it as `unsigned long`.

Homework 2.30: Modify the C code in Program 2.12 so that it doesn't use recursion.

Homework 2.31: Give an approximation of $\sqrt{2}$ using the decimal fixed-point ($\Delta = 0.001$) format.

Homework 2.32: Give an approximation of $\sqrt{2}$ using the binary fixed-point ($\Delta = 2^{-8}$) format.

Homework 2.33: Assume you have a 12-bit signed ADC. Let V_{in} be the analog voltage in volts and N be the digital ADC output. The input range is $-5 \leq V_{in} \leq +5V$. The ADC digital output range is $-2048 \leq N \leq +2047$. First, write a linear equation that relates V_{in} as a function of N. Next, rewrite the equation in fixed-point math assuming V_{in} is represented as a decimal fixed-point number with $\Delta = 0.001$ V.

Homework 2.34: Assume you have a 10-bit signed ADC. Let V_{in} be the analog voltage in volts and N be the digital ADC output. The input range of $-10 \leq V_{in} \leq +10$ V. The ADC digital output range is $-512 \leq N \leq +511$. First, write a linear equation that relates V_{in} as a function of N. Next, rewrite the equation in fixed-point math assuming V_{in} is represented as a decimal fixed point number with $\Delta = 0.01$ V.

Homework 2.35: First, rewrite the following digital filter using decimal fixed-point math. Assume the inputs are unsigned 8-bit values (0 to 255). Then, rewrite it so that it can be calculated with integer math using the fact that 0.11111 is about 1/9 and 0.088889 is about 4/45 and 0.8 is 4/5. In both cases, the calculations are to be performed in 16-bit unsigned integer form without overflow.

$$y = 0.11111 \bullet x + 0.08889 \bullet x_1 + 0.80000 \bullet y_1$$

Homework 2.36: Give short real floating point representation of $\sqrt{2}$.

Homework 2.37: Give short real floating point representation of -134.4.

Homework 2.38: Give short real floating point representation of -0.0123.

Homework 2.39: Perform the operation $10 + \pi$ in short real floating-point format. Determine the difference between what you got and what you should have gotten ($10 + \pi$). This error has two components: truncation error that results in the approximation π itself and roundoff error that occurs during the addition.

Homework 2.40: Perform the operation 0.1*0.1 in short real floating-point format. Determine the difference between what you got and what you should have gotten (0.01). This error has two components: truncation error that results in the approximation of 0.1 itself and roundoff error that occurs during the multiplication.

Homework 2.41: Perform the operation $\pi * \pi$ in short real floating-point format. Determine the difference between what you got and what you should have gotten (π^2). This error has two components: truncation error that results in the approximation of π itself and roundoff error that occurs during the multiplication.

3 Microcomputer Architecture

Chapter 3 objectives are to

❑ present the basic microcomputer architecture,
❑ list the available 6811/6812 microcomputers and their memory configurations,
❑ define some of the machine level instructions available on the 6811/6812,
❑ explain how the computer uses addressing modes to access memory,
❑ introduce I/O ports.

The overall objective of this book is to develop the hardware and software components of an embedded system using the **TExaS** simulator. In this chapter, we start with the general concepts of a computer, then present the details of the Motorola 6811 and 6812. For more information concerning a microcomputer, refer to the respective Motorola manual. Data sheets for the microcomputers we will use can be found in PDF format on the CD that accompanies this book. A free PDF reader can be downloaded for `adobe.com.` Given this basic knowledge, we will use **TExaS** to select the microcomputer and memory configuration for our simulation. The architecture of a computer includes all its hardware components and how the pieces are connected together. The more we know the strengths and weaknesses of our computer, the better programmers we will be. The performance of an embedded system depends on both its hardware and software components. When developing in assembly language, it is difficult to separate hardware architecture (e.g., wires, chips, memory etc.) from software architecture (machine language instructions.) Therefore, an introduction to assembly language programming is presented in this chapter alongside the traditional architecture issues like registers, busses, memory, and I/O connections.

3.1 Introduction

A computer system combines a processor, RAM, ROM, input devices and output devices, as shown in Figure 3.1. In a system with *memory mapped I/O*, the I/O devices are connected to the processor in a manner similar to memory. I/O devices are assigned addresses, and the software accesses I/O using reads and writes to the specific I/O address. The software inputs from an input device using the same instructions as it would if it were reading from memory. Similarly, the software outputs from an output device using the same instructions as it would if it were writing to memory.

Figure 3.1
A memory-mapped
computer system.

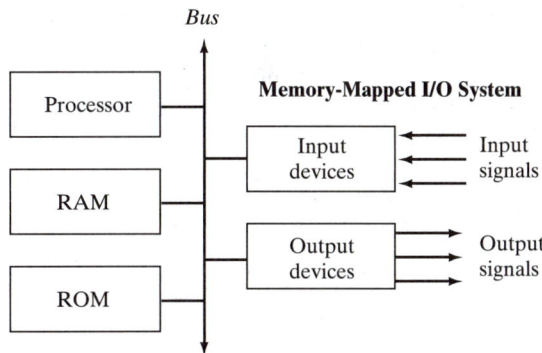

Two of the microcomputers used often in this book are MC68HC711E9 and MC68HC812A4. The memory map of the MC68HC711E9 is as follows:

$0000 to $01FF RAM
$1000 to $103F I/O
$B600 to $B7FF ROM (can be erased and reprogrammed electrically)
$D000 to $FFFF ROM (can be erased by UV and reprogrammed electrically)

The following is the memory map of the MC68HC812A4:

$0000 to $01FF I/O
$0800 to $0BFF RAM
$F000 to $FFFF ROM (can be erased by UV and reprogrammed electrically)

Input/output devices are important in all computers, but they are especially significant in an embedded system. In a computer system with *isolated I/O,* the control bus signals that activate the I/O are separate from those that activate the memory devices, as shown in Figure 3.2. These systems have a separate address space and instructions to access the I/O devices. The Intel x86 family has four control bus signals: MEMR, MEMW, IOR, and IOW. MEMR is active during memory read cycles, MEMW is active during memory write cycles, IOR is active during I/O read cycles, and IOW is active during I/O write cycles. The advantages of isolated I/O are that software does not inadvertently access I/O when it thinks it is accessing memory, and the I/O hardware interfaces are simpler. On the IBM-PC, there are 32 memory address lines, but only 11 of those 32 lines are used to access I/O devices. The other address lines are not used during an I/O bus cycle. Limiting the system to only 2048 I/O addresses was a design decision made in the early 1980s. Rather than use the regular memory access instructions, the Intel x86 computer uses special in and out instructions to access the I/O devices.

Figure 3.2
An isolated-I/O
computer system.

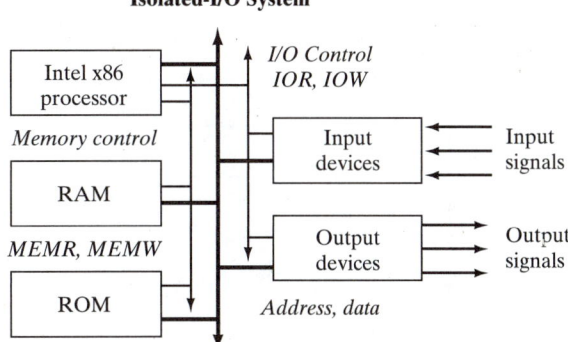

The *bus* contains address, data, and control information that provides data transfer between the various modules in the system. The address specifies which module (input, output, RAM or ROM) will communicate with the processor. The data contain the information that is being transferred. Control signals specify the direction of the transfer. We call a complete data transfer a *bus cycle* In a simple computer system, like the 6811 and 6812, the two types of transfers are shown in Table 3.1. In this simple system, the processor always controls the address (where to access), the direction (whether read or write), and the control (when to access.) The 6811 data bus contains 8 bits of data, while the 6812 data bus allows either 8 or 16 bits. Since each individual byte on most computers has a unique address, the maximum memory size in bytes is equal to the number of different addresses. The **TExaS** simulator allows you to observe bus activity during execution.

Table 3.1
Simple computers generate two types of cycles.

Type	Address Driven by	Data Driven by	Transfer
Read Cycle	processor	RAM, ROM or Input	Data copied to processor
Write Cycle	processor	processor	Data copied to Output or RAM

Checkpoint 3.1: The 6811 and 6812 have a 16-bit address bus. How many locations can it address?

During a read cycle (Figure 3.3), data flow from memory or an input device into the processor.

Figure 3.3
A read cycle copies data from RAM, ROM, or an input device into the processor.

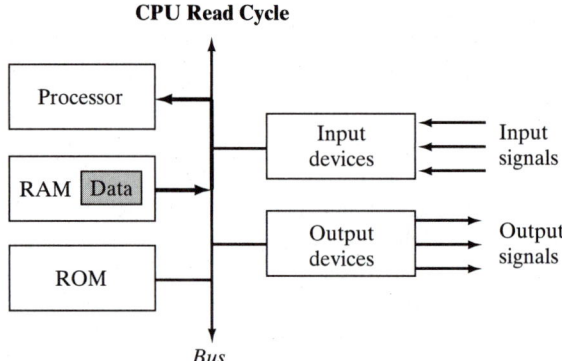

During a write cycle (Figure 3.4), data flow from the processor into memory or an output device.

Figure 3.4
A write cycle copies data from the processor into RAM, or output device.

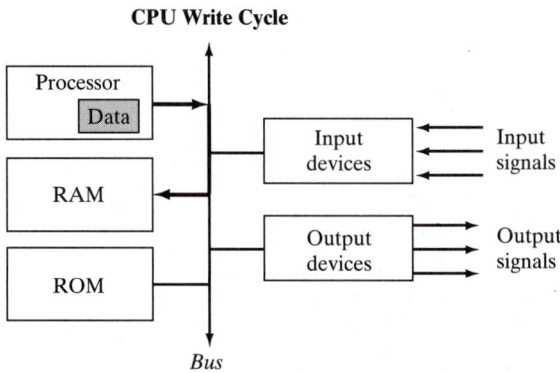

Checkpoint 3.2: The 6812 has a 16-bit address bus and a 16-bit data bus, but can still only address 65536 bytes of memory. Why?

You see that if you wish to transfer data from an input device into RAM, you must first transfer it from input to the processor, then from the processor into RAM. In more complex systems like the Motorola 6808, Motorola 68340 and IBM-PC, we will be able to transfer data directly from input to RAM or RAM to output using **D**irect **M**emory **A**ccess, DMA. The *bandwidth* of an I/O device is the number of bytes/sec that can be transferred. Because DMA is faster, we will use this method to interface high bandwidth devices like disks and networks. During a read DMA cycle (Figure 3.5) data flows directly from the memory to the output device. Many systems support DMA that transfers data from memory to memory.

Figure 3.5
A DMA read cycle copies data from RAM, ROM or input device into an output device.

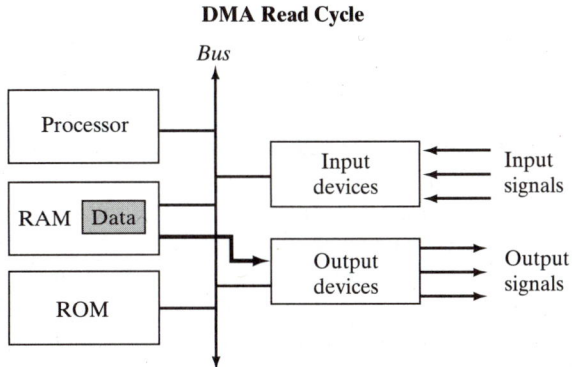

DMA Read Cycle

During a write DMA cycle (Figure 3.6) data flow directly from the input device to memory.

Figure 3.6
A DMA write cycle copies data from the input device into RAM, or output device.

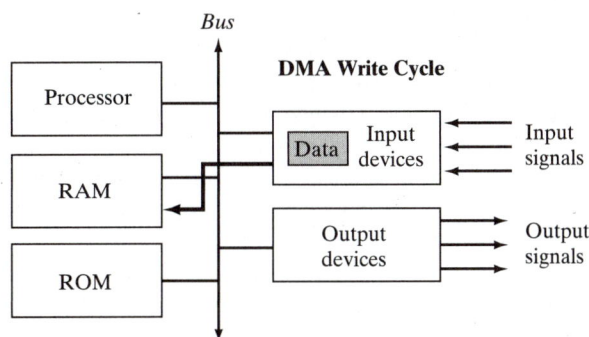

DMA Write Cycle

A simple processor like the 6811 or 6812 has four major components, as illustrated in Figure 3.7. The **bus interface unit** (BIU) reads data from the bus during a read cycle, and writes data onto the bus during a write cycle. The 6811 and 6812 have a single processor and do not support DMA. Therefore, the BIU always drives the address bus and the control signals of the bus. The **effective address register** (EAR) contains the data address for the current instruction. The **TExaS** simulator allows you to observe the EAR during execution.

Figure 3.7
The four basic components of 6811/6812 processor.

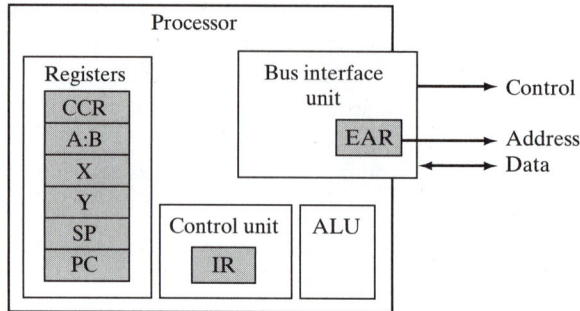

The **control unit** (CU) orchestrates the sequence of operations in the processor. The CU issues commands to the other three components. The **instruction register** (IR) contains the op code for the current instruction. Most 6811/6812 op codes are 8 bits wide, but some are 16 bits. The **TExaS** simulator allows you to observe the IR during execution.

The **registers** are high-speed storage devices located in the processor. Registers do not have addresses like regular memory, but rather they have specific functions explicitly defined by the instruction. **Accumulators** are registers that typically contain data. **Index registers** typically contain addresses. The **program counter (PC)** points to the memory containing the instruction to execute next. In an embedded system, the PC usually points into nonvolatile memory like ROM, EPROM, or EEPROM. The information stored in nonvolatile memory (e.g., the program) is not lost when power is removed. The **stack pointer** (SP) points to the RAM, and defines the stack. The stack is an extremely important component of software development, which can be used to pass parameters, save temporary information, and implement local variables. The internal RAM of the 6811/6812 is volatile memory, meaning its information is lost when power is removed. It is possible to connect external RAM to the 6811/6812 powered by a separate battery, so that one has nonvolatile RAM. The **condition code register** (CCR) contains the status of the previous operation, as well as some operating mode flags such as the interrupt enable bit. This register is called the **flag register** on Intel computers.

The **arithmetic logic unit** (ALU) performs arithmetic and logic operations. Addition, subtraction, multiplication, and division are examples of arithmetic operations. And, or, exclusive or, and shift are examples of logical operations.

Checkpoint 3.3: For what do the acronyms CU, DMA, BIU, and ALU stand?

In general, the execution of an instruction goes through four phases. First, the computer fetches the machine code for the instruction by reading the value in memory pointed to by the program counter (PC). After each byte of the instruction is fetched, the PC is incremented. At this time, the instruction is decoded, and the effective address is determined (EAR). Many instructions require additional data, and during phase 2 the data is retrieved from memory at the effective address. Next, the actual function for this instruction is performed. Often the computer bus is idle at this time, because no additional data are required.

During the last phase, the results are written back to memory. All instructions have a phase 1, but the other three phases may or may not occur for any specific instruction. Each of the phases may require one or more bus cycles to complete. Each bus cycle reads or writes one piece of data.

Phase	Function	R/W	Address	Comment
1	Instruction fetch	read	PC++	Put into IR,
2	Data read	read	EAR	Data passes through ALU,
3	Operation	none		ALU operations, set CCR
4	Data store	write	EAR	Results stored in memory

3.2 Common Architecture of the 6811 and 6812

In this section common features of the 6811 and 6812 are presented. In subsequent sections, information specific to each processor will be presented. From an assembly language perspective, the 6812 is a superset of the 6811. In other words, assembly language programs written for the 6811 can be reassembled and run on the 6812. The stack pointer operates slightly differently on the two computers, but this difference will not prevent most 6811 programs from running on the 6812. Software written for embedded systems like the 6811 and 6812 is tightly coupled to their I/O devices. Although the 6812 has more I/O ports than the 6811, the 6812 I/O structure is not a superset of the 6811. Consequently, a significant effort, translating I/O interface software, will be required to convert a 6811 system to run on a 6812. The machine code for the two processors is different, so software developed on one must be reassembled to run on the other.

> **Observation:** *Motorola should have developed versions of the 6812 with I/O ports and pinouts identical to its most popular versions of the 6811.*

3.2.1 Registers

The 6811/6812 registers are depicted in Figure 3.8. Registers A and B concatenated together form a 16-bit accumulator, Register D, with Register A containing the most significant byte. Typically Registers A and B contain data (numbers) while Registers X and Y contain addresses (pointers.) As stated earlier, the stack pointer operates slightly differently on the 6811 and 6812. On the 6811, SP+1 points to the top element of the stack. On the 6812, SP points to the top element of the stack. Register PC points to the current instruction.

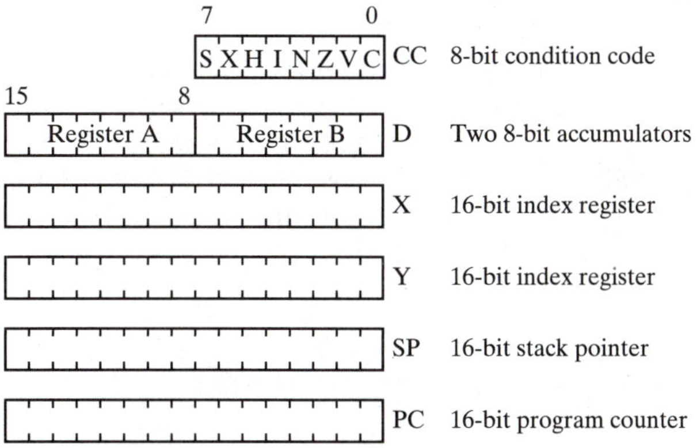

Figure 3.8
The 6811 and 6812
have 6 registers.

S X H I N Z V C CC	8-bit condition code
Register A | Register B D	Two 8-bit accumulators
X	16-bit index register
Y	16-bit index register
SP	16-bit stack pointer
PC	16-bit program counter

The Condition Code bits are shown in Figure 3.9. Many instructions set these bits to signify the result of the operation. When S=1, the stop instruction is disabled. When X=0, XIRQ interrupts are allowed. Once X is set to zero, the software can not set it back to 1. The H bit is used for BCD addition (see the `adda` and `daa` instructions.) When I=0, IRQ interrupts are enabled. Interrupts will be discussed in Chapter 11. The N, Z, V, and C bits signify the status of the previous ALU operation.

Figure 3.9
The 6811/6812
condition code bits.

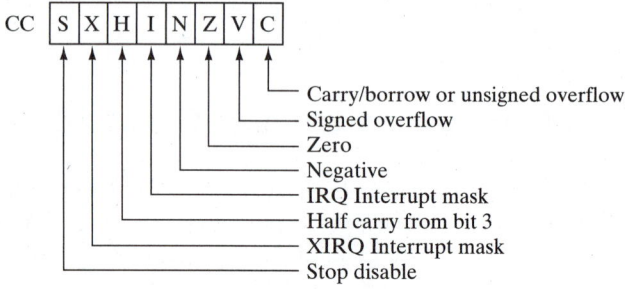

Checkpoint 3.4: To prevent (disable) interrupts, what value should the I bit be?

3.2.2 Terminology

This chapter focuses on 6811/6812 architecture, but now we introduce simple instructions in order to understand how the microcomputer works. The **TExaS** simulator that accompanies this book can be used to visualize program execution by activating the *CycleView* mode. Recall from chapter 2, the following notation.

```
w    is a signed 8-bit -128 to +127 or unsigned 8-bit 0 to 255
n    is a signed 8-bit -128 to +127
u    is a unsigned 8-bit 0 to 255
W    is a signed 16-bit -32768 to +32767 or unsigned 16-bit 0 to 65535
N    is a signed 16-bit -32768 to +32767
U    is a unsigned 16-bit 0 to 65535
=[addr]  specifies an 8-bit read from addr
={addr}  specifies a 16-bit read from addr using "big endian"
=<addr>  specifies a 32-bit read from addr using "big endian"
[addr] =  specifies an 8-bit write to addr
{addr} =  specifies a 16-bit write to addr using "big endian"
<addr> =  specifies a 32-bit write to addr using "big endian"
```

Assembly language instructions have four fields. The label field is optional, starts in the first column, and is used to identify the position in memory of the current instruction. You must choose a unique name for each label. The opcode field specifies the microcomputer command to execute. The 6812 includes all the 6811 instructions. The operand field specifies where to find the data to execute the instruction. We will see opcodes have 0, 1, 2, or 3 operands. The comment field is also optional and is ignored by the computer, but allows you to describe the software making it easier to understand. Good programmers add comments to explain the software.

label	opcode	operand	comment
here	ldaa	100	RegA=[$0064]

Observation: The TExaS assembler does not allow spaces in the operand field. Multiple operands are separated by commas.

As we will learn in later in the book, it is much better to add comments to explain how or even better why we do the action. But for now we are learning what the instruction is doing, so in this chapter comments will describe what the instruction does. The assembly language instructions (like the above example) are translated into machine instructions. The `ldaa 100` instruction is translated into 2 bytes of machine code:

```
object code      instruction      comment

$96 $64          ldaa 100         RegA=[$0064]
```

Machine code, object code, and machine instructions are identical terms defining the exact operations to be performed by the computer. Note that the numbers that start with $ (e.g., $64) are specified in hexadecimal. In C, we start hexadecimal numbers with 0x (e.g., 0x64). Intel assembly language adds a "H" at the end to specify hexadecimal (e.g., 64H). Texas Instruments uses "h" (e.g., 64h).

Observation: TExaS will accept hexadecimal numbers in $64 or 0x64 format.

**3.2.3
Addressing Modes**

A fundamental issue in program development is the differentiation between data and address. It is in assembly language programming in general and addressing modes in specific that this differentiation becomes clear. When we put the number 1000 into register X, whether this is data or address depends on how the 1000 is used. Most instructions access memory to fetch parameters or save results. The addressing mode is the format the instruction uses to specify the memory location to read or write data. All instructions begin by fetching the machine instruction (op code and operand) pointed to by the PC. Some instructions operate completely within the processor and require no memory data fetches. These instructions have no operand and are classified as **inherent.** If the data is found in the instruction itself, the instruction uses **immediate** addressing mode. If the instruction uses the absolute address to specify the memory data location, the instruction uses either **direct** or **extended** addressing mode. There are many more addressing modes, but for now, these five addressing modes, as illustrated in Table 3.2, are enough to get us started. Notice that the assembly code for the 6811 and 6812 are the same, but the machine code is slightly different. Therefore, when converting a 6811 program to run on a 6812, the original software must be reassembled using a 6812 assembler.

Table 3.2   Simple addressing modes.

6811 code	6812 code	opcode	operand	comment
$4F	$87	clra		Reg A = 0 (inherent)
$86 24	$86 24	ldaa	#36	Reg A = $24 (immediate)
$96 24	$96 24	ldaa	36	Reg A = [$0024] (direct)
$B6 08 01	$B6 08 01	ldaa	$0801	Reg A = [$0801] (extended)
$20 FE	$20 FE	bra	$F000	Assume this instruction is at $F000

Checkpoint 3.5: What is the addressing mode used for?

Many computers, including the 6811 and 6812, use **PC-relative** addressing mode to encode branch instructions. PC-relative addressing makes the object code smaller and relocatable. In this section, these five addressing modes are introduced. These simple addressing modes will be sufficient to understand most of the software presented in this book. The more complex addressing modes, presented later in the chapter, will be required to implement local variables and to build the data structures presented in Chapters 7 and 9.

1. Inherent addressing mode has no operand field. Sometimes there is no data for the instruction at all. For example, the `stop` instruction halts execution. Sometimes the data for the instruction is implied. For example, the `clra` instruction sets register A to zero. In this case, the data value of zero is implied. On the other hand, sometimes the data must be fetched from memory, but the address of the data is implied. For example, the `pula` instruction will pop an 8-bit data from the stack and store it in register A. In particular, the data value pointed to by the SP is read from memory and stored into register A.

2. Immediate addressing mode uses a fixed data constant. The data itself is included in the machine code. For example, the `ldaa #36` instruction will store a data value of 36 into register A (Figure 3.10). Notice that the "36" itself is encoded in the machine code for the `ldaa #36` instruction. In assembly code, this mode is signified by the # sign.

Figure 3.10
Example of the immediate addressing mode.

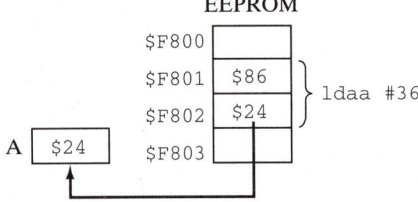

Observation: *With immediate mode addressing, the information is stored in the machine code.*

Common Error: *It is illegal to use the immediate addressing mode with instructions that store data into memory (e.g., `staa stab std stx sty` and `sts`).*

The **TExaS** 6812 simulation of this instruction shows the following two cycles. The first cycle reads the op code, and the second reads the operand.

```
Opcode fetch    R 0xF801 0x86 from EEPROM
Operand fetch   R 0xF802 0x24 from EEPROM
```

Checkpoint 3.6: *What is the difference between* `ldaa #36` *and* `ldaa #$24`*?*

3. Direct Page addressing mode uses an 8-bit address to access from addresses 0 to $00FF. In many computer systems outside the Motorola family this addressing mode is called *zero-page*. These addresses include the 6811 single chip RAM, but on the 6812 they reference the I/O ports. Some 6811 instructions can be used with both direct and extended addressing while others can be used with extended, but not direct mode. In assembly language, the < operator forces direct addressing. Figure 3.11 illustrates the execution of the `ldaa 36` instruction. Notice the address, $24, is encoded into the machine code.

Figure 3.11
Example of the direct-page addressing mode.

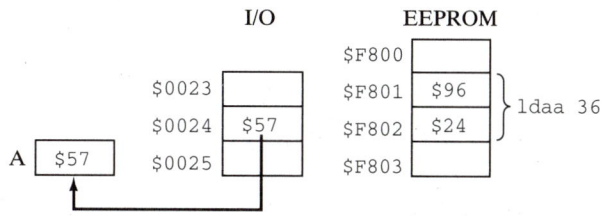

Observation: *With direct and extended mode addressing, a fixed pointer to the information is stored in the machine code. The data itself may change dynamically, but its location is fixed.*

The **TExaS** 6812 simulation of the `ldaa 36` instruction shows the following three cycles. The first cycle reads the op code, the second reads the operand, and the third cycle reads the data.

```
Opcode fetch   R 0xF801 0x96 from EEPROM
Operand fetch  R 0xF802 0x24 from EEPROM
Fetch using EARR 0x0024 0x57 from I/O
```

Checkpoint 3.7: *What is the difference between* `ldaa #36` *and* `ldaa 36`*?*

4. Extended addressing mode uses a 16-bit address allowing access to all memory and I/O devices. In many computer systems outside the Motorola family this addressing mode is called *direct*, because it can directly access all of memory. In this book, we will adhere to the Motorola terminology. Some 6811 instructions, which can be used with direct and indexed addressing, can not be used with extended addressing. The > operator forces extended addressing. In general, when the address happens to fall in the 0 to $00FF range, the assembler will automatically use direct addressing; otherwise it uses extended addressing. Figure 3.12 illustrates the execution of the `ldaa $0801` instruction. Notice the address, $0801, is encoded into the machine code.

Figure 3.12
Example of the extended addressing mode.

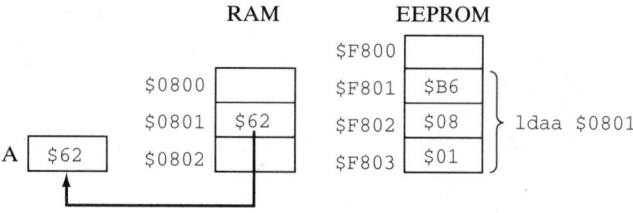

The **TExaS** 6812 simulation of the `ldaa $0801` instruction shows the following four cycles. The first cycle reads the op code, the second/third cycles read the operand, and the fourth cycle reads the data.

```
Opcode fetch   R 0xF801 0xB6 from EEPROM
Operand fetch  R 0xF802 0x08 from EEPROM
Operand fetch  R 0xF803 0x01 from EEPROM
Fetch using EARR 0x0801 0x62 from RAM
```

Common Error: *It is wrong to assume the < and > operators affect the amount of data that is transferred. The < and > operators will affect the addressing mode; that is, how the address is represented.*

Observation: Accesses to Registers A, B, and CC transfer 8 bits, while accesses to Registers D, X, Y, SP, and PC transfer 16 bits regardless of the addressing mode.

Checkpoint 3.8: What is the difference between `ldx #$0801` and `ldx $0801`?

Checkpoint 3.9: What is the difference between the direct mode instruction `ldx <$12` and the extended mode instruction `ldx >$0012`?

5. PC Relative addressing mode is used for the branch and branch to subroutine instructions. Stored in the machine code is not the absolute address of where to branch, but the 8-bit signed offset relative distance from the current PC value. When the branch address is being calculated, the PC already points to the next instruction. Calculating relative offsets gives first-time programmers a lot of trouble, but luckily, the assembler calculates these offsets for us. The concept is explained here in order to better understand how the computer works, rather than as an actual programming responsibility. The address of the next instruction is (location of instruction) + (number of bytes in the machine code). In the following example, assume the branch instruction is located at address $F880. The destination address is before the current instruction, which is called a backward jump.

```
bra $F840
```

The operand field for PC relative addressing is an 8-bit value called **rr**, which is calculated using the equation (destination address) − (location of instruction) − (size of the instruction). Since the `bra` op code is one byte ($20) and the operand is one byte, this instruction requires two bytes and the **rr** field is

```
$F840 - $F880 - 2 = -$42 = $BE
```

and the object code for this instruction will be `$20BE`. Again assume the branch instruction is located at address $F880. This time the destination address is after the current instruction, which is called a forward jump.

```
bra $F8C8
```

The **rr** field is

```
$F8C8 - $F880 - 2 = $46
```

and the object code for this instruction will be `$2046`.

Common Error: Since not every instruction supports every addressing mode, it would be a mistake to use an addressing mode not available for that instruction.

Observation: The number of cycles required to execute instructions on the 6811 is fixed for each instruction. The time for each 6811 cycle is also fixed.

Observation: Some of the conditional branch instructions on the 6812 require a different number of cycles to execute depending on whether or not the branch is taken. The cycle time when accessing external memory on a 6812 depends on the speed of the external memory. It also depends on whether the address is an even number or an odd number. These facts complicate the task of predetermining how long a 6812 program will take to execute.

Observation: Relative addressing within a program block is essential for implementing relocatable code.

Checkpoint 3.10: Give the object code of the assembly code that branches to itself, causing an infinite loop, `loop bra loop`.

3.2.4
6811 Machine
Language
Execution

In this section we will discuss the sequence of events that occur during 6811 program execution. Understanding computers at this hardware level is important for both the hardware engineer who builds new computers as well as the software engineer who programs them. The first step of every instruction is to fetch the op code. As you can see from Figure 3.7, the BIU allows the processor to read and write from memory. To fetch an op code, the processor places its PC on the address bus and issues a read cycle, the value read from memory is placed in the IR. The PC is then incremented. Most instructions on the 6811 and 6812 are one byte long, but some op codes are two bytes. Motorola uses the terms page-2, page-3, page-4 for op codes that require two bytes. The PC is automatically incremented after each fetch.

The **TExaS** simulator executes all the 6811 instructions and implements all its addressing modes. **TExaS** maintains an accurate cycle count for the MC68HC11 microcomputers. The assembler will give the number of cycles and cycle sequence in the listing. When showing the cycle-by-cycle execution, the actual 6811 bus cycles are shown.

In this section we present how the 6811 creates cycle-by-cycle accesses. There are two types of bus cycles (the 6811 does not support DMA) that the processor uses to communicate with memory. For both types of cycles, the processor drives the address bus and the R/W signal. The 16-bit address bus selects which memory location (or I/O device) to access. The R/W signal specifies whether the cycle is a read or a write. If R/W is one, then data is transferred from memory (or I/O device) into the processor. If R/W is zero, then data is transferred from the processor to memory (or I/O device).

During a read cycle (R/W=1), the memory (or I/O device) at the specified address puts the information on the data bus, and the processor transfers the information (8 bits at a time for a 6811) into the appropriate place within the processor. The processor has seven types of read cycles (we will label the last three types of cycles as free cycles that occur when the 6811 processor needs to perform internal calculations, but no additional memory information is required):

instruction fetch. The address is the PC and the 8-bit data are loaded into the instruction register, IR.

operand fetch. The address is also the PC, but the 8-bit data are used to calculate the effective address.

data fetch. The address is the EAR, and the 8-bit data are loaded into a register or sent to the ALU.

stack pull. First, the stack pointer is incremented (SP=SP+1). Then 8-bit data are read from memory, pointed to by SP, and stored inside the processor, usually into a register.

null cycle. The address is the $FFFF, and the 8-bit data are ignored.

dummy stack read. The address is the SP, the stack pointer is not changed, and the 8-bit data are ignored.

dummy PC read. The address is the PC, the program counter is not changed, and the 8-bit data are ignored.

During a write cycle (R/W=0), the processor puts the information on the data bus, and the memory (or I/O device) transfers the information into the specified location. The write cycles can be grouped into two types:

data write. The 8-bit data from a register or ALU are stored in memory at the address specified by the EAR.

stack write. The 8-bit data from a register are stored in memory at the address specified by the SP. Then the stack pointer is decremented (SP=SP-1).

The 6811 execution has six phases:

Phase	Function	R/W	Address	Comment
1	Op code fetch	read	PC++	Put data into IR,
2	Operand fetch	read	PC++	Immediate or calculate EA
3	Free cycle	read	PC/SP/$FFFF	Calculate EA
4	Data read	read	SP,EAR	Data passes through ALU,
5	Free cycle	read	PC/SP/$FFFF	ALU operations, set CCR
6	Data store	write	SP,EAR	Results stored in memory

Phase 1. **Op code fetch.** The execution of 6811 instruction begins with fetching the op code and putting it in the IR. The page 2,3,4 instructions have a prebyte ($18,$1A,$CD), so these instructions require two 8-bit cycles to obtain the entire op code. All instructions on the 6811 that involve Reg Y will require a prebyte. Having to fetch a 16-bit op code is the reason it is slower to use Reg Y compared to using Reg X.

Phase 2. **Operand fetch.** Inherent mode instructions require no additional information (no operand fetches). Immediate addressing mode requires 1 or 2 bytes of operand data, which is transferred to either a register (e.g., ldaa #10) or to the ALU (e.g., eora #$44). Direct addressing mode instructions will fetch 1 operand byte, which is used to create the effective address and stored in the EAR (with the most significant byte set to zero.) Extended addressing mode instructions will fetch 2 operand bytes that are used to create the effective address and stored in the EAR. Indexed addressing mode instructions will fetch exactly 1 operand byte, which is used to create the effective address and stored in the EAR. The index register is not modified during indexed addressing mode instructions on the 6811. PC relative addressing mode instructions (e.g., bra beq, etc.) will fetch 1 operand byte that is added to the PC forming the target branch address. Since the PC is incremented after fetching the op code and incremented again after fetching the operand, the PC is pointing to the next instruction at the time the target branch address is calculated. That is why the PC relative offset is the difference

$$\mathbf{rr} = \text{target} - \text{current} - \text{n}$$

where n is the number of bytes (usually 2 but sometimes 4 or 5) in the instruction.

Phase 3. **Free cycles.** After the instruction is fetched, and before the data can be accessed, the processor must calculate the effective address. The effective address register (EAR) in the BIU is set during this phase. All indexed addressing mode instructions have one null cycle during phase 3, during which time the effective address is calculated. For example, the effective address of the instruction ldaa 5,X is X+5. This X+5 addition operation will occur in this phase. The push and pull instructions also require additional time to calculate the effective address, and thus have free cycles between the instruction fetch and data access phases. On the other hand, even though the direct and extended addressing modes do generate an effective address, no additional time is required, therefore no phase 3 cycles are required for these instructions.

Phase 4. **Data read.** If the instruction requires data from memory, it will use the EAR to fetch 1 or 2 bytes. These cycles will be either data fetches or a stack pulls.

Phase 5. **Free cycles.** Any ALU functions occur next. The ALU is used in all instructions that set the CCR. The ALU calculation may require addition time to execute (e.g., idiv, mul). Since no data need to be read/written at this time, the 6811 will generate null cycles. In a real 6811 computer, these null cycles look like real memory accesses (R/W, Address, Data), but the data are ignored. The **TExaS** simulation of the 6811 realistically shows these do-nothing cycles.

Phase 6. **Data store.** If required, the last step involves writing data to memory. It will use the EAR to store 1 or 2 bytes. These cycles will be either data writes or stack pushes.

The actual 6811 cycle-by-cycle execution can be studied by first listing the cycle sequence code for the instruction, which can be found in the Motorola MC68HC11 Reference Manual. When the **ShowCycleTypes. . .** option is enabled via the **Assembly->Options** command, the cycle types will be included in the assembly listing. When the **Mode->CycleView** function is enabled, then the cycle types are displayed in the `TheLog.rtf` file during execution.

The following paragraphs explain the cycle code letters used. Each single letter code represents a single CPU cycle. All 6811 accesses are 8 bits, therefore lower case letters are used (the 6812 allows both 8- and 16-bit accesses, see 6812 cycle types.) There are cycle codes for each addressing mode variation of each instruction. The number of code letters specifies the execution time of an instruction.

f - Free cycle. This indicates a cycle where the CPU does not require use of the system buses. A **f** cycle is always one cycle of the system bus clock. The address will be the program counter, and the data will be ignored. The program counter, PC, is not incremented like other memory fetches that use the PC.

n - Null cycle. This indicates a cycle where the CPU does not require use of the system buses. A **n** cycle can be one or more cycles of the system bus clock. The address will be $FFFF, and the data will be ignored.

x - Dummy stack cycle. This indicates a cycle where the CPU does not require use of the system buses. An **x** cycle is always one cycle of the system bus clock. The address will be the stack pointer, and the data will be ignored. The SP is not incremented like other stack pulls that use the SP.

p - Program byte access. Program (opcode and operand) information is fetched as 8-bit bytes. The address will be the PC. The PC will be incremented immediately after each program-byte access. Opcode data will be stored in the instruction register. Immediate operand data will go directly to a register and/or the ALU (sets the condition code). Direct, extended, and indexed-mode operand data will be used to generate the effective address

r - 8-bit data read. The address is the effective address register. For 16-bit reads, two 8-bit data-read cycles are required. The most significant byte is read first, the EAR is incremented, and then the least significant byte is read. The 8-bit data go from the memory into the microcomputer.

s - 8-bit stack data. The stack push operation generates this type of cycle. First, the data are written at the current value of the stack pointer, SP. Then, the SP is decremented. For 16-bit pushes, two 8-bit stack data cycles are required. The least significant byte is written first, and then the most significant byte is written. In this way the 16-bit data exist in memory defined in the usual Big Endian format.

w - 8-bit data write. The address is the effective address register. For 16-bit writes, two 8-bit data write cycles are required. The most significant byte is written first, the EAR is incremented, and then the least significant byte is written. The 8-bit data go from the microcomputer to memory.

u - 8-bit unstack data. Unstack data cycles are generated during a stack pull operation. First, the stack pointer, SP, is incremented. Then, the data are read at the address of the SP. For 16-bit pulls, two 8-bit unstack data cycles are required. The most significant byte is read first, and then the least significant byte is read. In this way the 16-bit data exist in memory defined in the usual Big Endian format.

v - Vector fetch. The 6811 interrupt and reset vectors exist at the last addresses in ROM. These cycles can be observed when an interrupt is being serviced or during the execution of the software interrupt instruction (`swi`).

We can illustrate this process with an example. **TExaS** assembles the 6811 instruction `ldx $0800` as the following. The first field, `$F000`, is the memory location that will hold the instruction. The second field, `FE0800`, is the object code in hexadecimal that the assembler generates and is loaded into memory. The next field, `[5]`, specifies the total number of cycles required to execute this instruction. The fourth field, `{ppprr}`, gives the cycle codes explaining the details of the execution. The last field is the original source code.

```
$F000 FE0800 [ 5] {ppprr } ldx $0800
```

When the execution is simulated by TExaS it generates the 5 read cycles that would have occurred in a real 6811. Assume the 16-bit data at $8000 is $1234 (big endian.)

```
Read $F000 FE Op code fetch
Read $F001 08 Operand fetch
Read $F002 00 Operand fetch
Read $0800 12 Effective address data fetch
Read $0801 34 Effective address data fetch
```

The following shows the memory cycles generated by **TExaS** when the instruction `ldaa $F001` is executed by a 6811. The object code for this instruction at $F000 is $B6F001. The first cycle fetches the op code. The next two cycles fetch the operand ($F001), and the last cycle fetches the data.

```
Opcode fetch    R 0xF000 0xB6 from ROM
Operand fetch   R 0xF001 0xF0 from ROM
Operand fetch   R 0xF002 0x01 from ROM
Fetch using EARR 0xF001 0xF0 from ROM
```

3.2.5
Simulated 6812
Machine
Language
Execution

The **TExaS** simulator executes all the 6812 instructions and implements all its addressing modes. **TExaS** maintains an accurate cycle count for the 6812 running in single chip mode. A real 6812 can run in expanded mode with cycle stretching. For this situation, the cycle time will depend on the speed of the memory, but **TExaS** does not simulate this complex behavior. For more information on 6812 expanded mode and cycle stretching read Chapter 9 of *Embedded Microcomputer Systems: Real Time Interfacing* by Jonathan W. Valvano, published by Brooks/Cole, copyright (c) 2000. The assembler will give the num-

Table 3.3
Differences between a
real 6812 and the **TExaS**
bus cycle simulation.

actual 6812 cycle-by-cycle	simplified cycle-by-cycle
sometimes 8-, sometimes 16-bit data	sequence of 8-bit accesses
special case for misaligned 16-bit-access	even and odd addresses treated the same
pipeline enhances execution speed	simple fetch-execute sequence
fetches op codes for later execution	fetches op codes for immediate execution
fetches op codes that are never executed	fetched op codes are always executed
MC68HC812A4 allows for 22-bit address	always 16-bit address
variable length off-chip accesses	all accesses exactly same cycle period

ber of cycles and cycle sequence in the listing. When showing cycle-by-cycle execution, a simplified pseudo-cycle is displayed instead of the actual 6812 bus cycles. The major differences are shown in Table 3.3. Assuming the computer is running in single chip mode, with no external slow memory, **TExaS** properly simulates the action of each instruction and the amount of time each instruction takes to execute.

In this section we present how the **TExaS** 6812 simulator creates cycle-by-cycle accesses. There are two types of bus cycles (the 6812 does not support DMA) that the processor uses to communicate with memory. For both types of cycles, the processor drives the address bus and the R/W signal. The 16-bit address bus selects which memory location (or I/O device) to access. The R/W signal specifies read or write.

During a read cycle (R/W=1), the memory at the specified address puts the information on the data bus, and the processor transfers the information (8 bits in this simplified simulation) into the appropriate place within the processor. The processor has four types of read cycles:

instruction fetch. The address is the PC and the 8-bit data is loaded into the instruction register, IR.

operand fetch. The address is also the PC, but the 8-bit data is used to calculate the effective address.

data fetch. The address is the EAR, and the 8-bit data is loaded into a register or sent to the ALU.

stack pull. First, the 8-bit data is read from memory pointed to by SP and stored in a register, then the stack pointer is incremented SP=SP+1.

During a write cycle (R/W=0), the processor puts the information on the data bus, and the memory transfers the information into the specified location. The write cycles can be grouped into two types:

data write. The 8-bit data from a register or ALU is stored in memory at the address specified by the EAR.

stack write. First, the stack pointer is decremented SP=SP−1, then the 8-bit data from a register is stored in memory at the address specified by the SP.

The simplified execution has five phases:

Phase	Function	R/W	Address	Comment
1	Op code fetch	read	PC++	Put data into IR,
2	Operand fetch	read	PC++	Immediate or calculate EA
3	Data read	read	SP,EAR	Data passes through ALU,
4	Free cycle	read	PC/SP/$FFFF	ALU operations, set CCR
5	Data store	write	SP,EAR	Results stored in memory

Phase 1. **Op code fetch.** The execution of 6812 instruction begins with fetching the op code and putting it in the IR. The page 2 instructions have a prebyte ($18), so these instructions require two 8-bit cycles to obtain the entire op code.

Phase 2. **Operand fetch.** Inherent mode instructions require no additional information. Immediate addressing mode requires 1 or 2 bytes of operand data, which are transferred either to a register (e.g., `ldaa #10`) or to the ALU (e.g., `eora #$44`). Direct addressing mode instructions will fetch 1 operand byte, which is used to create the effective address and stored in the EAR (with the most significant byte set to zero). Extended addressing mode instructions will fetch 2 operand bytes that are used to create the effective address and stored in the EAR. Indexed addressing mode instructions will fetch 1 2 or 3 operand bytes, which are used to create the effective address and stored in the EAR. The index register may or may not be modified. PC relative addressing mode instructions will fetch 1 or 2 operand bytes that are added to the PC, forming the target branch address. Since the PC is incremented after fetching the op code and incremented again after fetching the operand, the PC is pointing to the next instruction at the time the target branch address is calculated. That is why the PC relative offset is the difference

$$rel = target - current - n$$
where n is the number of bytes (2, 3, or 4) in the instruction.

Phase 3. **Data read.** If the instruction requires data from memory, it will use the EAR to fetch 1 or 2 bytes. These cycles will be either data fetches or stack pulls.

Phase 4. **Free cycles.** Any ALU functions occur next. The ALU is used in many instructions to set the CCR. The ALU calculation may require addition time to execute (e.g., `idiv, mem`). Since no data need to be read/written at this time, the actual 6812 will generate null cycles or free cycles (f). In the real computer, these free cycles look like real memory accesses (R/W, Address, Data), but the data are ignored. The simplified cycle-by-cycle simulation counts these cycles, but does not generate a cycle access for these do-nothing cycles.

Phase 5. **Data write.** If required, the last step involves writing data to memory. It will use the EAR to store 1 or 2 bytes. These cycles will be either data writes or stack pushes.

TExaS assembles the 6812 instruction `ldx $0800` as the following. The first field, $F000, is the memory location that will hold the instruction. The second field, FE0800, is the object code in hexadecimal that the assembler generates and is loaded into memory. The next field, [3], specifies the total number of cycles required to execute this instruction. The fourth field, {ROP}, gives the cycle codes explaining the details of the execution of the real 6812. The last field is the original source code.

```
$F000 FE0800 [ 3]( 0){ROP } ldx $0800
```

When a real 6812 executes this instruction, it will create three 16-bit memory cycles. Execution of the real 6812 is discussed in the next section. The **TExaS** simulator will correctly count the three cycles, but show five 8-bit read cycles as it simulates this instruction. Assume the 16-bit data at $8000 is $1234 (big endian.)

```
Opcode fetch    R 0xF000 0xFE from EEPROM    Phase 1
Operand fetch   R 0xF001 0x08 from EEPROM    Phase 2
Operand fetch   R 0xF002 0x00 from EEPROM    Phase 2
Fetch msb @ EARR 0x0800 0x12 from RAM        Phase 3
Fetch lsb @ EARR 0x0801 0x34 from RAM        Phase 3
```

3.2.6
*Actual 12
Machine
Language
Execution

There are eight types of bus cycles that the 6812 uses to communicate with memory. For all types of cycles, the processor drives the address bus, the R/W signal, and the LSTRB signal. The 16-bit address bus selects which memory location R/W signal specifies read (R/W=1) or write (R/W=0). The LSTRB signal specifies the type of cycle as shown in Table 3.4.

Table 3.4
Eight types of memory cycles on a 6812.

LSTRB	A0	R/W	Type of Access
1	0	1	8-bit read of an even address
0	1	1	8-bit read of an odd address
1	0	0	8-bit write of an even address
0	1	0	8-bit write of an odd address
0	0	1	16-bit read of an even address
1	1	1	16-bit read of an odd address (low/high data swapped)
0	0	0	16-bit write to an even address
1	1	0	16-bit write to an odd address (low/high data swapped)

During a read cycle, the memory at the specified address puts the information on the data bus, and the processor transfers the information (8 or 16 bits) into the appropriate place within the processor. During a write cycle, the processor puts the information on the data bus, and the memory transfers the information into the specified location. The 8-bit transfers are straightforward because the 16-bit address explicitly defines the memory location. The 16-bit transfers to even addresses are also simple, because the top 15 bits of the address determine the two locations of interest. The tricky situation is a 16-bit access to an odd address. For example, a 16-bit read of location $08FF must access both locations $08FF (most significant byte) and $0900 (least significant byte) in a single memory cycle. In this case, the $08FF data will arrive at the processor as the lower 8 bits of the data bus, and the $0900 data will arrive as the upper 8 bits. The two bytes will be swapped inside the processor before it is used.

Another complicating feature of a real 6812 is its pipeline. The 6812 BIU reads op codes, operands and memory data in a sequence similar to the simple execution of the 6811 explained previously in Section 3.2.4. First, it reads the op code, then it reads the operand, and finally it reads and writes memory data as required. The pipeline is a hardware first-in-first-out queue, placed between the BIU and the CU. All data read from memory passes through the pipeline, which can hold up to three 16-bit values. Writes from the processor to memory do not pass through the pipeline. Using a pipeline allows the 6812 to fetch the op code and operands of the next instruction while it is executing the current instruction. This explains the curious behavior of 6812 instructions, which seem to perform its data transfers first, followed by the op code fetches. Actually, the op code fetches specified as part of an instruction execution are reading the op codes for the next instruction. A conditional branch instruction has the tricky question of which op codes should it fetch. It has not yet performed the conditional test, so it doesn't know the answer. The 6812 assumes a conditional branch will not occur, and will simply fetch the next op code. If the branch is to occur, then its pipeline will be filled with the wrong op codes, and the pipeline will have to be flushed and refilled. This is the reason conditional branch instructions take three or one cycle to execute, depending on whether the branch is taken or not. When faced with a conditional branch instruction, the Intel Pentium processor fills its pipeline with both possible branch paths, so it is efficient regardless of whether or not the branch is to occur.

The actual 6812 cycle-by-cycle execution can be studied by first listing the cycle sequence code for the instruction. The sequence code can be found in the Motorola CPU12 Reference Manual (on the CD) or by assembling the instruction using **TExaS**. When the **ShowCycleTypes. . .** option is enabled via the **Assembly->Options** command, the cycle types will be included in the assembly listing. Although **TExaS** doesn't show the real cycles during simulated execution, its assembler will show the cycle sequence code in the assembly listing. A single letter code represents a single CPU cycle. Upper case letters indicate 16-bit access cycles. Lower case letters mean 8-bit accesses. (The 6811 allows only eight accesses; see 6811 cycle types.) There are cycle codes for each addressing mode variation of each instruction. Simply count code letters to determine the execution time of an instruction in a best-case system. An example of a best-case system is a single-chip 16-bit system with no 16-bit off-boundary data accesses to any locations other than on-chip RAM. Many conditions can cause one or more instruction cycles to be stretched, but the CPU is not aware of the stretch delays because the clock to the CPU is temporarily stopped during these delays. The following paragraphs explain the cycle code letters used and note conditions that can cause each type of cycle to be stretched.

f - Free cycle. This indicates a cycle where the CPU does not require use of the system buses. An **f** cycle is always one cycle of the system bus clock. These cycles can be used by a queue controller or the background debug system to perform single-cycle accesses without disturbing the CPU.

g - Read 8-bit PPAGE register. These cycles are used only with the `call` instruction to read the current value of the PPAGE register, and are not visible on the external bus. Since the PPAGE register is an internal 8-bit register, these cycles are never stretched.

I - Read indirect pointer. Indexed indirect instructions use this 16-bit pointer from memory to address the operand for the instruction. These are always 16-bit reads but they can be either aligned or misaligned. These cycles are extended to two bus cycles if the MCU is operating with an 8-bit external data bus and the corresponding data are stored in external memory. There can be additional stretching when the address space is assigned to a chip-select circuit programmed for slow memory. These cycles are also stretched if they correspond to misaligned access to a memory that is not designed for single-cycle misaligned access.

i - Read indirect PPAGE value. These cycles are used only with indexed indirect versions of the `call` instruction, where the 8-bit value for the memory expansion page register of the `call` destination is fetched from an indirect memory location. These cycles are stretched only when controlled by a chip-select circuit that is programmed for slow memory.

n - Write 8-bit PPAGE register. These cycles are used only with the `call` and `rtc` instructions to write the destination value of the PPAGE register and are not visible on the external bus. Since the PPAGE register is an internal 8-bit register, these cycles are never stretched.

O - Optional cycle. Program information is always fetched as aligned 16-bit words. When an instruction consists of an odd number of bytes, and the first byte is misaligned, an **O** cycle is used to make an additional program word access (**P**) cycle that maintains queue order. In all other cases, the **O** cycle appears as a free (**f**) cycle. The $18 prebyte for page two opcodes is treated as a special one-byte instruction. If the prebyte is misaligned, the **O** cycle is used as a program word access for the prebyte; if the prebyte is aligned, the **O** cycle appears as a free cycle. If the remainder of the instruction consists of an odd number of bytes, another **O** cycle is required some time before the instruction is completed. If the **O** cycle for the prebyte is treated as a **P** cycle, any subsequent **O** cycle in the same instruction is treated as an **f** cycle; if the **O** cycle for the prebyte is treated as an **f** cycle, any subsequent **O** cycle in the same instruction is treated as a **P** cycle. Optional cycles used for program word accesses can be extended to two bus cycles if the MCU is operating with an 8-bit external data bus and the program is stored in external memory. There can be additional stretching when the address space is assigned to a chip-select circuit programmed for slow memory. Optional cycles used as free cycles are never stretched.

P - Program word access. Program information is fetched as aligned 16-bit words. These cycles are extended to two bus cycles if the MCU is operating with an 8-bit external data bus and the program is stored externally. There can be additional stretching when the address space is assigned to a chip-select circuit programmed for slow memory.

r - 8-bit data read. These cycles are stretched only when controlled by a chip-select circuit programmed for slow memory.

R - 16-bit data read. These cycles are extended to two bus cycles if the MCU is operating with an 8-bit external data bus and the corresponding data is stored in external memory. There can be additional stretching when the address space is assigned to a chip-select circuit programmed for slow memory. These cycles are also stretched if they correspond to misaligned accesses to memory that is not designed for single-cycle misaligned access.

s - 8-bit stack data. These cycles are stretched only when controlled by a chip-select circuit programmed for slow memory.

S - 16-bit stack data. These cycles are extended to two bus cycles if the MCU is operating with an 8-bit external data bus and the SP is pointing to external memory. There can be additional stretching if the address space is assigned to a chip-select circuit programmed for slow memory. These cycles are also stretched if they correspond to misaligned accesses to a memory that is not designed for single-cycle misaligned access. The internal RAM is designed to allow single-cycle misaligned word access.

w - 8-bit data write. These cycles are stretched only when controlled by a chip-select circuit programmed for slow memory.

W - 16-bit data write. These cycles are extended to two bus cycles if the MCU is operating with an 8-bit external data bus and the corresponding data is stored in external memory. There can be additional stretching when the address space is assigned to a chip-select circuit programmed for slow memory. These cycles are also stretched if they correspond to misaligned access to a memory that is not designed for single-cycle misaligned access.

u - 8-bit unstack data. These cycles are stretched only when controlled by a chip-select circuit programmed for slow memory.

U - 16-bit unstack data. These cycles are extended to two bus cycles if the MCU is operating with an 8-bit external data bus and the SP is pointing to external memory. There can be additional stretching when the address space is assigned to a chip-select circuit programmed for slow memory. These cycles are also stretched if they correspond to misaligned accesses to a memory that is not designed for single-cycle misaligned access. The internal RAM is designed to allow single-cycle misaligned word access.

V - Vector fetch. Vectors are always aligned 16-bit words. These cycles are extended to two bus cycles if the MCU is operating with an 8-bit external data bus and the program is stored in external memory. There can be additional stretching when the address space is assigned to a chip-select circuit programmed for slow memory.

t - 8-bit conditional read. These cycles are either data read cycles or free cycles, depending upon the data and flow of the `revw` instruction. These cycles are stretched only when controlled by a chip-select circuit programmed for slow memory.

T - 16-bit conditional read. These cycles are either data read cycles or free cycles, depending upon the data and flow of the `rev` or `revw` instruction. These cycles are extended to two bus cycles if the MCU is operating with an 8-bit external data bus and the corresponding data is stored in external memory. There can be additional stretching when the address space is assigned to a chip-select circuit programmed for slow memory. These cycles are also stretched if they correspond to misaligned accesses to a memory that is not designed for single-cycle misaligned access.

x - 8-bit conditional write. These cycles are either data write cycles or free cycles, depending upon the data and flow of the `rev` or `revw` instruction. These cycles are stretched only when controlled by a chip-select circuit programmed for slow memory.

PPP/P - Short branches require three cycles if taken, one cycle if not taken. Since the instruction consists of a single word containing both an op code and an 8-bit offset, the not-taken case is simple: The pipeline advances, another program word fetch is made, and execution continues with the next instruction. The taken case requires that the pipeline be refilled so that execution can continue at a new address. First, the effective address of the destination is determined, then the CPU performs three program word fetches from that address.

OPPP/OPO - Long branches require four cycles if taken, three cycles if not taken. Optional cycles are required because all long branches are page two op codes, and thus include the $18 prebyte. The CPU12 treats the prebyte as a special 1-byte instruction. If the prebyte is misaligned, the optional cycle is used to perform a program word access; if the prebyte is aligned, the optional cycle is used to perform a free cycle. As a result, both the taken and not-taken cases use one optional cycle for the prebyte. In the not-taken case, the pipeline must advance so that execution can continue with the next instruction, and another optional cycle is required to maintain the pipeline. The taken case requires that the pipeline be refilled so that execution can continue at a new address. First, the effective address of the destination is determined, then the CPU performs three program word fetches from that address.

Because of the pipeline, one must study a sequence of instructions to determine the actual memory bus cycles. As an example, we will analyze the tutorial example from Appendix 1. First, we look at the listing file generated by the **TExaS** assembler. The first field, e.g., $F000, is the memory location that will hold the instruction. The second field, e.g., 8600, is the object code in hexadecimal that the assembler generates and is loaded into memory. The next field, e.g., [1], specifies the total number of cycles required to execute this instruction. The fourth field, e.g., {P}, gives the cycle codes explaining the details of the execution. The last field is the original source code.

```
$F000 8600      [ 1] {P        }main  ldaa  #$00
$F002 5A06      [ 2] {Pw       }      staa  DDRC
$F004   86FF    [ 1] {P        }      ldaa  #$FF
$F006 5A03      [ 2] {Pw       }      staa  DDRB
$F008 9604      [ 3] {rfP      }loop  ldaa  PORTC
$F00A 7A0800    [ 3] {wOP      }      staa  Data
$F00D 5A01      [ 2] {Pw       }      staa  PORTB
$F00F   20F7    [ 3] {PPP      }      bra   loop
```

The next step is to string the cycle codes together: PPwPPw,rfPwOPPwPPP. On a 8 MHz 6812, each cycle takes 125ns. The sequence after the comma will be repeated. At the beginning of each instruction, we assume the entire instruction is loaded in the pipeline. In other words, the end of each instruction must fetch enough data to capture the entire next instruction. The reset sequence will set the PC using the reset vector (16-bit value from $FFFE), and load the pipeline with the $8600,$5A06 machine codes. We also assume the Port C inputs are set such that their data will be $0A. Table 3.5 shows the actual 6812 cycles during execution. Notice that the front of the pipeline contains the instruction being executed. The [**] means irrelevant data.

Table 3.5

Real memory cycles generated by a 6812 executing the Chapter 1 tutorial program.

Pipeline before	cycle	address	R/W	LSTRB	data	executing
$8600,$5A06	P	$F004	Read	0	$86FF	ldaa #$00
$5A06,$86FF	P	$F006	Read	0	$5A03	staa DDRC
$5A06,$86FF,$5A03	w	$0006	Write	1	$00	
$86FF,$5A03	P	$F008	Read	0	$9604	ldaa #$FF
$5A03,$9604	P	$F00A	Read	0	$7A08	staa DDRB
$5A03,$9604,$7A08	w	$0003	Write	0	$FF	
$9604,$7A08	r	$0004	Read	0	$0A	ldaa PORTC
$9604,$7A08	f	$FFFE	Read	0	****	
$9604,$7A08	P	$F00C	Read	0	$005A	
$7A08,$005A	w	$0800	Write	1	$0A	staa Data
$7A08,$005A	O	$FFFE	Read	0	****	
$7A08,$005A	P	$F00E	Read	0	$0120	
$5A,$0120	P	$F010	Read	0	$F7[**]	staa PORTB
$5A,$0120,$F7[**]	w	$0001	Write	1	$0A	
$20,$F7[**]	P	$F012	Read	0	****	bra loop
$20,$F7[**],****	P	$F008	Read	0	$9604	
$20,$F7[**],$9604	P	$F00A	Read	0	$7A08	

Table 3.6 shows the simulated 6812 cycles generated by **TExaS**. The 6812 simulation is similar to a real 6811, with the null cycles removed.

Table 3.6
Simulated memory cycles generated by a 6812 executing the Chapter 1 tutorial program.

cycle type	address	R/W	data	executing
Opcode fetch	$F000	Read	$86	ldaa #$00
Operand fetch	$F001	Read	$00	
Opcode fetch	$F002	Read	$5A	staa DDRC
Operand fetch	$F003	Read	$06	
Write at EAR	$0006	Write	$00	
Opcode fetch	$F004	Read	$86	ldaa #$FF
Operand fetch	$F005	Read	$FF	
Opcode fetch	$F006	Read	$5A	staa DDRB
Operand fetch	$F007	Read	$03	
Write at EAR	$0003	Write	$FF	
Opcode fetch	$F008	Read	$96	ldaa PORTC
Operand fetch	$F009	Read	$04	
Read at EAR	$0004	Read	$0A	
Opcode fetch	$F00A	Read	$7A	staa Data
Operand fetch	$F00B	Read	$08	
Operand fetch	$F00C	Read	$00	
Write at EAR	$0800	Write	$0A	
Opcode fetch	$F00D	Read	$5A	staa PORTB
Operand fetch	$F00E	Read	$01	
Write at EAR	$0001	Write	$0A	
Opcode fetch	$F00F	Read	$20	bra loop
Operand fetch	$F010	Read	$F7	

Checkpoint 3.11: *Give the actual memory cycles created when the following program is executed by a 6812. Assume the pipeline contains the data* B608, 0006. *Assume location $0800 contains the data $55.*

```
$F000 B60800 [ 3](  0){rOP }loop ldaa $0800
$F003 06F000 [ 3](  3){PPP }     jmp  loop
```

3.3   Introduction to Programming

A few simple instructions are presented in this section. The subset of instructions provided here in this section will be sufficient to understand many of the examples in this book. For a complete list of instructions, refer to the CD enclosed with this text, the help system on **TExaS,** or to the corresponding Motorola programming manual.

**3.3.1
Memory and
Register Transfer
Operations**

The 8-bit load instructions transfer data from memory into a register. Since the original data in memory exists, there are now two copies of the information. If the address U is between 0 and $00FF, then direct addressing mode will be used; otherwise extended addressing mode will be used.

```
ldaa  #w     RegA=w
ldaa  U      RegA=[U]
ldab  #w     RegB=w
ldab  U      RegB=[U]
```

Condition code bits are set with R equal to the 8-bit memory contents loaded into the register.

N: result is negative N=R7
Z: result is zero Z=not(R7)·not(R6)·not(R5)·not(R4)·not(R3)·not(R2)·not(R1)·not(R0)
V: signed overflow V=0

The 16-bit load instructions also transfer information from memory into a register. Although D usually contains data and X,Y usually contain addresses, it is acceptable programming practice to place address or data information in any of these three registers. The stack pointer (S or SP) will always contain an address specifying the top of the stack. The program counter (PC) will always contain an address specifying the next instruction to execute.

```
ldd #W  RegD=W
ldd U   RegD={U}
lds #W  RegSP=W
lds U   RegSP={U}
ldx #W  RegX=W
ldx U   RegX={U}
ldy #W  RegY=W
ldy U   RegY={U}
```

Condition code bits are set with R equal to the 16-bit memory contents loaded into the register.

N: result is negative N=R15
Z: result is zero Z=not(R15)·not(R14)·. . .·not(R1)·not(R0)
V: signed overflow V=0

> **Observation:** The `lds #W` instruction is used to initialize the stack, where W is the last RAM address.

The 6812 has a couple of very convenient memory to memory move instructions, which set no flags.

```
movb #w,addr      [addr]=w
movb addr1,addr2  [addr2]=[addr1]
movw #W,addr      {addr}=W
movw addr1,addr2  {addr2}={addr1}
```

The 8-bit store instructions move data from a register to memory. The data in the register remain intact, so after execution of one of these instructions, there are two copies of the data.

```
staa U [U]=RegA
stab U [U]=RegB
```

Condition code bits are set with R equal to the 8-bit register contents stored into memory.

N: result is negative N=R7Z: result is zero
Z=not(R7)·not(R6)·not(R5)·not(R4)·not(R3)·not(R2)·not(R1)·not(R0)
V: signed overflow V=0

The 16-bit store instructions move data from a register to memory.

```
std U {U}=RegD
sts U {U}=RegSP
stx U {U}=RegX
sty U {U}=RegY
```

Condition code bits are set with R equal to the 16-bit register contents stored into memory.

N: result is negative N=R15
Z: result is zero Z=not(R15)·not(R14)·...·not(R1)·not(R0)
V: signed overflow V=0

The following transfer operations use inherent addressing.

```
xgdx    swap RegD and RegX
xgdy    swap RegD and RegY
clc     clear carry bit, C=0
cli     clear interrupt mask bit, enable interrupts, I=0
clv     clear overflow bit, V=0
sec     set carry bit, C=1
sei     set interrupt mask bit, disable interrupts, I=1
sev     set overflow bit, V=1
tap     transfer A to CC, (can not change X bit from 0 to 1)
tpa     transfer CC to A
```

**3.3.2
Arithmetic
Operations**

The `adda` and `addb` instructions add an 8-bit value from memory to the corresponding register. These instructions work for both signed and unsigned data.

```
adda #w    RegA=RegA+w
adda U     RegA=RegA+[U]
addb #w    RegB=RegB+w
addb U     RegB=RegB+[U]
```

Condition code bits are set after R=X+M; X is initial register value, M is the memory data, and R is the final register value.

N: result is negative N=R7Z: result is zero
Z=not(R7)·not(R6)·not(R5)·not(R4)·not(R3)·not(R2)·not(R1)·not(R0)
V: signed overflow V=X7·M7·not(R7)+not(X7)·not(M7)·R7
C: unsigned overflow C=X7·M7+M7·not(R7)+not(R7)·X7

Let N and M be 8-bit unsigned locations. The following assembly code implements M=N+10.

```
ldaa N
adda #10 RegA=N+10, error if C is set
staa M
```

The `addd` instruction adds a 16-bit value from memory to Register D. This instruction works for both signed and unsigned data.

```
addd #W  RegD=RegD+W
addd U   RegD=RegD+{U}
```

Condition code bits are set after R=D+M; where D is the initial value in Register D.

N: result is negative N=R15
Z: result is zero Z=not(R15)·not(R14)·...·not(R2)·not(R1)·not(R0)
V: signed overflow V=D15·M15·not(R15)+not(D15)·not(M15)·R15
C: unsigned overflow C=D15·M15+M15·not(R15)+not(R15)·D15

Let N and M be 16-bit unsigned locations. The following assembly code implements M=N+1000.

```
ldd   N
addd  #1000 RegD=N+1000, error if C is set
std   M
```

Checkpoint 3.12: *Write assembly code that adds a constant 100 to Register X.*

These instructions subtract an 8-bit memory value from a register. The operation works for both signed and unsigned values. The compare instructions do not change the register value. The condition code bits can be tested by a conditional branch instruction to compare the two values. If the numbers represent unsigned values, then follow the subtraction with an unsigned conditional branch: `beq`, `bne`, `bhi`, `bhs`, `blo`, or `bls`. If the numbers represent signed values, then follow the subtraction with a signed conditional branch: `beq`, `bne`, `bgt`, `bge`, `blt`, or `ble`.

```
cmpa #w RegA-w
cmpa U  RegA-[U]
cmpb #w RegB-w
cmpb U  RegB-[U]
suba #w RegA=RegA-w
suba U  RegA=RegA-[U]
subb #w RegB=RegB-w
subb U  RegB=RegB-[U]
tsta    RegA-0
tstb    RegB-0
```

Condition code bits are set after $R = X - M$; X is initial register value, and R is the final register value.

N: result is negative $N = R7Z$: result is zero
$Z = \text{not}(R7) \cdot \text{not}(R6) \cdot \text{not}(R5) \cdot \text{not}(R4) \cdot \text{not}(R3) \cdot \text{not}(R2) \cdot \text{not}(R1) \cdot \text{not}(R0)$
V: signed overflow $V = X7 \cdot \text{not}(M7) \cdot \text{not}(R7) + \text{not}(X7) \cdot M7 \cdot R7$
C: unsigned overflow $C = \text{not}(X7) \cdot M7 + M7 \cdot R7 + R7 \cdot \text{not}(X7)$

Let N and M be 8-bit unsigned locations. The following assembly code implements $M = N-10$.

```
ldaa N
suba #10 RegA=N-10, error if C is set
staa M
```

These instructions subtract a 16-bit memory value from a register. Just like the 8-bit subtraction operators, these operators works for both signed and unsigned values. Again, the condition code bits can be tested by a conditional branch instruction to compare the two values.

```
cpd  #W  RegD-W
cpd  U   RegD-{U}
cpx  #W  RegX-W
cpx  U   RegX-{U}
cpy  #W  RegY-W
cpy  U   RegY-{U}
subd #W  RegD=RegD-W
subd U   RegD=RegD-{U}
```

Condition code bits are set after $R = X - M$; X is initial register value, and R is the final register value.

N: result is negative N=R15
Z: result is zero Z=not(R15)·not(R14)·. . .·not(R2)·not(R1)·not(R0)
V: signed overflow V=X15·not(M15)·not(R15)+not(X15)·M15·R15
C: unsigned overflow C=not(X15)·M15+M15·R15+R15·not(X15)

Let N and M be 16-bit unsigned locations. The following assembly code implements M=N−1000.

```
ldd   N
subd #1000 RegD=N-1000, error if C is set
std  M
```

There are a full set of increment and decrement instructions, which operate properly on either signed or unsigned values. These instructions use inherent addressing. The Z bit is set if the result is zero.

```
deca RegA=RegA-1
decb RegA=RegA-1
dex  RegX=RegX-1
dey  RegY=RegY-1
inca RegA=RegA+1
incb RegB=RegB+1
inx  RegX=RegX+1
iny  RegY=RegY+1
```

The `mul` instruction performs an 8-bit by 8-bit into 16-bit unsigned multiply, giving RegD=RegA*RegB, as shown in Figure 3.13. No overflow is possible. It can also be used to perform an 8 by 8-bit into 8-bit signed multiply, giving RegB=RegA*RegB.

Figure 3.13
The `mul` instruction takes two 8-bit inputs and generates a 16-bit product.

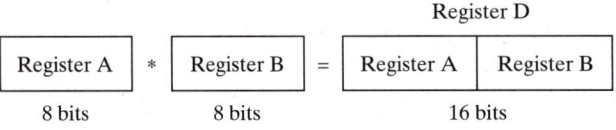

Condition code bits are set after R=A*B.

C: R7, set if bit 7 of 16-bit result is one

> **Checkpoint 3.13:** Prove the `mul` instruction can't overflow.

Let N and M be 8-bit unsigned locations. The following assembly code implements M=3*N.

```
ldaa  N
ldab  #3
mul        RegD=3*N, error if RegA is not zero
stab  M
```

The `idiv` instruction performs a 16-bit by 16-bit unsigned divide with remainder, giving RegX=RegD/RegX, as shown in Figure 3.14. Register D is the remainder.

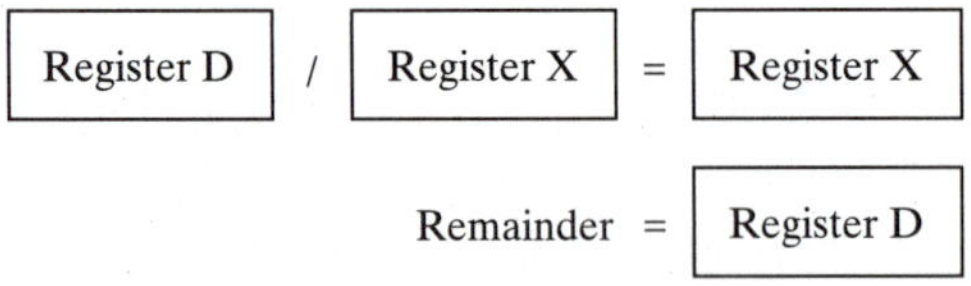

Figure 3.14
The `idiv` instruction takes two 16-bit inputs and generates a 16-bit quotient and a 16-bit remainder.

Condition code bits are set after quotient=dividend/divisor or Q=D/X.

Z: result is zero, Z=not(Q15)·not(Q14)·. . .·not(Q2)·not(Q1)·not(Q0)
V: 0
C: divide by zero, C=not(X15)·not(X14)·. . .·not(X2)·not(X1)·not(X0)

Let N and M be 16-bit unsigned locations. The following assembly code implements M=N/15. Notice that this operation can not overflow because the result will be less than or equal to 4369.

```
ldd    N    D=N (between 0 and 65535)
ldx    #15
idiv        X=N/15 (between 0 and 4369)
stx    M
```

Checkpoint 3.14: *Give a single mathematical equation relating the dividend, divisor, quotient, and remainder. This equation gives a unique solution as long as you assume the remainder is strictly less than the divisor.*

The `fdiv` instruction also performs a 16-bit by 16-bit unsigned divide with remainder. In contrast, this instruction calculates RegX=$(2^{16}$*RegD)/RegX, as shown in Figure 3.15. RegD is the remainder.

Figure 3.15
The `fdiv` instruction takes two 16-bit inputs and generates a 16-bit quotient and a 16-bit remainder.

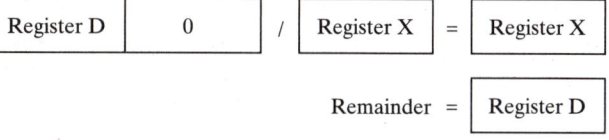

Register D | 0 / Register X = Register X

Remainder = Register D

Condition code bits are set after R=(65536*D)/X.

Z: result is zero, Z=not(R15)·not(R14)·. . .·not(R2)·not(R1)·not(R0)
V: overflow if RegX is less than or equal to RegD, result >\$FFFF
C: divide by zero, C=not(X15)·not(X14)·. . .·not(X2)·not(X1)·not(X0)

Let N and M be 16-bit unsigned locations. The following assembly code implements M=12.34*N. We approximate 12.34 by 65536/5311.

```
ldd    N
ldx    #5311
fdiv        RegX=(65536*N)/5311
stx    M
```

Checkpoint 3.15: *Let N and M be 8-bit unsigned locations. Write assembly code to implement M=(7*N)/31.*

3.3.3
Shift Operations
The shift instructions use inherent addressing. The N bit is set if the result is negative. The Z bit is set if the result is zero. The V bit is set on a signed overflow, and detected by a change in the sign bit. The C bit is the carry out after the shift.

```
asla RegA=RegA*2 (signed, but the same as lsla)
aslb RegB=RegB*2 (signed, but the same as lslb)
asld RegD=RegD*2 (signed, but the same as lsld)
lsla RegA=RegA*2 (unsigned but the same as asla)
lslb RegB=RegB*2 (unsigned but the same as aslb)
lsld RegD=RegD*2 (unsigned but the same as asld)
asra RegA=RegA/2 (signed)
asrb RegB=RegB/2 (signed)
asrd RegD=RegD/2 (signed)
lsra RegA=RegA/2 (unsigned)
lsrb RegB=RegB/2 (unsigned)
lsrd RegD=RegD/2 (unsigned)
rola RegA=rol(RegA)    (C←A7←. . .←A0←C)
rolb RegB=rol(RegB)    (C←B7←. . .←B0←C)
rora RegA=ror(RegA)    (C→A7→. . .→A0→C)
rorb RegB=ror(RegB)    (C→B7→. . .→B0→C)
```

Let N and M be 16-bit unsigned locations. The following assembly code implements M=N>>3.

```
ldd  N
lsrd
lsrd
lsrd
std  M
```

Checkpoint 3.16: Let N and M be 8-bit signed locations. Write assembly code to implement M=4*N.

3.3.4
Logical
Operations

Most 8-bit logical instructions take two inputs, one from a register and the other from memory. All but the `bita bitb` instructions put the result back in the register. The N bit will be set if the result is negative. The Z bit will be set if the result is zero. These logical instructions will clear the V bit and leave the C bit unchanged.

```
anda  #w   RegA=RegA&w
anda  U    RegA=RegA&[U]
andb  #w   RegB=RegB&w
andb  U    RegB=RegB&[U]
bita  #w   RegA&w
bita  U    RegA&[U]
bitb  #w   RegB&w
bitb  U    RegB&[U]
coma       RegA=$FF-RegA, or RegA=~RegA
comb       RegB=$FF-RegB, or RegB=~RegB
eora  #w   RegA=RegA ^ w
eora  U    RegA=RegA ^ [U]
eorb  #w   RegB=RegB ^ w
eorb  U    RegB=RegB ^ [U]
oraa  #w   RegA=RegA | w
oraa  U    RegA=RegA | [U]
orab  #w   RegB=RegB | w
orab  U    RegB=RegB | [U]
```

Condition code bits are set, where R is the result of the operation.

N: result is negative N=R7
Z: result is zero Z=not(R7)·not(R6)·not(R5)·not(R4)·not(R3)·not(R2)·not(R1)·not(R0)
V: signed overflow V=0

Let N be an 8-bit location. The following assembly code implements N=$FC&(N|$10), which sets bit 4 and clears bits 1 and 0.

```
ldaa  N
oraa  #$10 RegA = N|$10 (set bit 4)
anda  #$FC RegA = $FC&(N|$10) (clears bits 1,0)
staa  P
```

Checkpoint 3.17: Write assembly code that implements RegD=RegD&$0F3C.

Checkpoint 3.18: Write assembly code that implements RegX=RegX|$1234.

Checkpoint 3.19: Let N be an 8-bit location. Write assembly code that clears bit 4.

3.3.5 Subroutines and the Stack

We begin this section with a general description of the stack, and introduce the basic concepts common to both the 6811 and 6812. Sophisticated use of the stack will be developed later in Chapter 8. In general, we initialize the stack pointer (SP) into RAM using the `lds` instruction, which is usually done once at the beginning of the program. In the classical definition of the stack, there are just two operations one can perform: **push** and **pull.** Some computers define the two stack operations as push and pop. The push function saves data on the top of the stack, and the pull function removes data from the top of the stack. For example, the `psha` instruction will push the value in register A onto the stack, leaving register A unchanged. The `pula` instruction will pull (or pop) a value off the stack bringing it into register A. The pull operation does modify the stack such that the pulled data is no longer on the stack. The stack implements last in first out (LIFO) behavior. The following code pushes the numbers 1, 2, and 3 in that order:

```
ldaa  #1
psha              ; push 1 on the stack
ldaa  #2
psha              ; push 2 on the stack
ldaa  #3
psha              ; push 3 on the stack
```

After these three push operations, the stack would contain these numbers with the 3 on the top, as shown in Figure 3.16. The top entry of the stack contains the newest data, i.e., the data pushed last. On the 6811, SP points to the entry just above the top element. On the 6812, SP points to the top element. The white boxes represent empty stack positions, while the shaded boxes represent values saved on the stack.

Figure 3.16
The stack holding three elements, with the 3 on top.

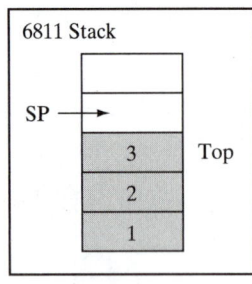

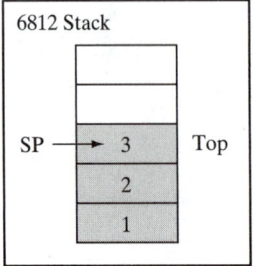

At this point if one were to pull from the stack (e.g., execute `pula`), the 3 would be returned, and 2 would now be on the top of the stack, as shown in Figure 3.17.

Figure 3.17
The stack holding two elements, with the 2 on top.

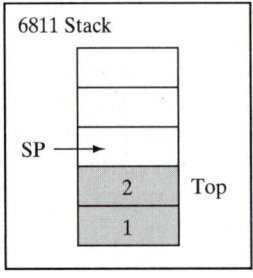

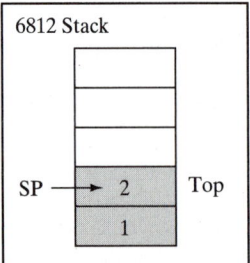

The push and pull instructions use inherent addressing and do not modify the condition code. The push instructions produce two copies of the data, one on the stack and the other still in the register. The pull instructions remove the data from the stack, so there will be only one copy of the data left, which is in the register.

```
psha    push Register A on the stack
pshb    push Register B on the stack
pshx    push Register X on the stack
pshy    push Register Y on the stack
des     S=S-1 (reserve space on the stack)
pula    pull value from stack, put in Register A
pulb    pull value from stack, put in Register B
pulx    pull value from stack, put in Register X
puly    pull value from stack, put in Register Y
ins     S=S+1 (discard top of stack)
```

The stack is used for many purposes. A common use is temporary storage. If a piece of information is important, we can push it onto the stack. Later, when we wish to retrieve the data, we pull it off the stack.

Checkpoint 3.20: Assume you have two 8-bit global variables M and N. Write assembly code that switches the values in M and N using just the `ldaa staa psha` and `pula` instructions.

Later in Chapter 8, we will learn the additional stack operations of **stack read** and **stack write.** The stack read operation allows you to retrieve data previously pushed onto the stack without modifying the data or the stack pointer. The stack write operation allows you to change a value previously pushed onto the stack without changing the stack pointer. Even though these two stack operations are not part of the classical definition of a stack, they will be essential for implementing parameter passing and local variables. The following are important instructions that greatly facilitate the use of the stack.

```
tsx    transfer  S to X,  6811  does  X=SP+1,  6812  does  X=SP
tsy    transfer  S to Y,  6811  does  Y=SP+1,  6812  does  Y=SP
txs    transfer  X to S,  6811  does  SP=X-1,  6812  does . SP=X
tys    transfer  Y to S,  6811  does  SP=Y-1,  6812  does  SP=Y
```

The instruction `tsx` will move a copy of the stack pointer into Register X. Although `tsx` and `tsy` work a little differently on the 6811 versus the 6812, in both cases the register points to the top element of the stack.

Procedures and functions are programs that can be called to perform specific tasks. Some programming environments differentiate functions (return a value) from procedures (do not return a value.) In assembly language, we use the term **subroutine** for all subprograms whether or not they return a value. Subroutines allow us to develop modular software. In assembly language, we will use either the `bsr` or `jsr` instruction to call a subroutine, and we will use the `rts` instruction to return from the subroutine. The `bsr` and `jsr` instructions will push the return address on the stack. The return address is the address of the instruction immediately after the branch to subroutine instruction. The `rts` will pull the return address from the stack, returning the program to the place from which the subroutine was called.

Observation: Since the `bsr` instruction uses relative addressing, it can only be used to call a subroutine near the current instruction. Since the `jsr` instruction allows extended addressing, it can be used to call a subroutine anywhere in memory.

We will study the concept of subroutine using the simple example shown in Program 3.1. The input parameter to the subroutine is passed in using Register A. The subroutine adds one, and returns the result also in Register A. The main program calls the subroutine using the `bsr` instruction. The subroutine returns to the main program using the `rts` instruction.

Program 3.1
Simple program showing how to use the bsr and rts instructions to implement a subroutine.

```
        org  $f800
main lds   #$00FF ;on 6812 change $00FF to $0C00
        clra
loop bsr   sub     ; branch to subroutine
        bra  loop
; Input: RegA, Output: RegA=Input+1
sub  inca             ; adds one to RegA
        rts
        org $fffe
        fdb main
```

We begin the study by looking at the listing file generated by the assembler, shown as Program 3.2.

Program 3.2
6811 Assembly listing of Program 3.1.

```
$F800                                   org  $f800
$F800  8E00FF  {ppp    }main lds  #$00ff
$F803  4F      {pf     }     clra
$F804  8D02    {ppnfss }loop bsr  sub ;branch to subroutine
$F806  20FC    {pfn    }     bra  loop
                              ; Input: RegA, Output: RegA=Input+1
$F808  4C      {pf     }sub  inca        ;adds one to RegA
$F809  39      {pfxuu  }     rts
$FFFE                        org $fffe
$FFFE  F800                  fdb main
```

Figure 3.18 shows the stack before and after the `bsr` instruction is executed. We can also understand the execution of `bsr` by looking at the cycles it generates `ppnfss`. The 6812 runs faster and more efficiently, but performs basically the same sequence of operations, so we'll study 6811. The 6811 needs six cycles to execute `bsr`. During the first two cycles it fetches the op code and operand. At this point the PC is $F806, which will be the return location. During the next two cycles, the effective address ($F808) is calculated. The read $FFFF and read $F808 cycles perform no useful work, except to use up time while the processor is performing internal calculations. The last two cycles push the return address on the stack.

```
Opcode fetch   R   0xF804   0x8D   from   ROM
Operand fetch  R   0xF805   0x02   from   ROM
Null Cycle     R   0xFFFF   0x00   from   ROM
Dummy PC fetch R   0xF808   0x4C   from   ROM
Stack store lsbW   0x00FF   0x06   to     RAM
Stack store msbW   0x00FE   0xF8   to     RAM
```

Figure 3.18
The stack before and
after execution of the
`bsr` instruction.

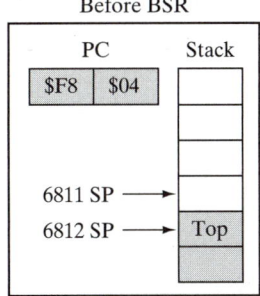

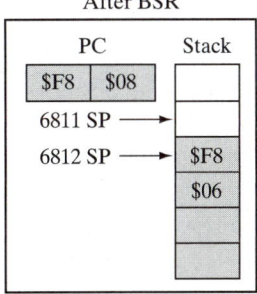

The `rts` instruction will return to the program that called the subroutine. Figure 3.19 shows the stack before and after the `rts` instruction is executed. From the assembly listing, we see that the `rts` generates the five cycles `pfxuu`. Again, the 6812 runs faster and more efficiently, but performs basically the same sequence of operations. During the first cycle it fetches the op code. The read $F80A and read $00FD cycles perform no useful work, except to use up time while the processor is performing internal calculations. The last two cycles pull the return address from the stack.

```
Opcode  fetch   R   0xF809   0x39   from   ROM
Dummy   PC fetch R  0xF80A   0xFF   from   ROM
Dummy   SP read  R  0x00FD   0x00   from   RAM
Stack   read msb R  0x00FE   0xF8   from   RAM
Stack   read lsb R  0x00FF   0x06   from   RAM
```

Figure 3.19
The stack before and
after execution of the
`rts` instruction.

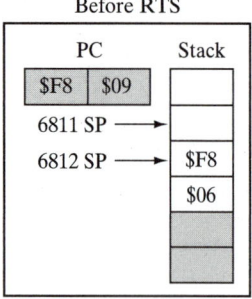

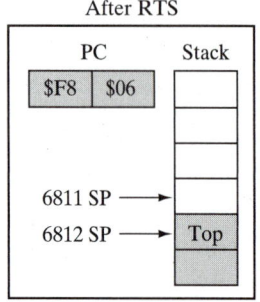

3.3.6
Branch Operations

Normally the computer executes one instruction after another in a linear fashion. In particular, the next instruction to execute is found immediately following the current instruction. We use branch instructions to deviate from this straight line path. Simple control structures are presented here, and more complex control structures can be found later in Section 7.3.

```
bcc     branch if C=0
bcs     branch if C=1
beq     branch if Z=1
bne     branch if Z=0
bmi     branch if N=1
bpl     branch if N=0
bra     branch always
brn     branch never
bvc     branch if V=0
bvs     branch if V=1
jmp     branch always, extended addressing
```

The following branch instructions must follow a subtract, compare, or test instruction, such as `suba subb sbca sbcb subd cba cmpa cmpb cpd tsta tstb tst`.

```
bge    Branch if signed greater than or equal to,
            if (N^V)=0, or (~N•V+N•~V)=0
bgt    Branch if signed greater than,
            if (Z+N^V)=0, or (Z+~N•V+N•~V)=0
ble    Branch if signed less than or equal to,
            if (Z+N^V)=1, or (Z+~N•V+N•~V)=1
blt    Branch if signed less than,
            if (N^V)=1, or (~N•V+N•~V)=1
bhs    Branch if unsigned greater than or equal to,
            if C=0, same as bcc
```

```
bhi    Branch if unsigned greater than,
            if C+Z=0
blo    Branch if unsigned less than,
            if C=1, same as bcs
bls    Branch if unsigned less than or equal to,
            if C+Z=1
```

Conditional execution is an important aspect of software programming. Two values are compared and certain blocks of program are executed or skipped depending on the results of the comparison. In assembly language it is important to know the precision (e.g., 8-bit, 16-bit) and the format of the two values (e.g., unsigned, signed). It takes three steps to perform a comparison. You begin by reading the first value into a register. For 8-bit values you can use either Register A or Register B. 16-bit values can be loaded into Register D, Register X or Register Y. The second step is to compare the first value with the second value. You can use either a subtract instruction (suba subb subd) or a compare instruction (cmpa cmpb cpd cpx cpy). These instructions set the condition code bits. The last step is a conditional branch. Table 3.7 lists some simple comparisons. When testing for equal or not equal it doesn't matter whether the numbers are signed or unsigned. In the following examples, we assume that G1 and G2 are 8-bit values while H1 and H2 are 16-bit variables.

Table 3.7
Conditional structures that test for equality.

C code	assembly code

`if(G2 == G1) {`
`   isEqual();`
`}`
	`ldaa G2`
	`cmpa G1`
	`bne next        ; skip if not equal`
	`jsr isEqual    ; G2==G1`
	`next`

`if(G2 != G1){`
`   isNotEqual();`
`}`
	`ldaa G2`
	`cmpa G1`
	`beq next        ; skip if equal`
	`jsr isNotEqual ; G2!=G1`
	`next`

`if(H2 == H1){`
`   isEqual();`
`}`
	`ldd H2`
	`cpd H1`
	`bne next        ; skip if not equal`
	`jsr isEqual    ; H2==H1`
	`next`

`if(H2 != H1){`
`   isNotEqual();`
`}`
	`ldd H2`
	`cpd H1`
	`beq next        ; skip if equal`
	`jsr isNotEqual ; H2!=H1`
	`next`

Common error. *It is an error to use an 8-bit comparison to test two 16-bit values.*

Checkpoint 3.21: *Assume you have an 8-bit global variable* N. *Write assembly code that implements* `if(N==25)isEqual();`

When testing for greater than or less than, it does matter whether the numbers are signed or unsigned. Table 3.8 lists some 8-bit unsigned comparisons. When comparing unsigned values, the instructions `bhi`, `blo`, `bhs`, and `bls` should follow the subtraction or comparison instruction. To convert these examples to 16 bits, change the `ldaa G2` to `ldd H2` and the `cmpa G1` to `cpd H1`.

Table 3.8
Unsigned conditional
structures.

C code	assembly code

`if(G2 > G1){`
`   isGreater();`
`}`
	`ldaa G2`
	`cmpa G1`
	`bls next        ; skip if G2<=G1`
	`jsr isGreater  ; G2>G1`
	`next`

`if(G2 >= G1){`
`   isGreaterEq();`
`}`
	`ldaa G2`
	`cmpa G1`
	`blo next        ; skip if G2<G1`
	`jsr isGreaterEq ; G2>=G1`
	`next`

`if(G2 < G1){`
`   isLess();`
`}`
	`ldaa G2`
	`cmpa G1`
	`bhs next        ; skip if G2>=G1`
	`jsr isLess      ; G2<G1`
	`next`

`if(G2 <= G1){`
`   isLessEq();`
`}`
	`ldaa G2`
	`cmpa G1`
	`bhi next        ; skip if G2>G1`
	`jsr isLessEq    ; G2<=G1`
	`next`

Table 3.9 lists some 8-bit signed comparisons. When comparing signed values, the instructions `bgt`, `bls`, `bge`, and `ble` should follow the subtraction or comparison instruction. To convert these examples to 16 bits, change the `ldaa G2` to `ldd H2` and the `cmpa G1` to `cpd H1`.

Table 3.9
Signed conditional
structures.

C code	assembly code

`if(G2 > G1){`
`   isGreater();`
`}`
	`ldaa G2`
	`cmpa G1`
	`ble next        ; skip if G2<=G1`
	`jsr isGreater  ; G2>G1`
	`next`

`if(G2 >= G1){`
`   isGreaterEq();`
`}`
	`ldaa G2`
	`cmpa G1`
	`blt next        ; skip if G2<G1`
	`jsr isGreaterEq ; G2>=G1`
	`next`

`if(G2 < G1){`
`   isLess();`
`}`
	`ldaa G2`
	`cmpa G1`
	`bge next        ; skip if G2>=G1`
	`jsr isLess      ; G2<G1`
	`next`

```
if(G2 <= G1){                    ldaa G2
   isLessEq();                   cmpa G1
}                                bgt  next       ; skip if G2>G1
                                 jsr  isLessEq ; G2<=G1
                              next
```

Checkpoint 3.22: *When implementing* `if(N>25)isGreater();` *why is it important to know if* N *is signed or unsigned?*

Common error. *It is an error to use an unsigned conditional branch when comparing two signed values. Similarly, it is a mistake to use a signed conditional branch when comparing two unsigned values.*

Quite often the microcomputer is asked to wait for events or to search for objects. Both of these operations are solved using the `while` or `do-while` structure. A simple example of the while loop is illustrated in Figure 3.20 and presented in Program 3.3. The operation is defined by the C code:

```
while(G2 > G1) body();
```

Figure 3.20
Flowchart of a while structure.

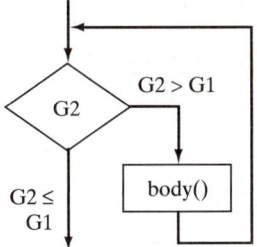

The program begins with a test of `G2>G1`. If `G2<=G1` then the body of the while loop is skipped.

Program 3.3
A while loop structure.

```
loop ldaa G2
     cmpa G1
     bls  next    ; stop when G2<=G1
     jsr  body    ; execute body of the while loop
     bra  loop
next
```

Observation: *Because the load and store instructions clear the V bit, it is acceptable to follow these instructions with one of the signed conditional branch instructions:* `bge bgt ble blt`.

Checkpoint 3.23: *Assume you have a 16-bit global variable N. Write assembly code that implements* `while(N!=25)body();`

3.4 6811 Architecture

3.4.1
MC68HC11 Family

The 6811 is a powerful yet inexpensive family of microcomputer chips. The 6811 can operate in one of four modes. In the **single chip** operating mode, the 6811 chip includes all the program memory, I/O devices and data memory (RAM). In the **expanded** operating mode, some of the memory and I/O devices are external to the 6811 chip. The other two modes are **bootstrap** mode, which can be used to load programs into RAM, and **test** mode, which is used by Motorola to verify the chip is operational.

In many applications, the MC68HC11 provides a single-chip solution with mask-programmed ROM or user-programmable EPROM. All family derivatives are also expandable for the incorporation of external memory in the design. A four-channel Direct Memory Access (DMA) unit on some devices permits fast data transfer between two blocks of memory (including externally mapped memory in expanded mode), between registers or between registers and memory. Within the MC68HC11 Family, there are nine major series of microcontroller units. The following are examples of the features the MC68HC11 can offer, as shown through specific devices within each series:

The **A-series** includes the basic 6811, featuring 40 I/O pins, timer, serial port and 8-bit ADC.

The **D-series** offers an economical alternative for applications that require fewer peripherals and less memory.

The **E-series** comes in a wide range of I/O capabilities. It combines EEPROM and EPROM on a single chip. The E-series offers multiple memory sizes in a pin compatible package.

The **F-series** combines high-speed, expanded memory, extra I/O ports, and 1K RAM.

The **G-series** offers 10-bit ADC resolution and the most sophisticated timer systems in the family.

The **K-series** offers high speed, large memories, a memory management unit, pulse width modulation along with standard I/O ports.

The **L-series** is a high-speed, low-power family with a multiplexed bus. It has an additional bidirectional port. Its fully static design allows operation at frequencies down to DC.

The **M-series** includes enhanced, high-performance microcontrollers derived from the 68HC11K4 and include large memory modules, a 16-bit math coprocessor, and 4 channels of DMA.

The **P-series** offers a power-saving programmable phase-lock loop (PLL)-based clock circuit along with many I/O pins, large memory and 3 SCI ports.

TExaS supports four 6811 configurations: MC68HC811E2, MC68HC711E9, MC68HC711D3, and MC68HC11A1 EVB. The first three operate in single-chip mode, where all the memory and I/O devices are included on a single 6811 chip, as shown in Figure 3.21. The EVB system operates in expanded mode, where some of the memory and I/O devices are external to the 6811 chip. This 6811 system controls a DC motor. **TExaS** was first used to simulate this motor controller, then the hardware/software were converted to physical devices.

Figure 3.21
ADAPT-11 Single Chip
6811 Module from
Technological Arts used
in a motor controller.

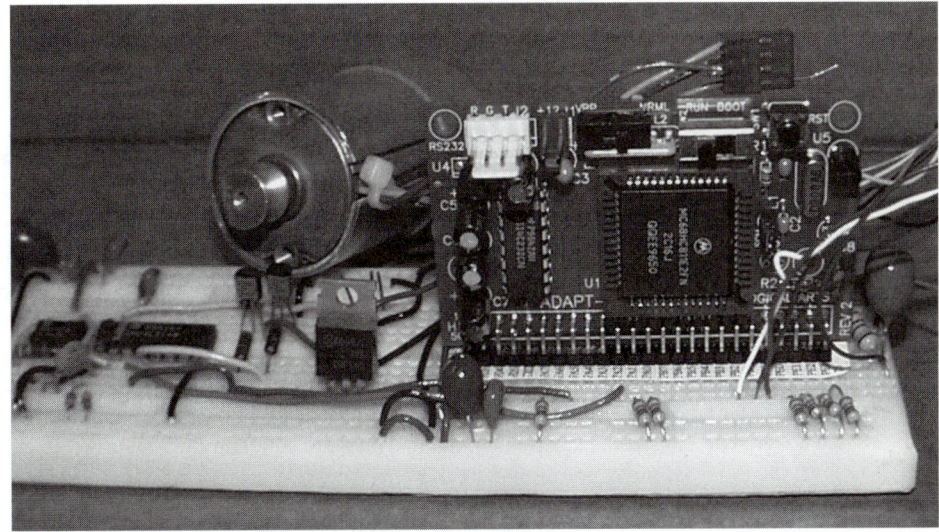

The digit following the HC specifies the type of program memory. No digit means ROM, 7 means EPROM, 8 means EEPROM and 9 means flash EEPROM. Most 6811 systems have both program memory and EEPROM. The extra EEPROM can be used to store calibration constants or other information that might vary from one system to the next. It takes about 10 ms to program or erase EEPROM, so it will not be used instead of RAM to store data that change frequently. Various versions of the 6811 are shown in Table 3.10.

Table 3.10
Some memory
configurations available
for the 6811.

chip	program memory	EEPROM size	RAM size	I/O pins
MC68HC11A1	-	512	256	40
MC68HC11A8	8K ROM	512	256	40
MC68HC11D0	-	-	192	32
MC68HC11D3	4K ROM	-	192	32
MC68HC711D3	4K EPROM	-	192	32
MC68HC11E0	-	-	512	36
MC68HC11E1	-	512	512	36
MC68HC811E2	-	2048	256	36
MC68HC11E9	12K ROM	512	512	36
MC68HC11E20	20K ROM	512	768	36
MC68HC711E9	12K EPROM	512	512	36
MC68HC711E20	20K EPROM	512	768	36
MC68HC11F1	-	512	1024	54
MC68HC11K4	24K ROM	640	768	62

Observation: *The locations of the I/O, RAM, and EEPROM can be remapped to other addresses on a real 6811, but the **TExaS** simulator does not allow this flexibility, fixing their addresses to the default positions.*

The final product uses single-chip mode, where the application software is programmed into the ROM EPROM or EEPROM. Typically during hardware and software development, we operate the 6811 in expanded mode on a development system like the Motorola EVB, so that the edit/assemble/load/run software development cycle is short. Once the system is debugged, the program can be burned into ROM, and the 6811 single-chip computer can be embedded into the system.

3.4.2 The E Series 6811

The Motorola E series is one of the most popular families of the 6811, as shown in Figure 3.22. The "8" in its number refers to the fact that the program memory is implemented with EEPROM. This feature makes it very convenient to use during the development stage of a project where many erase/program/test cycles are required. The "7" in its number refers to the fact that the program memory is implemented with EPROM, which can be erased with UV light and reprogrammed. The EPROM devices are convenient to use during the development stage of a project that will eventually go into production. There are one-time-programmable (OTP) devices that are electrically the same as other EPROM devices, but do not have a glass window, so they can not be erased. As the name implies, these devices can be programmed once, and are suitable for the final phase of small production projects. ROM devices are programmed by Motorola and are used exclusively for mass production. Motorola uses the "9" designation in its number to specify the program memory is made with flash EEPROM.

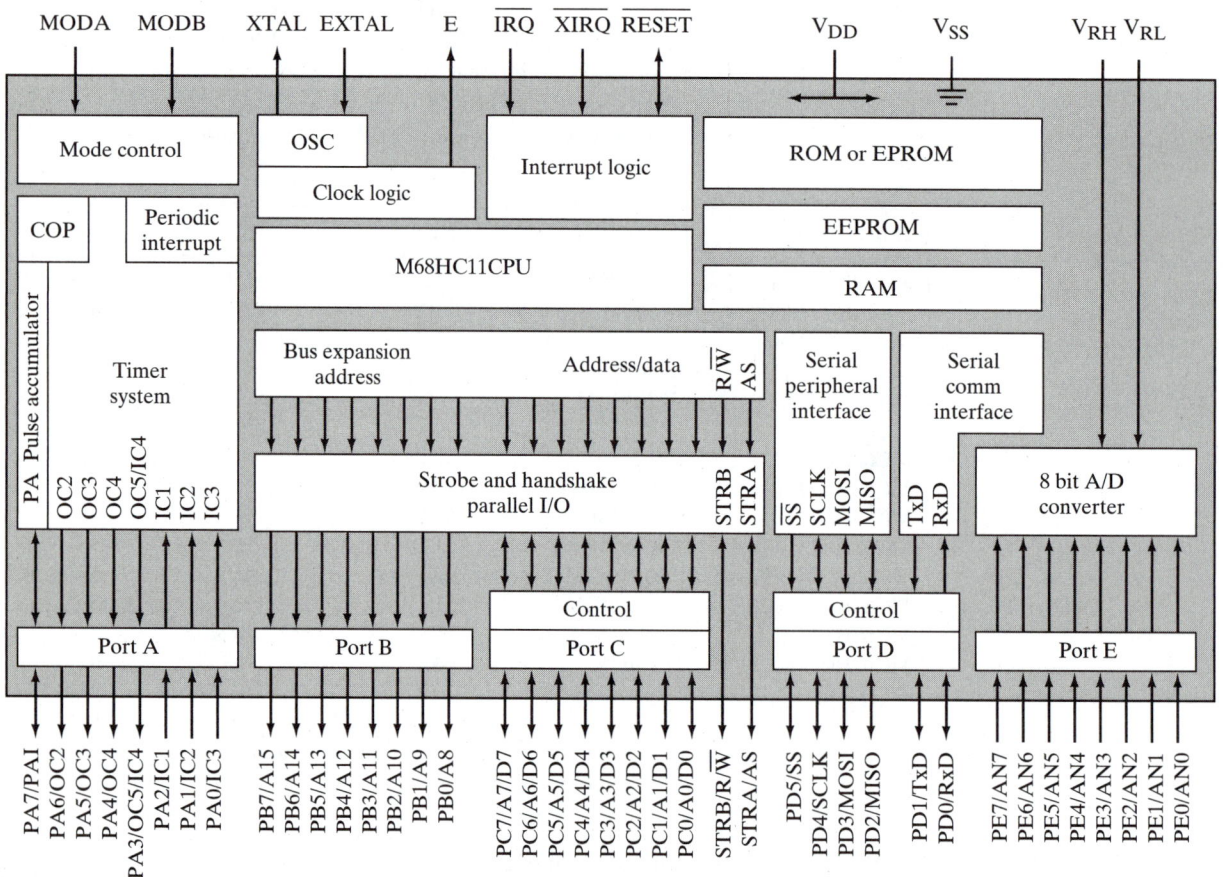

Figure 3.22   Block diagram of the Motorola 6811 E series.

The I/O port addresses range from $1000 to $103F. Each has a Boot ROM from $BF00 to $BFFF. The RAM and ROM configurations of members of the E series 6811 are presented in Table 3.11.

Table 3.11
The 6811 E series comes in a wide range of memory configurations.

Device	RAM	Program memory
MC68HC11E0	512 from 0 to $01FF	none
MC68HC11E1	512 from 0 to $01FF	512 EEPROM from $B600 to $B7FF
MC68HC11E8	512 from 0 to $01FF	12K ROM from $D000 to $FFFF
MC68HC11E9	512 from 0 to $01FF	512 EEPROM from $B600 to $B7FF 12K ROM from $D000 to $FFFF
MC68HC711E9	512 from 0 to $01FF	512 EEPROM from $B600 to $B7FF 12K EPROM from $D000 to $FFFF
MC68HC11E20	768 from 0 to $02FF	8K ROM from $9000 to $AFFF 512 EEPROM from $B600 to $B7FF 12K ROM from $D000 to $FFFF
MC68HC711E20	768 from 0 to $02FF	8K EROM from $9000 to $AFFF 512 EEPROM from $B600 to $B7FF 12K EPROM from $D000 to $FFFF
MC68HC811E2	256 from 0 to $00FF	2K EEPROM from $F800 to $FFFF

The 6811 implements memory-mapped I/O; therefore, each I/O device is assigned an address. Some of the MC68HC711E9 addresses are included in Table 3.12. Refer to the CD enclosed with this text for a complete list.

Table 3.12
The MC68HC711E9 has five external I/O ports and an internal timer.

Port	Address	Usage	Alternate Usage
PORTA	$1000	Bits 7,3 I/O, Bits 6-4 out, Bits 2-0 in	Timer
PORTB	$1004	8-bit Output Port	A15-A8 in expanded mode
PORTC	$1003	8-bit Input/Output Port	A7-A0/D7-D0 in expanded mode
DDRC	$1007	Data Direction Register	
PORTD	$1008	6-bit Input/Output Port	Serial port
DDRD	$1009	Data Direction Register	
PORTE	$100A	8-bit Input Port	ADC
TCNT	$100E	16-bit Timer	

3.4.3
MC68HC711D3

Compare the block diagram of the D3 computer with the other 6811 products and notice that it has fewer I/O devices, but is available in a smaller, less expensive package. The onboard 192-byte RAM is initially located at $0040 after reset, but can be placed at any other 4K boundary ($x040) by writing an appropriate value to the INIT register. The 64-byte register block originates at $0000 after reset, but can be placed at any other 4K boundary ($x000) after reset by writing an appropriate value to the INIT register. There is no EEPROM or ADC on this version of the 6811. The 4-Kbyte EPROM is located at $F000 through $FFFF.

The MC68HC711D3 has four 8-bit I/O ports: A, B, C, and D, as shown in Table 3.13. In the 40-pin package, port A bits 4 and 6 are not connected to pins. In single-chip and bootstrap modes, all ports are parallel I/O data ports. In expanded-multiplexed and test modes, ports B and C and lines AS and R/W are a memory expansion bus with port B serving as the high-order address bus, port C as the multiplexed address and data bus, AS as the demultiplexing signal, and R/W as the data bus direction control.

Table 3.13
The MC68HC711D3 has
four external I/O ports.

Port	Input Pins	Output Pins	Bidirectional Pins	Shared Functions
Port A	3	3	2	Timer
Port B	—	—	8	High Order Address
Port C	—	—	8	Low Order Address and Data Bus
Port D	—	—	8	SCI, SPI, AS, and R/W

The address map of a single chip MC68HC711D3, OTP or UV erasable EPROM, is shown in Table 3.14.

Table 3.14
The MC68HC711D3 has
4K of EPROM and 192
bytes of RAM.

Address (hex)	Size	Device	Device	Contents
0000 to 003F	64	I/O	Input/output devices	
0040 to 00FF	192	RAM	Random Access Memory	Variables and stack
F000 to FFFF	4K	EPROM	OTP or UV erasable	Programs and fixed constants

3.4.4
Expanded Mode
6811

Figure 3.23 shows an expanded mode system built with an MC68HC11A1. The A1 is one of the first 6811s made and has no ROM. Expanded mode systems have external RAM and ROM modules. During the hardware/software development, we can use a 6811 in the ex-

Figure 3.23
Bus interconnection for
a 6811 expanded mode
system.

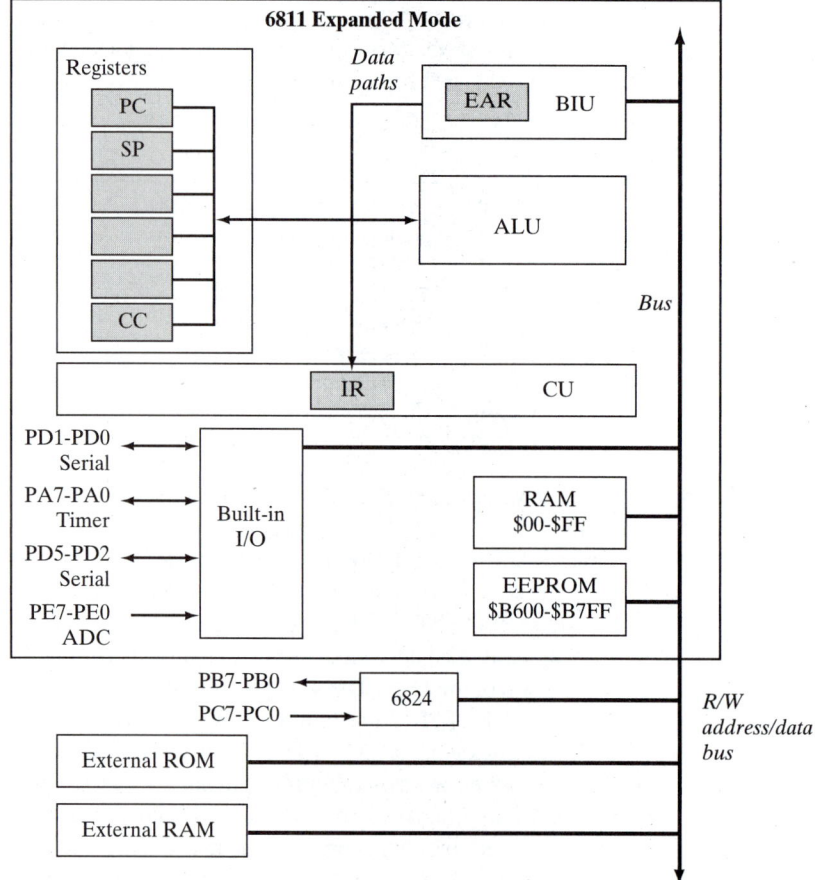

panded mode because the edit/compile/download time is short and because we can utilize a debugger. In expanded mode, 6811 ports B and C are used for the address and data bus. The Peripheral Recovery Unit (PRU, 6824) is added to reconstruct I/O ports B and C. External ROM and RAM are used, and the 6811 acts like a microprocessor. The address map of a Motorola 6811 EVB development board is shown in Table 3.15.

Table 3.15
The 6811 EVB board has 8K of PROM, 16K of PROM/RAM and 256 bytes of RAM.

Address (hex)	Device	Description
0000 to 00FF	RAM	Internal to 6811 can used by you
1000 to 103F	I/O	Internal I/O of the 6811/6824 system
4000	switch	output a 1 to this location to use the HOST Serial Port 0 to use PD0 for other purposes
6000 to 7FFF	8K RAM/PROM	You put your software here
9800 to 9801	6850	Universal Asynchronous Receiver/Transmitter
B600 to B7FF	512 bytes EEPROM	Internal to 6811, Electrically erasable PROM
C000 to DFFF	8K RAM/PROM	You put your software here
E000 to FFFF	8K PROM	Bit User's Fast Friendly Aid to Logical Operation

3.4.5
6811 Indexed Addressing Mode

6811 indexed addressing mode uses an 8-bit unsigned offset with either RegX or RegY. The 8-bit unsigned offset is included even if the offset is zero. Notice that instructions using RegX take less object code and execute faster than those using RegY do. Indexed mode is useful when addressing the 6811 I/O ports located from $1000 to $103F. Indexed mode is also useful when addressing the data structures and information on the stack. In each case, the 16-bit register is used as a pointer (index) and is unmodified by the instruction. The instruction `staa 4,x` has an object code $A704. The $A7 specifies the staa instruction with indexed mode addressing. The $04 is the index. The effective address will be X+4.

```
staa    4,x    [X+4]=RegA
```

Assuming Register X=$0023, the instruction `staa 4,X` will store a copy of the value in Register A at $0027 leaving Register X unchanged as shown in Figure 3.24. The effective address (EA) is $0023+4=$0027.

Figure 3.24
Example of the 6811 indexed addressing mode.

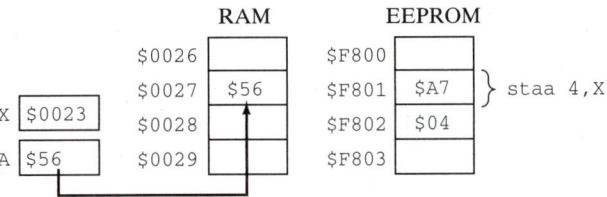

More indexed mode examples:

```
machine code    opcode    operand    comment
$A664           ldaa      100,x      RegA = [X+100]
$18A420         anda      $20,y      RegA = RegA&[Y+32]
$1C0301         bset      $03,x,#1   [X+3] = [X+3]|1
$18ED05         std       5,y        {Y+5} = RegD
```

Observation: With indexed mode addressing, the pointer to the information is calculated at run time, so the data and its location may change dynamically.

Observation: 6811 instructions involving Register Y require an extra byte of machine code.

Observation: The 6811 instructions `bset`, `bclr`, `brset`, and `brclr` do not operate with extended addressing. You have to use indexed addressing mode when using these instructions with the I/O ports.

Common error: A space will define the end of the operand. For example, ldaa 100g X will give an error because the instruction ldaa 100, has no meaning.

3.5 6812 Architecture

The MC68HC12 is a highly integrated, general-purpose family of microcomputers with a 16-bit microcontroller architecture specifically designed for low power consumption. It is source-code compatible with the popular 68HC11 8-bit microcontroller family, but some of the I/O ports operate slightly differently. Features include:

Low power consumption and low voltage operation at full bus speed
Single wire Background Debug™ Mode for non-intrusive in-circuit programming and debugging
High-level language optimization
Flash EEPROM and byte-erasable EEPROM integrated on a single device
Fuzzy logic instructions
Enhanced arithmetic instructions over the 68HC11

Observation: The locations of the I/O, RAM, and EEPROM can be remapped to other addresses on a real 6812, but the **TExaS** simulator does not allow this flexibility, fixing their addresses to the default positions.

3.5.1 MC68HC812A4

The MC68HC812A4 microcontroller unit (MCU) is a 16-bit device composed of standard on-chip peripheral modules connected by an intermodule bus, as shown in Figure 3.25. Modules include a 16-bit central processing unit (CPU12), a Lite integration module (LIM), two asynchronous serial communications interfaces (SCI0 and SCI1), a serial peripheral interface (SPI), a timer and pulse accumulation module, an 8-bit analog-to-digital converter (ATD), 1-Kbyte RAM, 4-Kbyte EEPROM, and memory expansion logic with chip selects, key wakeup ports, and a phase-locked loop (PLL).

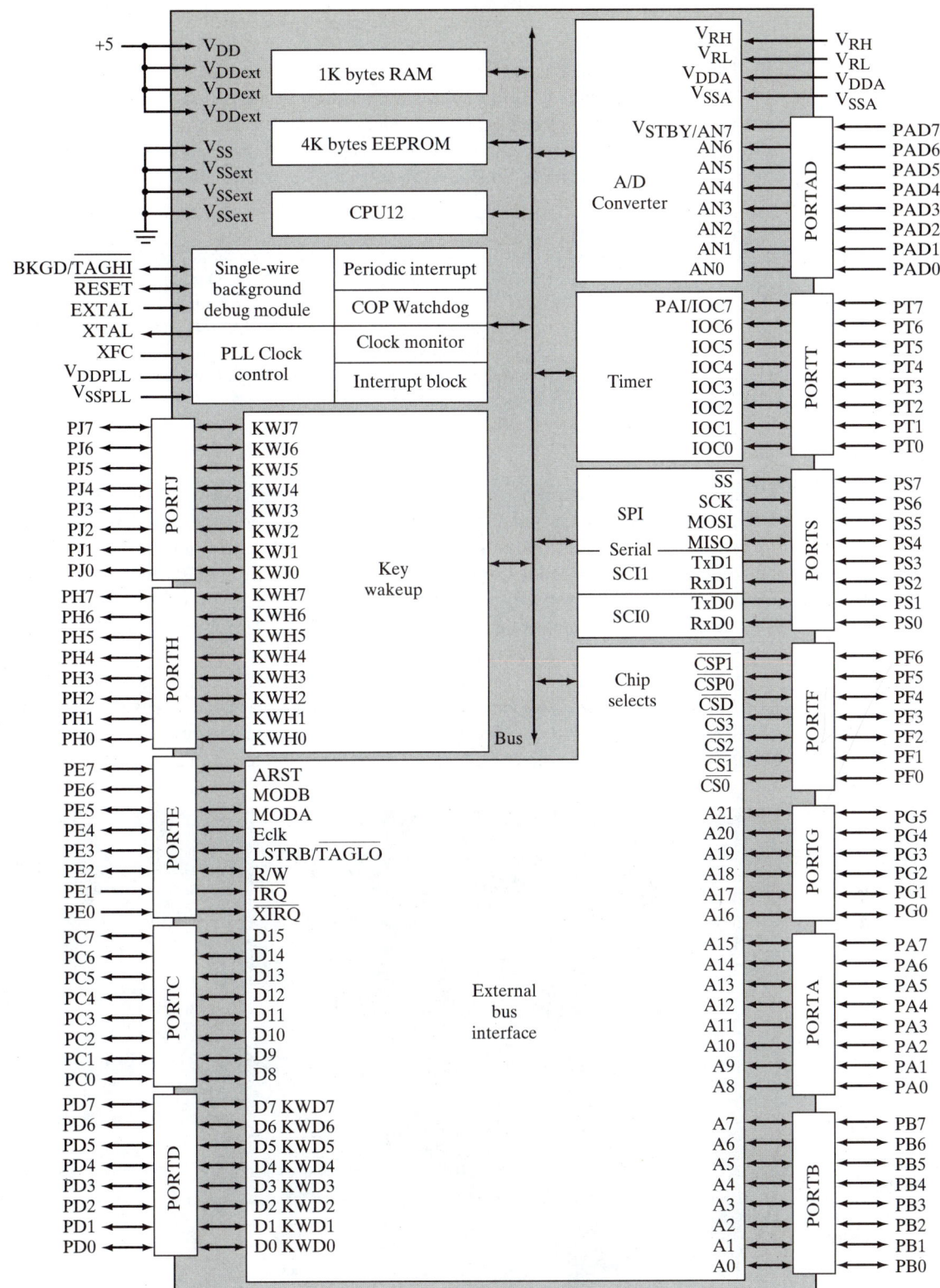

Figure 3.25 Block diagram of a Motorola MC68HC812A4.

The 68HC812A4 can operate in one of seven modes as specified by the three input signals BKGD, MODA, and MODB, as shown in Table 3.16. In single chip mode, the 6812 contains the four major building blocks required to make a complete computer system: processor, I/O, RAM, and (P)ROM. In this mode, all ports are available for input/output, as shown in Figure 3.25. In expanded mode some of the ports are used for the address and data bus.

Table 3.16
The MC68HC812A4 has seven operating modes.

BKGD	MODB	MODA	Mode	Port A	Port B	Port C	Port D
0	0	0	Special Single Chip	In/Out	In/Out	In/Out	In/Out
0	0	1	Special Expanded Narrow	A15-8	A7-A0	D7-D0	In/Out
0	1	0	Special Peripheral	A15-8	A7-A0	D15-D8	D7-D0
0	1	1	Special Expanded Wide	A15-8	A7-A0	D15-D8	D7-D0
1	0	0	Normal Single Chip	In/Out	In/Out	In/Out	In/Out
1	0	1	Normal Expanded Narrow	A15-8	A7-A0	D7-D0	In/Out
1	1	1	Normal Expanded Wide	A15-8	A7-A0	D15-D8	D7-D0

We use expanded mode during development because it provides a fast edit/assemble/download cycle, and we use single-chip mode to embed the microcomputer into our final product. Once the system is debugged, the program can be burned into ROM or programmed into PROM. In the **single-chip operating mode**, the MC68HC812A4 has 4096 bytes of EEPROM (electrically erasable programmable read-only memory), and 1024 bytes of RAM. Other versions of the 6812 have differing amounts of these types of memory. In single-chip mode, the 6812 implements a complete microcomputer, where all its I/O ports are available. This mode is used for the final product with the application software programmed into the ROM. The address space of the Input/Output Devices and the address space of the RAM are mappable to any 2K space. The EEPROM is mappable to any 4K space. The standard address map of a single chip MC68HC812A4 is shown in Table 3.17. Figure 3.26 shows a prototype of an embedded system.

Figure 3.26
ADAPT-812
MC68HC812A4 Module
from Technological Arts.

Table 3.17
The MC68HC812A4 has
4K of EEPROM and 1K
bytes of RAM.

Address (hex)	Size	Device	Device	Contents
0000 to 01FF	512	I/O	Input/output devices	
0800 to 0BFF	1028	RAM	Random Access Memory	Variables and stack
F000 to FFFF	4096	EEPROM	Electrically erasable PROM	Programs and fixed constants

The 6812 implements memory-mapped I/O, therefore each I/O device is assigned an address. Some of the MC68HC812A4 addresses are included in the Table 3.18. Refer to the CD enclosed with this text for a complete list.

Table 3.18   The MC68HC812A4 has twelve external I/O ports and an internal timer.

Port	Address	Usage	Alternate Usage
PORTA	$0000	General Purpose Input/Output Port	Address A15-A8 in expanded mode
PORTB	$0001	General Purpose Input/Output Port	Address A7-A0 in expanded mode
DDRA	$0002	Data Direction Register	
DDRB	$0003	Data Direction Register	
PORTC	$0004	General Purpose Input/Output Port	Data D15-D8 in expanded wide mode
PORTD	$0005	General Purpose Input/Output Port	Data D7-D0 in expanded wide mode
DDRC	$0006	Data Direction Register	
DDRD	$0007	Data Direction Register	
PORTE	$0008	General Purpose Input/Output Port	Mode and bus control
DDRE	$0009	Data Direction Register	
PORTH	$0024	General Purpose Input/Output Port	Key wakeup, interrupt
DDRH	$0025	Data Direction Register	
PORTJ	$0028	General Purpose Input/Output Port	Key wakeup, interrupt
DDRJ	$0029	Data Direction Register	
PORTF	$0030	General Purpose Input/Output Port	Chip select logic
PORTG	$0031	General Purpose Input/Output Port	Address A21-A16 Memory expansion
DDRF	$0032	Data Direction Register	
DDRG	$0033	Data Direction Register	
PORTAD	$006F	Analog input	General Purpose Input Port
TCNT	$0084	16-bit Timer	
PORTT	$00AE	Timer Input/Output Port	General Purpose Input/Output Port
DDRT	$00AF	Data Direction Register	
PORTS	$00D6	Serial Input/Output Port	General Purpose Input/Output Port
DDRS	$00D7	Data Direction Register	

Observation: When the 6812 operates in expanded mode with external memory, the memory cycle time (time to fetch data) may be slower for data accesses in external memory. This fact makes the task of counting cycles impossible.

Observation: The **TExaS** application does not currently simulate any 6812 expanded mode configurations.

3.5.2
MC68HC912B32

The MC68HC912B32 microcontroller unit (MCU) is a 16-bit device composed of standard on-chip peripherals including a 16-bit central processing unit (CPU12), 32 Kbyte flash EEPROM, 1 Kbyte RAM, 768 byte EEPROM, an asynchronous serial communications interface (SCI), a serial peripheral interface (SPI), an 8-channel timer and 16-bit pulse accumulator, an 8-bit analog-to-digital converter (ADC), a four-channel pulse-width modulator (PWM), and a J1850-compatible byte data link communications module (BDLC). Table 3.19 presents its address map. The MC68HC912B32 has full 16-bit data paths throughout;

however, the multiplexed external bus can operate in an 8-bit narrow mode so single 8-bit wide memory can be interfaced for lower cost systems.

Table 3.19
The MC68HC912B32 has 32K of flash EEPROM and 1K bytes of RAM.

Address (hex)	Size	Device	Device	Contents
0000 to 01FF	512	I/O	Input/output devices	
0800 to 0BFF	1028	RAM	Random Access Memory	Variables and stack
0D00 to 0FFF	768	EEPROM	Electrically erasable PROM	Programs and fixed constants
8000 to FFFF	32768	EEPROM	Electrically erasable PROM	Programs and fixed constants

Like the MC68HC812A4, the MC68HC912B32 operates in one of seven modes as specified by the three input signals BKGD, MODA, and MODB, as shown in Table 3.20. In single-chip mode, the 6812 contains the four major building blocks required to make a complete computer system: processor, I/O, RAM, and (P)ROM. In this mode, all ports are available for input/output. In expanded mode, some of the ports are used for the address and data bus. Refer to the CD enclosed with this text for a complete list of I/O devices.

Table 3.20
The MC68HC912B32 has seven operating modes.

BKGD	MODB	MODA	Mode	Port A	Port B
0	0	0	Special Single Chip	In/Out	In/Out
0	0	1	Special Expanded Narrow	Addr[15:8]/Data[7:0]	Addr[7:0]
0	1	0	Special Peripheral	Addr[15:8]/Data[15:8]	Addr[7:0]/Data[7:0]
0	1	1	Special Expanded Wide	Addr[15:8]/Data[15:8]	A7-A0
1	0	0	Normal Single Chip	In/Out	In/Out
1	0	1	Normal Expanded Narrow	Addr[15:8]/Data[7:0]	Addr[7:0]
1	1	1	Normal Expanded Wide	Addr[15:8]/Data[15:8]	Addr[7:0]/Data[7:0]

3.5.3 MC68HC12D60/MC68HC912D60

The MC68HC912D60 microcontroller unit is a 16-bit device composed of standard on-chip peripherals including a 16-bit central processing unit (CPU12), 60 Kbyte of flash EEPROM, 1 Kbyte of EEPROM, 2 Kbyte of RAM, two asynchronous serial communications interfaces (SCI), a serial peripheral interface (SPI), an enhanced capture timer (ECT), two (one on 80QFP) 8-channel, 10-bit analog-to-digital converters (ATD), a four-channel pulse-width modulator (PWM), and a CAN 2.0 A, B software-compatible module (MSCAN12). Table 3.21 presents its address map. The MC68HC12D60 has factory programmed ROM in place of the flash EEPROM. This microcontroller unit is available in two package options, 80-pin QFP and 112-pin TQFP. The addresses of the I/O, RAM, and 1K EEPROM can be remapped to other locations.

Table 3.21
The MC68HC912D60 has 60K of flash EEPROM and 2K bytes of RAM.

Address (hex)	Size	Device	Device	Contents
0000 to 01FF	512	I/O	Input/output devices	
0200 to 0BFF	2048	RAM	Random Access Memory	Variables and stack
0C00 to 0FFF	1024	EEPROM	Electrically erasable PROM	Fixed constants
1000 to FFFF	60K	EEPROM	Electrically erasable PROM	Programs and fixed constants

3.5.4
MC68HC12DG128/
MC68HC912DG128

The MC68HC912DG128 microcontroller unit is a 16-bit device composed of standard on-chip peripherals including a 16-bit central processing unit (CPU12), 128 kbyte of flash EEPROM, 2 kbyte of EEPROM, 8 kbyte of RAM, two asynchronous serial communications interfaces (SCI), a serial peripheral interface (SPI), an inter-IC interface (I2C), an enhanced capture timer (ECT), two 8-channel, 10-bit analog-to-digital converters (ATD), a four-channel pulse-width modulator (PWM), and a CAN 2.0 A, B software-compatible module (MSCAN12). Table 3.22 presents its address map. The MC68HC12D60 has factory-programmed ROM in place of the flash EEPROM. This microcontroller unit is available in a 112-pin TQFP package. Regular memory exists from $4000-$7FFF and $C000-$FFFF, but 96 kbytes of storage is squeezed into 16 kbyte of address space using the paged memory scheme.

Table 3.22
The MC68HC912DG128 has 128K of flash EEPROM and 2K bytes of RAM.

Address (hex)	Size	Device	Device	Contents
0000 to 03FF	1024	I/O	Input/output devices	
2000 to 3FFF	8192	RAM	Random Access Memory	Variables and stack
0800 to 0FFF	2048	EEPROM	Electrically erasable PROM	Fixed constants
4000 to 7FFF	16K	EEPROM	Electrically erasable PROM	Programs and fixed constants
8000 to BFFF	96K	EEPROM	Electrically erasable PROM	Paged program space
C000 to FFFF	16K	EEPROM	Electrically erasable PROM	Programs and fixed constants

Observation: The **TExaS** simulator does not currently simulate the *MC68HC912DG128.*

3.5.5
Additional
Addressing Modes
on the 6812

The 6812 instruction set has ten addressing modes. Five of the modes were presented earlier (inherent, immediate, direct, extended, and PC relative), and the remaining modes are presented next.

1. Indexed addressing mode uses a fixed offset with the 16-bit registers: X, Y, SP, or PC. On the 6811, instructions that use register Y take more memory and run slower than the equivalent instruction using register X. The 6812 eliminates this speed and memory cost of using Reg Y. In addition, the 6812 indexing modes can be used with the stack pointer (SP) and the program counter (PC). The offset can be 5-bit (-16 to $+15$), 9-bit (-256 to $+127$), or 16-bit. Five bit (-16 to $+15$) index mode requires one machine byte to encode the operand. In the first example that uses 5-bit indexed mode, $6A is the `staa` instruction and $**5C** is the index mode operand. Tables A.3 and A.4 of the Motorola CPU12 Reference Manual show the machine codes for the indexed instructions.

```
machine code      opcode      operand      comment

$6A5C             staa        -4,Y         [Y-4] = RegA
```

Assuming Register Y=$0823, the instruction `staa -4,Y` will store a copy of the value in Register A at $081F leaving Register Y unchanged, as shown in Figure 3.27. The effective address (EA) is 0823-4=$$081F.

Figure 3.27
Example of the 6812 indexed addressing mode.

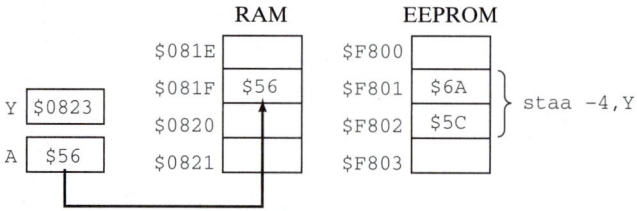

Nine-bit (−256 to +255) indexed mode requires two machine bytes to encode the operand.

machine code	opcode	operand	comment
$6AE840	staa	$40,Y	[Y+$40] = RegA

Assuming Register Y=$0823, the instruction `staa $40,Y` will store a copy of the value in Register A at $0863 leaving Register Y unchanged, as shown in Figure 3.28. The effective address (EA) is $0823+$40=$0863.

Figure 3.28
Another example of the 6812 indexed addressing mode.

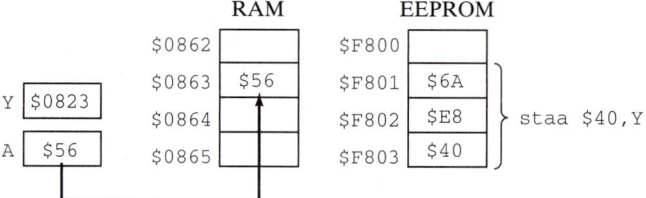

Sixteen-bit indexed mode requires three machine bytes to encode the operand.

machine code	opcode	operand	comment
$6AEA0200	staa	$200,Y	[Y+$200] = RegA

Assuming Register Y=$0823, the instruction `staa $200,Y` will store a copy of the value in Register A at $0A23 leaving Register Y unchanged, as shown in Figure 3.29. The effective address (EA) is $0823+$200=$0A23.

Figure 3.29
A third example of the 6812 indexed addressing mode.

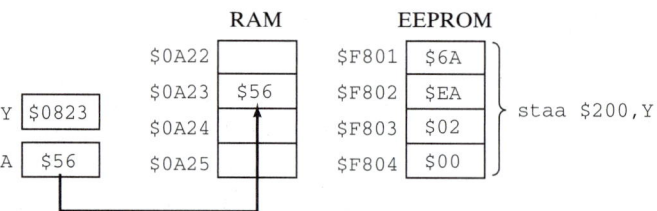

Due to the properties of 16-bit addition, the 16-bit offset can be interpreted either as unsigned (0 to 65535) or signed (−32768 to +32767.) Indexed mode is useful when addressing the data structures and information on the stack. In each case, the 16-bit register used as a pointer (index) is not modified by the instruction. More indexed mode examples:

machine code	opcode	operand	comment
$A6E064	ldaa	100,x	RegA = [X+100]
$A4E820	anda	$20,y	RegA = RegA&[Y+32]
$0C0301	bset	$03,x,#1	[X+3] = [X+3]|1
$6C45	std	5,y	{Y+5} = RegD

Common Error: *SP relative indexed addressing with a negative constant is usually defined as an illegal stack access.*

2. Auto Pre/Post Decrement/Increment Indexed Addressing uses the 16-bit registers: X, Y, or SP. The PC can not be used with these index modes that modify the index register. In each case, the 16-bit register used as a pointer (index) is modified either before (pre) or after (post) the memory access. These modes are useful when addressing the data structures. The 6812 allows the programmer to specify the amount added to (subtracted from) the index register from 1 to 8. In each case assume Reg Y is initially $2345.

Post-increment examples:

```
staa 1,Y+ [$2345]=Reg A, then Reg Y=$2346
staa 4,Y+ [$2345]=Reg A, then Reg Y=$2349
```

Pre-increment examples:

```
staa 1,+Y Reg Y=$2346, then [$2346]=Reg A
staa 4,+Y Reg Y=$2349, then [$2349]=Reg A
```

Post-decrement examples:

```
staa 1,Y- [$2345]=Reg A, then Reg Y=$2344
staa 4,Y- [$2345]=Reg A, then Reg Y=$2341
```

Pre-decrement examples:

```
staa 1,-Y Reg Y=$2344, then [$2344]=Reg A
staa 4,-Y Reg Y=$2341, then [$2341]=Reg A
```

Observation: *Usually we would add/subtract one when accessing an 8-bit value and add/subtract two when accessing a 16-bit value.*

Common Error: *The improper use of these index modes with the SP can result in an illegal stack access or unbalanced stack.*

3. Accumulator Offset Indexed Addressing uses two registers. The offset is located in one of the accumulators A, B or D, and the index (memory address) the 16-bit registers: X, Y, SP, or PC. In each case, the accumulator used for the offset and the index register used as a pointer (index) are not modified by the instruction.

Examples:

```
ldab #4
ldy  #$2345
staa B,Y   [$2349]=Reg A, (B & Y unchanged)
```

4. Indexed Indirect Addressing mode uses a fixed offset with the 16-bit registers; X, Y, SP, or PC. The fixed offset is always 16 bits. The fixed 16-bit value is added to the index register (X, Y, SP, or PC), and used to fetch a second 16-bit big endian address from memory. The load or store is performed at this second address, as shown in Figure 3.30. Indexed indirect mode is useful when data structures contain pointers. In each case, the 16-bit index register and the memory pointer are not modified by the instruction. For example,

```
ldy  #$2345
staa [-4,Y] fetch 16-bit address from $2341, store $56 at $1234
```

Figure 3.30
Example of the 6812 indexed-indirect addressing mode.

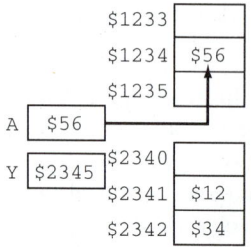

5. Accumulator D Offset Indexed Indirect Addressing uses two registers. The offset is located in accumulator D, and the index (memory address) is in one of the 16-bit registers: X, Y, SP, or PC. The value in D is added to the index register (X, Y, SP, or PC), and used to fetch a second 16-bit big endian address from memory, as shown in Figure 3.31. The load or store is performed at this second address. This mode is also useful when data structures contain pointers. In each case, accumulator D and the index register used as a pointer (index) are not modified by the instruction. For example

```
ldd  #4
ldy  #$2341
stx [D,Y]    [$1234]=Reg X, (D & Y unchanged)
```

Figure 3.31
Example of the 6812 accumulator-offset indexed-indirect addressing mode.

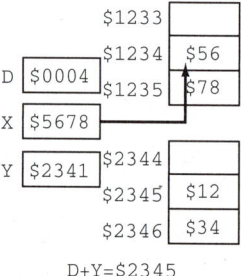

Table 3.23 shows the object code for `rr,` which specifies X, Y, SP, or PC.

Table 3.23
The MC68HC812A4 indexed-register code.

rr	register
00	X
01	Y
10	SP
11	PC

The expression `ff` refers to the 8-bit offset needed for IDX1 modes, and `eeff` refers to the 16-bit offset needed for IDX2 modes. Table 3.24 gives the object code for indexed-mode instructions.

Table 3.24
Postbyte values for the 6812 indexed-addressing modes.

postbyte(xb)	syntax	mode	explanations
rr000000	,r	IDX	5-bit constant, n=0
rr00nnnn	n,r	IDX	5-bit constant, n from 0 to +15
rr01nnnn	-n,r	IDX	5-bit constant, n from -16 to -1
rr100nnn	n,+r	IDX	pre-increment, n from 1 to 8
rr101nnn	n,-r	IDX	pre-decrement, n from 1 to 8
rr110nnn	n,r+	IDX	post-increment, n from 1 to 8
rr111nnn	n,r-	IDX	post-decrement, n from 1 to 8
111rr100	A,r	IDX	Reg A accumulator offset
111rr101	B,r	IDX	Reg B accumulator offset
111rr110	D,r	IDX	Reg D accumulator offset
111rr000 ff	n,r	IDX1	9-bit constant, n 16 to 255
111rr001 ff	-n,r	IDX1	9-bit constant, n -256 to -16
111rr010 ffee	n,r	IDX2	16-bit constant, any 16-bit n
111rr111	[D,r]	[D,IDX]	Reg D offset, indirect
111rr011 ffee	[n,r]	[IDX2]	16-bit constant, indirect

The 6812 load-effective address instructions can be used only with indexed addressing mode. These instructions are very useful for manipulating the 16-bit registers. Let idx represent one of the above index addressing modes. They do not affect any condition code bits.

```
leax idx    RegX=EA
leay idx    RegY=EA
leas idx    RegS=EA
```

The basic idea is that the effective address is calculated in the usual manner. But rather than fetching the memory contents at that address as a regular load instruction (`ldaa`, `ldab`, `ldd`, `ldx`, `ldy`, or `lds`) would, this instruction puts the effective address itself into the register. In each of the following cases, the effective address, `EA`, is loaded into Register X.

```
leax m,r    #IDX 5-bit index, EA=r+1m (-16 to 15)
leax v,+r   #IDX pre-increment r=r+v, EA=r (1 to 8)
leax v,-r   #IDX pre-decrement r=r-v, EA=r (1 to 8)
leax v,r+   #IDX post-increment, EA=r, r=r+v (1 to 8)
leax v,r-   #IDX post-decrement, EA=r, r=r-v (1 to 8)
leax A,r    #IDX Reg A offset EA=r+A, zero padded
leax B,r    #IDX Reg B offset EA=r+B, zero padded
leax D,r    #IDX Reg D offset EA=r+D
leax q,r    #IDX1 9-bit index EA=r+q (-256 to 255)
leax W,r    #IDX2 16-bit index EA=r+W (-32768 to 65535)
```

where r is Reg X, Y, SP, or PC, and the fixed constants are

m is any signed 5-bit -16 to $+15$

q is any signed 9-bit -256 to $+255$

v is any unsigned 3 bit 1 to 8

W is any signed 16-bit -32768 to $+32767$ or any unsigned 16-bit 0 to 65535

> ***Observation:*** *The* `leas -4,sp` *instruction subtracts four from the stack pointer.*
>
> ***Checkpoint 3.24:*** *Write 6812 assembly code that sets Register Y equal to X+10.*

3.6 Memory Allocation

Memory allocation is the decision of where in memory we put the various pieces of our software. The memory on a PC-compatible computer is physically configured as a simple linear array. In other words, if you have 256 Mbytes of RAM, then this memory exists as a continuous linear object with no fundamental difference in the behavior of one memory cell to the next. Although the memory itself forces no structure in the way it is used, the Intel x86 processors have implemented a memory access scheme that requires the programmer to separate various components of their program (machine code, local variables, and global variables) into different parts of memory. The term **x86** refers to any Intel processor from the 8086 through the current Pentiums. There can be more than three segments, but three is enough to illustrate the point. The mechanism to access these segments is called **segmentation**. Figure 3.32 shows a simple view of the memory allocation on the Intel x86 family.

Figure 3.32
The Intel x86 uses segmented memory allocation.

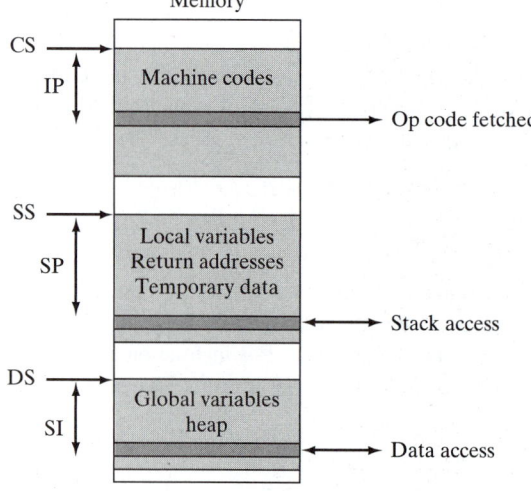

In particular, when the Pentium fetches a machine code, it uses two registers. The code segment selector (CS) points to the beginning of the code segment, and the instruction pointer (IP) contains the offset within this segment of the op code to fetch. Similarly, when the Pentium accesses a global variable, it uses two different registers. The data segment selector (DS) points to the beginning of the data segment, and a data pointer (e.g., DI) contains the offset within this segment of global variable. Lastly, when the Pentium accesses a local variable, it uses a stack segment selector (SS) and either the stack pointer (SP) or the base pointer (BP). The stack segment selector (SS) points to the beginning of the stack segment, and a stack pointer (SP or BP) contains the offset within this segment of local variable. Segmentation forces the programmer to allocate in memory information that has similar properties. In other words, all the machine codes are placed in one group, the global variables are in another group, and the stack is in a third group. This allocation scheme provides for protection so that the errors or stack overflow, stack underflow, accessing an illegal pointer do not modify machine codes.

We will allocate memory on our embedded system in a fashion similar to segmentation, but for a different reason. Because different types of memory on an embedded computer behave in different fashions, it makes sense to group together in memory information that has similar properties or usage. Typical examples of this grouping include global variables, the heap, local variables, fixed constants, and machine instructions. Figure 3.33 shows a typical memory allocation scheme for an embedded system. We place variables in RAM and programs in ROM on an embedded system.

Figure 3.33
We place variables in RAM and programs in ROM on an embedded system.

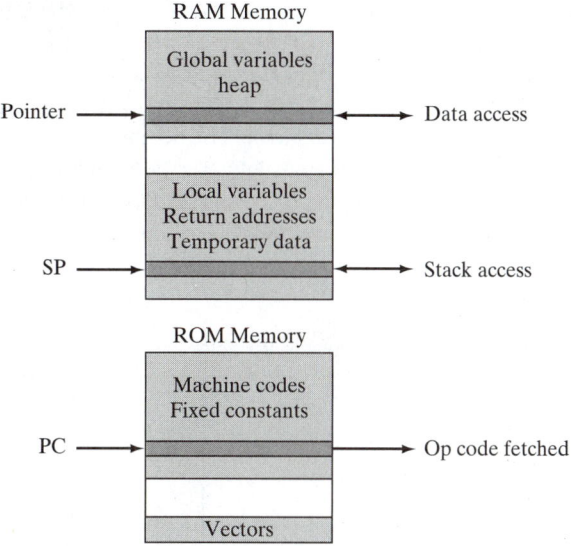

Global variables are permanently allocated and usually accessible by more than one program. We must use global variables for information that must be permanently available, or for information that is to be shared by more than one module. We will see many applications later in this book of the first-in-first-out (FIFO) queue that is a global data structure shared by more than one module. Some software systems use a heap to dynamically allocate and release memory. This information can be shared or not shared depending on which modules have pointers to the data. The heap is efficient in situations where storage is needed for only a limited amount of time. Local variables are usually allocated on the stack at the beginning of the subroutine/function, used within the subroutine/function, and deallocated at the end of the subroutine/function. Local variables are not shared with other modules. Fixed constants do not change and include information such as numbers, strings, sounds and pictures. Just like the heap, the fixed constants can be shared or not shared depending on which modules have pointers to the data. As we saw in the previous chapter, the assembler or compiler translates our software into machine instructions (op codes and operands) that, when executed, perform the intended operations. For single-chip microcomputers, there are three types of memory. The RAM contains temporary information that is lost when the power is shut off (i.e., volatile). Thus, all variables allocated in RAM must be explicitly initialized at run time by the software. Some C compilers initialize all RAM-based global variables to zero, and others do not. It is good software development practice to set globals to the desired initial value explicitly. As we saw earlier in this chapter, many Motorola microcomputers have a little bit of EEPROM. The ROM is a low-cost nonvolatile storage that can be programmed only once.

In an embedded application, we usually put global variables, the heap, and local variables in RAM because these types of information can change during execution. When software is to be executed on a regular computer, the machine instructions are usually read from a mass storage device (like a disk) and loaded into memory. Because the embedded system usually has no mass storage device, the machine instructions and fixed constants must be stored in nonvolatile memory. If there is both EEPROM and ROM on our microcomputer, we put some fixed constants in EEPROM and some in ROM. If it is information that we may wish to change in the future, we could put it in EEPROM. Examples include language-specific strings, calibration constants, finite-state machines, and system ID numbers. This allows us to make minor modifications to the system by reprogramming the EEPROM without throwing the chip away. If our project involves producing a small number of devices, the program can be placed in EPROM or EEPROM. For a project with a large volume, it will be cost effective to place the machine instructions in ROM.

Program 3.4 is a simple illustration of how we allocate various sections of our software using the `org` pseudo-op. The program outputs to Port C the sequence 0,5,10,15, ... The global variable is placed at the start of the RAM, the stack is initialized to the top of RAM (and grows down), and the program is placed in ROM. The constants are placed in the 6811 EEPROM.

Program 3.4   Memory allocation places variables in RAM and programs in ROM.

```
; MC68HC11A8                        ; MC68HC812A4
      org  $0000 ;RAM                     org  $0800 ;RAM
cnt   rmb  1    ;global         cnt   rmb  1    ;global

      org  $B600 ;EEPROM                  org  $F000 ;EEPROM
const fcb  5    ;amount to add   const fcb  5    ;amount to add

      org  $E000 ;ROM            init  movb #$FF,DDRC ;outputs
init  ldaa #$FF                        clr  cnt
      staa DDRC  ;outputs              rts
      clr  cnt                   main  lds  #$0C00 ;sp=>RAM
      rts                              bsr  init
main  lds  #$00FF ;sp=>RAM       loop  ldaa cnt
      bsr  init                        staa PORTC  ;output
loop  ldaa cnt                         adda const
      staa PORTC ;output               staa cnt
      adda const                       bra  loop
      staa cnt
      bra  loop                        org  $FFFE ;EEPROM
                                       fdb  main  ;reset vector
      org  $FFFE ;ROM
      fdb  main ;reset vector
```

Checkpoint 3.25: How would you change Program 3.4 if you were to use a MC68HC711E9 and MC68HC912B32?

3.7 Tutorial 3. Building a Microcomputer and Executing Machine Code

In this example, we will create a 4-input 4-output NOT gate using an embedded micro-computer (see Figure 3.34). The four input signals will be connected to an input port of the microcomputer and the 4 output signals will be connected to an output port. Some pins of Port C are used as input and some pins are used as output.

Figure 3.34
External inputs are connected to PC3-PC0 and PC7-PC are the outputs.

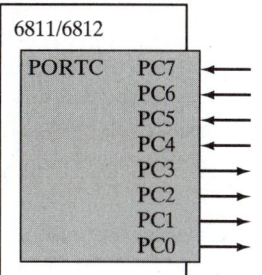

The first software step is to write an initialization function that specifies Port C bits PC7-PC4 will be outputs and Port C bits PC3-PC0 will be inputs (see Program 3.5). This function, which we usually execute once at the start of our program, is called a **ritual**. The direction register, DDRC, specifies whether each pin is an output (1) or an output (0). The main program will input from Port C, perform the logical complement, shift the data into the proper position, then output it back to Port C. The input operation performed on the output pins will return the previous value written, while the output operation to the input pins has no effect. In each case, the stack pointer is initialized to the top of RAM.

Program 3.5 Assembly program that implements the 4-bit NOT gate

```
; MC68HC711E9                          ; MC68HC812A4
PORTC equ  $1003                       PORTC equ  $0004
DDRC  equ  $1007                       DDRC  equ  $0006
      org  $D000                             org  $F000
main  lds  $01FF  ;Initialize stack    main  lds  #$0C00 ;Initialize stack
      bsr  init   ;ritual                    bsr  init   ;ritual
loop  ldaa PORTC  ;input               loop  ldaa PORTC  ;input
      coma         ;logical not               coma         ;logical not
      lsla         ;shift                     lsla         ;shift
      lsla                                    lsla
      lsla                                    lsla
      lsla                                    lsla
      staa PORTC  ;output                    staa PORTC  ;output
      bra  loop   ;repeat                     bra  loop   ;repeat
init  ldaa #$F0   ;PC7-PC3 out         init  ldaa #$F0   ;PC7-PC3 out
      staa DDRC   ;PC3-PC0 in                staa DDRC   ;PC3-PC0 in
      rts                                    rts
      org                                    org  $FFFE
      fdb  main   ;place to start           fdb  main   ;place to start
```

We could implement these operations in C (see Program 3.6). In these C programs we will assume the compiler will handle the segmentation (placing variables in RAM, programs in ROM/EEPROM), initializing the stack, and setting the reset vector.

Program 3.6
C program that implements the 4-bit NOT gate.

```
void init(void){
  DDRC=0xF0;   // PC7-PC4 are outputs, PC3-PC0 are inputs
}
void main(void){ unsigned char data;
  init();              // call ritual once
  while(1){
    data=PORTC;        // input
    data=(~data)<<4;   // complement and shift
    PORTC=data;        // output
  }
}
```

Question 3.1. Choose the processor you wish to study and look up its RAM and ROM locations. Look up the locations of PORTC and DDRC.

Action: Copy the three files `Chap3.rtf`, `Chap3.uc`, and `Chap3.io` from the CD onto your hard drive. They can be found on the CD in the tutorial folder. Select the 6811 or 6812 version as appropriate. Start a fresh copy of **TExaS** and open the `Chap3.rtf` file; **TExaS** should open the other two files. Execute the **Mode->Processor. . .** command and select the processor you identified in Question 3.1. Save these three documents. Figure 3.35 shows the MC68HC812A4 being selected.

Figure 3.35
Mode->Processor dialog.

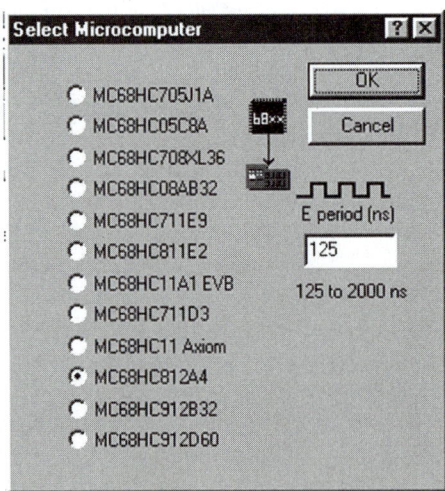

Action: Click on the Program window and execute the **Assemble->Options. . .** command. Make sure all the check boxes except *Automatically create a *.S19 file* are checked. Make sure the *Complete Assembly (small programs)* radio button is selected, as shown in Figure 3.36.

Figure 3.36
Assemble->Options
dialog.

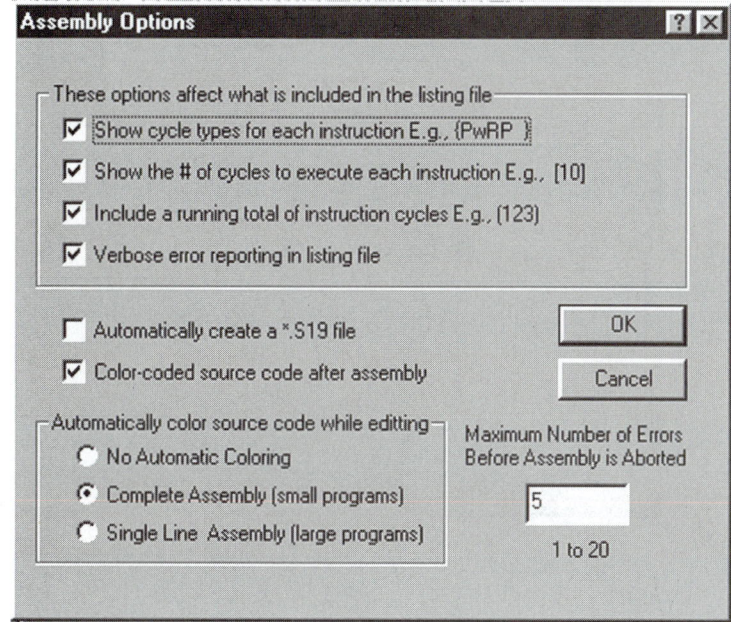

Action: Edit the assembly source code, `Chap3.rtf`. Adjust the $0004, $0006, $F000, and $0C00 in the first four lines to match the processor you selected in Question 3.1. Assemble the program by executing the **Assemble->Assemble** command.

Question 3.2. Execute this program by hand using paper and pencil up to and including the `bra` instruction. For each instruction show the memory cycles generated and the values of Registers A, PC, and SP after each instruction. For the 6811, show the actual cycles. For the 6812, show the simplified cycles as described in section 3.2.5.

Action: Activate the **FollowPC CycleView InstructionView** and **LogRecord** modes using the commands in the Mode menu. Single step (F10) the program up to and including the `bra` instruction. Verify the answers you gave for Question 3.2.

Action: Bring the `Chap3.io` window to the top and notice that the switches connected to PC1 and PC0 are activated, as shown in Figure 3.37. Start the simulation (F12). Toggle each of the switches by clicking on it with the mouse. Notice that the corresponding LED is the logical complement of the switch.

Figure 3.37
I/O window.

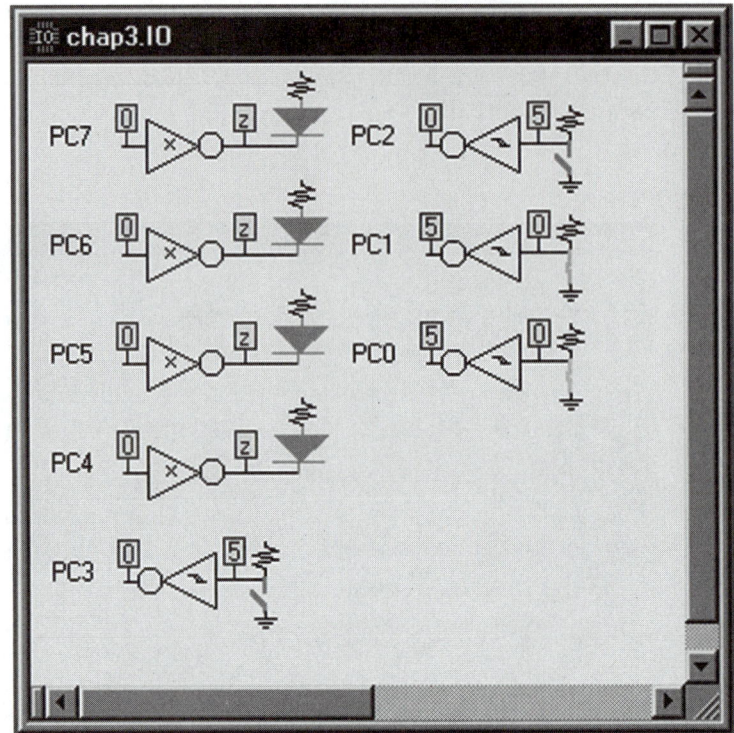

3.8   Homework Assignments

Homework 3.1: What is the difference between memory-mapped and isolated I/O?

Homework 3.2: Name the four main components of the processor. Give a brief description of their functions.

Homework 3.3: What does the effective address register contain?

Homework 3.4: What is the purpose of the following registers: CCR, SP, PC, IR, and EAR?

Homework 3.5: Write assembly code that exchanges the values in Register X and Y.

Homework 3.6: Let N, M, and P be three 8-bit unsigned locations. Write assembly code to implement P=2*N+M.

Homework 3.7: Let N, M, and P be three 8-bit unsigned locations. Write assembly code to implement P=(5*N+M)/17.

Homework 3.8: Let N, M, and P be three 8-bit locations. Write assembly code to implement P=(5|N)&M.

Homework 3.9: Let N be an 8-bit location. Write assembly code to set bit 7 and clear bit 0.

Homework 3.10: Let N be an 8-bit location. Write assembly code to toggle bit 7 and set bit 0.

Homework 3.11: Assume you have a 16-bit global variable H. Write assembly code that implements `if(H>1234)isGreater();`

Homework 3.12: Assume you have an 8-bit global variable G. Write assembly code that implements `while(G&0x80)body();`

Homework 3.13: Assume Register X contains the address $2000, Register Y contains the address $2080, Register A contains $45, and Register B contains $67. For each of the following instructions, specify the effective address and the resulting operation. In particular specify what value(s) is stored into what memory location(s). Give all your answers in hexadecimal.

```
staa  40,x
stab  $40,x
std   $66,y
staa  25,y
stab  $FF,y
std   $CD,x
```

Homework 3.14: Assume Register X contains the address $2000, and Register Y contains the address $2080. Assume memory contains the following initial values $2000=0, $2001=1,...,$20FF=$FF. For each of the following instructions, specify the effective address and the resulting operation. Give all your answers in hexadecimal.

```
ldaa  40,x
ldab  $40,y
ldaa  $66,x
ldaa  25,y
ldd   $FE,x
ldd   $0D,y
```

Homework 3.15: Show the 6811 bus cycles generated by the execution of the following program. For the 6812, give the simplified cycles generated by the **TExaS** simulator. The first step is to find the object code for the three instructions, and the second step is to break each instruction into individual bus cycles required to execute it.

```
org   $F000
ldaa  #44
ldy   #$0010
staa  4,y
```

Homework 3.16: Show the 6811 bus cycles generated by the execution of the following program. For the 6812, give the simplified cycles generated by the **TExaS** simulator. The first step is to find the object code for the three instructions, and the second step is to break each instruction into individual bus cycles required to execute it.

```
org   $F000
ldab  #$55
ldx   #$0020
stab  5,x
```

Homework 3.17: The following data are stored in sequential memory locations. Determine the sequence of memory instructions these data represent. Each value is in hexadecimal.

On the 6811: `86,55,CE,00,20,F6,F0,00,97,05,E7,05,20,FA`
On the 6812: `86,55,CE,00,20,F6,F0,00,5A,05,6B,05,20,FA`

Homework 3.18: The following data are stored in sequential memory locations. Determine the sequence of memory instructions these data represent. Each value is in hexadecimal.

On the 6811: `4F,CE,00,40,F6,F0,01,97,08,54,E7,05,20,FB`
On the 6812: `87,CE,00,40,F6,F0,01,5A,08,54,6B,05,20,FB`

Homework 3.19: You will write four assembly language versions of the following C code

`n=100; while(n!=0){n—; body();}`

a) Assume the variable `n` is implemented as a 16-bit global variable.

b) Assume the variable `n` is implemented as an 8-bit global variable.

c) Assume the variable `n` is implemented as a 16-bit variable in Register D.

d) Assume the variable `n` is implemented as an 8-bit variable in Register A.

Homework 3.20: You will write four assembly language versions of the following C code

`n=0; while(n<100){n++; body();}`

a) Assume the variable `n` is implemented as a 16-bit global variable.

b) Assume the variable `n` is implemented as an 8-bit global variable.

c) Assume the variable `n` is implemented as a 16-bit variable in Register D.

d) Assume the variable `n` is implemented as an 8-bit variable in Register A.

Homework 3.21: This question only applies to the 6812. Assume Register X contains the address $0800, Register Y contains the address $0900, Register A contains $02, and Register B contains $67. Assume locations $0802 and $0803 contain the 16-bit value $0A00. For each of the following instructions, specify the effective address and the resulting operation. In particular specify what value(s) is stored into what memory location(s). Give all your answers in hexadecimal.

```
staa  b,x
stab  -$40,y
std   [2,x]
stx   d,y
stab  1,-y
std   2,x+
```

Homework 3.22: This question only applies to the 6812. Assume Register X contains the address $0800, Register Y contains the address $0900, Register A contains $03, and Register B contains $67. Assume locations $0804 and $0805 contain the 16-bit value $0B12. For each of the following instructions, specify the effective address and the resulting operation. In particular specify what value(s) is stored into what memory location(s). Give all your answers in hexadecimal.

```
stab  a,x
staa  -1,y
std   [4,x]
sty   d,x
staa  1,+x
std   2,y+
```

Homework 3.23: Write assembly code that adds 10 to Register X and subtracts 100 from Register Y.

Homework 3.24: Write assembly code that sets Register X equal to Register Y plus 100.

Homework 3.25: How would you change Program 3.4 if you were to use a MC68HC711D3 and MC68HC912D60?

Homework 3.26: Look up the machine code created by the following instructions. Explain the basic function of each instructon. The first one is completed.

machine code	instruction	comment
$860A	ldaa #10	RegA = 10
	ldaa 10	
	ldaa 10,x	
	ldaa 10,y	

Homework 3.27: Look up the machine code created by the following 6812 instructions. Explain the basic function of each instructon. The first one is completed.

machine code	instruction	comment
$A602	ldaa 2,x	RegA = [x+2]
	ldaa -2,x	
	ldaa 2,+x	
	ldaa 2,x+	
	ldaa 2,-x	
	ldaa 2,x-	
	ldaa [2,x]	

Homework 3.28: Assume RegA contains an ASCII character. Write assembly code that converts any upper case (A-Z) to lower case letters (a-z). For example, if RegA is initially 'G', convert it to 'g'. Leave all other characters unchanged.

Homework 3.29: Assume RegA contains an ASCII character. Write assembly code that converts any upper case letters (A-Z) to lower case (a-z). For example, if RegA is initially 'G', convert it to 'g'. Leave all other characters unchanged.

3.9   Laboratory Assignments

Lab 3.1   Start a fresh copy of **TExaS** and create new Program and Microcomputer windows. Execute the **Mode->Processor . . .** command and select either the MC68HC711E9 or the MC68HC812A4. Save these two documents with names having the same root name, e.g., Lab3.rtf, Lab3.uc. Click on the Program window and execute the **Assemble->Options . . .** command. Make sure the check boxes make the configuration shown in Figure 3.36.

Step 1. Type the assembly source code from Program 3.16. Assemble the program by executing the **Assemble->Assemble** command.

Step 2. Execute this program by hand up to and including the "stop" instruction. For each instruction show the memory cycles generated and the values of any registers that change. For the 6811, show the actual cycles. For the 6812, show the simplified cycles as described in section 3.2.5.

Step 3. Activate the **FollowPC CycleView InstructionView** and **LogRecord** modes using the commands in the Mode menu. Single step the program up to and including the "stop" instruction. Verify the answers you gave for Step 2.

Program 3.16
Assembly program used in Lab 3.1.

```
; MC68HC711E9              ; MC68HC812A4
  org   $D000                org   $F000
  ldx   #$0000               ldx   #$0800
  ldy   #$0080               ldy   #$0880
  ldd   #$4567               ldd   #$4567
  staa  40,x                 staa  40,x
  stab  $40,x                stab  $40,x
  std   $66,y                std   -$66,y
  staa  25,y                 staa  25,y
  stab  $0F,y                stab  -2,y
  std   $CD,x                std   $CD,x
  stop                       stop
  org   $FFFE                org   $FFFE
  fdb   $D000                fdb   $F000
```

Lab 3.2 Start a fresh copy of **TExaS** and create new Program and Microcomputer windows. Execute the **Mode->Processor . . .** command and select either the MC68HC711E9 or the MC68HC812A4. Save these two documents with names having the same root name, e.g., `Lab3.rtf`, `Lab3.uc`. Click on the Program window and execute the **Assemble->Options . . .** command. Make sure the check boxes make the configuration shown in Figure 3.36.

Step 1. Type the assembly source code from Program 3.17. Assemble the program by executing the **Assemble->Assemble** command.

Step 2. Execute this program by hand up to and including the "stop" instruction. For each instruction show the memory cycles generated and the values of any registers that change. For the 6811, show the actual cycles. For the 6812, show the simplified cycles as described in section 3.2.5.

Step 3. Activate the **FollowPC CycleView InstructionView** and **LogRecord** modes using the commands in the Mode menu. Single step the program up to and including the "stop" instruction. Verify the answers you gave for Step 2.

Program 3.17
Assembly program used in Lab 3.2.

```
; MC68HC711E9              ; MC68HC812A4
  org   $D000                org   $F000
  ldx   #$0000               ldx   #$0800
  ldy   #$0040               ldy   #$0900
  ldd   #$1234               ldd   #$1234
  ldaa  $50,x                ldaa  $50,x
  ldab  $1A,y                ldab  $1A,y
  ldaa  $AC,x                ldaa  $AC,x
  ldaa  13,y                 ldaa  -13,y
  ldd   $EF,x                ldd   $EF,x
  ldd   $0B,y                ldd   -$0B,y
  stop                       stop
  org   $FFFE                org   $FFFE
  fdb   $D000                fdb   $F000
```

Lab 3.3   Start a fresh copy of **TExaS** and create new Program and Microcomputer windows. Execute the **Mode->Processor . . .** command and select either the MC68HC711E9 or the MC68HC812A4. Save these two documents with names having the same root name, e.g., Lab3.rtf, Lab3.uc. Click on the Program window and execute the **Assemble->Options . . .** command. Make sure the check boxes make the configuration shown in Figure 3.36.

Step 1. Type the assembly source code from Program 3.18. Assemble the program by executing the **Assemble->Assemble** command.

Step 2. Execute this program by hand up to and including the "stop" instruction. For each instruction show the instruction executed, the memory cycles generated, and the values of any registers that change. For the 6811, show the actual cycles. For the 6812, show the simplified cycles as described in section 3.2.5.

Step 3. Activate the **FollowPC CycleView InstructionView** and **LogRecord** modes using the commands in the Mode menu. Single step the program up to and including the "stop" instruction. Verify the answers you gave for Step 2.

Program 3.18
Assembly program used in Lab 3.3.

```
; MC68HC711E9                    ; MC68HC812A4
  org   $D000                      org   $F000
  fcb   $86,$55,$CE,$00,$20        fcb   $86,$55,$CE,$00,$20
  fcb   $F6,$D0,$00,$97,$05        fcb   $F6,$F0,$00,$5A,$05
  fcb   $E7,$05,$20,$FA            fcb   $6B,$05,$20,$FA
  stop                             stop
  org   $FFFE                      org   $FFFE
  fdb   $D000                      fdb   $F000
```

Lab 3.4   Start a fresh copy of **TExaS** and create new Program and Microcomputer windows. Execute the **Mode->Processor . . .** command and select either the MC68HC711E9 or the MC68HC812A4. Save these two documents with names having the same root name, e.g., Lab3.rtf, Lab3.uc. Click on the Program window and execute the **Assemble->Options . . .** command. Make sure the check boxes make the configuration shown in Figure 3.36.

Step 1. Type the assembly source code from Program 3.19. Assemble the program by executing the **Assemble->Assemble** command.

Step 2. Execute this program by hand up to and including the "stop" instruction. For each instruction show the instruction executed, the memory cycles generated, and the values of any registers that change. For the 6811, show the actual cycles. For the 6812, show the simplified cycles as described in section 3.2.5.

Step 3. Activate the **FollowPC CycleView InstructionView** and **LogRecord** modes using the commands in the Mode menu. Single step the program up to and including the "stop" instruction. Verify the answers you gave for Step 2.

Program 3.19
Assembly program used in Lab 3.4.

```
; MC68HC711E9                    ; MC68HC812A4
  org   $D000                      org   $F000
  fcb   $4F,$CE,$00,$40,$F6        fcb   $87,$CE,$00,$40,$F6
  fcb   $D0,$01,$97,$08,$54        fcb   $F0,$01,$5A,$08,$54
  fcb   $E7,$05,$20,$FB            fcb   $6B,$05,$20,$FB
  stop                             stop
  org   $FFFE                      org   $FFFE
  fdb   $D000                      fdb   $F000
```

4 Assembly Language Programming

Chapter 4 objectives are to

❑ present an introduction to software development,
❑ discuss the special features of the **TExaS** editor and assembler,
❑ describe the basic approach to assembly language programming,
❑ define the pseudo-ops supported by **TExaS**.

In truth, we have been programming in assembly language since the beginning of the book. In this chapter, we will begin by presenting a general approach to software development. Then we will present a formal definition of the 6811/6812 assembly language as implemented by the **TExaS** application. An overview of the executable instructions was presented in the last chapter. A complete list of the 6811/6812 instructions can be found on the CD enclosed with this text. In this chapter the assembly directives or pseudo-ops are defined. There is also a brief discussion of how assemblers work.

4.1 Introduction

4.1.1
Assembly
Language
Development

To develop assembly language software, we first use an **editor** to create our **source code**. Source code contains specific commands in human-readable form. Next we use an **assembler** to translate our source code into **object code**. Object code contains the specific commands in machine-readable form. When developing software for a real microcomputer, a **loader** is used to place the object code into our computer's memory. We test our program with the aid of a **debugger**. In the final product, the power on reset is used to start software execution after power is supplied to the microcomputer. If the object code is to be stored in EPROM, we can use an EPROM programmer. Many microcomputers contain built-in features that assist in programming its EPROM, while other microcomputers require a separate apparatus to program. Some hardware development systems use RAM to hold the program. In these systems the loader downloads the object code into RAM. Since RAM is volatile, the programs must be loaded each time power is removed. Figure 4.1 outlines the assembly language development process.

Assemblers are programs that process assembly language source program statements and translate them into executable machine language object files. Cross assemblers (such as **TExaS**) allow source programs written and edited on one computer (the host) to generate executable code for another computer (the target). In our case the target is microcomputer simulator, but since **TExaS** generates S19 records, you could download the S19 file onto a real 6811/6812 for execution.

Figure 4.1
Assembly language
development process

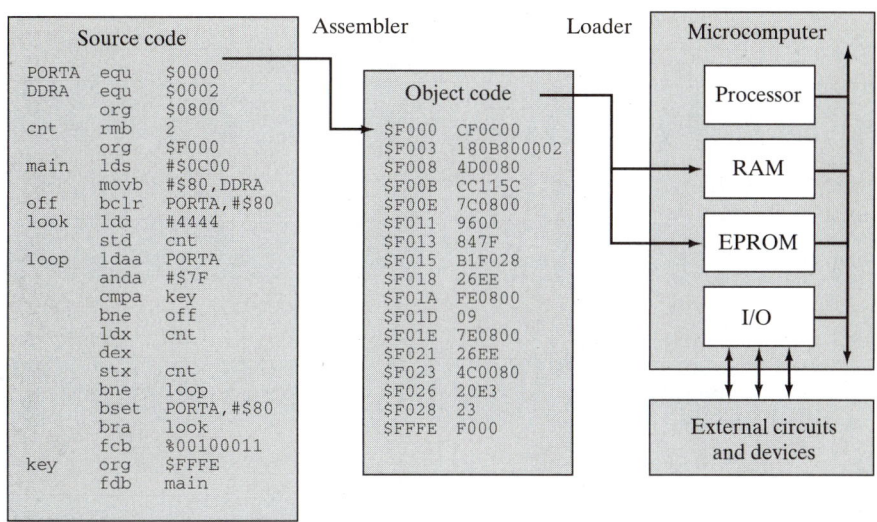

The symbolic language used to code source programs to be processed by the assembler is called assembly language. The language is a collection of mnemonic symbols representing operations (i.e., machine instruction mnemonics or directives to the assembler), symbolic names, operators, and special symbols. The assembly language provides mnemonic operation codes for all machine instructions in the instruction set. The instructions are defined and explained in the Programming Reference Manual for the 6811/6812, which can be found in the **pdf** directory of the CD. A brief overview of each instruction and example usage can be found by searching the **Contents** page of the help engine included with the **TExaS** application. The assembly language also contains mnemonic directives that specify auxiliary actions to be performed by the assembler. Most of these directives or pseudo-ops are not translated into machine language.

The **TExaS** assembler is a two-pass assembler. During the first pass, the source program is analyzed in order to develop the symbol table. A **symbol table** is a mapping between symbolic names (e.g., PORTC) and their numeric values (e.g., $1003.) During the second pass, the object file is created (assembled) using the symbol table developed in pass one. It is during the second pass that the source program listing is also produced. The symbol table is recreated in the second pass. A phasing error occurs if the symbol table values calculated in the two passes are different.

Errors that occur during the assembly process (e.g., undefined symbol, illegal op code, branch destination too far, etc.) are explained in the listing file. Program 4.1 contains the listing file for the short program shown in Figure 4.1. The listing file contains the source code, the object code, and the symbol table.

Program 4.1   Assembly listing file.

```
Copyright 2001-2002 Test EXecute And Simulate - Version 1.15
                                    ; *****lock.rtf*******
                                    ; activate solenoid (PA7=1)
                                    ; bounce is less than 10 ms
$0000                               PORTA equ   $0000
$0002                               DDRA  equ   $0002
$0800                                     org   $0800
$0800                               cnt   rmb   2
$F000                                     org   $F000
$F000 CF0C00          [ 2](  0){OP      }main  lds   #$0C00
$F003 180B800002      [ 4](  2){OPwP    }      movb  #$80,DDRA
$F008 4D0080          [ 4](  6){rPOw    }off   bclr  PORTA,#$80
$F00B CC115C          [ 2]( 10){OP      }look  ldd   #4444
$F00E 7C0800          [ 3]( 12){WOP     }      std   cnt
$F011 9600            [ 3]( 15){rfP     }loop  ldaa  PORTA
$F013 847F            [ 1]( 18){P       }      anda  #$7F
$F015 B1F028          [ 3]( 19){rOP     }      cmpa  key
$F018 26EE            [ 3]( 22){PPP/P   }      bne   off
$F01A FE0800          [ 3]( 25){ROP     }      ldx   cnt
$F01D 09              [ 1]( 28){O       }      dex
$F01E 7E0800          [ 3]( 29){WOP     }      stx   cnt
$F021 26EE            [ 3]( 32){PPP/P   }      bne   loop
                                    ; 7 switches match key code
$F023 4C0080          [ 4]( 35){rPOw    }      bset  PORTA,#$80
$F026 20E3            [ 3]( 39){PPP     }      bra   look
$F028 23                                 key   fcb   %00100011
$FFFE                                          org   $FFFE
$FFFE F000                                     fdb   main
***************Symbol Table********************
DDRA            $0002
PORTA          $0000
cnt            $0800
key            $F028
look           $F00B
loop           $F011
main           $F000
off            $F008
Assembly successful
```

The **source code** is a file of ASCII characters usually created with an editor. Each source statement (i.e., aline) is processed completely before the next source statement is read. As each statement is processed, the assembler examines the label, operation code, and operand fields. The operation code table is scanned for a match with a known op code. During the processing of a standard operation code mnemonic, the standard machine code is inserted into the object file. If an assembler directive is being processed, the proper action is taken.

Any errors that are detected by the assembler are displayed after the actual line containing the error is printed. If no source listing is being produced, error messages are still displayed in the TheLog.rtf to indicate that the assembly process did not proceed normally. **Object code** is the binary values (instructions and data) that, when executed by the computer, perform the intended function. The listing file contains the address, object code, and a copy of the source code. You can optionally include the number of cycles to execute the instruction, a running total of cycles, and explicit details of the cycle types required to execute the instruction. The listing file also provides a symbol table describing where in memory the program and data will be loaded. The symbol table is a list of all the names used in the program along with the values. A symbol is created when you put a label starting in column 1. `look`, `loop`, and `key` are examples of symbols in Program 4.1. The symbol table value for this type of symbol is the absolute memory address where the instruction, variable, or constant will reside in memory. The second type of label is created by the `equ` pseudo-op, e.g., `PORTA`. The value for this type of symbol is simply the number specified in the operand field. When the assembler processes an instruction with a symbol in it, it simply substitutes the fixed value in place of the symbol. Therefore we will use symbols to clarify (make it easier to understand) our programs. The symbol table for this example is given at the end of the listing file.

A compiler converts high-level language source code into object code. A cross-compiler also converts source code into object code and creates a listing file, except that the object code is created for a target machine that is different from the machine running the cross-compiler. The **TExaS** is a cross-assembler because it runs on an Intel computer but creates 6805/6808/6811/6812 object code. ImageCraft's ICC11 and ICC12 include both a cross-assembler and a cross-compiler because they run on the PC and create 6811/6812 object code.

Motorola uses S-records to represent object code. The 16-bit values are stored in memory with the most significant byte first. For more information about the specific format of the S-record see Section 4.5. The S-record for the above example is as follows. The actual object code is in **bold** face.

```
S121F000CF0C00180B8000024D008018030FA008009600847FB1F033260EFE0800091F
S119F01E7E080026EE4C008020E918030FA008004D008020DE23A9
S105FFFEF0000D
S903F0000C
```

Checkpoint 4.1: *What does the assembler do in pass 1?*

Checkpoint 4.2: *What does the assembler do in pass 2?*

4.1.2 *Running On an Evaluation Board

When debugging involves controlling real hardware, we must use an actual microcomputer running in real time. Because the programs are stored in ROM or EEPROM on the single-chip microcomputer they are difficult to debug. To solve this problem there are evaluation boards that run the microcomputer in expanded mode and include a debugger. These systems have RAM in the positions where the final system will have ROM. Some development systems load programs into EPROM or EEPROM. With this development system, we also use a cross-assembler or a cross-compiler to convert our source programs into a listing file and object file. The object file is transmitted via the serial port to the evaluation board and loaded into RAM.

One possibility for 6811 program development uses a evaluation board (EVB) and a resident debugger. "Resident" means the debugger exists as 6811 software that exists along with the regular user program on the system. A popular debugger, developed in the mid 1980s, is called BUFFALO (Bit User's Fast Friendly Aid to Logical Operation). The debugger allows us to test our software. Typical debugger features include

start execution at a specific address,
set breakpoint at a specific address,
read/modify microcomputer registers,
read/modify microcomputer I/O ports,
read/modify RAM,
read/program EPROM (erased with UV light),

read/erase/program EEPROM,
load object code into RAM/EPROM/EEPROM received from the HOST PC.

Sophisticated development environments integrate the editor, compiler, assembler, serial port communication and debugger into a single application running on the PC.

> **Checkpoint 4.3:** *In what ways is running your programs out of RAM different from running them out of EEPROM/EPROM/ROM?*

4.1.3
Simulation

The focus of this book is developing assembly language using the **TExaS** simulator. When developing software using a simulator, we use a cross-assembler or a cross-compiler running on a PC to convert our source programs into a listing file and object file. We then "run" our program on a simulator that emulates the microcomputer with its external components. Simulation in an embedded environment is more difficult than other computer applications because the software execution is tightly coupled to (extremely dependent on) the hardware. Simulation of an embedded system is effective only if all the software, computer, external mechanical, and external electrical components are modeled. Another complicating issue is the real-time nature of the external mechanical and electrical devices. Figure 4.2 outlines the software development process using the **TExaS** simulator. **TExaS** includes the editor, assembler, loader, simulated microcomputer, and simulated external devices.

Figure 4.2
Assembly language development process using **TExaS**.

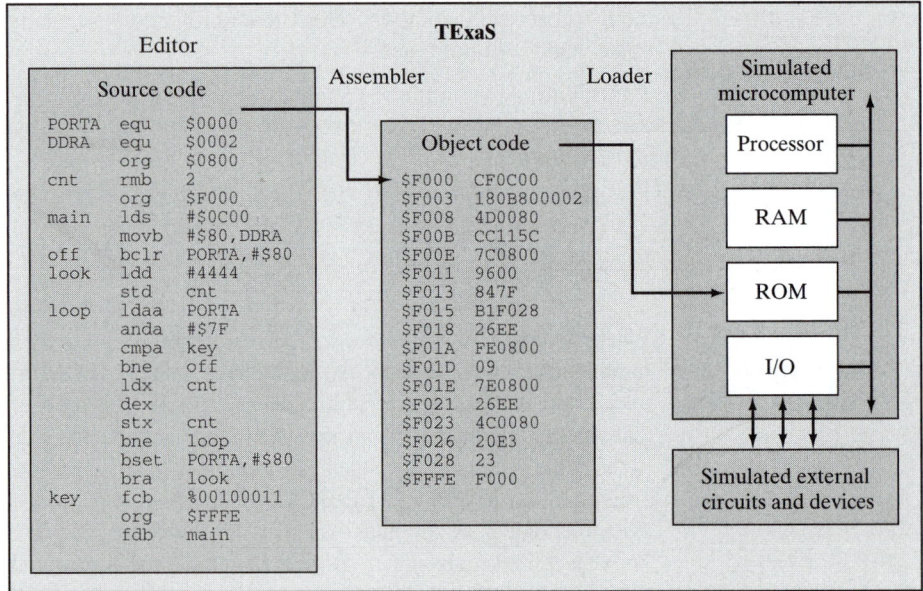

4.2   TExaS Help System

There are so many details about the microcomputer, it is important to have reference materials at your finger tips while writing assembly programs. There are three ways to find information using the interactive **TExaS**'s Help system. We can search by **Contents**, **Index** or **Find**. Contents is the best way to search for what the **TExaS** commands are and how they work.

For the first example, we will search for more information about the `Assemble->` `Assemble` command. One way to invoke the Help system is to highlight the command, as shown in Figure 4.3, and type F1. In particular, click on the `Assemble menu`, roll the mouse over the `Assemble->Assemble` command, then hit the F1 function key. For commands that have a dialog, specific help about the dialog can be found by typing F1 while in the dialog.

Figure 4.3
To get help on a command, highlight the command and type F1.

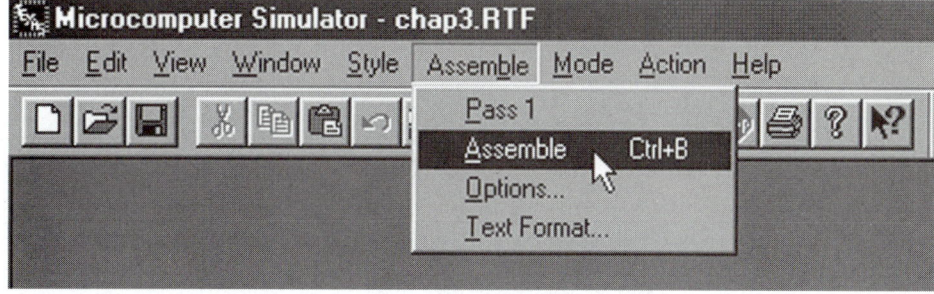

A second way to get help is to execute the `Help->HelpTopics` command. Within the help system, the fastest way to search for help about commands is to use the **Contents** search method, as shown in Figure 4.4. All the commands are located in the **Menu** section, and are organized by menu.

Figure 4.4
To get help on a command, search by contents in the help system.

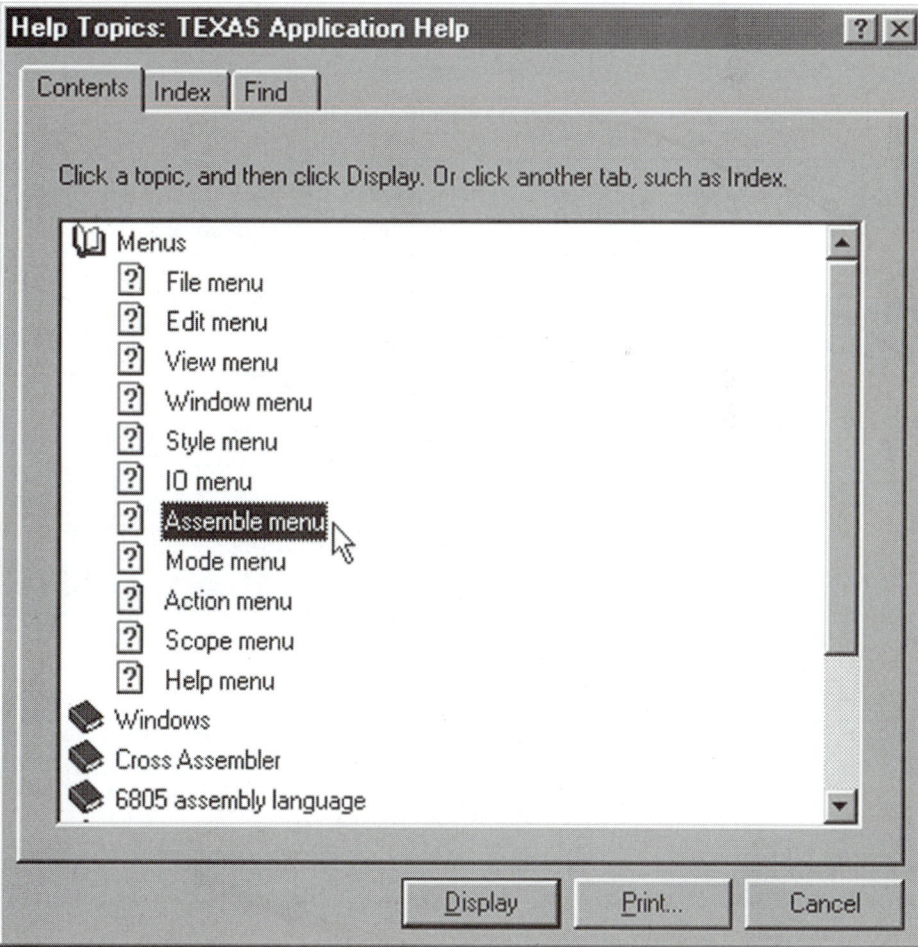

For the second example, we will search for details about the 6811 instruction `bset`. We start by executing the **`Help->HelpTopics`** command. A good way to search for help about instructions is to use the **Contents** search method, as shown in Figure 4.5. All the 6811 instructions are organized in the **6811 assembly language** section. `bset` can be found in the logical instructions. This method is OK if you are familiar with the organization of the help system.

Figure 4.5
To get help on an instruction, search by contents in the Help system.

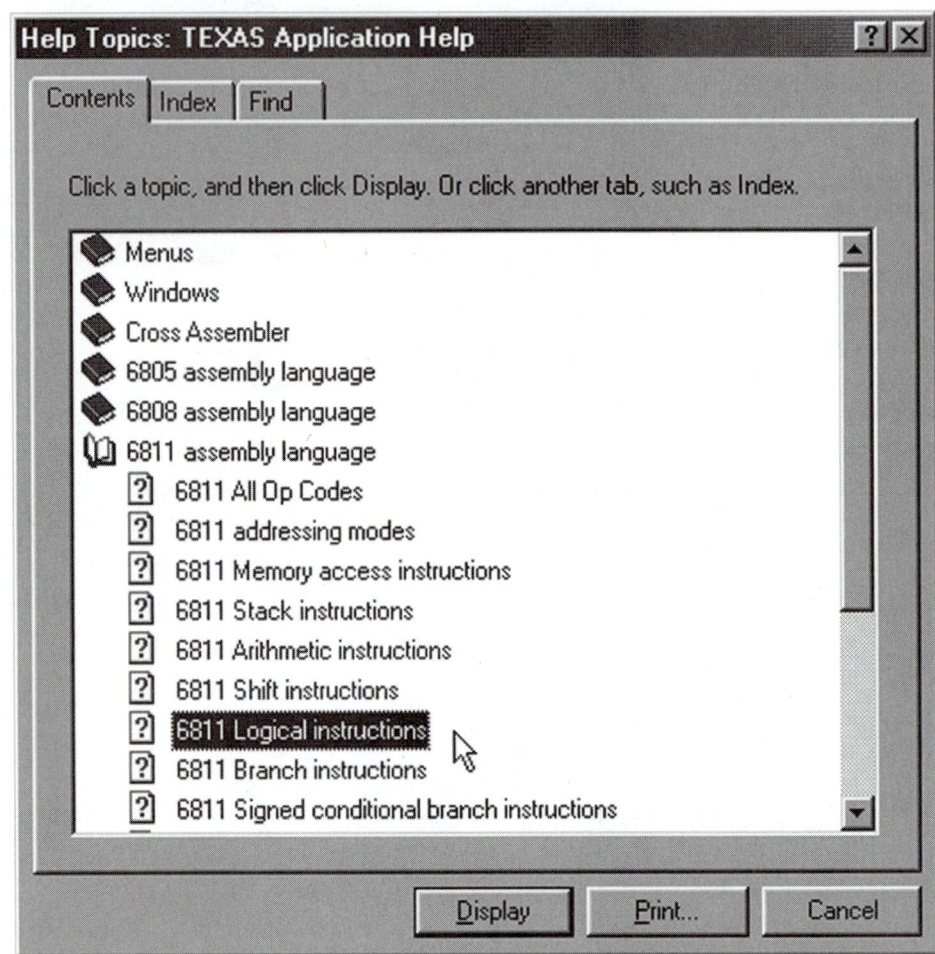

A more general search method is to use **Find**. Within the Help system, click on the **Find** tab, and enter `bset` into the search string box, as shown in Figure 4.6. Information about this instruction can be found at the *6811 bset instruction* entry.

Figure 4.6
To get help on an
instruction, search by
find in the Help system.

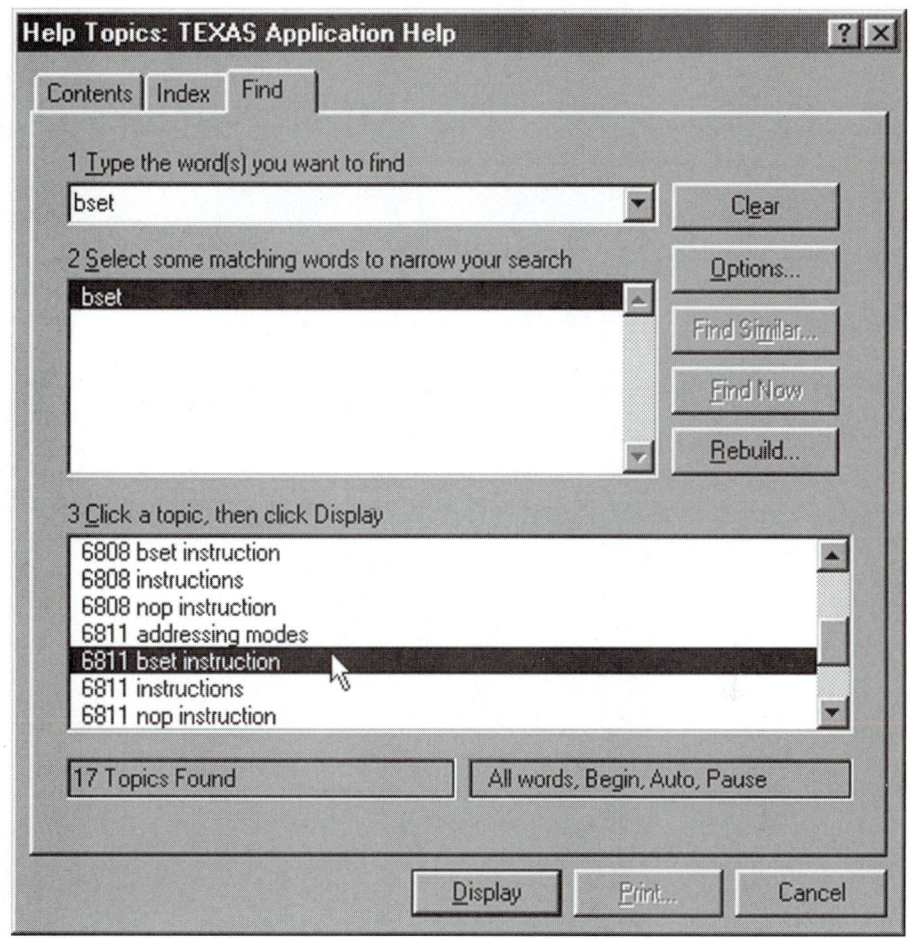

For the third example, we will search for information about how to write assembly code to implement `for loops`. We start by executing the **Help->HelpTopics** command. The best way to search for help about programming examples is to use the **Contents** search method, as shown in Figure 4.7. All the 6811 examples are organized in the **6811 programming examples** section. `For loops` can be found in the section called Control Structures.

Figure 4.7
To get help on programming, search by contents in the Help system.

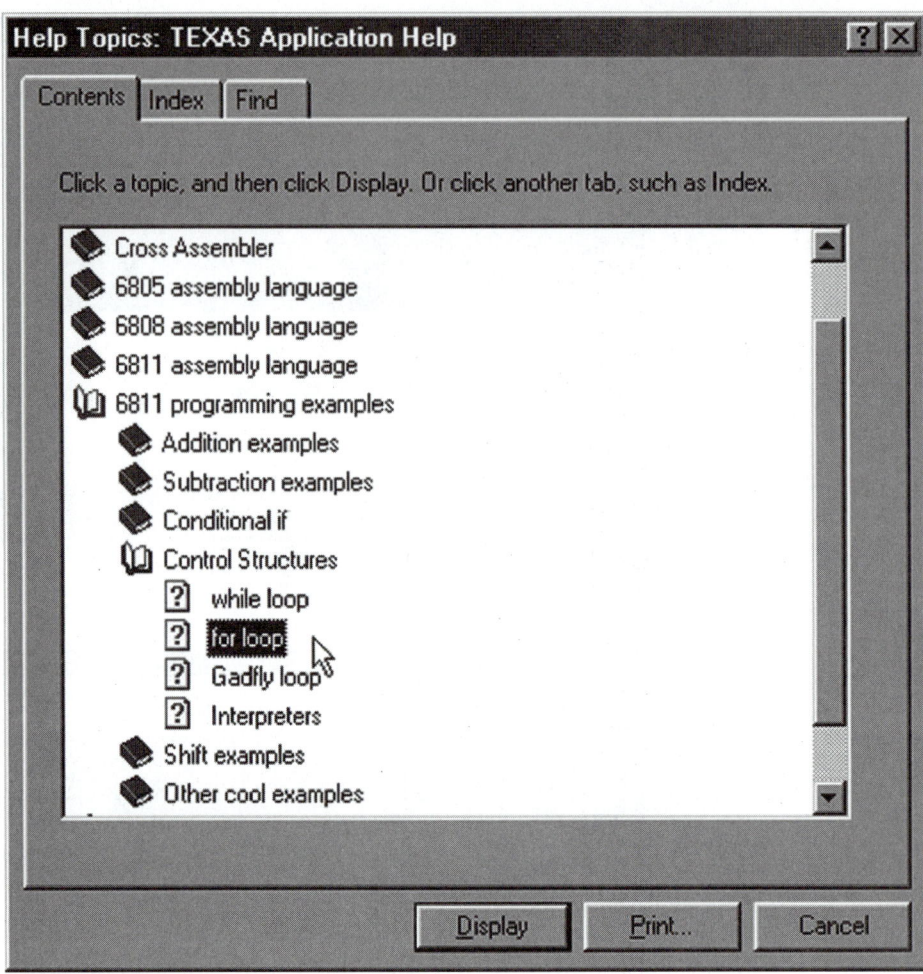

4.3 TExaS Editor

The editor is a simplified version of WordPad. You can specify fonts, sizes and colors. Embedded figures can be added to clarify the software. For example, you can add circuit diagrams, flowcharts, and speadsheet objects into your programs. These embedded objects are ignored by the assembler when creating the object code, but can be quite useful for documentation. The editor uses rich text format (RTF), so formatted text can be cut and pasted from other applications that support rich text format. The following list shows the default color settings of the **TExaS** editor:

the *labels* are shown in purple,
the *op codes* are shown in blue,
the *pseudo-op codes* are shown in gray,

the *numbers* are shown in dark blue,
the *strings* are shown in magneta,
the *operands* are shown in black,
the *comments* are shown in green, and
the *assembly errors* are shown in **bold red**.

These colors can be changed using the **Assemble->TextFormat**. . . command.

> *Checkpoint 4.4: When using the **TExaS** editor, how can you tell if an operation is an op code or pseudo-op?*

4.4 Assembly Language Syntax

4.4.1
Overall Structure

Programs written in assembly language consist of a sequence of source statements. Each source statement consists of a sequence of ASCII characters ending with a carriage return. Each source statement may include up to four fields: a label, an operation (instruction mnemonic or assembler directive), an operand, and a comment. We use pseudo-op codes in our source code to give instructions to the assembler itself. The `equ` is an assembly directive and the `ldaa` is a regular machine instruction.

```
PORTA   equ    $0000   ; Assembly time constant
Inp     ldaa   PORTA   ; Read data from fixed address I/O data port
```

An assembly language statement contains the following fields:

Label Field can be used to define a symbol,
Operation Field defines the operation code or pseudo-op,
Operand Field specifies either the address or the data,
Comment Field allows the programmer to document the software.

Sometimes not all four fields are present in an assembly language statement. A line may contain just a comment. The entire line is considered a comment if the first character of the line is an asterisk (*) or a semicolon (;). For example,

```
; This line is a comment
* This is a comment too
    * This line is a comment
```

Instructions with inherent mode addressing do not have an operand field. For example,

```
label clra comment
      deca comment
      cli  comment
      inca comment
```

> *Recommendation: For small programs, you should enable automatic assembly colors. The editor will then immediately and continuously color each field according to its type.*

> *Recommendation: For large programs, you should disable automatic assembly colors, because the system will run too slowly. Instead, use the assembler to color the source code explicitly each time the program is assembled.*

4.4.2
Label Field

The label field occurs as the first field of a source statement. The label field can take one of the following three forms:

A. An asterisk (*) or semicolon (;) as the first character in the label field indicates that the rest of the source statement is a comment. Comments are ignored by the assembler, and are printed on the source listing only for the programmer's information. Examples:

```
* This line is a comment
; This line is also a comment
```

B. A white-space character (blank or tab) as the first character indicates that the label field is empty. The line has no label and is not a comment. These assembly lines have no labels:

```
    ldaa 0
    rmb 10
```

C. A symbol character as the first character indicates that the line has a label. Symbol characters are the upper or lower case letters a–z, digits 0–9, and the special characters, period (.), dollar sign ($), and underscore (_). Symbols consist of at least one and at most 99 characters, the first of which must be alphabetic or the special characters period (.) or underscore (_). All characters are significant and upper and lower case letters are distinct.

A label may occur only once in the program. If a label does occur more than once in a program, then each reference to that label will be flagged with an error. The exception to this rule is the set pseudo-op that allows you to define and redefine the same label. We typically use set to define the stack offsets for the local variables in a subroutine. The set pseudo-op allows two separate subroutines to re-use the same name for their local variables.

With the exception of the equ = and set directives, a label is assigned the value of the program counter of the first byte of the instruction or data being assembled. The value assigned to the label is absolute. Labels may optionally be ended with a colon (:). If the colon is present, it is not part of the label but merely acts to set the label off from the rest of the source line. Thus, the following code fragments are equivalent:

```
here: deca
      bne    here
here  deca
      bne    here
```

A label may appear on a line by itself. The assembler interprets this as "Set the value of the label equal to the current value of the program counter." A label may occur on a line with a pseudo-op. The size of the symbol table depends on the available PC computer memory, but you are typically allowed to have thousands of labels.

4.4.3
Operation Field

The operation field occurs after the label field, and must be preceded by at least one white-space character. The operation field must contain a legal opcode mnemonic or an assembler directive. Upper case characters in this field are converted to lower case before being checked as a legal mnemonic. Thus nop, NOP, and NoP are recognized as the same mnemonic. Entries in the operation field may be op codes or directives.

Op codes correspond directly to the machine instructions. The operation code includes any register name associated with the instruction. These register names must not be separated from the op code with any white-space characters. Thus `clra` means clear accumulator A, but `clr a` means clear memory location identified by the label `a`. The available instructions depend on the microcomputer you are using.

Directives or *pseudo-ops* are special operation codes known to the assembler that control the assembly process rather than being translated into machine instructions. The directives that **TExaS** supports are described in detail later in this chapter.

4.4.4 Operand Field

The interpretation of the operand field is dependent on the contents of the operation field. The operand field, if required, must follow the operation field, and must be preceded by at least one white-space character. The operand field may contain a symbol, an expression, or a combination of symbols and expressions separated by commas. There can be no white spaces in the operand field. For example, the following two lines produce identical object code because of the space between "data" and "+" in the first line:

```
ldaa    data + 1
ldaa    data
```

Observation: *Some assemblers allow spaces within the operand field, and then require that a semicolon (;) be placed before each comment.*

The operand field of machine instructions is used to specify the addressing mode of the instruction, as well as the operand of the instruction. Table 4.1 summarizes the operand field formats on the 6811 and 6812. On the 6811 the index register, `idx`, is X or Y, but on the 6812 it can be X, Y, SP, or PC.

Table 4.1 Example operands for the 6811 and 6812.

Operand	Format	Example
no operand	inherent	`clra`
<expression>	direct, extended, or relative	`ldaa 4`
#<expression>	immediate	`ldaa #4`
<expression>,idx	indexed with address register	`ldaa 4,x`
<expr>,#<expr>	bit set or clear	`bset 4,#$01`
<expr>,#<expr>,<expr>	bit test and branch	`brset 4,#$01,there`
<expr>,idx,#<expr>,<expr>	bit test and branch	`brset 4,x,#$01,there`

The 6812 assembly language includes some additional operand formats, as shown in Table 4.2. The accumulator offset, `acc`, is A, B or D, and the index register, `idx`, is X, Y, SP, or PC. The PC is not allowed with any of the predecrement, postdecrement, preincrement, or postincrement addressing modes. The 6812 `movb` and `movw` instructions require two operands separated by a comma, e.g., `movb 2,y,1,x`.

Table 4.2
Additional example
operands for the 6812.

Operand	Format	Example
<expression>,idx+	indexed, post increment	ldd 2,SP+
<expression>,idx-	indexed, post decrement	ldaa 4,Y-
<expression>,+idx	indexed, pre increment	ldaa 4,+X
<expression>,-idx	indexed, pre decrement	staa 1,-SP
acc,idx	accumulator offset indexed	ldaa A,X
[<expression>,idx]	indexed indirect	ldaa [4,X]
[D,idx]	RegD indexed indirect	ldaa [D,Y]

The valid syntax of the operand field depends on the microcomputer. For a detailed explanation of the instructions and their addressing modes, see the Help system with the **TExaS** application.

4.4.5
Expressions

An expression is a combination of symbols, constants, algebraic operators, and parentheses. The expression is used to specify a value that is to be used as an operand. Expressions may consist of symbols, constants, or the character '*' (denoting the current value of the program counter) joined together by one of the operators: + - * / % & | ^.

+	add
−	subtract
*	multiply
/	divide
%	remainder after division
&	bitwise and
|	bitwise or
^	bitwise exclusive or

Expressions may include parentheses and other expressions. Expressions are evaluated using the standard arithmetic precedence. Evaluation occurs left to right for multiple operations with the same precedence. The precedence follows standard mathematic conventions, as shown in Table 4.3. Arithmetic is carried out in signed 32-bit twos-complement integer precision (on the IBM PC).

Table 4.3
Operator precedence.

Precedence	Operation
Highest	parentheses
2	unary + - ~
3	binary * / % &
lowest	binary + - ^ |

> *Maintenance Tip:* It is good programming practice to add parentheses even if they are not necessary in order to clarify the operation. E.g., (A&B)|(C&D) is clearer than A&B|C&D.

Each symbol is associated with a 16-bit integer value that is used in place of the symbol during the expression evaluation. The asterisk (*) used in an expression as a symbol represents the current value of the location counter (the first byte of a multi-byte instruction.)

Constants represent numbers that do not vary in value during the execution of a program. Constants may be presented to the assembler in one of four formats: decimal, hexadecimal, binary, or ASCII. The programmer indicates the number format to the assembler with the following prefixes:

`0x`	hexadecimal, C syntax
`$`	hexadecimal, assembly syntax
`%`	binary
`'c'`	ASCII code for a single letter 'c'

Unprefixed constants are interpreted as decimal. The assembler converts all constants to binary machine code and are displayed in the assembly listing as hexadecimal. A decimal constant consists of a string of numeric digits. The value of an 8-bit decimal constant ranges from -128 to 255. The value of a 16-bit decimal constant must fall in the range from -32768 to 65535. Some valid decimal constants are `12`, `1235`, and `-3200`. Table 4.4 shows some invalid decimal constants.

Table 4.4
Invalid decimal constants.

Invalid	Reason Invalid
`12,456`	illegal symbol
`12.3`	invalid character
`- 5`	spaces not allowed
`123456`	too many digits

A hexadecimal constant consists of a maximum of four characters from the set of digits (0–9) and the alphabetic letters (A–F), and is preceded by a dollar sign (`$`). Hexadecimal constants must be in the range $0000 to $FFFF. Some valid hexadecimal constants are `-$12`, `$ABCD`, and `$001f`. Table 4.5 shows some invalid hexadecimal constants.

Table 4.5
Invalid hexadecimal constants.

Invalid	Reason Invalid
`ABCD`	no preceding "$"
`$G2A`	invalid character
`$2F018`	too many digits

A binary constant consists of a maximum of 16 ones or zeros preceded by a percent sign (`%`). Some valid binary constants are `%00101`, `-%1`, and `%10100`. Table 4.6 shows some invalid binary constants.

Table 4.6
Invalid binary constants.

Invalid	Reason Invalid
`1010101`	missing percent
`%1001,1000,1010`	invalid character
`%210101`	invalid digit

A single ASCII character can be used as a constant in expressions. ASCII constants are surrounded by a single quotes (`'`). Any character, except the single quote, can be used as a character constant. Some valid character constants are `'*'`, `'a'`, and `'Q'`. Table 4.7 shows some invalid character constants.

Table 4.7
Invalid character
constants.

Invalid	Reason Invalid
`'Valid'`	too long
`'a`	unterminated
`"Q"`	wrong quote mark

The invalid cases will be identified as syntax errors by the assembler.

Checkpoint 4.5: *What is the value of* `2+4*6/5+1`*?*

Checkpoint 4.6: *The following two expressions evaluate to exactly the same result:* `$0F&'A'|$F0&'0'` *and* `($0F&'A')|$F0&'0')`*. Which is better and why?*

Checkpoint 4.7: *The following two assembly code sequences produce similar results:* `ldaa #5+6` *and* `ldaa #5 adda #6`*. How are they different?*

**4.4.6
Comment Field**

The last field of an assembler source statement is the comment field. This field is optional and is printed on the source listing only for documentation purposes. The comment field is separated from the operand field (or from the operation field if no operand is required) by at least one white-space character. The comment field can contain any printable ASCII characters.

Observation: *Some assemblers require that a semicolon* `(;)` *be placed before each comment.*

As software developers, our goal is to produce code that not only solves our current problem, but can serve as the basis of our future solutions. In order to reuse software we must leave our code in a condition such that future programmers (including ourselves) can easily understand its purpose, constraints, and implementation. Documentation is not something tacked onto software after it is done, but rather a discipline built into it at each stage of the development. We carefully develop a programming style providing appropriate comments. A comment that tells us *why* we perform certain functions is more informative than comments that tell us what the functions are. An example of bad comments would be

```
clr    Flag    Flag=0
sei            Set I=1
ldaa   $1003   Read PortC
```

These are bad comments because they provide no information to help us in the future to understand what the program is doing. An example of good comments would be

```
clr    Flag    Signifies no key has been typed
sei            The following code will not be interrupted
ldaa   $1003   Bit7=1 iff the switch is pressed
```

These are good comments because they make it easier to change the program in the future.

Self-documenting code is software written in a simple and obvious way, such that its purpose and function are self-apparent. To write wonderful code like this, we first must formulate the problem, organizing it into clear, well-defined subproblems. How we break a complex problem into small parts goes a long way making the software self-documenting.

Both the concept of abstraction and modular code address this important issue of software organization.

Maintaining software is the process of fixing bugs, adding new features, optimizing for speed or memory size, porting to new computer hardware, and configuring the software system for new situations. It is the *most important* phase of software development. Flowcharts are effective in the design phase of a project. Flowcharts and software manuals are good mechanisms for documenting programs only when these types of documentation are kept up to date when modifications are made.

We should use careful indenting, and descriptive names for variables, functions, labels, I/O ports. Effective use of `equ` provides explanation of software function without cost of execution speed or memory requirements. A disciplined approach to programming is to develop patterns of writing that you consistently follow. Software developers are unlike short story writers. It is OK to use the same *subroutine outline* over and over again. In Program 4.2, notice the following style issues:

1. Begins and ends with a line of *s
2. States the purpose of the subroutine
3. Gives the input/output parameters, what they mean and how they are passed
4. Different phases (submodules) of the code delineated by a line of –s

Program 4.2
An example use of comments.

```
***************** Max *****************************
*   Purpose: returns the maximum of two 16-bit numbers
*   Inputs: RegX and RegY are two 16-bit unsigned numbers
*   Output: RegX is the maximum of the two inputs
*   Destroyed: CCR
*   Calling sequence
*     ldx  #100    first number
*     ldy  #200    second number
*     jsr  Max
Max       psha           Save registers, that will be modified
          pshb
          pshy
* - - - - - - - - - - - - - - - - - - - - - - - - - - - -
          pshx             first number on the stack
          xgdy             RegD is second number
          tsx              access the stack
          cpd  0,x         which is bigger
          bhs  second      go if second>=first
first     pulx             RegX =first
          bra  end
second    pulx
          xgdx             RegX = second
end
* - - - - - - - - - - - - - - - - - - - - - - - - - - - -
          puly    Restore registers
          pulb
          pula
          rts
***************** End of Max ****************************
```

4.4.7
Assembly Listing

The assembler output includes a listing containing the source program, the object code, and any assembly errors. The listing file by **TExaS** is created when the `TheList.rtf` file is open. Each line of the listing contains a reference line number, the address and bytes assembled, and the original source input line. If an input line causes more than 8 bytes to be output (e.g., a long `fcc` directive), the additional bytes are included in the object code (S19 file or loaded into memory) but not shown in the listing. There are three assembly options; each can be toggled on/off using the **Assembly->Options** command.

(4)	cycles	shows the number of cycles to execute this instruction
[100]	total	gives a running cycle total since last `org` pseudo-op
{PPP}	type	gives the cycle type

The codes used in the cycle type were presented in Chapter 3. The end of the assembly listing contains a symbol table. The symbol table contains the name of each symbol, along with its defined value. Since the `set` pseudo-op can be used to redefine the symbol, the value in the symbol table is the last definition.

4.4.8
Assembly Errors

Programming errors fall into two categories. Simple typing/syntax error will be flagged by the **TExaS** assembler as an **error** when the assembler tries to translate source code into machine code. The more difficult programming errors to find and remove are functional bugs that can be identified during execution, when the program does not perform as expected. Error messages are meant to be self-explanatory. The assembler has a **verbose** (see **Assembler->Options** command) mode that provides more details about the error and suggests possible solutions. The assembler error types are as follows:

1. *Label previously defined error:* the same label occurs multiple times
   How to fix: check spelling of all the labels
2. *Undefined opcode error:* operation does not exist
   How to fix: check the spelling/availability of the instruction, verify the correct processor is being used
3. *Operand error:* syntax error within the operand
   How to fix expression error: check parentheses, start with a simpler expression
   How to fix undefined symbol: check spelling of both the definition and access
   How to fix addressing mode error: look up the addressing modes available for the instruction
4. *Phasing error:* the value of a symbol changes from pass1 to pass2
   How to fix: first remove any undefined symbols, then remove forward references
   If you really need a forward reference: use > and < to force extended or direct addressing
5. *Can't program address error*
   How to fix: use the `org` pseudo-op to match available memory
6. *Branch too far error:* Destination address is too far away to use 8-bit PC-relative addressing
   How to fix on the 6811: rework code to use the `jmp` or `jsr` instruction
   How to fix on the 6812: switch to long branch version of the instruction

Error diagnostic messages are placed in the listing file just after the line containing the error. If there is no `TheList.rtf` file, then assembly errors are reported in `TheLog.rtf` file. If neither `TheList.rtf` or `TheLog.rtf` exist, then assembly errors are not reported.

A phasing error occurs during Pass 2 of the assembler when the address of a label is different from what it was when it was previously calculated. The purpose of Pass 1 of the assembler is to create the symbol table. In order to calculate the address of each assembly line, the assembler must be able to determine the exact number of bytes each line will take during Pass 1. For most instructions, the number of bytes required is fixed and easy to calculate, but for other instructions, the number of bytes can vary. A phasing errors occur when the assembler calculates the size of an instruction different in Pass 2 than previously calculated in Pass 1. Sometimes a phasing error occurs on a line further down in the program than where the mistake occurs. A phasing error usually results from the use of forward references. In this first example, the symbol "size" is not available at the time of assembling the ldaa size. The assembler incorrectly chooses the extended addressing mode version rather than the correct direct mode. One solution is to move the variables to the top, and a second solution is to force direct mode using ldaa <size.

```
     ldaa size
;  ...
     org  0
size fcb  5
```

In this 6812 example, the symbol "index" is not available at the time of assembling the ldaa index,x. The assembler incorrectly chooses the 2-byte IDX addressing mode version rather than the correct 3-byte IDX1 mode.

```
       ldaa  index,x
index  equ 100
;  ...
loop   ldaa #0
```

The listing shows the phasing error

```
$0000 A6E064         ldaa  index,x
$0064              index equ   100
         ;  ...
$0003 8600    loop  ldaa  #0
#####
Phasing error
This line was at address $0002 in pass 1, now in pass 2 it is
$0003
***************Symbol Table********************
index  $0064
loop   $0002
##### Assembly failed, 1 errors!
```

When the assembler gets to loop, the Pass 1 and Pass 2 values are off by one, causing a phasing error at the loop ldaa #0 instruction. The solution here to simply put the index equ 100 first.

> *Observation: The assembler must be able to accurately determine the object code size of each instruction during pass 1.*

4.4.9 Assembler Pseudo-ops

Pseudo-ops are specific commands to the assembler that are interpreted during the assembly process. An alternative name for pseudo-op is **assembly directive**. A few of them create object code, but most do not. There are many assemblers available for developing

Motorola assembly code. Although they all use the standard Motorola op codes, the spelling of the pseudo-op codes varies. The **TExaS** assembler supports many of the various dialects. The pseudo-op codes supported by this assembler are shown in Table 4.8. If you plan to export software developed with **TExaS** to another application, then you should limit your use to only the pseudo-ops compatible with that application. Group A is supported by Motorola's MCUez, and HiWare (now Metroworks). Group B is supported by Motorola's DOS level AS05, AS08, AS11 and AS12. Group C are used by ImageCraft's ICC11 and ICC12.

Table 4.8 Assembly directives supported by **TExaS**.

Group A	Group B	Group C	Meaning
org	org	.org	Specific absolute address to put subsequent object code
=	equ		Define a constant symbol
	set		Define or redefine a constant symbol
dc.b db	fcb	.byte	Allocate byte(s) of storage with initialized values
	fcc		Create an ASCII string (no termination character)
dc.w dw	fdb	.word	Allocate word(s) of storage with initialized values
dc.l dl		.long	Allocate 32-bit long word(s) of storage with initialized values
ds ds.b	rmb	.blkb	Allocate bytes of storage without initialization
ds.w		.blkw	Allocate bytes of storage without initialization
ds.l		.blkl	Allocate 32-bit words of storage without initialization
end	end	.end	Signifies the end of the source code (**TExaS** ignores these)

Equate Symbol to a Value

```
<label> equ <expression> (<comment>)
<label> = <expression> (<comment>)
```

The equ (or =) directive assigns the value of the expression in the operand field to the label. The equ directive assigns a value other than the program counter to the label. The label cannot be redefined anywhere else in the program. The expression cannot contain any forward references or undefined symbols. Equates with forward references are flagged as a phasing error.

Program 4.3
A constant implemented with equ might make the program easier to change.

```
; 6811
        org   0
size equ   5
data rmb   size
        org   $F000
sum  ldaa #size
     ldx  #data
     clrb
loop addb 0,x
     inx
     deca
     bne  loop
     rts
```

```
; 6812
        org   $800
size equ   5
data rmb   size
        org   $F000
sum  ldaa #size
     ldx  #data
     clrb
loop addb 1,x+
     dbne A,loop
     rts
```

The `equ` pseudo-op is used to define the I/O ports, and to access the elements of a data structure. More information about data structures can be found in Chapter 10.

> **Programming tip:** *Use* `equ` *definitions only if it makes the program easier to understand, to debug, or to change.*

Redefinable Equate
Symbol to a Value

```
<label> set <expression> (<comment>)
```

The `set` directive assigns the value of the expression in the operand field to the label. The `set` directive assigns a value other than the program counter to the label. Unlike the `equ` pseudo-op, the label can be redefined within the program. Although allowed, it is probably a mistake to use forward references. The use of this pseudo-op with forward references will not be flagged with a phasing error. In Program 4.4, the local variable names created with the `set` directive could be reused in another subroutine. More information about local variables can be found in Chapter 8.

Program 4.4 Simple functions with local variables using `set`.

```
; 6811                                   ; 6812
; *****binding phase**************       ; *****binding phase**************
I    set   0                            I    set   0
J    set   1                            J    set   1
; ******allocation phase *********      ; ******allocation phase *********
function pshx    save old Reg X         function
     des         allocate                    leas -2,sp allocate I,J
     des         allocate J             ; *******access phase ***********
     tsx         create stack frame pointer   clr  I,sp   Clear I
; *******access phase ***********            ldab I,sp   Reg B is a copy of I
     clr  I,x    Clear I                      staa J,sp   store into J
     ldab I,x    Reg B is a copy of I   ; *******deallocation phase *****
     staa J,x    store into J                leas 2,sp   deallocate J,I
; ********deallocation phase *****            rts
     ins         deallocate J
     ins         deallocate I
     pulx        restore old X
     rts
```

Form Constant Byte

```
(<label>) fcb   <expr>(,<expr>, . . . ,<expr>) (<comment>)
(<label>) dc.b  <expr>(,<expr>, . . . ,<expr>) (<comment>)
(<label>) db    <expr>(,<expr>, . . . ,<expr>) (<comment>)
(<label>) .byte <expr>(,<expr>, . . . ,<expr>) (<comment>)
```

The `fcb` directive may have one or more operands separated by commas. The value of each operand is truncated to eight bits, and is stored in a single byte of object code. Multiple operands are stored in successive bytes. The operand may be a numeric constant, a character constant, a symbol, or an expression. If multiple operands are present, one or more of them can be null (two adjacent commas), in which case a single byte of zero will be assigned for that operand. If an operand is larger than the range of an 8-bit number (-128 to $+255$), the result is truncated without a warning, and the least significant 8 bits are used.

A string can be included, which is stored as a sequence of ASCII characters. The delimiters supported by **TExaS** are ″, ′, and \. The string does not include a null-termination, so if desired, the programmer must explicitly terminate it. The following three examples produce identical null-terminated strings:

```
str1 fcb "Hello World",0
str2 fcb 'Hello World',0
str3 fcb \Hello World\,0
```

The stepper motor controller shown in Program 4.5 uses the `fcb` definitions to store the four stepper motor output values.

Program 4.5   A stepper motor controller using `fcb`.

```
; 6811                                   ; 6812
size  equ  4                             size  equ  4
PORTB equ $1004     PB3-PB0 to stepper   PORTB equ $0001     PB3-PB0 to stepper
      org $F000                          DDRB  equ $0003
main  ldaa #size                               org $F000
      ldx  #steps                        main  movb #$FF,DDRB    PB3-PB0 outputs
step  ldab 0,x                           run   ldaa #size
      inx                                      ldx  #steps
      stab PORTB       step the motor    step  movb 1,x+,PORTB  step the motor
      deca                                     dbne A,step
      bne  step                                bra  run
      bra  main                          steps fcb  5,6,10,9    output sequence
steps fcb  5,6,10,9  output sequence           org  $FFFE
      org  $FFFE                               fdb  main
      fdb  main
```

Form Constant Character String

```
(<label>) fcc <delimiter><string><delimiter> (<comment>)
```

The `fcc` directive is used to store ASCII strings into consecutive bytes of memory. The byte storage begins at the current program counter. The label is assigned to the first byte in the string. Any of the printable ASCII characters can be contained in the string. The string is specified between two identical delimiters. The first nonblank character after the `fcc` directive is used as the delimiter. The delimiters supported by **TExaS** are ", ', and \. The following are examples:

```
LABEL1  FCC  'ABC'
LABEL2  fcc  "Jon Valvano "
LABEL4  fcc  /Welcome to FunCity!/
```

The first line creates the ASCII characters **ABC** at location LABEL1. Be careful to position the `fcc` code away from executable instructions. The assembler will produce object code as it would for regular instructions, one line at a time. For example, the following would crash because, after executing the `ldx` instruction, the 6811 would try to execute the ASCII characters "Trouble" as instructions:

```
      ldaa 100
      ldx  #Strg
Strg fcc  "Trouble"
```

Typically we collect all the `fcc`, `fcb`, `fdb` together and place them at the end of our program, so that the microcomputer does not try to execute the constant data. For example, we might have

```
Loop  ldaa Con8
      ldy  Con16
      ldx  #Strg
      bra  loop
* Since the bra loop is unconditional,
* the computer won't go beyond this point.
Strg  fcc  "No Trouble"
Con8  fcb  100
Con16 fdb  1000
```

The ASCII string generated by `fcc` is not null-terminated, so if a termination is needed, you must add it explicitly using either

```
Strg1 fcc  "happy"
      fcb  0
```

or

```
Strg2 fcb  "happy",0
```

Form Double Byte

```
(<label>) fdb   <expr>(,<expr>,...,<expr>) (<comment>)
(<label>) dc.w  <expr>(,<expr>,...,<expr>) (<comment>)
(<label>) dw    <expr>(,<expr>,...,<expr>) (<comment>)
(<label>) .word <expr>(,<expr>,...,<expr>) (<comment>)
```

The `fdb` directive may have one or more operands separated by commas. The 16-bit value corresponding to each operand is stored into two consecutive bytes of the object program. The storage begins at the current program counter. The label is assigned to the first 16-bit value. Multiple operands are stored in successive bytes. The operand may be a numeric constant, a character constant, a symbol, or an expression. If multiple operands are present, one or more of them can be null (two adjacent commas), in which case two bytes of zeros will be assigned for that operand. The `fdb` has been used many times so far in the book to define the reset vector.

```
 org  $FFFE
 fdb  main
```

Define 32-bit Constant

```
(<label>) dc.l  <expr>(,<expr>,...,<expr>) (<comment>)
(<label>) dl    <expr>(,<expr>.,<expr>) (<comment>)
(<label>) .long <expr>(,<expr>,...,<expr>) (<comment>)
```

The `dl` directive may have one or more operands separated by commas. The 32-bit value corresponding to each operand is stored into four consecutive bytes of the object program (big endian). The storage begins at the current program counter. The label is assigned to the first 32-bit value. Multiple operands are stored in successive bytes. The operand may be a numeric constant, a character constant, a symbol, or an expression. If multiple operands are present, one or more of them can be null (two adjacent commas), in which case four bytes of zeros will be assigned for that operand. In the following finite state machine, the `dl` definitions are used to define 32-bit constants:

```
S1  dl    100000,$12345678
S2  .long 1,10,100,1000,10000,100000,1000000,10000000
S3  dc.l  -1,0,1
```

Set Program
Counter Origin

```
org  <expression> (<comment>)
.org <expression> (<comment>)
```

The `org` directive changes the program counter to the value specified by the expression in the operand field. Subsequent statements are assembled into memory locations starting with the new program counter value. If no `org` directive is encountered in a source program, the program counter is initialized to zero. Expressions cannot contain forward references or undefined symbols. The `org` statements in Program 4.4 place the variables in RAM and the programs in EEPROM of a MC68HC812A4. The `org` statement is also used to set the reset vector.

Reserve Multiple Bytes

```
(<label>) rmb   <expression> (<comment>)
(<label>) ds    <expression> (<comment>)
(<label>) ds.b  <expression> (<comment>)
(<label>) .blkb <expression> (<comment>)
```

The `rmb` directive causes the location counter to be advanced by the value of the expression in the operand field. This directive reserves a block of memory the length of which in bytes is equal to the value of the expression. The block of memory reserved is not initialized to any given value. The expression cannot contain any forward references or undefined symbols. This directive is commonly used to reserve a scratchpad or table area for later use.

Checkpoint 4.8: Why can't you use a forward reference in an `rmb` directive?

Reserve Multiple Words

```
(<label>) ds.w  <expression> (<comment>)
(<label>) .blkw <expression> (<comment>)
```

The `ds.w` directive causes the location counter to be advanced by 2 times the value of the expression in the operand field. This directive reserves a block of memory the length of which in words (16-bit) is equal to the value of the expression. The block of memory reserved is not initialized to any given value. The expression cannot contain any forward references or undefined symbols. This directive is commonly used to reserve a scratchpad or table area for later use.

ds.l Reserve Multiple
32-bit Words

```
(<label>) ds.l  <expression> (<comment>)
(<label>) .blkl <expression> (<comment>)
```

The `ds.l` directive causes the location counter to be advanced by 4 times the value of the expression in the operand field. This directive reserves a block of memory the length of which in words (32-bit) is equal to the value of the expression. The block of memory reserved is not initialized to any given value. The expression cannot contain any forward references or undefined symbols. This directive is commonly used to reserve a scratchpad or table area for later use.

End of Program
(Optional)

```
end  (<comment>)
.end (<comment>)
```

This directive signifies the end of the source code. The **TExaS** assembler will ignore these pseudo operation codes. Some other assemblers require one of these directives at the end of every program.

4.5 S-19 Object Code

The S-19 record encodes program and data object modules into a printable format. This allows viewing of the object file with standard tools and allows display of the module while transferring from one computer to the next or during loads between a host and target. The S-record format also includes information for use in error checking to insure the integrity of data transfers.

S-Records are character strings made of several fields that identify the record type, record length, memory address, code/data, and checksum. Each byte of binary data is encoded as a two-character hexadecimal number: the first character representing the high-order 4 bits, and the second the low-order 4 bits of the byte.

The five fields that make up an S-record are

1. Type S0, S1 or S9,
2. Record Length,
3. Address,
4. Code/Data,
5. Checksum.

Eight types of S-records have been defined to accommodate various encoding, transportation, and decoding needs, but only three **types** are used in most Motorola microcontrollers. The *S0 record* is a title record containing the ASCII name of the file in the Code/Data field. The address field of this type is usually 0000. The *S1 record* is a data record containing the information to be loaded sequentially starting at the specified address. The *S9 record* is an end-of-file marker, and sometimes contains the starting address to begin execution. In an embedded microcomputer environment, the starting address must be programmed at the appropriate place. The **Record Length** contains the count of the character pairs in the length record, excluding the type and record length. For S0, S1, S9 record types, the **Address** field is a 4-byte value. For the S1 record type the address specifies where the data field is to be loaded into memory. There are from 0 to n bytes in the **Code/Data** field. This information contains executable code, memory loadable data, or descriptive information. The **Checksum** field is two ASCII characters used for error checking. The least significant byte of the one's complement of the sum of the values represented by the pairs of characters making up the record length, address, and the code/data fields. When generating a checksum, one adds (call the result **sum**) the record length, address and code/data field using 8-bit modulo arithmetic (ignoring overflows).

The checksum is calculated

```
checksum = $FF - sum.
```

When verifying a checksum, one adds (call the result **sum**) the record length, address code/data field and checksum using 8-bit modulo arithmetic (ignoring overflows). The **sum** should be $FF. Each record may be terminated with a CR/LF/NULL. The following is a typical S-record module:

```
S1130000285F245F2212226A000424290008237C2A
S11300100002000800082629001853812341001813
S113002041E900084E42234300182342000824A952
S107003000144ED492
S9030000FC
```

The module consists of four code/data records and an S9 termination record. The first S1 code/data record is explained as follows:

> S1 S-record type S1, indicating a code/data record to be loaded/verified at a 2-byte address.
> 13 Hex 13 (decimal 19), indicating 19 character pairs, representing 19 bytes of binary data, follow.
> 0000 Four-character 2-byte address: hex address 0000, indicates location where the data are to be loaded.
> The next 16 character pairs are the ASCII bytes of the actual program code/data
> 2A Checksum of the first S1 record.

The second and third S1 code/data records each also contain $13 character pairs and are ended with checksums. The fourth S1 code/data record contains 7 character pairs. The S9 termination record is explained as follows:

> S9 S-record type S9, indicating a termination record.
> 03 Hex 03, indicating three character pairs (3 bytes) to follow.
> 0000 Four-character 2-byte address field, zeroes.
> FC Checksum of S9 record.

Checkpoint 4.9: What loader operation is caused by the following S19 record?

S107F026F08020FE54

Checkpoint 4.10: Is the `checksum` correct in the following S19 record?

S10CF0DD80CCF0E616F4BC20B876

Checkpoint 4.11: Create an S19 record that stores the value $1234 into location $5678.

4.6

Tutorial 4. Editing and Assembling

To illustrate the assembly process, we will hand assemble the software given in Program 4.6.

Program 4.6
Simple program used to study the assembly process.

```
; MC68HC11A8                       ; MC68HC812A4
PORTC   equ  $1003                 PORTC equ  $0004
DDRC    equ  $1007                 DDRC  equ  $0006
        org  $E000   ;ROM                org  $F000  ;EEPROM
main    lds  #$00FF ;sp=>RAM       main  lds  #$0C00 ;sp=>RAM
        bsr  init                        bsr  init
loop    staa PORTC   ;output       loop  staa PORTC   ;output
        inca                             inca
        bra  loop                        bra  loop
init    ldaa #$FF                  init  ldaa #$FF
        staa DDRC    ;outputs            staa DDRC     ;outputs
        clra                             clra
        rts                              rts
        org  $FFFE ;ROM                  org  $FFFE  ;EEPROM
        fdb  main  ;reset vector         fdb  main  ;reset vector
```

During pass 1, we need to create the symbol table. The symbol table is a mapping between symbols and their values. In this particular example, the values of PORTC and DDRC are obvious. The values of main, loop, and init will be the memory address of the corresponding line. The org pseudo-op specifies the address of the next line. Otherwise, the instructions will be allocated in order one after another. To determine the line addresses, we need to know the size of each line. The first step is to determine the addressing modes. Recall that direct addressing can be used when accessing information at addresses 0 to $00FF, while extended addressing is required for the other locations.

Question 4.1. Determine the addressing mode for each instruction in Program 4.6.

Question 4.2. Determine the size of each instruction in Program 4.6. We can find the size of an instruction in the Motorola instruction manuals on the CD. At this point, all we need to know is the number of bytes required to encode each instruction for that particular addressing mode. Some pseudo-ops, like equ and org, do not create object code, so these lines do not have a size. On the other hand, other pseudo-ops, like fdb, do create object code, so they will have sizes.

Question 4.3. Finish pass 1 by creating the symbol table. The value for the lines with the equ pseudo-op is simply the value of the operand field. The value of the other lines is the address of that line.

During pass 2, we create the object code for each line. We can also find object codes in the Motorola instruction manuals which can be found on the CD.

Question 4.4. Determine the object code for each instruction in Program 4.6.

Question 4.5. What trouble would we have during pass 1 if the equ pseudo-ops were placed at the end of the program?

4.7 Homework Assignments

The first four homework assignments in this chapter involve hand assembly. Pass1 contains three steps. The first step is to determine addressing mode for each instruction. Next, you calculate the object code size for the instruction. The third step is to create the symbol table. Pass2 contains two steps. The first step is to determine the object code for each instruction, and the second step is to calculate the S19 record.

Homework 4.1: Assemble the assembly language program shown in Program 4.7. Include the symbol table, the address data for each instruction, and the S19 record for the entire program.

Program 4.7 Program for Homework 4.1.

```
;MC68HC11A8                      ;MC68HC812A4
PPROG = 0x103B                   TCNT    = $0084
    org $E000  ;ROM              EEPROT  = $00F1
; **********Erase EEPROM***      EEPROG  = $00F3
EraseAll:                        EraseAll = $06  ;bulk erase all
    ldab #0x06                   EraseWord= $16  ;erase 16-bit word
    stab PPROG                   EraseRow = $0E  ;erase one row
    stab 0xB600                  ProgWord = $02  ;prgram 16-bit word
    incb                             org $F000  ;EEPROM
    stab PPROG                   ; **********program EEPROM********
```

Continued on p. 168

Continued from p. 167

```
        bsr  Wait10ms                    ; RegX =address to change
        clr  PPROG                       ; RegY =data, returns data written
        rts                              ; RegA has command $06,$16,$0E,$02
; **********program 1 byte***           Prog:  ldab  EEPROT
; RegX points into $B600-$B7FF                 pshb
; RegA =data, returns data written             clr   EEPROT
Prog: ldab #0x02                               staa  EEPROG
      stab PPROG                               sty   0,x
      staa 0,x                                 inca
      incb                                     staa  EEPROG
      stab PPROG                               bsr   Wait10ms
      bsr  Wait10ms                            clr   EEPROG
      clr  PPROG                               pulb
      ldaa 0,x                                 stab  EEPROT
      rts                                      ldy   0,x
; *****Wait for 10 ms******                    rts
Wait10ms: pshx                           ; *****Wait for 10 ms******
          ldx  #3334                     Wait10ms:
wloop: dex                                      ldd   TCNT
       bne  wloop                               addd  #10000
       pulx                              wloop: cpd   TCNT
       rts                                      bpl   wloop
                                                rts
```

Homework 4.2: Assemble the assembly language program shown in Program 4.8. Include the symbol table, the address data for each instruction, and the S19 record for the entire program.

Program 4.8
Program for Homework 4.2.

```
; MC68HC11A8                          ; MC68HC812A4
PORTB equ  $1004                      PORTB equ  $0001
      org  $0000   ;RAM               DDRB  equ  $0003
last  rmb  1       ;global                  org  $0800   ;RAM
      org  $E000   ;ROM               last  rmb  1       ;global
main  lds  #$00FF ;sp=>RA                   org  $F000   ;EEPROM
      ldaa #1                         main  lds  #$0C00 ;sp=>RAM
      staa last                             movb #1,last
loop  bsr  out                              movb #$FF,DDRB ;make output
      bra  loop                       loop  bsr  out
out   ldaa last                             bra  loop
      rora                            out   ldaa last
      staa PORTB ;output                    rora
      staa last                             staa PORTB ;output
      rts                                   staa last
      org $FFFE ;start vector               rts
      fdb main                              org $FFFE ;start vector
                                            fdb main
```

Homework 4.3: Assemble the assembly language program shown in Program 4.9. Include the symbol table, the address data for each instruction, and the S19 record for the entire program.

Program 4.9
Program for Homework
4.3.

```
; MC68HC11A8
      org  $E000 ;ROM
;output 8 bits, Reg A=data
out8  ldab #8         ;8 bits
      ldx  #$1000
clop  bclr 8,x,#$10 ;CLK=0
      lsra            ;lsb first
      bcc  set0
      bset 8,x,#$08 ;DQ=1
      bra  next
set0  bclr 8,x,#$08 ;DQ=0
next  bset 8,x,#$10 ;CLK=1
      decb
      bne  clop
      rts
```

```
; MC68HC812A4/MC68HC912B32
PORTS equ  $00D6  ;I/O port
DDRS  equ  $00D7  ;Direction Reg
      org  $F000  ;EEPROM
out8  ldab #8          ;8 bits
clop  bclr PORTS,#$40 ;CLK=0
      lsra            ;lsb first
      bcc  set0
      bset PORTS,#$20 ;DQ=1
      bra  next
set0  bclr PORTS,#$20 ;DQ=0
next  bset PORTS,#$40 ;CLK=1
      decb
      bne  clop
      rts
```

Homework 4.4: Assemble the assembly language program shown in Program 4.10. Include the symbol table, the address data for each instruction, and the S19 record for the entire program.

Program 4.10   Program for Homework 4.4.

```
; MC68HC11A8
      org  $E000 ;ROM
;input 8 bits, Reg A=data
in8   ldy  #8         ;8 bits
      ldx  #$1000
      bclr 9,x,#$08 ;DQ input
ilop  bclr 8,x,#$10 ;CLK=0
      lsra            ;lsb first
      brclr 8,x,#$08,inext
      oraa #$80 ;DQ=1
inext bset 8,x,#$10 ;CLK=1
      dey
      bne  ilop
      bset 9,x,#$08 ;DQ output
      rts
```

```
; MC68HC812A4/MC68HC912B32
PORTS equ  $00D6  ;I/O port
DDRS  equ  $00D7  ;Direction Reg
      org      $F000  ;EEPROM
;input 8 bits, Reg A=data
in8   ldx  #8         ;8 bits
      bclr DDRS,#$20 ;DQ input
ilop  bclr PORTS,#$40 ;CLK=0
      lsra            ;lsb
      brclr PORTS,#$20,inext
      oraa #$80       ;DQ=1
inext bset PORTS,#$40 ;CLK=1
      dex
      bne  ilop
      bset DDRS,#$20 ;DQ output
      rts
```

Homework 4.5: Write a subroutine to convert a null-terminated string to upper case. In particular, convert all lower case ASCII characters to upper case. The calling sequence is

```
ldx  #string   ; pointer to ASCII string
jsr  UpperCase
```

Homework 4.6: Write a subroutine to convert a null-terminated string to lower case. In particular, convert all upper case ASCII characters to lower case. The calling sequence is

```
ldx  #string   ; pointer to ASCII string
jsr  LowerCase
```

Homework 4.7: Write a subroutine that compares two null-terminated strings. Register A will be 0 if the strings do not match and nonzero if the strings match. The calling sequence is

```
ldx  #string1  ; pointer to first string
ldy  #string2  ; pointer to second string
jsr  StringCompare
```

4.8   Laboratory Assignments

Lab 4.1
Logic Function

Purpose
The purpose of this laboratory is to familiarize you with the software development steps using the **TExaS** Simulator. The specific function you will implement is logical AND, i.e., P=M&N.

Description
Part A.   Begin by running the `ReadMe.exe` instructional movie that you can find on the CD. In particular, you should learn how to start the **TExaS** simulator, access on-line Help, open, edit, print, and save source files, interface switches and LEDs to the microcomputer, assemble source files into listing files, run the simulator, observe registers memory and stack, single step, set breakpoints, stop execution, back up your files on floppy disks and quit.

Part B.   Use the simulator to create three files. `Lab4.rtf` will contain the assembly source code. `Lab4.uc` will contain the microcomputer configuration. `Lab4.io` will define the external connections. Specify the microcomputer, and connect two switches and one LED as specified in Figure 4.8. You will connect switches to Port A bits 0 and 1, and interface a LED to PA3. PA0 and PA1 will be input ports, and PA3 will be an output port. The switches should be labeled N and M, and the LED should be labeled P. When N switch is "off" or open position, PA0 will be "0". For this situation, your software will consider N to be false. When the N switch is "on" or closed position, PA0 will be "1". In this case, your software will consider N to be true. The M switch, which is connected to PA1, will operate in a similar fashion. When your software writes a "1" to PA3, the LED will activate.

Figure 4.8
Hardware circuit for
Lab 4.1.

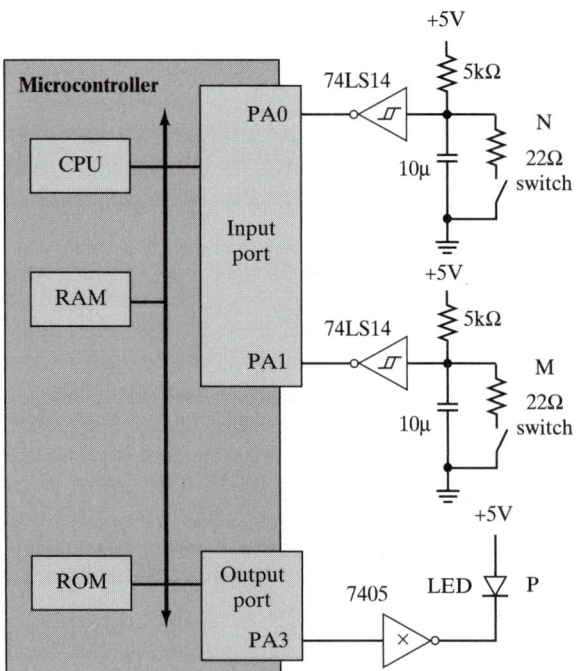

Part C.   You will write assembly code that inputs from PA0 and PA1, and outputs to PA3. Your software will calculate P to be the logical AND of N and M. If P is true, then the LED should be turned on, by writing a "1" to PA3. Otherwise the LED should be turned off, by writing "0" to PA3. The C program shown as Program 4.11 describes the software algorithm.

Program 4.11 The first C program to illustrate Lab 4.1.

```
unsigned char N,M,P;
void main(void){     // make PA3 is output, PA1,PA0 are inputs
  DDRA=0x08;          // this step is not needed for a 6811
  while(1){
    if(PORTA&0x01)
      N=1;            // N is true if PA0 is 1
    else
      N=0;            // N is false if PA0 is 0
    if(PORTA&0x02)
      M=1;            // M is true if PA1 is 1
    else
      M=0;            // M is false if PA1 is 01
    P=N&M;            // logical AND
    if(P)
      PORTA=0x08;     // P is 1, so turn LED on
    else
      PORTA=0;        // P is 0, so turn LED off
  }
}
```

There are many ways to implement this system. You are free to implement the system in any way you wish, as long as your software is easy to understand and well documented. In particular, consider the approach shown in Program 4.12.

Program 4.12 The second C program to illustrate Lab 4.1.

```
void main(void){   // make PA3 is output, PA1,PA0 are inputs
  DDRA=0x08;        // this step is not needed for a 6811
  while(1){
    if((PORTA&0x03)==0x03)
      PORTA=0x08; // turn on
    else
      PORTA=0;     // turn off
  }
}
```

This third approach requires no conditional statements. Software without conditional branches is easier to debug, because it has fewer cases to test. See Program 4.13.

Program 4.13 The third C program to illustrate Lab 4.1.

```c
unsigned char N,M,P;
void main(void){  // make PA3 is output, PA1,PA0 are inputs
  DDRA=0x08;      // this step is not needed for a 6811
  while(1){
    N=(PORTA&0x01)<<3;  // N is 8 iff N switch is one
    M=(PORTA&0x02)<<2;  // M is 8 iff M switch is one
    P=N&M;              // P is 8 iff both N and M switches are one
    PORTA=P;            // turn on or off, as needed
  }
}
```

Part D. During the demonstration, you will be asked to run your program to verify proper operation. You should be able to single step your program and explain what your program is doing and why. You need to know how to set and clear breakpoints. You will be asked to look up the meaning of commands like **Mode->FollowPC** using the on-line Help. Be prepared to make changes to Lab4.io, such as changing the names and colors of the switches and LEDs.

Lab 4.2
Logic Function

Same as Lab 4.1, except connect to Port C, and implement the *OR* function, P=M|N. Both the 6811 and 6812 will need to execute DDRC=0x08.

Lab 4.3
Logic Function

Same as Lab 4.1, except connect to Port D, and implement the *EXCLUSIVE OR* function, P=M^N. This is equivalent to the not equals function. P=1 iff M!=N. Both the 6811 and 6812 will need to execute DDRD=0x08.

Lab 4.4
Logic Function

Same as Lab 4.1, except connect to Port C, and implement the *NOR* function, P=not(M|N). Both the 6811 and 6812 will need to execute DDRC=0x08.

Lab 4.5
Logic Function

Same as Lab 4.1, except connect to Port D, and implement the *EXCLUSIVE NOR* function, P=not(M^N). This is equivalent to the equals function. P=1 iff M==N. Both the 6811 and 6812 will need to execute DDRD=0x08.

Lab 4.6
Input/Output Interface to a Stepper Motor

Purpose

The purpose of this laboratory is to develop a microcomputer system that spins a stepper motor.

Description

Part A. Use the simulator to create three files. Stepper.rtf will contain the assembly source code. Stepper.uc will contain the microcomputer configuration. Stepper.io will define the external connections, as shown in Figure 4.9. You should specify the microcomputer and attach one switch and four LEDs. You will connect a switch to Port A bit 0, and interface a stepper motor to Port B (4 signals: PB3, PB2, PB1, PB0). The four stepper motor signals are called B, B', A, and A'.

Part B. You will write assembly code that inputs from Port A, and outputs to Port B. When the input switch is "off" or open position, Port A bit 0 will be "0". For this situation, your software will not change the Port B stepper motor outputs. When the input switch is "on" or closed position, Port A bit 0 will be "1". In this case, your software will output the sequence 5, 6, 10,

Figure 4.9
Hardware circuit for
Lab 4.6.

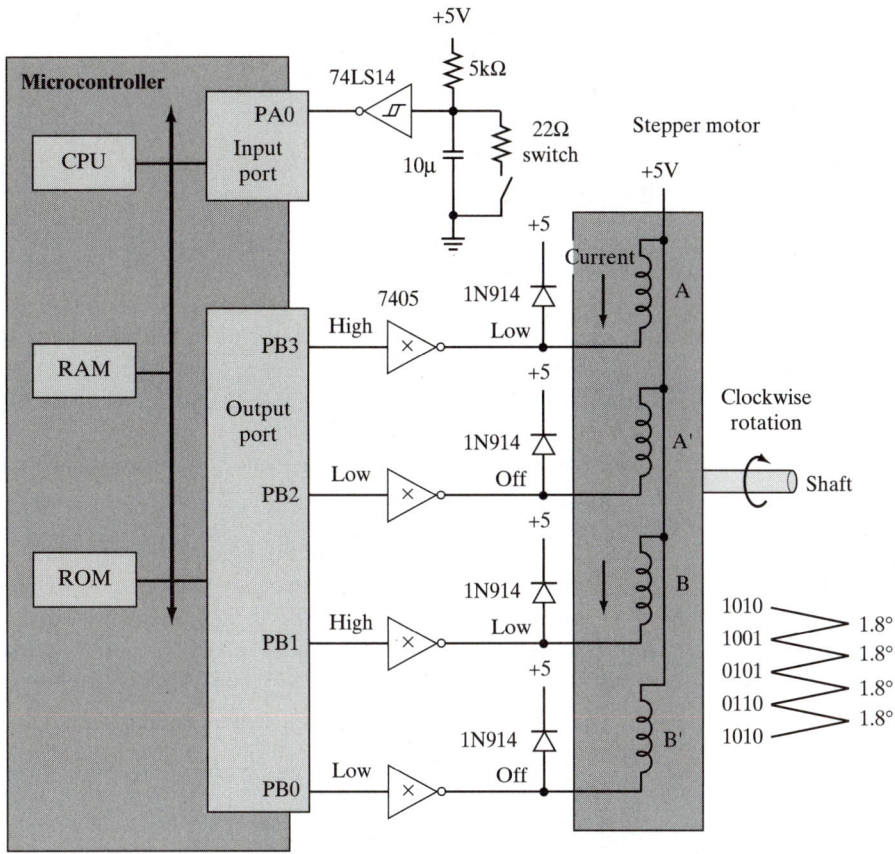

9, 5, 6, 10, 9, . . . over and over again to the stepper motor. The motor will turn 1.8° for every
new output to Port B. The four outputs will be connected to a simulated stepper motor. The
following C program describes the software algorithm.

Program 4.14
The C program to
illustrate Lab 4.6.

```c
unsigned char Angle; // ranges from 0 to 199
void main(void){
  Angle=0;          // initialize global
  DDRA=0;           // make Port A inputs (not needed for a 6811)
  DDRB=0xFF;        // make Port B outputs (not needed for a 6811)
  while(1){
    while((PORTA&0x01)==0){};   // stop if PA0=0, continue if PA0=1
    PORTB=5;   Angle++;
    PORTB=6;   Angle++;
    PORTB=10;  Angle++;
    PORTB=9;   Angle++;
    if(Angle==200)
      Angle=0;
  }
}
```

The software variable `Angle` varies from 0 to 199 as the stepper motor angle varies from
0 to 358°.

Part C.   During the demonstration, you will be asked to run the program to verify proper operation. Be prepared to use the debugger to determine how fast the simulated motor is spinning. Each output to Port B causes a 1.8° step.

Lab 4.7
Hand Assembly
and Execution

Purpose

In this lab you will learn how to hand-assemble source code. During Pass 1 you will create the symbol table. During Pass 2 you will create the object code. Another objective is to understand how the microcomputer executes instructions. For each memory cycle during execution you will predict the R/W line, the 16-bit address, and the 8-bit data bus.

Description

In preparation for this assignment, you should familiarize yourself with the format of the Microcomputer Programming Reference Manual. In particular, you should understand the addressing modes. You need to be able to look up op codes for each instruction. For each instruction you need to determine the object code and CPU execution cycles. Many instructions have multiple addressing modes; each addressing mode has a distinct object code and execution cycles.

Part A.   Pretend you are Pass 1 of the **TExaS** cross-assembler and create the symbol table for the Program 4.15. Labels start in column 1. A symbol table is a list of symbols and their 16-bit unsigned values. There will be an entry in the symbol table for each label. For all labels except equ[1] or set, the value of a symbol is the beginning address of that line. For labels with equ or set, the value of the symbol is the 16-bit value of the operand.

Program 4.15

The assembly program for Lab 4.7.

```
;MC68HC11A8                          ;MC68HC812A4
        org   $0    ; RAM                    org   $900    ; RAM
Result  rmb   2                      Result  rmb   2
Index   rmb   1                      Index   rmb   1
        org   $E000  ; ROM                   org   $F800    ; EEPROM
Main    lds   #$00FF                 Main    lds   #$0C00
        ldy   #data                          ldy   #data
        ldaa  #2                             ldaa  #2
        bsr   Sum                            bsr   Sum
        std   Result                         std   Result
        stop                                 stop
Sum     pshy                         Sum     pshy
        staa  Index                          staa  Index
        ldd   #0                             ldd   #0
SLoop   addd  0,y                    SLoop   addd  2,y+
        Iny                                  dec   Index
        Iny                                  bne   SLoop
        dec   Index                          puly
        bne   Sloop                          rts
        puly                                 org   $FC00
        rts                          data    fdb   13,9
        org   $B600  ; EEPROM                org   $FFFE
data    fdb   13,9                           fdb   Main
        org   $FFFE
        fdb   Main
```

---

[1] In **TExaS**, the pseudo-ops equ and = are identical.

Part B. Pretend you are Pass 2 of the **TExaS** cross-assembler and create the object code for the Program 4.15. Include four fields for each line of assembly code:

the **address** is the 16-bit unsigned hexadecimal location of the start of this line
the **object code** is a group of 8-bit unsigned hexadecimal values
the number of cycles to execute this line (called **Cycles** in the manual)
the execution pattern is called **Access Detail** in the CPU manual

Every line has an address. Some pseudo-ops will create object code (e.g., `fcb`[2] `fdb`[3].) Since pseudo-ops are not executed, no pseudo-op will have values for the **Cycles** or **Access Detail** entries. For example, the 6812 yields

Address	Object Code(s)	Cycles	Access Detail	Source code
$F800				`org  $F800`
$F800	B6 08 00	[3]	rOP	`ldaa $0800`

Part C. Type the source code into the system and run the cross-assembler. Please correct your Part B with a red pen. Please do Parts A and B on paper first, then run the machine.

Part D. Pretend you are the microcomputer and hand execute this program up until the `stop` instruction. Perform the pseudo-execution showing the R/W, 16-bit Address, and 8-bit Data in hexadecimal for each cycle. On the 6811, the simulated execution matches exactly the actual 6811. On the 6812 the pseudo-execution will not match the actual 6812 execution. This is because the 6812 has an instruction pipeline and can actually fetch 16 bits at a time. **TExaS** does not simulate the 6812 instruction pipeline and will always fetch 8 bits. **TExaS** properly simulates the software timing on all its microcomputers. For example, **TExaS** will show the 6812 instruction `ldaa $0800` as the following four pseudo cycles:

```
Read $F800 B6    fetch opcode
Read $F801 08    fetch operand
Read $F802 00    fetch operand
Read $0800 xx    memory read, xx is the contents at $800
```

Nonetheless, the simulated time will be correctly incremented by 3. In fact, all timing aspects of the simulation will be accurate. Add in the comment field at the start of each instruction which instruction is being executed.

Part E. Run this program with the simulator and verify your answers to Part D. Correct any mistakes with a red pen. Please do Part D on paper first, then run the machine.

---

[2] In **TExaS**, the pseudo-ops `fcb` `dc.b` and `dc` are identical.

[3] In **TExaS**, the pseudo-ops `fdb` and `dc.w` are identical.

5 I/O Programming

Chapter 5 objectives are to

❑ describe I/O ports available on the 6811 and 6812,
❑ explain the fundamentals of I/O programming,
❑ give simple examples of I/O device drivers.

An embedded system uses its input/output devices to interact with the external world. Input devices allow the computer to gather information, and output devices can display information. Output devices also allow the computer to manipulate its environment. The tight coupling between the computer and external world distinguishes an embedded system from a regular computer system. The interactions between the computer and its environment are critical for the success of an embedded system. In this chapter, we will cover I/O ports and I/O programming, which represent the first half of the connection between the computer and its environment. I/O ports are the specific components of a microcomputer that allow it to interact with its environment. A **device driver** is a collection of software functions that allow higher-level software to utilize an I/O device. In other words, the set of low-level functions that input/output directly with the hardware are grouped together in a single module and called a device driver. Advances in the number and sophistication of the I/O ports have contributed greatly to the long-term growth of applications of embedded systems. The next chapter contains the second half of the connection, which is the interface between the microcomputer and the external electrical and mechanical devices.

5.1 Basic Concept of an I/O Port Address and Direction Register

On most embedded microcomputers, the I/O ports are *memory mapped*. This means the software accesses an input/output port simply by reading from or writing to the appropriate address. Usually an input port is read only. That means a read-cycle access from the port address returns the current values existing on the inputs. In particular, the tristate driver (triangle-shaped circuit shown in Figure 5.1) will drive the input signals onto the data bus during a read cycle from the port address. A write-cycle access to an input port usually produces no effect.[1] To make our software more readable, we usually include symbolic definitions for the I/O ports. This chapter covers just some of the ports for the 6811 and 6812 microcomputers. For a complete list of I/O ports, refer to the respective data sheets. Since I/O ports exist at specific addresses within the computer, we use assembly language pseudo-

---

[1] An input port on the 8051 is created by writing a "1" to the port address; therefore, writing a "0" to the port will switch the input to an output.

ops to define symbolic names for the ports. Program 5.1 defines some of the parallel I/O ports for the 6811. A full list of I/O ports can be found in the 6811 reference manual. The 6811 ports simulated by **TExaS** are defined in the file `Port11.rtf`.

Program 5.1
Assembly definitions of some of the MC68HC11 I/O ports.

```
PORTA   equ   $1000   PA7 input/output, PA6-3 outputs, PA2-0 inputs
PACTL   equ   $1026   Bit 7(DDRA7) specifies the direction of PA7
PORTB   equ   $1004   PB7-PB0 are all read able outputs
PORTC   equ   $1003   PC7-PC0 can be input or output
DDRC    equ   $1007   Direction register for Port C
PORTD   equ   $1008   PD5-PD0 can be input or output
DDRD    equ   $1009   Direction register for Port D
PORTE   equ   $100A   PE7-PE0 are all inputs
```

Program 5.2 defines these 6811 I/O port addresses in C. The `volatile` qualifier means the contents at that address can change by means other than explicit software action of the current module. It is necessary to add the `volatile` qualifier to prevent the compiler from optimizing a C program that accesses I/O ports. A full list of I/O ports can be found in the file `HC11.h` on the CD.

Program 5.2
C definitions of some of the 6811 I/O ports.

```
#define   PORTA   *(unsigned   char   volatile   *)   (0x1000)
#define   PACTL   *(unsigned   char   volatile   *)   (0x1026)
#define   PORTB   *(unsigned   char   volatile   *)   (0x1004)
#define   PORTC   *(unsigned   char   volatile   *)   (0x1003)
#define   DDRC    *(unsigned   char   volatile   *)   (0x1007)
#define   PORTD   *(unsigned   char   volatile   *)   (0x1008)
#define   DDRD    *(unsigned   char   volatile   *)   (0x1009)
#define   PORTE   *(unsigned   char   volatile   *)   (0x100A)
```

Checkpoint 5.1: On the 6811, TCNT is a 16-bit unsigned counter at address $100E. Write the C code that defines a symbol for this counter.

Program 5.3 defines some of the I/O ports for the MC68HC812A4. A full list of I/O ports can be found in the 6812 reference manual. The 6812 ports simulated by **TExaS** are defined in the file `Port12.rtf`.

Program 5.3
Assembly definitions of some of the 6812 I/O ports.

```
PORTA   equ   $0000   General Purpose Input/Output Port
PORTB   equ   $0001   General Purpose Input/Output Port
DDRA    equ   $0002   Data Direction Register
DDRB    equ   $0003   Data Direction Register
PORTC   equ   $0004   General Purpose Input/Output Port
PORTD   equ   $0005   General Purpose Input/Output Port
DDRC    equ   $0006   Data Direction Register
DDRD    equ   $0007   Data Direction Register
PORTE   equ   $0008   General Purpose Input/Output Port
DDRE    equ   $0009   Data Direction Register
```

Program 5.4 defines these 6812 I/O port addresses in C. A full list of I/O ports can be found in the file `HC12.h` on the CD.

Program 5.4
C definitions of some of the 6811 I/O ports.

```
#define   PORTA  *(unsigned   char   volatile *)  (0x0000)
#define   PORTB  *(unsigned   char   volatile *)  (0x0001)
#define   DDRA   *(unsigned   char   volatile *)  (0x0002)
#define   DDRB   *(unsigned   char   volatile *)  (0x0003)
#define   PORTC  *(unsigned   char   volatile *)  (0x0004)
#define   PORTD  *(unsigned   char   volatile *)  (0x0005)
#define   DDRC   *(unsigned   char   volatile *)  (0x0006)
#define   DDRD   *(unsigned   char   volatile *)  (0x0007)
#define   PORTE  *(unsigned   char   volatile *)  (0x0008)
#define   DDRE   *(unsigned   char   volatile *)  (0x0009)
```

> **Checkpoint 5.2:** On the 6812, TCNT is a 16-bit unsigned counter at address $0086.
> Write the C code that defines a symbol for this counter.

A simple fixed input port behaves like the circuit shown in Figure 5.1. The digital values existing on the input pins are copied into the microcomputer when the software executes a read from the port address.

Figure 5.1
A read-only input port allows the software to read external digital signals.

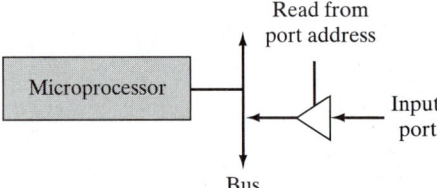

A latched input port behaves like the circuit shown in Figure 5.2. The digital values existing on the input pins are copied into an internal latch on an appropriate edge of the external control signal. At a later time, the data are transferred to the microcomputer when the software executes a read from the latch address. Notice that this latched input port also supports the regular input function. In other words, the software has the option of reading the port address to get information directly from the input port pins or from the latch address to get information that existed at the time of the previous active edge of the external control signal.

Figure 5.2
A latched input port allows the software to read external digital signals that are captured via the external control signal.

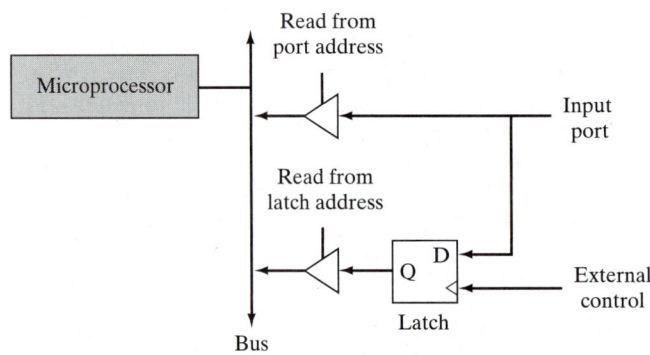

Often the latched input port allows the software to select the strategic edge to affect the latch function. In other words, the software specifies whether the rise or fall of an external control signal latches the input data. It is important to remember that when the software reads from the latch address, it obtains the values of the input signals occurring at the time of the active edge of the external control signal. In this way, the external device can provide the data at the input, issue an edge on the control signal latching into the computer, and the software can process the data at a later time without requiring the external device to maintain the valid data at the input. One of the limitations of this particular configuration is that it does not include a way for the software to signal back (acknowledge) to the external hardware that the previous data has been read by the software.

While an input device usually involves just the software reading the port, an output port can participate in both the read and write cycles very much like a regular memory. Figure 5.3 describes a "readable output port." For an 8-bit output port, there will be eight D flip-flops to hold the values on the output pins. A write cycle to the port address will affect the values on the output pins. In particular, the microcomputer places information on the data bus and that information is clocked into the D flip-flops. Since it is a readable output, a read-cycle access from the port address returns the current values existing on the port pins. A "write-only output port" does not allow software to read the current values.

Figure 5.3
A readable output port allows the software to generate external digital signals.

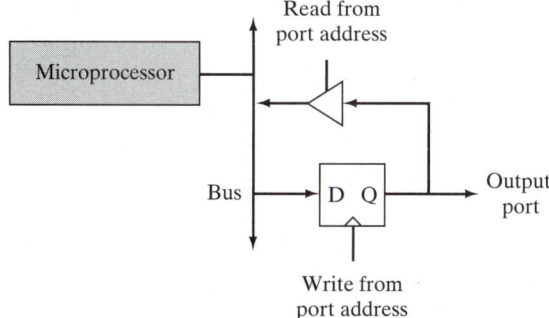

Some ports are fixed as inputs or outputs, while others can be specified by the software to be inputs or outputs. Figure 5.4 shows a bidirectional port. Motorola uses the concept of a *direction register* to determine whether a pin is an input (direction register bit is 0) or an

Figure 5.4
A bidirectional port can be configured as a read-only input port or a readable output port.

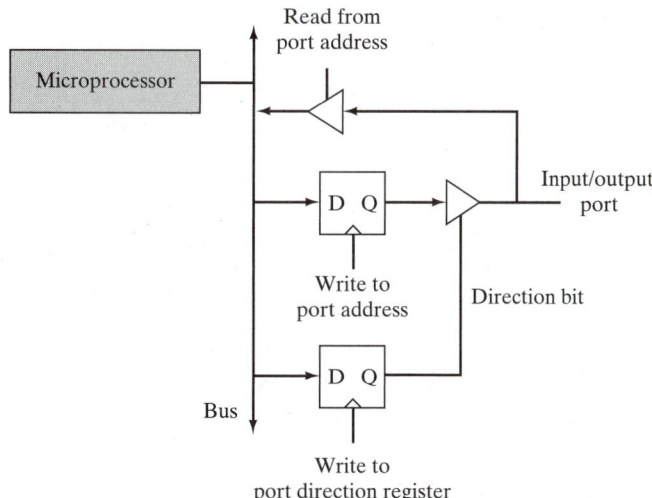

output (direction register bit is 1). We define a *ritual* as a program executed during startup that initializes hardware and software. If the direction bit is zero, the port behaves like a simple input, and if the direction bit is one, it becomes a readable output port.

> **Common Error:** *Many program errors can be traced to confusion between I/O ports and regular memory. For example, you should not write to an input port, and sometimes we can not read from an output port.*

> **Observation:** *If a port pin is configured as a readable output, but external loading causes the pin voltage to be different from the value written by the software, then a read from the port will return the current voltage level at the pin and not the value last written by the software. This fact can be used by the software to detect excess loading by the external circuits.*

Both the 6811 and 6812 can provide address and data bus signals for connecting external memory and I/O devices. In this expanded mode, some of the I/O port signals are used for the address/data bus and thus not available for I/O.

> **Checkpoint 5.3:** *List the I/O pins of Ports A, B, C, D, and E on the MC68HC711E9, classifying them as input only, output only or bidirectional.*

> **Checkpoint 5.4:** *There are many ports on the 6812, but not all are 8 bits and not all are bidirectional. List the ports on the MC68HC812A4 and the MC68HC912B32 that are a full 8 bits wide and bidirectional.*

5.2 ## Parallel Port I/O Programming

Port C on the 6811/6812 is a single 8-bit I/O port that can be used as either read-only inputs or readable outputs. Three hardware configurations for Port C are shown in Figure 5.5.

Figure 5.5
The input/output direction of a bidirectional port is specified by its direction register.

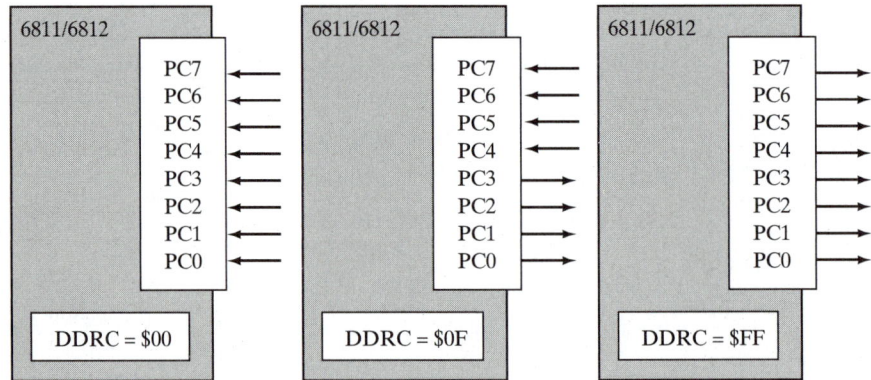

The first step to access a bidirectional I/O port is to set the direction register. The direction register specifies bit for bit whether the corresponding pins are input (0) or output (1). To make all pins input, you should clear the direction register:

```
clr   DDRC
```

or in C you could execute

```
DDRC=0x00;
```

The following software reads a regular input and saves the information into a global variable called `Happiness`.

```
ldaa   PORTC
staa   Happiness
```

In C, an I/O port name on the right side of an assignment operator is implemented with an input port read access. In actuality, an input port read access is produced whenever the I/O port name exists anywhere except on the right side of an assignment operator (e.g., `if` `while` `for` `switch`, etc.):

```
Happiness=PORTC;
```

During the execution of the `ldaa` instruction, the computer will perform a read bus cycle to the address of the port. It is during this exact time that the current value of the input is transferred into the computer. If a pin is programmed as an input, a write to that port has no effect on that pin.

If a pin is programmed as an output, a write to that port will set/clear that pin, and a read from that port will return the value that we had previously written. To make pins 7-4 input and pins 3-0 output, you should set the direction register to $0F.

```
ldaa   #$0F
staa   DDRC
```

or in C you could execute

```
DDRC=0x0F;
```

The following software will set the outputs (pins 3-0) high, but have no effect on the inputs (pins 7-4)

```
ldaa   #$FF
staa   PORTC
```

or in C you could execute

```
PORTC=0xFF;
```

We could write general subroutines to access the parallel port as shown in Program 5.5. The advantage to developing subroutines for parallel I/O access is that changes can be easily made. When you develop a layered approach to the I/O ports, you are creating an I/O abstraction, or hardware abstraction layer (HAL). The separation of the policy (how parameters are passed to/from `Init Set Read`) from the implementation (`staa PORTC`), makes it easy to debug and maintain. For example, if you had a system implemented on the 6811 using Port C, and wished to redesign it on the 6812 using Port J, then it should be a simple manner of changing these three subroutines. These assembly language subroutines pass the data in and out using Register A. The term "software driver" is used to describe a set of programs that perform a task. In particular, an **I/O driver** or **device driver** is a set of programs that facilitate the use of an I/O device. When `Set` is called, only the output pins are affected. One the other hand, the `Read` subroutine reads from the port and gets the current value of all pins, both inputs and outputs.

Program 5.5 Simple
I/O port driver.

```
* RegA specifies input/output        void Init(unsigned char value) {
Init staa DDRC                         DDRC = value;
     rts                             }
* Reg A is value output to port      void Set(unsigned char value){
Set  staa PORTC                        PORTC = value;
     rts                             }
* Reg A is returned with input       unsigned char Read(void) {
Read ldaa PORTC                        return(PORTC);
     rts                             }
```

On an isolated I/O system like the Intel x86, if we wish to bring into Register AL the current values of an input port that is located at $10, we could

    in AL, $10[2] Copy the values of Port $10 into Register AL

On an isolated I/O system like the Intel x86, if we wish to send the value in Register AL to an output port, that is located at $11, we could

    out AL, $11   copy the values of Register AL out to Port $11

When there is more than one initialization routine, one initialization routine may affect the operation of the other. For example, assume there are two independent modules, one uses Port C bit 0 as an output, and the other uses Port C bit 1 as an output. If the initialization routine for the first module sets DDRC to 1 and the second initialization sets DDRC to 2, then an error will occur when both initialization routines are executed. In particular, the execution of the second one overrides the first.

To illustrate this important point, we will write a software driver for just Port C bit 0. The example in Program 5.6 is friendly because it does not modify the other seven bits. The initialization will define PC0 as an output, the function `SetPC0` will make it high, and the function `ClrPC0` will make it low.

Program 5.6
A friendly I/O port driver.

```
* Make PC0 an output pin          // Make PC0 an output pin
InitPC0 ldaa DDRC                 void InitPC0(void){
        oraa #$01                   DDRC |= 0x01;
        staa DDRC                 }
        rts                       // Make PC0 high
* Make PC0 high                   void SetPC0(void){
SetPC0  ldaa PORTC                  PORTC |= 0x01;
        oraa #$01                 }
        staa PORTC                // Make PC0 low
        rts                       void ClrPC0(void){
* Make PC0 low                      PORTC &= ~0x01;
ClrPC0  ldaa PORTC                }
        anda #~$01
        staa PORTC
        rts
```

---

[2] Hexadecimal numbers on the Intel assemblers are actually specified as 10H.

Observation: An initialization routine is called "friendly" if it sets just the mode bits that are needed and leaves the other bits as is.

Checkpoint 5.5: Write the initialization routines for two independent modules. One uses Port C bit 1 as an output, and the other uses Port C bit 2 as an output.

5.3 Serial Communications Interface, SCI

Serial transmission involves sending one bit a time, where the data are spread out over time. The total number of bits transmitted per second is the *baud rate.* Most of the Motorola-embedded microcomputers support at least one *Serial Communications Interface* or SCI. Before discussing the detailed operation of particular devices, we will begin with general features common to all devices. Each SCI module has a *baud rate* control register, which we use to select the transmission rate. There is a mode bit, **M**, which selects 8-bit (M=0) or 9-bit (M=1) data frames. Each device is capable of creating its own serial port clock with a period that is an integer multiple of the E clock period. The programmer will select the baud rate by specifying the integer divide-by used to convert the E clock into the serial port clock. A *frame* is the smallest complete unit of serial transmission. Figure 5.6 plots the signal versus time on a serial port, showing a single frame, which includes a start bit (0), 8 bits of data (least significant bit first) and a stop bit (1). This protocol is used for both transmitting and receiving. The information rate, or *bandwidth,* is defined as the amount of data or useful information transmitted per second. From Figure 5.6, we see that 10 bits are sent for every byte of useful data. Therefore the bandwidth of the serial channel (in bytes/second) is the baud rate (in bits/sec) divided by 10.

Figure 5.6
A serial data frame with M=0.

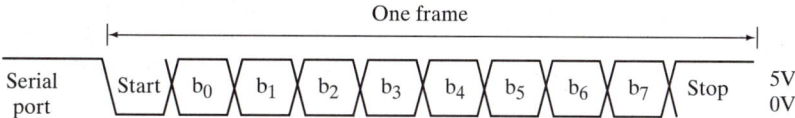

Common Error: If you change the E clock frequency without changing the baud rate register, the SCI will operate at an incorrect baud rate.

**5.3.1
Transmitting in
Asynchronous
Mode**

We will begin with transmission, because it is simpler than reception. The transmitter portion of the SCI includes a TxD data output pin, with TTL voltage levels. (See Figure 5.7.) The transmitter has a 10- or 11-bit shift register, which can not be directly accessed by the programmer, and this shift register is separate from the receive shift register. To output data using the SCI, the software will write to the Serial Communications Data Register. On the 6811, the data register is called SCDR; on the 6812, it is called SC0DRL or SCI1DRL. The transmit data register is write only, which means the software can write to it (to start a new transmission), but can not read from it. Even though the transmit data register is at the same address as the receive data register, the transmit and receive data registers are two separate registers. When using 9-bit data mode (M=1), we first set the T8 bit, then we write to the transmit data register to start transmission.

Figure 5.7
Data and shift registers
implement the serial
transmission.

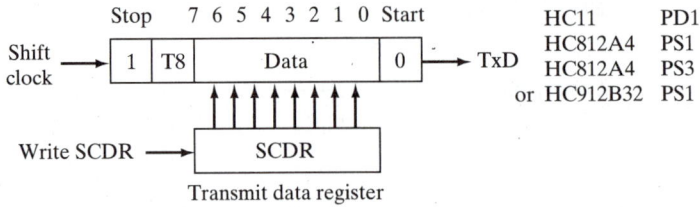

Four control bits affect transmission. We initialize the Transmit Enable control bit, TE, to 1 to enable the transmitter. We set the Send Break control bit, SBK, to 1 to send blocks of 10 (or 11 if M=1) zeros. We arm for interrupts by setting the Transmit Interrupt Enable control bit, TIE. The Transmit Complete Enable control bit, TCIE, allows the TC flag to interrupt. Interrupts will be covered in Chapter 11. There are two status bits generated by transmitter activity. The Transmit Data Register Empty flag, TDRE, is set when the transmit SCDR is empty. The TDRE bit is cleared by reading the TDRE flag (with it set), then writing to the SCDR. The Transmit Complete flag, TC, is set when the transmit shift register is done shifting. The TC is cleared by reading the TC flag (with it set), then writing to the SCDR.

When new data (8 bits) is loaded into the SCDR, it is copied into the 10- or 11-bit transmit shift register. Next, the start bit, T8 (if M=1) and stop bits are added. Then, the frame is shifted out one bit at a time at a rate specified by the baud rate register. If there is already data in the shift register when the SCDR is written, it will wait until the previous frame is transmitted, before it too is transferred. The serial port hardware is actually controlled by a clock that is 16 times faster than the baud rate. The digital hardware in the SCI counts 16 times in between changes to the TxD output line.

In essence, the SCDR and transmit shift register behave together like a two-element first-in-first-out queue (FIFO). In other words, the software can actually write two bytes to the SCDR, and the hardware will send them both one at a time. In fact, the serial port interface chip used in most PC computers has a 16-byte hardware FIFO between the data register and the shift register. A PC that has a 16C550-compatible UART supports this hardware FIFO function. This FIFO reduces the software response time requirements of the operating system to service the serial port hardware.

5.3.2 Receiving in Asynchronous Mode

Receiving data frames is a little trickier than transmission because we have to synchronize the receive shift register with the incoming data. The receiver portion of the SCI includes a RxD data input pin, with TTL voltage levels. (See Figure 5.8.) There is also a 10- or 11-bit shift register, which can not be directly accessed by the programmer. Again the receive shift register is separate from the transmit shift register. The receiver has a Serial Communications Data Register. Again, this register is called SCDR on the 6811, and SC0DRL/SCI1DRL on the 6812. The receive data register is read only; thus, write operations to this address have no effect on this register. When operating in 9-bit mode (M=1), the ninth data bit is saved in the R8 bit.

There are four control bits that affect the receiver. We will set the Receiver Enable control bit, RE, to 1 to enable the receiver. If we set the Receiver Wakeup control bit, RWU, to 1, then a receiver input will wake up the computer out of a low-power sleep mode. There are two interrupt arm bits for the receiver. The Receiver Interrupt Enable control bit, RIE, enables the RDRF flag to request interrupts. The Idle Line Interrupt Enable control bit, ILIE, enables the IDLE flag to request interrupts. There are five status bits generated by receiver activity. The Receive Data Register Full flag, RDRF, is set when new input data are available. The RDRF bit is cleared by reading the RDRF flag (with it set) then reading the SCDR. The Receiver Idle flag, IDLE, is set when the receiver line becomes idle. The IDLE bit is cleared by reading the IDLE flag (with it set) and then reading the SCDR. The Overrun flag, OR, is set when input data are lost because previous data frames had not been read. The OR bit is cleared by reading the OR flag (with it set) then reading the SCDR. The Noise flag, NF, is set when the input is noisy. The NF bit is cleared by reading the NF flag reading the SCDR. Each bit is sampled three times by the receiver, and the NF flag is cleared if all bits yielded unanimous decisions. The NF bit is set when any of the groups of three samples did not all agree. NF errors can occur if there is indeed noise on the line, but more likely it is caused by a mismatch between the transmitter and receiver baud rates. The Framing Error, FE, is set when the stop bit is incorrect. The FE bit is cleared by reading the FE flag (with it set) and then reading the SCDR. Framing errors are also probably caused by a mismatch in baud rate.

Figure 5.8
Data register shift registers implement the receive serial interface.

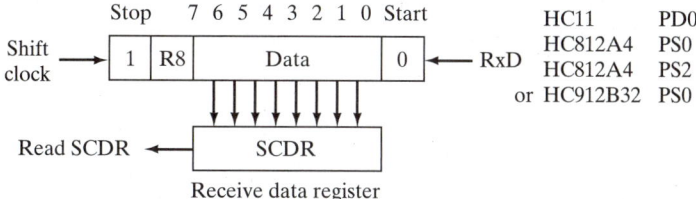

The receiver waits for the 1 to 0 edge signifying a start bit, then shifts in 10 or 11 bits of data one at a time from the RxD line. The start and stop bits are removed (checked for noise and framing errors), the 8 bits are loaded into the SCDR, the ninth data bit is put in R8 (if $M=1$), and the RDRF flag is set. If data already are in the SCDR when the shift register is finished, it will wait until the previous frame is read by the software before it is transferred.

An overrun occurs when there is one receive frame in the SCDR, one receive frame in the receive shift register, and a third frame comes into RxD. In order to avoid overrun, we can design a real-time system (i.e., one with a maximum latency). The latency of an SCI receiver is the delay between the time when new data arrive in the receiver SCDR, and the time the software reads the SCDR. If the latency is always less than 10 (11 if $M=1$) bit times, then overrun will never occur.

> **Observation:** *With a serial port that has a shift register and one data register (no additional FIFO buffering), the latency requirement of the input interface is the time it takes to transmit one data frame.*

In the example illustrated in Figure 5.9, assume the SCI receive shift register and receive data register are initially empty. Three incoming serial frames occur one right after another, but the software does not respond. At the end of the first frame, the $31 goes into the receive SCDR and the RDRF flag is set. In this scenario, the software is busy doing other things, and does not respond to the setting of RDRF. Next, the second frame is entered into the receive shift register. At the end of the second frame, there is the $31 in the SCDR and the $32 in the shift register. If the software were to respond at this point, then both characters would be properly received. If the third frame begins before the first is read by the software, then an overrun error occurs and a frame is lost. We can see from this worst case scenario that the software must read the data from SCDR within 10 bit times of the setting of RDRF.

Figure 5.9
Three receive data
frames result in an
overrun (OR) error.

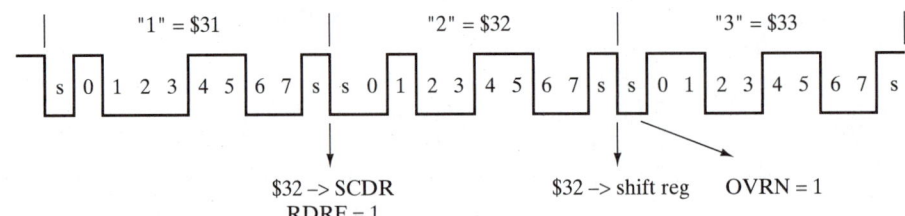

$32 \rightarrow$ SCDR $32 \rightarrow$ shift reg   OVRN = 1
RDRF = 1

Next we will overview the specific SCI functions on particular Motorola microcomputers. This section is intended to supplement rather than replace the Motorola manuals. When designing systems with a SCI, please also refer to the reference manual of your specific Motorola microcomputer.

5.3.3
6811 SCI Details

Most versions of the 6811 have one asynchronous serial port using Port D bits 1,0. Table 5.1 shows some of the 6811 I/O ports that implement the SCI functions. The full details can be found in the Motorola data sheets. We will assume the 6811 SCI is running in the default 8-bit data mode. Only some of the details are presented here; the rest can be found in Chapter 11.

Table 5.1   6811 SCI ports.

address	Bit 7	6	5	4	3	2	1	Bit 0	Name
$102B	TCLR	SCP2	SCP1	SCP0	RCKB	SCR2	SCR1	SCR0	BAUD
$102D	TIE	TCIE	RIE	ILIE	TE	RE	RWU	SBK	SCCR2
$102E	TDRE	TC	RDRF	IDLE	OR	NF	FE	0	SCSR
$102F	R7/T7	R6/T6	R5/T5	R4/T4	R3/T3	R2/T2	R1/T1	R0/T0	SCDR

BAUD

We use the BAUD register to select the baud rate. There are two integer factors that divide the E clock to get the baud rate, shown in Table 5.2. The 2 bits **SCP1** and **SCP0** determine the prescale factor. **SCP2** adds divide by 39 to SCI prescaler and is present only in MC68HC(7)11E20. The 3 bits **SCR2**, **SCR1** and **SCR0** specify the other factor. The baud rate can be calculated using the equation

$$\text{SCI Baud Rate} = \frac{\text{ECLK}}{(16 \bullet P \bullet BR)}$$

Table 5.2
Baud rate selection bits
for the 6811.

SCP1	SCP0	prescaler, P
0	0	1
0	1	3
1	0	4
1	1	13

SCR2	SCR1	SCR0	divider, BR
0	0	0	1
0	0	1	2
0	1	0	4
0	1	1	8
1	0	0	16
1	0	1	32
1	1	0	64
1	1	1	128

Table 5.3 shows the BAUD register values for common baud rates when the E clock is 2 MHz.

Table 5.3
6811 baud rate selection bits to create common protocols.

BAUD	baud rate (bits/sec)
$30	9600
$31	4800
$32	2400
$33	1200

Checkpoint 5.6: What is the baud rate on a 2 MHz 6811 if BAUD equals $23?

Checkpoint 5.7: What value should BAUD be for a baud rate of 2604 bits/sec on a 2 MHz 6811?

The **SCCR2** control register contains the bits that turn on the SCI, and contains the interrupt arm bits. Interrupts will be discussed in Chapter 11. **TE** is the Transmitter Enable bit. You turn this bit off to disable the transmitter, and set this bit to enable the transmitter. **RE** is the Receiver Enable bit. This bit should be set to one to activate the receiver. If RE is 0, then the receiver is disabled.

The flags in the **SCSR** register can be read by the software, but can not be modified by writing to this register. **TDRE** is the Transmit Data Register Empty Flag. TDRE is set if transmit data can be written to SCDR. If TDRE is zero, the transmit data register contains previous data that has not yet been moved to the transmit shift register. TDRE is cleared by SCSR read with TDRE set followed by SCDR write. **RDRF** is the Receive Data Register Full Flag. It is set if a received character is ready to be read from SCDR. RDRF is cleared by a SCSR read with RDRF set followed by a SCDR read.

5.3.4
6812 SCI Details

The MC68HC812A4 has two asynchronous serial ports using Port S bits 3,2 and bits 1,0. Only one of the devices is discussed in this section. The MC68HC912B32 and MC68HC912D60 have only one serial port using Port S bits 1,0. Each serial port has a 16-bit baud rate register. Table 5.4 shows some of the port names and addresses for the 6812 SCI. The full details can be found in the Motorola data sheet. We will assume the SCI is running in the default 8-bit data mode.

Table 5.4 6812 SCI ports.

Address	Bit 7	6	5	4	3	2	1	Bit 0	Name
$0047	LCK	PLLON	PLLS	BCSC	BCSB	BCSA	**MCSB**	**MCSA**	CLKCTL
$00C0	BTST	BSPL	BRLD	**SBR12**	**SBR11**	**SBR10**	SBR9	SBR8	SC0BD
$00C1	**SBR7**	**SBR6**	SBR5	SBR4	SBR3	SBR2	SBR1	SBR0	SC0BD
$00C3	TIE	TCIE	RIE	ILIE	**TE**	**RE**	RWU	SBK	SC0CR2
$00C4	**TDRE**	TC	**RDRF**	IDLE	OR	NF	FE	PF	SC0SR1
$00C7	**R7T7**	**R6T6**	**R5T5**	**R4T4**	**R3T3**	**R2T2**	**R1T1**	**R0T0**	SC0DRL

SC0BD   The **SC0BD** register is 16 bits, occupying two bytes. The least significant 13 bits of SC0BD determine the baud rate for the SCI port. If BR is the value written to bits 12:0, and MCLK is the module clock, then the baud rate is

$$\text{SCI Baud Rate} = \frac{\text{MCLK}}{(16 \bullet \text{BR})}$$

The MCLK frequency is determined by the 2 bits MCSB and MCSA in the clock control (CLKCTL) register, as shown in Table 5.5. The PCLK is the internal bus rate clock, typically 8 MHz.

Table 5.5
MCLK rate selection bits for the 6812.

MCSB	MCSA	MCLK Rate
0	0	MCLK = PCLK
0	1	divide by 2
1	0	divide by 4
1	1	divide by 8

Checkpoint 5.8: Assume MCLK is 8 MHz. What is the baud rate if SC0BD equals $23?

Checkpoint 5.9: Assume MCLK is 8 MHz. What value should SC0BD be for a baud rate of 2604 bits/sec?

The **SC0CR2** control register contains the bits that turn on the SCI, and contains the interrupt arm bits. Interrupts will be covered in Chapter 11. **TE** is the Transmitter Enable bit, and **RE** is the Receiver Enable bit. We set both TE and RE equal to 1 in order to activate the SCI device.

The flags in the **SC0SR1** register can be read by the software, but can not be modified by writing to this register. **TDRE** is the Transmit Data Register Empty Flag. It is set by the SCI hardware if transmit data can be written to SC0DRL. If TDRE is zero, transmit data register contains previous data that has not yet been moved to the transmit shift register. Writing into the SC0DRL when TDRE is set will result in a loss of data. On the other hand, when this bit is set, the software can begin another output transmission by writing to SC0DRL. This flag is cleared by first reading SC0SR1 with TDRE set followed by a SC0DRL write. **RDRF** is the Receive Data Register Full bit. RDRF is set if a received character is ready to be read from SC0DRL. We clear the RDRF flag by reading SC0SR1 with RDRF set followed by reading SC0DRL.

The **SC0DRL** register contains the data transmitted out and received in by the SCI device. Even though there are separate transmit and receive data registers, these two registers exist at the same I/O port address. Reads to SC0DRL access the eight bits of the read-only SCI receive data register (RDR). Writes to SC0DRL access the eight bits of the write-only SCI transmit data register (TDR).

5.3.5
SCI I/O
Programming

Software that sends and receives data must implement a mechanism to synchronize the software with the hardware. In particular, the software should read data from the input device only when data is indeed ready. Similarly, software should write data to an output device only when the device is ready to accept new data. *Busy waiting, gadfly,* or *polling* are three equivalent names for the same synchronization method. In this scheme the software continuously checks the hardware status waiting for it to be ready. In this section, we will use busy-waiting to write I/O programs that send and receive data using the SCI port. When a new 8-bit character is received into the serial port, RDRF is set and the 8-bit data is available in the serial data register, SCDR. To get new data from the serial port, the software first waits for the flag to be set, then reads the result. Recall that when the software reads SCDR (SC0DRL on the 6812) it accesses the receive data register. This operation is illustrated in Figure 5.10 and shown in Programs 5.7 and 5.8. In a similar fashion, when the software wishes to output via the serial port, it first waits for TDRE to be set, then performs the output. When the software writes SCDR (SC0DRL on the 6812) it sets the transmit data register. An interrupt synchronization method will be presented in Chapter 11.

Figure 5.10
Flowcharts of InChar
and OutChar using
busy-waiting.

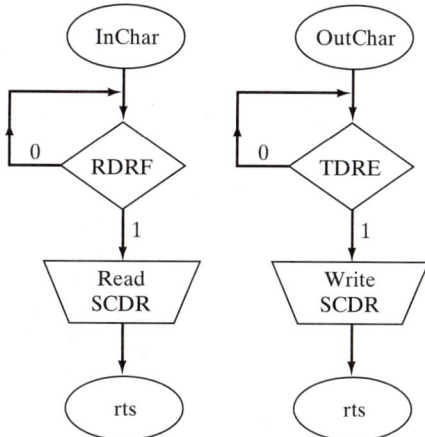

The initialization program, `SCI_Init`, enables the SCI device and selects the fastest baud rate possible.

Program 5.7   Assembly functions that implement serial I/O.

```
; Initalize 6811 SCI              * Initialize 6812 SCI
; Inputs: none Outputs: none      * Inputs: none      Outputs: none
SCI_Init                          SCI_Init
     ldaa #$0C                          movb #$0c,SC0CR2   enable SCI
     staa SCCR2   enable SCI            movw #0,SC0BD      500000 bps
     ldaa #0                            rts
     staa BAUD    125000 bps
     rts
; Read 8 bits from serial port    ; Read 8 bits from serial port
; Return 8-bit byte in RegA       ; Return 8-bit byte in RegA
RDRF equ $20                      RDRF equ  $20
SCI_InChar                        SCI_InChar
     ldaa SCSR    serial status        ldaa SC0SR1 ; serial status
     anda #RDRF   available?           anda #RDRF   ; available?
     beq  SCI_InChar wait for RDRF     beq SCI_InChar ; wait
     ldaa SCDR    read ASCII           ldaa SC0DRL ; read ASCII
     rts                                rts
; Write 8 bits to serial port     ; Write 8 bits to serial port
; Input 8-bit byte in RegA        ; Input 8-bit byte in RegA
TDRE equ $80                      TDRE equ  $80
SCI_OutChar                       SCI_OutChar
     ldab SCSR     serial status       ldab SC0SR1   ; serial status
     andb #TDRE output ready?          andb #TDRE    ; output ready?
     beq  SCI_OutChar wait for TDRE    beq SCI_OutChar ; wait
     staa SCDR write ASCII             staa SC0DRL   ; write ASCII
     rts                                rts
```

Program 5.8   C functions that implement serial I/O.

```
// 6811 initialize SCI                      // 6812 initialize SCI
void SCI_Init(void){                        void SCI_Init(void){
   BAUD = 0;     // 125000 bits/sec            SC0BD  = 0;     // 500000 bits/sec
   SCCR2= 0x0C; // enable                      SC0CR2 = 0x0C; // enable
}                                           }
#define RDRF 0x20                           #define RDRF 0x20
// Wait for new input,                      // Wait for new input,
// then return ASCII code                   // then return ASCII code
char SCI_InChar(void){                      char SCI_InChar(void){
   while((SCSR&RDRF) == 0){};                  while((SC0SR1&RDRF) == 0) {};
   return(SCDR);                               return(SC0DRL);
}                                           }
#define TDRE 0x80                           #define TDRE 0x80
// Wait for buffer to be empty,             // Wait for buffer to be empty,
// then output                              // then output
void SCI_OutChar(char data){                void SCI_OutChar(char data){
   while((SCSR&TDRE) == 0){};                  while((SC0SR1&TDRE) == 0){};
   SCDR = data;                                SC0DRL = data;
}                                           }
```

The **TExaS** simulator has example serial port software for the 6811 and 6812 written in both assembly and C. Once **TExaS** is installed, the busy-waiting SCI functions can be found in the TUT2.* files. The TUT4.* files implement similar serial port operations using interrupt synchronization.

5.4   Synchronous Peripheral Interface, SPI

5.4.1
SPI Fundamentals

Many of the Motorola embedded microcomputers include a Serial Peripheral Interface or SPI. Typically, we use the SPI to attach additional I/O devices, or peripherals, to the microcomputer. The fundamental difference between a SCI, which implements an asynchronous protocol, and a SPI, which implements a synchronous protocol, is the manner in which the clock is handled. Two devices communicating with asynchronous serial interfaces (SCI) operate at the same frequency (baud rate) but have two separate (not synchronized) clocks. Two devices communicating with synchronous serial interfaces (SPI) operate the same clock (synchronized). Typically, the master device creates the clock, and the slave device(s) uses the clock to latch the data (in or out.) Before discussing the detailed operation of particular devices, we will begin with general features common to all devices. The Motorola SPI includes four I/O lines. The slave select ($\overline{SS}$) is an optional negative logic control signal from master to slave signal signifying the channel is active. The second line, SClK, is a 50% duty cycle clock generated by the master. The MOSI (master out slave in) is a data line driven by the master and received by the slave. The MISO (master in slave out) is a data line driven by the slave and received by the master. In order to work properly, the transmitting device uses one edge of the clock to change its output, and the receiving device uses the other edge to accept the data. Table 5.6 lists the I/O port locations of the synchronous serial ports for the various microcomputers discussed in this book.

Table 5.6

Synchronous serial port
pins on various Motorola
microcomputers.

Microcomputer	pin for $\overline{SS}$	pin for SClK	pin for MOSI	pin for MISO
MC68HC11	PD5	PD4	PD3	PD2
MC68HC812A4	PS7	PS6	PS5	PS4
MC68HC912B32	PS7	PS6	PS5	PS4

Although it could be used to connect microcomputers together, we will use the serial peripheral interface external devices. In this situation, the SPI system in the microcomputer operates as the master and the external devices are slaves. In the SPI system the 8-bit data register, SPDR, in the master and the 8-bit data register in the slave are linked to form a distributed 16-bit register. Figure 5.11 illustrates communication between a microcomputer and a single external device. The SS signal is used to activate the external I/O device, the SClk shifts both shift registers.

Figure 5.11

A synchronous serial
interface between a
microcomputer and an
I/O device.

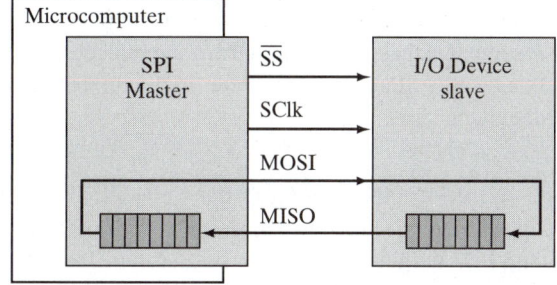

When a data transfer operation is performed, this 16-bit register is serially shifted 8 bit positions by the SClk clock from the master so the data are effectively exchanged between the master and the slave. Data written to the SPDR register of the master are transmitted to the slave. A baud rate control register is used to select the transmission rate. It is possible to implement output only using just MOSI or to implement input only using just MISO. It is also possible to concatenate multiple slaves in a large loop.

> ***Observation:*** *Because the clocks are shared, if you change the E clock frequency, the transfer rate will change, but the SPI should still operate properly.*

The SPI timing is shown in Figure 5.12. There are three mode control bits that affect the transmission protocol. If the device is a master it generates the SClk. Data are output on the MOSI pin and input on the MISO pin. If the device is a slave, the SClk is an input. Data are input on the MOSI pin and output on the MISO pin. The CPOL control bit specifies the polarity of the SClk. The CPHA bit affects the timing of the first bit transferred and received. In this diagram, the data are shown with MSB transferred first, but the 6812 offers an option for bit transfer in the other order (LSB first).

Figure 5.12   Synchronous serial modes of the Motorola SPI interface.

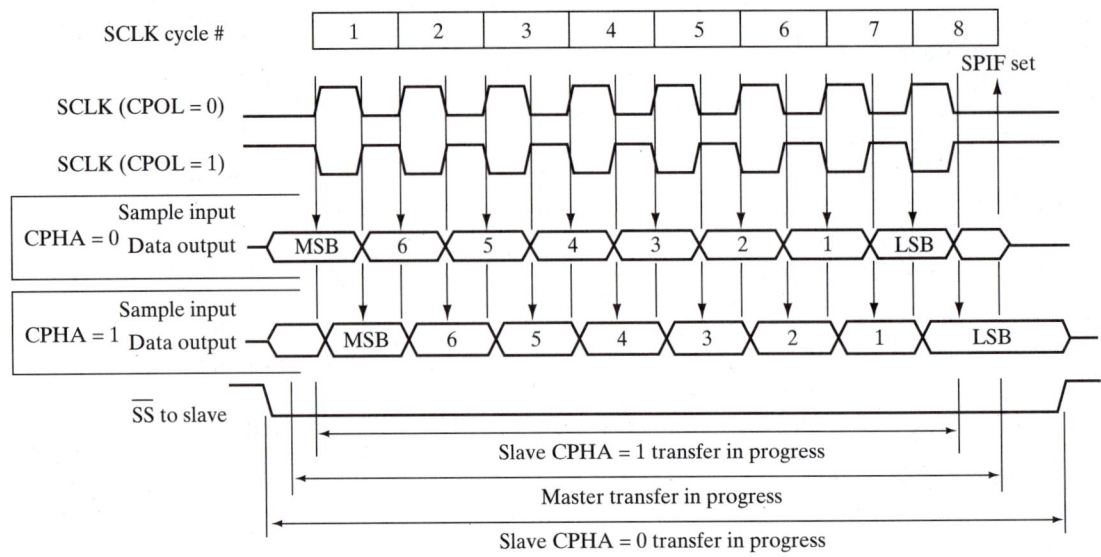

Next we will overview the specific SPI functions on particular Motorola microcomputers. This section is intended to supplement rather than replace the Motorola manuals. When designing systems with a SPI, please also refer to the reference manual of your specific Motorola microcomputer.

5.4.2 6811 SPI Details

The SPI port on the MC68HC11A8 uses the 4 pins PD5= $\overline{SS}$, PD4=SClk, PD3=MOSI, and PD2=MISO. If the 6811 is the master, then we should set the DDRD register to make PD5, PD4, and PD3 outputs. PD2 should be an input. Table 5.7 shows the 6811 SPI ports.

Table 5.7   6811 SPI ports.

address	Bit 7	6	5	4	3	2	1	Bit 0	Name
$1009	0	0	DDRD5	DDRD4	DDRD3	DDRD2	DDRD1	DDRD0	DDRD
$1028	SPIE	SPE	DWOM	MSTR	CPOL	CPHA	SPR1	SPR0	SPCR
$1029	SPIF	WCOL	0	MODF	0	0	0	0	SPSR
$102A	Bit 7	6	5	4	3	2	1	Bit 0	SPDR

SPCR   The **SPCR** register contains the configuration bits that determine the mode of operation. SPIE is the Serial Peripheral Interrupt Enable bit. We will turn this bit off to use busy-waiting synchronization. SPE is the Serial Peripheral System Enable bit. We will turn this bit on to enable the SPI device. The **DWOM** bit in the SPCR register is used to activate the Port D wired-OR mode. This one bit affects all six port D pins. If DWOM is 0, then PD5-PD0 are normal CMOS signals. A normal CMOS output has two states: low (0V or logic zero) or high (+5V or logic one). If DWOM is 1, then any port D output is open drain. "Open drain" means the output is either driven low or floats. Alternative names for the same behavior are "open collector" and "wire or mode." MSTR is the Master Mode

Select bit. We will set this bit to one to make the 6811 the master. **CPOL** and **CPHA** determine the SPI Clock Polarity and Clock Phase. These two bits are used to specify the protocol, as shown in Figure 5.12. The SPR1 and SPR0 control bits determines the transfer rate. Assuming an E clock frequency of 2 MHz, Table 5.8 shows the possible transmission rates.

Table 5.8
Bit rate selection for the synchronous serial port on the MC68HC11A8.

SPR1	SPR0	divisor	transfer frequency	bit time
0	0	2	1 MHz	1 μs
0	1	4	500 kHz	2 μs
1	0	16	125 kHz	8 μs
1	1	32	62.5 kHz	16 μs

SPSR   The **SPSR** register contains the SPI status bits. SPIF is the SPI Transfer Complete Flag. It is set when an SPI transfer is complete. SPIF is cleared by reading SPSR with SPIF set followed by a SPDR access. **WCOL** is the Write Collision bit. It is set when SPDR is written while transfer is in progress. It is cleared by SPSR with WCOL set followed by a SPDR access. **MODF** is the mode fault bit. A mode fault will terminates SPI operation. It is set when $\overline{SS}$ is pulled low while MSTR = 1. It is cleared by SPSR read with MODF set followed by SPCR write.

5.4.3
6812 SPI Details

A simplified version of the 6812 SPI port is given. For more details see Chapter 7 of *Embedded Microcomputer Systems: Real Time Interfacing* by Jonathan W. Valvano. We will consider only the case where the 6812 is the master. The SPI port on the 6812 uses the 4 pins PS7= $\overline{SS}$, PS6=SClk, PS5=MOSI and PS4=MISO. When the SPI is enabled, all pins that are defined by the configuration as inputs will be inputs regardless of the state of the DDRS bits for those pins. All pins that are defined as SPI outputs will be outputs only if the DDRS bits for those pins are set. If the 6812 is the master, then we should set the DDRS register to make PS7, PS6, and PS5 outputs. PS4 will automatically be an input. Table 5.9 shows the 6812 SPI ports.

Table 5.9   6812 SPI ports.

Address	Bit 7	6	5	4	3	2	1	Bit 0	Name
$00D0	SPIE	SPE	SWOM	MSTR	CPOL	CPHA	SSOE	LSBF	SP0CR1
$00D2	0	0	0	0	0	SPR2	SPR1	SPR0	SP0BR
$00D3	SPIF	WCOL	0	MODF	0	0	0	0	SP0SR
$00D5	Bit 7	6	5	4	3	2	1	Bit 0	SP0DR
$00D6	PS7	PS6	PS5	PS4	PS3	PS2	PS1	PS0	PORTS
$00D7	Bit 7	6	5	4	3	2	1	Bit 0	DDRS

The M Clock is input to a divider series and the resulting SPI clock rate may be selected to be E divided by 2, 4, 8, 16, 32, 64, 128 or 256. Three bits in the SP0BR register control the SPI clock rate. The SP0BR register determines the transfer rate. Assuming an M clock frequency of 8 MHz, Table 5.10 shows the possible transmission rates.

Table 5.10
Bit rate selection for the synchronous serial port on the 6812.

SPR2	SPR1	SPR0	divisor	transfer frequency	bit time
0	0	0	2	4 MHz	250 ns
0	0	1	4	2 MHz	500 ns
0	1	0	8	1 MHz	1 μs
0	1	1	16	500 kHz	2 μs
1	0	0	32	250 kHz	4 μs
1	0	1	64	125 kHz	8 μs
1	1	0	128	62.5 kHz	16 μs
1	1	1	256	31.25 kHz	32 μs

We use the **SP0CR1** register to specify the SPI mode of operations. The **SPIE** bit will be cleared because interrupts are not needed. **SPE** is the SPI System Enable bit. We will turn this bit on whenever we wish to use the SPI. We will set the MSTR to one so that the 6812 becomes the master. **CPOL** and **CPHA** determine the SPI Clock Polarity and Clock Phase. These two bits are used to specify the protocol, as shown in Figure 5.12. All other bits not specifically mentioned will be cleared. **SPIF** is the SPI Interrupt Request bit. Even though we won't be using interrupts, this bit gets set after each byte is transferred, and will be used to implement the busy-waiting synchronization. In particular, SPIF is set after the eighth SCIK cycle in a data transfer and it is cleared by reading the **SP0SR** register (with SPIF set) followed by an access (read or write) to the SPI data register. The **SP0DR** 8-bit register is both the input and output register. Reads of this register are double buffered but writes cause data to be written directly into the serial shifter.

5.4.4
SPI DAC Interface

This first example shows the synchronous serial interface between the computer and a Maxim MAX550A digital-to-analog converter, as shown in Figure 5.13. A digital-to-analog converter accepts a digital input (in our case a number between 0 and 255) and creates an analog output, Out, which in our case will be a voltage between 0 and 5V. Detailed discussion of ADC and DAC converters can be found in Chapter 11 of *Embedded Microcomputer Systems: Real Time Interfacing* by Jonathan W. Valvano. In this section, we will focus on the digital hardware and software aspects of the serial interface.

Figure 5.13
An 8-bit DAC interfaced to the SPI port.

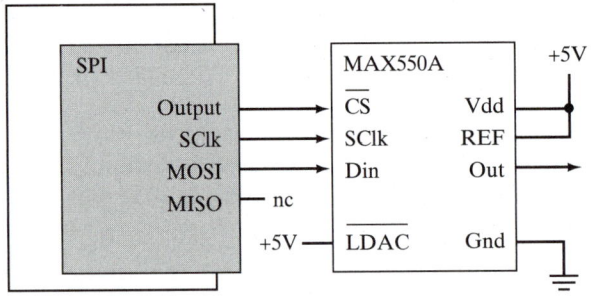

As with any SPI interface, there are basic interfacing issues to consider:

1. *Word size.* In this case we need to transmit two 8-bit packets to the DAC. The first will be a command code ($09), which will tell the DAC to output the

following data right away. The second 8-bit packet will be the digital data 0 to 255.

2. *Bit order.* The MAX550A requires the most significant bits first. This is the normal mode of the SPI.

3. *Clock phase, clock polarity.* There are two issues to resolve. Since the MAX550A samples its serial input data on the rising edge of the clock, the SPI must change the data on the falling edge. The mode with CPOL=CPHA=0 satisfies this requirement.

4. *Bandwidth.* We look at the timing specifications of the MAX550A, which can handle a clock period as short as 80ns. The fast SPI mode will therefore be used in each case.

The SPI/MAX550A timing is shown in Figure 5.14. First CS is made to fall, followed by two output frames (command and data), and then CS is made to rise again.

Figure 5.14
MAX550A DAC serial timing.

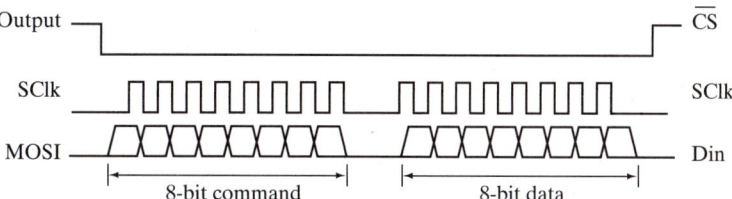

Because we want the CS signal to remain low for the entire 16-bit transfer, then pulse low, we will implement it using the regular I/O pin functions. The ritual, shown in Programs 5.9 and 5.10, initializes the direction register, SPI mode, and bandwidth.

Program 5.9   Assembly initialization for a MAX550A DAC interface using the SPI.

```
; 6811                                  ; 6812
DAC_Init                                DAC_Init(void){
   ldaa  #$38   PD3=Din                   movb #$E0,DDRS   PS6=SCLK
   staa  DDRD   PD4 = SCLK                movb #$80,PORTS PS7=CS=1
   ldaa  #$20                           ; bit SP0CR1      PS5=Din
   staa  PORTD PD5=CS=1                 ; 7 SPIE = 0    no interrupts
; bit SPCR                              ; 6 SPE  = 1    enable SPI
; 7 SPIE = 0      no interrupts         ; 5 SWOM = 0    regular outputs
; 6 SPE  = 1      enable SPI            ; 4 MSTR = 1    master
; 5 DWOM = 0      regular outputs       ; 3 CPOL = 0    changes on fall,
; 4 MSTR = 1      master                ; 2 CPHA = 0    clock normally low
; 3 CPOL = 0      changes on fall,      ; 1 SSOE = 0    PS7 regular output
; 2 CPHA = 0      clock normally low    ; 0 LSBF = 0    MSB first */
; 1 SPR1 = 0      1 MHz operation         movb #$50,SP0CR1
; 0 SPR0 = 0      */                      movb #0,SP0CR2 regular mode
   ldaa  #$50                             movb #0,SP0BR    4MHz
   staa  SPCR                             rts
   rts
```

Program 5.10 C language initialization for a MAX550A DAC interface using the SPI.

```
// 6811                              // 6812
void DAC_Init(void){                 void DAC_Init(void){
  DDRD  =0x38;    // PD4=SCLK          DDRS =0xE0;   // PS6=SCLK
  PORTD =0x20;    // PD5=CS=1          PORTS=0x80;   // PS7=CS=1
/* bit SPCR        PD3=Din           /* bit SP0CR1      PS5-Din
  7 SPIE = 0  no interrupts           7 SPIE = 0   no interrupts
  6 SPE  = 1  enable SPI              6 SPE  = 1   enable SPI
  5 DWOM = 0  regular outputs         5 SWOM = 0   regular outputs
  4 MSTR = 1  master                  4 MSTR = 1   master
  3 CPOL = 0  changes on fall,        3 CPOL = 0   changes on fall,
  2 CPHA = 0  clock normally low      2 CPHA = 0   clock normally low
  1 SPR1 = 0  1 MHz operation         1 SSOE = 0   PS7 regular output
  0 SPR0 = 0     */                   0 LSBF = 0   MSB first */
  SPCR=0x50;                          SP0CR1=0x50;
}                                     SP0CR2=0x00; // regular mode
                                      SP0BR=0x00;} // 4MHz
```

To change the DAC output, two 8-bit transmissions are sent, as shown in Programs 5.11 and 5.12. Writing to the SPI data register starts an 8-bit transmission, and the software waits for each byte to be transmitted.

Program 5.11 Assembly function for a MAX550A DAC interface using the SPI.

```
; 6811                              // 6812
;Register A contains new data       ;Register A contains new data
DAC_out                             DAC_out
    ldab PORTD                          bclr PORTS,#$20   PS7=CS=0
    andb #$DF      PD5=CS=0             movb #$09,SP0DR   send command
    stab PORTD                      11 brclr SP0SR,#$80,11     wait
    ldab #$09                           tst  SP0DR        clear SPIF
    stab SPDR    send command           staa SP0DR        send data
11 tst  SPSR   wait                 12 brclr SP0SR,#$80,12     wait
    bpl  11                             tst  SP0DR        clear SPIF
    tst  SPDR    clear SPIF             bset PORTS,#$20   PS7=CS=1
    staa SPDR    send data              rts
12 tst  SPSR   wait
    bpl  12
    tst  SPDR    clear SPIF
    ldab PORTD
    orab #$20    PD5=CS=1
    stab PORTD
    rts
```

Program 5.12   C language function for a MAX550A DAC interface using the SPI.

```
// 6811
#define SPIF 0x80
void DAC_out(unsigned char code){
unsigned char dummy;
  PORTD &= ~0x20; // PD5=CS=0
  SPDR=0x09;        // command
  while((SPSR&SPIF)==0); // wait
  dummy=SPDR;       // clear SPIF
  SPDR=code;        // data
  while((SPSR&SPIF)==0); // wait
  dummy=SPDR;       // clear SPIF
  PORTD |= 0x20;} // PD5=CS=1
```

```
// 6812
#define SPIF 0x80
void DAC_out(unsigned char code) {
unsigned char dummy;
  PORTS &= ~0x80; // PS7=CS=0
  SP0DR=0x09;        // command
  while((SP0SR&SPIF)==0); // wait
  dummy=SP0DR;        // clear SPIF
  SP0DR=code;         // data
  while((SP0SR&SPIF)==0); // wait
  dummy=SP0DR;        // clear SPIF
  PORTS |= 0x80;} // PS7=CS=1void
```

Checkpoint 5.10: Derive an equation for the analog voltage out versus digital value passed to the function `DAC_out`.

5.5   ADC Converters

The Motorola 6811/6812 microcomputers have built-in ADC converters, as shown in Figure 5.15. An analog-to-digital converter accepts an analog input, which in our case will be a voltage between 0 and 5V, and creates a digital output (in our case a number between 0 and 255). We will begin by discussing the particular I/O registers used to interface analog signals to the individual microcomputers. The common features include 8-channel operation, 8-bit resolution, and successive approximation conversion technique. In its two operation modes, the converter can perform a single sequence of conversions and then stop, or it can continuously perform analog-to-digital conversions. It can convert the same channel multiple times, or it can perform conversions on a group of channels. For example, it can perform ADC conversions on channels 0,1,2,3 with one action. The ADC uses external V_{RH}, V_{RL} analog high/low references. Normally V_{RH} is tied to +5V and V_{RL} is tied to ground. In both the 6811 and 6812, the software can read the port directly, using it as a regular 8-bit digital input port.

Figure 5.15
The 6811 shares the analog inputs with the Port E input port, while the 6812 uses Port AD.

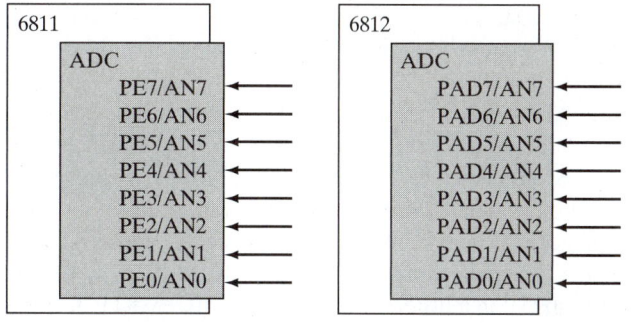

The Table 5.11 shows the straight binary format.

Table 5.11
Straight binary format used by the Motorola internal 8-bit ADCs.

Analog Input Voltage (V)	Digital Output
0.00	%00000000 $00 0
0.02	%00000001 $01 1
2.50	%10000000 $80 128
3.75	%11000000 $C0 192
5.00	%11111111 $FF 255

5.5.1
6811 ADC Details

The 6811 ADC system is enabled by setting **ADPU** equal to 1 in the **OPTION** register. When ADPU is zero, Port E operates as a normal digital input port. The **CSEL** control bit selects the clock for the charge pump. CSEL should be 0 for normal E clock frequencies. For E clock frequencies below 750 kHz, CSEL should be 1. Port E could also be used as an 8-bit digital input port. Table 5.12 shows the 6811 ADC ports.

Table 5.12 6811 ADC ports.

address	Bit 7	6	5	4	3	2	1	Bit 0	Name
$100A	PE7	PE6	PE5	PE4	PE3	PE2	PE1	PE0	PORTE
$1030	CCF	0	SCAN	MULT	CD	CC	CB	CA	ADCTL
$1031	Bit 7	6	5	4	3	2	1	Bit 0	ADR1
$1032	Bit 7	6	5	4	3	2	1	Bit 0	ADR2
$1033	Bit 7	6	5	4	3	2	1	Bit 0	ADR3
$1034	Bit 7	6	5	4	3	2	1	Bit 0	ADR4
$1039	ADPU	CSEL	IRQE	DLY	CME	0	CR1	CR0	OPTION

ADCTL   Bits 5-0 of the ADCTL register are used to set the operation mode and start the conversion. Each sequence of four conversions requires 128 E clock cycles to complete. For single sequence mode, we write to the ADCTL with SCAN=0 to start a new sequence. This register also contains a status bit, CCF, which we use to poll for the ADC conversion completion. The **CCF** flag is cleared by writing data into the ADCTL (i.e., starting a new conversion). To start a continuous conversion, we write to the ADCTL with SCAN=1. CCF is the Conversions Complete Flag, which is read only. 0 means the conversions are in progress, and 1 means the conversion is complete. When CCF is 1, the ADR1, ADR2, ADR3, ADR4 have valid data. **SCAN** is the Continuous Scan Control bit. If you set SCAN equal to 0, then the ADC will perform a single sequence of conversions and then stop. If you set SCAN equal to 1, then the ADC will perform continuous conversions. **MULT** is the Multiple Channel/Single Channel Control bit. We set MULT equal to 0 when we want a sequence of conversions on a single channel. We set MULT equal to 1 when we want a sequence of conversions on multiple channels, e.g., (0,1,2,3) (4,5,6,7) (8,9,10,11) or (12,13,14,15). In multiple-channel mode (MULT=1), the multiplexer control bits CD and CC specify the group of channels, and the multiplexer control bits CB and CA are ignored. CD,CC,CB,CA bits select which channel to convert, as described in Table 5.13.

Table 5.13
Multiplexer control for
the 6811 ADC.

CD,CC,CB,CA	Channel
0000	PE0 Analog Channel 0
0001	PE1 Analog Channel 1
0010	PE2 Analog Channel 2
0011	PE3 Analog Channel 3
0100	PE4 Analog Channel 4
0101	PE5 Analog Channel 5
0110	PE6 Analog Channel 6
0111	PE7 Analog Channel 7
1000	reserved
1001	reserved
1010	reserved
1011	reserved
1100	V_{RH} analog reference high
1101	V_{RL} analog reference low
1110	$(V_{RH}+V_{RL})/2$
1111	reserved

The ADC result registers contain the digital outputs of the ADC conversions. In single sequence mode (MULT=0), the specified single channel (CD,CC,CB,CA) is converted four times and the results placed in **ADR1**, **ADR2**, **ADR3**, and **ADR4**. In multiple sequence mode (MULT=1), the specified group of four channels (CD, CC) is each converted once and the results placed in ADR1, ADR2, ADR3, and ADR4. For example, if we write $14 into ADCTL, the 6811 will convert analog channels 4, 5, 6, 7 and put the digital results in ADR1, ADR2, ADR3, ADR4 respectively.

Performance Tip: *If we are interested in a single conversion, we could start the ADC wait 32 cycles, then read the result in ADR1.*

5.5.2
6812 ADC Details

Both the MC68HC812A4 and MC68HC912B32 have built in ADC converters. Port AD could also be used as an 8-bit digital input port. Table 5.14 shows the 6812 ADC ports. This section provides a brief introduction to the ADC, and additional details can be found in the Motorola data sheets.

Table 5.14 6812 ADC ports.

Address	Bit 7	6	5	4	3	2	1	Bit 0	Name
$0062	**ADPU**	AFFC	AWAI	0	0	0	ASCIE	ASCIF	ATDCTL2
$0065	0	S8CM	SCAN	MULT	CD	CC	CB	CA	ATDCTL5
$0066	SCF	0	0	0	0	CC2	CC1	CC0	ATDSTAT
$0067	CCF7	CCF6	CCF5	CCF4	CCF3	CCF2	CCF1	CCF0	ATDSTAT
$006F	PAD7	PAD6	PAD5	PAD4	PAD3	PAD2	PAD1	PAD0	PORTAD
$0070	Bit 7	6	5	4	3	2	1	Bit 0	ADR0H
$0072	Bit 7	6	5	4	3	2	1	Bit 0	ADR1H
$0074	Bit 7	6	5	4	3	2	1	Bit 0	ADR2H
$0076	Bit 7	6	5	4	3	2	1	Bit 0	ADR3H
$0078	Bit 7	6	5	4	3	2	1	Bit 0	ADR4H
$007A	Bit 7	6	5	4	3	2	1	Bit 0	ADR5H
$007C	Bit 7	6	5	4	3	2	1	Bit 0	ADR6H
$007E	Bit 7	6	5	4	3	2	1	Bit 0	ADR7H

ATDCTL2 and ATDCTL5   The 6812 ADC system is enabled by setting ADPU equal to 1 in the ATDCTL2 register. The ADCTL5 register is used to set the operation mode and start the conversion. S8CM determines if the sequence will be 4 or 8 conversions. For single sequence mode, we write to the ADCTL5 with SCAN=0 to start a new sequence. To start a continuous conversion, we write to the ADCTL5 with SCAN=1. S8CM is the Select 8 channel mode bit. If S8CM is 0, then the ADC will perform a four-conversion sequence. If it is 1, then an eight-conversion sequence is taken. SCAN is the Continuous Scan Control bit. When SCAN is 0, the ADC will take a single sequence of conversions and then stop. When it is 1, then the ADC performs continuous conversions. MULT is the Multiple Channel/Single Channel Control bit. If MULT is 0, the ADC performs a sequence of conversions on a single channel. When it is 1, the sequence of conversions is performed on multiple channels. For example, if S8CM = 1, then CD specifies the sequence (0,1,2,3,4,5,6,7) or (8,9,10,11,12,13,14,15). If S8CM = 0 then CD,CC specifies the sequence (0,1,2,3) or (4,5,6,7) or (8,9,10,11) or (12,13,14,15). The CD,CC,CB,CA bits select which channel to convert, as shown in Table 5.15.

Table 5.15
Multiplexer control for the 6812 ADC.

CD,CC,CB,CA	Channel
0000	PAD0 Analog Channel 0
0001	PAD1 Analog Channel 1
0010	PAD2 Analog Channel 2
0011	PAD3 Analog Channel 3
0100	PAD4 Analog Channel 4
0101	PAD5 Analog Channel 5
0110	PAD6 Analog Channel 6
0111	PAD7 Analog Channel 7
1000	reserved
1001	reserved
1010	reserved
1011	reserved
1100	V_{RH} analog reference high
1101	V_{RL} analog reference low
1110	$(V_{RH}+V_{RL})/2$
1111	reserved

ATDSTAT   This 16-bit status register contains a status bit, SCF, which we use to poll for the ADC conversion completion. The SCF flag is cleared by writing data into the ADCTL5 (i.e., starting a new conversion). The CC2,CC1,CC0 bits are the sequence counter as the ADC steps through a conversion sequence. The CCFn bits are individual flags for each of the conversions. Recall that the number of conversions in a sequence is determined by S8CM. SCF is the Sequence Complete Flag, which is read only. 0 means conversions are in progress, and 1 means the sequence of conversions is complete. When SCF is 1, all result registers have valid data. Each conversion in a sequence has its own status flag. CCF7, CCF6, CCF5, CCF4, CCF3, CCF2, CCF1, CCF0 are the Conversion Complete Flags, which are read only. When a flag is 0, that particular conversion has not finished. When one of these flags is 1, that particular conversion is complete, and the associated result register has valid data.

The ADC result registers contain the digital outputs of the ADC conversions. In single sequence mode (MULT=0), the specified single channel (CD,CC,CB,CA) is converted multiple times (four times if S8CM=0 or eight times if S8CM=1) and the results placed in the ADRnH, registers. In multiple sequence mode (MULT=1), the specified group of channels is each converted once and the results placed in the ADRnH, registers. For example, if we write $14 into ADCTL5, the 6812 will convert analog channels 4, 5, 6, 7 and put the digital results into ADR0H, ADR1H, ADR2H, ADR3H respectively. Similarly, if we write $50 into ADCTL5, the 6812 will convert analog channels 0 through 7 and put the digital results into ADR0H through ADR7H respectively. If we add the SCAN bit and write $70 to ADCTL5, the 6812 will repeatedly convert analog channels 0 through 7 and put the digital results into ADR0H through ADR7H respectively.

Performance Tip: *If we are interested in a single conversion, we could start the ADC, wait for CCF0, then read the result in ADR0H.*

5.5.3
ADC Software

In this example assume that an analog signal, ranging from 0 to +5 volts, is connected to analog input channel 1. The subroutine, ADC_In in Program 5.14, will return an unsigned value from 0 to 255 that represents the analog input. Program 5.15 is a version written in C. The analog input range is 0 to +5V, the analog input **resolution** is 5V/256 that is about 20 mV, and the **precision** is 8 bits. In this case, resolution is defined as the smallest change in analog input voltage that can be reliably detected by the system. Precision is defined as the number of distinguishable measurements. The 6811 and 6812 ADC will perform multiple conversions on the same analog signal. This subroutine returns only the first one.

Program 5.14 Assembly software to sample data using the ADC.

```
; MC68HC11                          ; MC68HC812A4/MC68HC912B32
; *****Initialize ADC*******        ;*****Initialize ADC*******
;Inputs: None                       ;Inputs: None
;Outputs: None                      ;Outputs: None
ADC_Init                            ADC_Init
    ldab  #$80                          movb #$80,TSCR    ;enable timer
    stab  OPTION ;enables ADC           movb #$80,ATDCTL2 ;enables ADC
    rts                                 rts
****Input sample from ADC****       ****Input sample from ADC*****
*Inputs: RegB is channel 0-7        *Inputs: RegB is channel 0-7
*Outputs: RegB is ADC sample        *Outputs: RegB is ADC sample
ADC_In                              ADC_In
    stab ADCTL   ;start                 stab  ATDCTL5 ;start
wait ldab ADCTL                         brclr ATDSTAT,#$80,* ;SCF
    andb #$80                           ldab  ADR0H
    beq  wait    ; wait for SCF          rts
    ldab ADR1
    rts
```

In the following C versions, the channel number (0 to 7) is passed in as an input parameter.

Program 5.15   C software to sample data using the ADC.

```
// MC68HC11A8                        // MC68HC812A4/MC68HC912B32
void ADC_Init(void){                 void ADC_Init(void){
  OPTION = 0x80;} //  Activate ADC      TSCR = 0x80;       // enable TCNT
#define CCF 0x80                        ATDCTL2 = 0x80;}  // Activate ADC
unsigned char ADC_In(unsigned char chan){  #define SCF 0x8000
  ADCTL=chan;      // Start ADC        unsigned char ADC_In(unsigned char chan){
  while ((ADCTL & CCF) == 0){};          ATDCTL5=chan;     // Start ADC
  return(ADR1); }                        while ((ATDSTAT & SCF) == 0) {};
                                         return(ADR0H); }
```

Checkpoint 5.11: *Derive an equation for the digital value returned by the function* `ADC_In` *versus the analog input voltage.*

5.6   16-bit Timer

Both the 6811 and 6812 have a 16-bit timer called TCNT. This counter is incremented at a fixed rate, and the software can read its value to know the current time. It can also be used to create pulses, squarewaves and pulse-width modulated waves. It can be configured to measure the period, pulse-width, or frequency of an input signal. For information on creating waveforms and measuring events see Chapter 6 of *Embedded Microcomputer Systems Interfacing* by Jonathan W. Valvano. In this chapter, we will use TCNT to create fixed-time software delays. In chapter 11, we will use TCNT to create a periodic interrupt.

5.6.1
6811 Timer Details

On the 6811, TCNT is located at address $100E. The rate at which TCNT is incremented is determined by two bits (PR1 and PR0) in the TMSK2 register ($1024) as shown in Tables 5.16 and 5.17. Every time the TCNT register overflows from $FFFF to 0, the TOF flag in the TFLG2 register is set. The TOF condition will cause an interrupt if the mask TOI=1. Table 5.16 shows the 6811 timer ports.

Table 5.16   6811 timer ports.

address	Bit 7	6	5	4	3	2	1	Bit 0	Name
$100E	**Bit 15**	14	13	12	11	10	9	**Bit 8**	TCNT
$100F	**Bit 7**	6	5	4	3	2	1	**Bit 0**	TCNT
$1024	TOI	RTII	PAOVI	PAII	0	0	**PR1**	**PR0**	TMSK2
$1025	**TOF**	RTIF	PAOVF	PAIF	0	0	0	0	TFLG2

The flags in the TFLG2 register are cleared by writing a 1 into the specific flag bit we wish to clear. For example, writing a $80 into TFLG2 will clear the TOF flag.

Table 5.17
Assuming a 2MHz E
clock on the 6811, PR1
and PR0 define the
TCNT clock period.

PR1	PR0	Divide by	TCNT clock
0	0	1	500 ns
0	1	2	1 μs
1	0	4	2 μs
1	1	8	4 μs

5.6.2
6812 Timer Details

Table 5.18 shows the 6812 timer ports. On the 6812, TCNT is a 16-bit unsigned counter that is incremented at a rate determined by three bits (PR2, PR1 and PR0) in the TMSK2 register ($008D) as shown in Table 5.19. Divide by 64 and divide by 128 modes are available in the TMSK2 register on the MC68HC912D60.

Table 5.18 6812 timer ports.

Address	Bit 7	6	5	4	3	2	1	Bit 0	Name
$0016	CME	FCME	FCM	FCOP	DISR	CR2	CR1	CR0	COPCTL
$0017	Bit 7	6	5	4	3	2	1	Bit 0	COPRST
$0084	Bit 15	14	13	12	11	10	9	Bit 8	TCNT
$0085	Bit 7	6	5	4	3	2	1	Bit 0	TCNT
$0086	TEN	TSWAI	TSBCK	TFFCA	PAOQE	T7QE	T1QE	T0QE	TSCR
$008D	TOI	0	TPU	TDRB	TCRE	PR2	PR1	PR0	TMSK2
$008F	TOF	0	0	0	0	0	0	0	TFLG2

Table 5.19
Assuming an 8MHz E
clock on the 6812, PR2,
PR1, and PR0 define the
TCNT clock period.

PR2	PR1	PR0	Divide by	TCNT clock
0	0	0	1	125 ns
0	0	1	2	250 ns
0	1	0	4	500 ns
0	1	1	8	1 μs
1	0	0	16	2 μs
1	0	1	32	4 μs
1	1	0	reserved	-
1	1	1	reserved	-

5.6.3
Fixed-Time Delay
Software Using
the Built-In Timer

In order to use TCNT on the 6812, you must first set bit 7 of the TSCR register. For both the 6811 and 6812, you can set the prescale bits to specify the TCNT count period. Program 5.16 shows how to use TCNT to create a fixed-time delay. The Init function is needed only for the 6812. The delay parameter can be any number from 1 to 32767.

Program 5.16 Timer function that implements a fixed time delay.

```
* Enable TCNT                                  void Init(void){
Init   movb #$80,TCSR                            TCSR = 0x80; /* enable TCNT */
       rts                                     }
* Reg D is the time to wait in cycles          void Wait(unsigned short delay){
Wait   addd  TCNT end of wait time              short Endt;           /* must be signed *
wloop  cpd   TCNT EndT-TCNT<0 when EndT<TCNT       Endt=TCNT+delay;   /* end of wait time */
       bpl   wloop                                while((Endt-(short)TCNT)>0);
       rts                                     }
```

Observation: *Assuming delay is less than or equal to 32767, then Endt-TCNT will switch from positive to negative only when TCNT counts past Endt.*

Checkpoint 5.12: *Write assembly code to initialize the timer so that it counts every 1 μs.*

Checkpoint 5.13: *If the TCNT counts every 4 μs, how long does it take for it to count all the way from 0 to $FFFF and back to 0 again?*

5.7 Pulse Accumulator

5.7.1
6811 Pulse
Accumulator
Details

The 6811 pulse accumulator is an 8-bit read/write counter that can operate in either of two modes. External event counting mode can be used for counting events or frequency measurement. We will use gated time accumulation mode for pulse width measurement. In the event counting mode, the 8-bit counter is incremented on either the rising edge or falling edge of PA7. The maximum clocking rate for the external event counting mode is the E clock frequency divided by two. In the gated time-accumulation mode, a free-running E clock/64 signal increments the 8-bit counter, but only while the PA7 input is enabled. For a 2-MHz 6811, the period of this clock will be 32μs. The I/O ports involved in the 6811 pulse accumulator are shown in Table 5.20.

Table 5.20   6811 I/O ports used by the pulse accumulator.

address	Bit 7	6	5	4	3	2	1	Bit 0	Name
$1000	PA7	PA6	PA5	PA4	PA3	PA2	PA1	PA0	PORTA
$1024	TOI	RTII	PAOVI	PAII	0	0	PR1	PR0	TMSK2
$1025	TOF	RTIF	PAOVF	PAIF	0	0	0	0	TFLG2
$1026	DDRA7	PAEN	PAMOD	PEDGE	DDRA3	I4/O5	RTR1	RTR0	PACTL
$1027	Bit 7	6	5	4	3	2	1	Bit 0	PACNT

DDRA7 is the Data Direction bit for PA7. Normally, the DDRA7 bit is cleared so PA7 is an input, but even if it is configured for output, PA7 still drives the pulse accumulator. **PAEN** is the Pulse Accumulator System Enable bit. Turn this bit on to activate the pulse accumulator. The PAMOD and PEDGE bits select the operation mode, as shown in Table 5.21.

Table 5.21
6811 pulse accumulator
operation modes.

PAMOD	PEDGE	Mode	Action on Clock
0	0	event counting	PA7 Falling Edge Increments the Counter
0	1	event counting	PA7 Rising Edge Increments the Counter
1	0	gated time accumulation	Counts when PA7=1
1	1	gated time accumulation	Counts when PA7=0

The **PAOVF** status bit is set each time the pulse-accumulator count rolls over from $FF to $00. To clear this status bit, we write a one to the TFLG2 register bit 5. The **PAOVI** will arm the device so that a pulse-accumulator interrupt is requested when PAOVF is set. When **PAOVI** is zero, pulse-accumulator overflow interrupts are disarmed. The **PAIF** status bit is automatically set each time a selected edge is detected at the PA7 pin (PAMOD=0 means falling edge, and PAMOD=1 means rising edge). To clear this status bit, write to the TFLG2 register bit 4. The **PAII** will arm the device so that a pulse-accumulator interrupt is requested when PAIF is set. When PAII is zero, pulse-accumulator input interrupts are disarmed.

5.7.2
6812 Pulse
Accumulator
Details

The 6812 pulse accumulator is a 16-bit read/write counter that can operate in either of two modes. External event-counting mode can be used for counting events or frequency measurement. We will use gated time-accumulation mode for pulse width measurement. In the event-counting mode, the 16-bit counter is incremented on either the rising edge or falling edge of PT7. The maximum clocking rate for the external event-counting mode is the E clock frequency divided by two. In the gated time-accumulation mode, a free-running clock increments the 16-bit counter, but only while the PT7 input is enabled. The I/O ports involved in the 6812 pulse accumulator are shown in Table 5.22.

Table 5.22   6812 I/O ports used by the pulse accumulator.

Address	Bit 7	6	5	4	3	2	1	Bit 0	Name
$0047	LCK	PLLON	PLLS	BCSC	BCSB	BCSA	MCSB	MCSA	CLKCTL
$0086	TEN	TSWAI	TSBCK	TFFCA	PAOQE	T7QE	T1QE	T0QE	TSCR
$00A0	0	PAEN	PAMOD	PEDGE	CLK1	CLK0	PAOVI	PAI	PACTL
$00A1	0	0	0	0	0	0	PAOVF	PAIF	PAFLG
$00A2	Bit 15	14	13	12	11	10	9	Bit 8	PACNT
$00A3	Bit 7	6	5	4	3	2	1	Bit 0	PACNT
$00AE	PT7	PT6	PT5	PT4	PT3	PT2	PT1	PT0	PORTT
$00AF	DDRT7	6	5	4	3	2	1	Bit 0	DDRT

DDRT7 is the Data Direction bit for PT7. Normally, the DDRT7 bit is cleared so PT7 is an input, but even if it is configured for output, PT7 still drives the pulse accumulator. **PAEN** is the Pulse Accumulator System Enable bit. Turn this bit on to activate the pulse accumulator. PAEN is independent from the timer enable TEN in the TSCR register. The PAMOD and PEDGE bits select the operation mode, as shown in Table 5.23.

Table 5.23
6812 pulse accumulator
operation modes.

PAMOD	PEDGE	Mode	Action on Clock
0	0	event counting	PT7 Falling Edge Increments the Counter
0	1	event counting	PT7 Rising Edge Increments the Counter
1	0	gated time accumulation	Counts when PT7=1
1	1	gated time accumulation	Counts when PT7=0

In gated accumulation mode, a free-running clock is used to count the PACNT while the PT7 input is active. Assuming the three bits BCSC, BCSB, BCSA in the CLKCTL register are zero, the 6812 PCLK has the same frequency as the E clock, typically 8 MHz. The MCSB and MCSA bits in the CLKCTL register determine the frequency of the module clock, MCLK, as shown in Table 5.24. The MCLK is also used by the SCI, TCNT, computer operating properly (COP) and the real-time interrupt (RTI). Therefore, changing the MCSB and MCSA bits will also affect these other devices. The PACLK is the MCLK divided by 64.

Table 5.24
6812 pulse accumulator
clock rates.

MCSB	MCSA	MCLK Rate	PACLK Rate	PACLK Period
0	0	8 MHz	125 kHz	8 μs
0	1	4 MHz	62.5 kHz	16 μs
1	0	2 MHz	31.25 kHz	32 μs
1	1	1 MHz	15.625 kHz	64 μs

The **PAOVF** status bit is set each time the pulse accumulator count rolls over from $FFFF to $0000. To clear this status bit, we write a one to the PAFLG register bit 1. The **PAOVI** will arm the device so that a pulse accumulator interrupt is requested when PAOVF is set. When **PAOVI** is zero, pulse-accumulator overflow interrupts are disarmed. The **PAIF** status bit is automatically set each time a selected edge is detected at the PT7 pin (PAMOD=0 means falling edge, and PAMOD=1 means rising edge). To clear this status bit, write to the PAFLG register bit 1. The **PAII** will arm the device so that a pulse accumulator interrupt is requested when PAIF is set. When PAII is zero, pulse-accumulator input interrupts are disarmed.

5.7.3
Frequency
Measurement

The goal is to measure frequency in Hz, which in this case is defined as the number of falling edges that occur in one second. The signal to be measured will be connected to the pulse accumulator input, which is PA7 on the 6811, and PT7 on the 6812. The frequency measurement function, shown in Program 5.17, enables the pulse accumulator and selects the event-counting mode. When measuring frequencys, it usually doesn't matter whether we count rising or falling edges. But, in this case, falling edges will be counted. The approach will be to initialize the pulse accumulator to event counting, clear the count, wait one second, then read the counter. Since frequency is defined as the number of edges in one second, the value in the PACNT after the one-second time delay will be frequency in Hz. The 6811 frequency-measurement range is 0 to 255 Hz, while the 6812 can measure 0 to 65535 Hz. In both cases, the frequency resolution (which is the smallest change in frequency that can be distinguished) will be 1 Hz. In general, the frequency resolution will be one divided by the fixed time during which counts are measured. The PAOVF bit will be set if the input frequency exceeds the measurement range.

Program 5.17   Frequency measurement using the pulse accumulator.

```
;6811 measures 0 to 255 Hz          ;6812 measures 0 to 65535 Hz
; returns Reg A = freq in Hz        ; returns Reg D = freq in Hz
Freq ldaa #$40    PA7 is input, PEN=1   Freq bclr DDRT,#$80  PT7 is input
     staa PACTL   count falling edges        movb #$40,PACTL count falling
     clr  PACNT                              movw #0,PACNT
     ldaa #$20                               movb #$02,PAFLG clear PAOVF
     staa TFLG2   clear PAOVF                bsr  Wait1sec
     bsr  Wait1sec                           brclr PAFLG,#$02,ok   check PAOVF
     ldaa TFLG2                         bad  ldd  #65535          too big
     bita #$20    check PAOVF                bra  out
     beq  ok                           ok   ldd  PACNT
bad  ldaa #255    too big             out  rts
     bra  out
ok   ldaa PACNT
out  rts
```

Checkpoint 5.14: *What is the frequency resolution of the system implemented in Program 5.17? What does it mean?*

Checkpoint 5.15: *How do you modify Program 5.17 so that it measures frequency in kHz?*

5.7.4 Pulse-Width Measurement

The goal is to measure pulse width, which in this case will be defined as the time the input signal is high. Again, the input signal will be connected to the pulse-accumulator input, which is PA7 on the 6811, and PT7 on the 6812. The pulse-width measurement function, shown in Program 5.18, enables the pulse accumulator and selects gated-accumulation mode. In this case, PEDGE is set to zero, so the PACNT will accumulate when the input is high. With PEDGE equal to zero, the PAIF will be set on the falling edge of the input, signaling that the pulse-width measurement is complete. The approach will be to initialize the pulse accumulator to gated-accumulation mode, clear the count, wait for PAIF to be set, then read the counter. Since PACNT counts while the input is high, the value in this counter will represent the width of the pulse. The pulse-width resolution is the smallest change in pulse width that can be distinguished. In general, the pulse-width resolution will be the period of the free-running clock used to increment the counter. For the 6811, the pulse-width resolution will be 32μs. Assuming the 6812 MCLK is the default value of 8 MHz, its pulse-width resolution will be 8μs. The 6811 pulse-width measurement range is 32μs to 8.16ms, while the 6812 can measure 8μs to 524.28ms. The PAOVF bit will be set if the input pulse width exceeds the measurement range. If the input signal has a pulse width of 1 ms, then the 6811 function will return a value of 1000/32 or 31, while the 6812 function will give a result of 1000/8 or 125.

Program 5.18 Pulse-width measurement using the pulse accumulator.

```
;6811 measures 32us to 8.16ms          ;6812 measures 8us to 524.28 ms
; returns Reg A = pulse width in 32us  ; returns Reg D = pulse width in 8us
Puls ldaa #$60   PA7 is input, PEN=1   Puls bclr DDRT,#$80   PT7 is input
     staa PACTL  measure when high          movb #$60,PACTL measure high
     clr  PACNT                             movw #0,PACNT
     ldaa #$20                              movb #$02,PAFLG clear PAOVF
     staa TFLG2  clear PAOVF           loop brclr PAFLG,#$01,loop
loop ldaa TFLG2                             brclr PAFLG,#$02,ok   check PAOVF
     bita #$10   check PAIF            bad  ldd  #65535          too big
     beq  loop   wait for falling edge      bra  out
     bita #$20   check PAOVF           ok   ldd  PACNT
     beq  ok                          out  rts
     ldaa #255   bad measurement
     bra  out
ok   ldaa PACNT
out  rts
```

Checkpoint 5.16: What is the pulse-width resolution of the system implemented in Program 5.18? What does it mean?

Checkpoint 5.17: What will be the output of Program 5.18 if the pulse width is 1234.5 μsec?

5.8 Tutorial 5. I/O Programming

The objective of this tutorial is to illustrate I/O programming of the serial port and ADC. An analog signal is sampled using the ADC, and the results are displayed on a remote CRT terminal using the serial port. More sophisticated data acquisition systems will be discussed later in Chapter 11. As our software becomes more complete, we need alternative methods to visualize its complexity. The first part of this tutorial will be to visualize the software in different formats.

Action: Copy the `Chap5.rtf`, `Chap5.uc`, `Chap5.scp`, and `Chap5.io.` files from the CD onto your hard drive. Start a fresh copy of **TExaS** and open the `Chap5.rtf` program file from within **TExaS**. This action should open the corresponding microcomputer, scope and IO Device windows.

Question 5.1. Flowcharts are a convenient way to describe computer algorithms. Look at the `Chap5.rtf` file and draw flowcharts of the `main ADC_In` and `SCI_OutCh` programs.

Question 5.2. Call-graphs are used to visualize software hierarchy. Look at the `Chap5.rtf` file and draw a call-graph of the software system. First, begin by defining the three software modules and two hardware devices, as shown in Figure 5.16. Ovals are software modules and rectangles are I/O devices. If there were any global data structures, they would also be shown as rectangles. Next, for each situation where one function calls another, draw a call arrow from the function that performs the call to the function it calls. If one function calls a second function more than once, show only one arrow. Finally, for each I/O device access (read or write) draw an arrow from the function to the I/O device.

Figure 5.16
The first step when drawing a call-graph is to list the functions and modules.

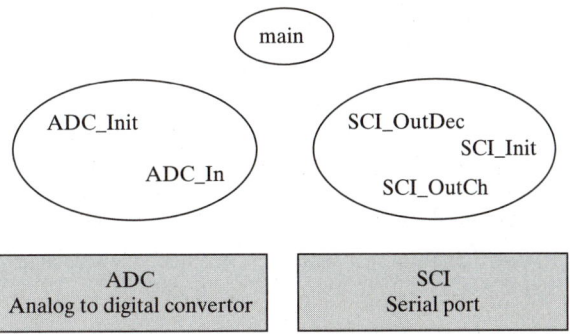

Question 5.3. A data flow graph illustrates the data as it flows from input to output. Draw a data-flow graph of this system, starting with the input hardware (ADC), going through each software function that handles the data, leading to the output hardware (SCI). Again, ovals are software modules and rectangles are I/O devices.

Action: Assemble and run `Chap5.rtf`. Observe the I/O device registers in the microcomputer window, the analog voltage versus time signal in the scope window and the status of the external hardware in the IO Device window. The results of the ADC sampling will be displayed on the `CRT.rtf` window.

Question 5.4. What is the TCNT clock period? In this system, the default value is used.

Question 5.5. How many TCNT cycles occur in 1 ms?

Action. You will measure the time from ADC sample to ADC sample. First, remove all entries in the Microcomputer **ViewBox** window except TCNT. Next, add a breakpoint at the assembly line that starts the ADC (first instruction of the `ADC_In` function.) Next, switch the breakpoint system from *Break* mode to *Scan* mode. In particular, execute `Mode->BreakMode` so that the check mark is absent. Run the program and observe the timing results in the `TheLog.RTF` window. You should get results like Table 5.25.

Table 5.25
TCNT times when the ADC is sampled.

TCNT=50	TCNT=78
TCNT=62	TCNT=707
TCNT=411	TCNT=1508
TCNT=1161	TCNT=2309
TCNT=1911	TCNT=3110
TCNT=2661	TCNT=3907
TCNT=3411	TCNT=4708

Question 5.6. Using the results from Question 5.4, what is the average time in between ADC samples? Ignore the first couple of samples. Give your results in cycles and in μsec. Calculate the approximate sample rate in samples/sec.

Question 5.7. Write a software delay function that waits the number of TCNT cycles that is the difference between the number of TCNT cycles in one millisecond (calculated in Question 5.5) and the number of TCNT cycles in the current software (measured in Question 5.6). You can use the approach developed in Program 5.16. The goal is to make the sampling rate equal to 1 kHz.

Action. Add this software delay function to the main program, and test it by repeating the timing experiment performed in the previous **Action**.

5.9   Homework Problems

Homework 5.1: Write an assembly language subroutine that initializes Port C bits 5,4,1,0 to outputs and bits 7,6,3,2 to inputs. Make all Port D bits inputs, and all Port B bits outputs.

Homework 5.2: Write an assembly language subroutine that initializes Port C bits 7,4,3,0 to outputs and bits 6,5,2,1 to inputs. Make all Port D bits outputs.

Homework 5.3: Write an assembly language subroutine that initializes Port C bit 3 to an output. All other bits in the DDRC should remain unchanged.

Homework 5.4: Write an assembly language subroutine that initializes Port D bit 1 to an input. All other bits in the DDRD should remain unchanged.

Homework 5.5: Assume the baud rate is 9600 bits/sec. Show the serial port output versus time waveform that occurs when the ASCII characters "ABC" are transmitted one right after another. What is the total time to transmit the three characters?

Homework 5.6: Assume the baud rate is 19200 bits/sec. Show the serial port output versus time waveform that occurs when the ASCII characters "125" are transmitted one right after another. What is the total time to transmit the three characters?

Homework 5.7: Assume the 6811 E clock is 2 MHz and the 6812 E clock is 8 MHz. Write an assembly language subroutine that initializes the serial port to communicate at 9600 bits/sec, 8-bit data, 1 start bit, and 1 stop bit.

Homework 5.8: Sometimes it is important for the software to know when the SCI transmission is complete. The transmit complete (TC) flag is set after the data in the shift register have been transmitted. Rewrite the `SCI_OutChar` subroutine so that it first writes to the data register, then waits for the TC flag to be set. The TC flag is cleared by first reading the status register with TC set and then writing into the transmit data register.

Homework 5.9: The Maxim MAX549 is a 2-channel 8-bit DAC similar to the MAX550. You can find the data sheet for both these chips on the CD as file MAX548.PDF. Develop DACinit and DACout functions similar to the MAX550 example in the chapter, except the DACout function includes a channel number in Register B. If Register B equals 0, then output the value in Register A to DAC channel A. If Register B equals 1, then output the value in Register A to DAC channel B.

Homework 5.10: Write an assembly language subroutine that samples ADC channel 2 four times, calculates the average of the four samples, and returns the result in Register A.

Homework 5.11: Write an assembly language subroutine that samples all eight ADC channels, calculates the average of the eight samples, and returns the result in Register A.

Homework 5.12: Write an assembly language subroutine that samples all eight ADC channels, calculates the minimum and maximum of the eight samples, and returns the range (maximum-minimum) in Register A.

Homework 5.13: Write an assembly language subroutine that samples ADC channels (0,1,2) calculates the median of the three samples, and returns the result in Register A.

Homework 5.14: Rewrite assembly language Program 5.14 so that it returns a voltage in Register D using decimal fixed point with $\Delta = 0.001$ V.

Homework 5.15: Rewrite assembly language Program 5.14 so that it returns a voltage in Register D using binary fixed point with $\Delta = 2^{-8}$ V.

Homework 5.16: Write an assembly language subroutine that implements a 1-second delay using the TCNT timer. Include both an initialization ritual and a delay function.

5.10 Laboratory Assignments

Each of the laboratory assignments in this chapter involves real-time periodic ADC sampling. The objectives of these labs are to

■ use the ADC to convert analog signals into digital form,
■ synchronize the ADC sampling to a hardware clock, using the TCNT clock,
■ study decimal fixed-point numbers,
■ understand how to convert binary data into an ASCII string, which can be displayed.

In each case you will write two subroutines, `Timer_Init` and `Timer_Sync`, that can be used to execute periodic events. One possible approach is illustrated in Program 5.19:

Program 5.19 Synchronization programs for Lab 5.1.

```
static short Period;    /* sampling period in cycles */
static short NextT;     /* it is important to case this as signed */
void Timer_Init(short thePeriod){
  TCSR = 0x80;          /* (6812 only) enable TCNT */
  Period = thePeriod;   /* TCNT cycles per period */
  NextT = TCNT;
}
void Timer_Sync(void){
  NextT = NextT+Period;          /*  TCNT at next sample */
  while((NextT-(short)TCNT)>0); /*  wait for TCNT to pass NextT */
}
void main(void){ unsigned char data;
  Timer_Init(1234);   /*  number of cycles per sample */
  ADC_Init();         /*  Activate ADC */
  while(1){
    Timer_Sync();
    data = ADC_In(0); /*  sample ADC */
/********* process data ************/
  }
}
```

Observation: *The addition of* `static` *to a variable that is otherwise already global makes the variable private, which means it is only accessible to functions within this file.*

If you develop the parts of this lab in a-b-c-d order, you performing **bottom-up** program development. To perform bottom-up development, you begin with a low-level function like part a, and use dummy main program (like the one shown in Program 5.19) to test the function. A dummy main program passes data into the function and evaluates the output results as needed. The operation is similar to the eventual main program, but is usually a lot simpler. After all low-level functions are designed and tested (parts a-b-c), the individual low-level functions are combined to build the next higher level. In this way, the components are first tested individually, then tested again each time the system conplexity is increased.

If you develop the parts of this lab in d-c-b-a order, you performing **top-down** program development. To perform top-down development, you begin with part d, and use snubs for each of the subroutines you haven't yet written. A **snub** is a dummy function that passes data in and out as needed, but doesn't actually perform any work. As you proceed further down the development process, you replace the snubs with actual programs.

Place strategic global variables in the **ViewBox** to assist in debugging. Use the debugger to set two breakpoints, one before and one immediately after important sections of code. Run your program and use the breakpoints to observe the inputs and outputs of the piece of code. Add labels to your program to simplify the process of adding breakpoints.

When the `TheCRT.rtf` window is active, most of the keys are passed from the PC keyboard to the SCI port. Some of the keystrokes may be processed by the simulator (e.g., <ctrl>B F12) and others are processed by the Windows operating system (e.g., <PrntScrn>). The CRT simulation is restricted to ASCII codes $08, $09, $0D and $20 to $7F.

Modular subroutines have one entry point (where to start) and one exit point. In particular, only one `rts` instructions allowed per subroutine. For clarity, make the first instruction the entry point and the last instruction `rts`. Save and restore any modified registers (obviously excluding the output parameters.)

Lab 5.1 Voltmeter

In Tutorial 5, an analog input voltage is measured, and the results are displayed as an integer on a CRT terminal, which is interfaced through the SCI serial port. In this lab you will extend this system in three ways. First, you will develop an accurate way to establish the 1 kHz sampling. In particular, the ADC should be started exactly every 1 ms. Second, you will convert the ADC sample (0 to 255) into a decimal fixed-point number, with a Δ of 0.01V. Lastly, you will develop a fixed-point display function, which will be used to display the sampled signal on the CRT.

Part A. Write two subroutines, `Timer_Init` and `Timer_Sync`, as described in Program 5.19.

Part B. Write a subroutine that converts an 8-bit binary ADC sample into unsigned fixed-point format. The input parameter to the subroutine will be passed in using Register A, and your subroutine will return the result in Register D. Table 5.26 shows some example results. Do not worry if your answers differ by ± 1, because of rounding.

Table 5.26
Example results of the conversion from ADC sample to fixed-point.

Analog input	ADC sample	Fixed-point Output
0.000 V	%00000000	0
1.234 V	%00111111	123
3.456 V	%10110000	346
5.000 V	%11111111	500

Part C. Write a subroutine that outputs the fixed-point number to the CRT interfaced to the SCI port. The input parameter to the subroutine will be passed using Register D. Table 5.27 shows some example results. The data to the CRT display must be in ASCII. You can call the `SCI_OutCh` function (from Tutorial 5) to send one ASCII character to the CRT.

Table 5.27
Example results of the
fixed-point display
subroutine.

Fixed-point	CRT display
0	0.00 V
123	1.23 V
346	3.46 V
500	5.00 V

Part D. Write the main program that first initializes the timer, SCI, and ADC. The main program runs in an infinite loop, taking an ADC sample every 1 ms, converting it to fixed point, then displaying the results on the CRT terminal.

Lab 5.2
Pressure Monitor

Assume analog signal connected to the microcomputer comes from a pressure sensor, such that the analog voltage ranges from 0 to +5V as the pressure ranges from 0 to 40 dynes/cm^2. In Tutorial 5, an analog input voltage is measured, and the results are displayed as an integer on a CRT terminal, which is interfaced through the SCI serial port. In this lab you will extend this system in three ways. First, you will develop an accurate way to establish 2 kHz sampling. In particular, the ADC should be started exactly every 0.5 ms. Second, you will convert the ADC sample (0 to 255) into a decimal fixed-point number, with a Δ of 0.1 dynes/cm^2. Lastly, you will develop a fixed-point display function, which will be used to display the sampled signal on the CRT.

Part A. Write two subroutines, `Timer_Init` and `Timer_Sync` as described in Program 5.19.

Part B. Write a subroutine that converts an 8-bit binary ADC sample into unsigned fixed-point format. The input parameter to the subroutine will be passed using Register A, and your subroutine will return the result also in Register A. Table 5.28 shows some example results. Do not worry if your answers differ by ± 1, because of rounding.

Table 5.28
Example results of the
conversion from ADC
sample to fixed-point.

Pressure	Analog input	ADC sample	Fixed-point Output
0.00 dynes/cm^2	0.000 V	%00000000	0
9.872 dynes/cm^2	1.234 V	%00111111	99
27.648 dynes/cm^2	3.456 V	%10110000	276
40.000 dynes/cm^2	5.000 V	%11111111	400

Part C. Write a subroutine that outputs the fixed-point number to the CRT interfaced to the SCI port. The input parameter to the subroutine will be passed using Register D. Table 5.29 shows some example results. The data to the CRT display must be in ASCII. You can call the `SCI_OutCh` function (from Tutorial 5) to send one ASCII character to the CRT.

Table 5.29
Example results of the
fixed-point display
subroutine.

Fixed-point	CRT display
0	0.0 dynes/cm**2
99	9.9 dynes/cm**2
276	27.6 dynes/cm**2
400	40.0 dynes/cm**2

Part D. Write the main program that first initializes the timer, SCI, and ADC. The main program runs in an infinite loop, taking an ADC sample every 0.5 ms, converting it to fixed point, then displaying the results on the CRT terminal.

Lab 5.3
Temperature
Monitor

Assume analog signal connected to the microcomputer comes from a temperature sensor, such that the analog voltage ranges from 0 to +5V as the temperature ranges from −30 to 30 °C. In Tutorial 5, an analog input voltage is measured, and the results are displayed as an integer on a CRT terminal, which is interfaced through the SCI serial port. In this lab you will extend this system in three ways. First, you will develop an accurate way to establish 2 kHz sampling. In particular, the ADC should be started exactly every 0.5 ms. Second, you will convert the ADC sample (0 to 255) into a decimal fixed-point number, with a Δ of 0.1 °C. Lastly, you will develop a fixed-point display function, which will be used to display the sampled signal on the CRT.

Part A. Write two subroutines, `Timer_Init` and `Timer_Sync` as described in Program 5.19.

Part B. Write a subroutine that converts an 8-bit binary ADC sample into signed fixed-point format. The input parameter to the subroutine will be passed in using Register A, and your subroutine will return the result in Register D. Table 5.30 shows some example results. Do not worry if your answers differs by ±1, because of rounding.

Table 5.30
Example results of the conversion from ADC sample to fixed-point.

Pressure	Analog input	ADC sample	Fixed-point Output
−30.000 °C	0.000 V	%00000000	−300
−15.192 °C	1.234 V	%00111111	−152
0.000 °C	2.500 V	%10000000	0
11.472 °C	3.456 V	%10110000	115
30.000 °C	5.000 V	%11111111	300

Part C. Write a subroutine that outputs the fixed-point number to the CRT interfaced to the SCI port. The input parameter to the subroutine will be passed using Register D. Table 5.31 shows some example results. The data to the CRT display must be in ASCII. You can call the `SCI_OutCh` function (from Tutorial 5) to send one ASCII character to the CRT.

Table 5.31
Example results of the fixed-point display subroutine.

Fixed-point	CRT display
−300	−30.0 C
−152	−15.2 C
0	0.0 C
115	11.5 C
300	30.0 C

Part D. Write the main program that first initializes the timer, SCI, and ADC. The main program runs in an infinite loop, taking an ADC sample every 0.5 ms, converting it to fixed point, then displaying the results on the CRT terminal.

6 Microcomputer Interfacing

Chapter 6 objectives are to

❏ discuss the basic approach to interfacing,
❏ present the external devices simulated by **TExaS**.

There are three components to microcomputer interfacing. Since many external devices have physical characteristics, the first step is the mechanical design of the physical components. Often the mechanical design is simply selecting the physical devices from a list of available components. The next step is the analog and digital electronics used to connect the physical devices to the computer. The input/output information may be encoded as simple digital signals or variable analog signals. More complex systems may use frequency, period, phase, or pulse width to represent the signals. The third component of interfacing is the low-level software that transforms the mechanical and electrical devices into objects that perform the desired tasks. The group of these low-level functions is often designated as an I/O device driver. The purpose of this chapter is to explain the various external devices available with the **TExaS** simulator. Since the basic hardware circuits and software interfaces are given, this chapter can also serve as an introduction to interfacing. For more detailed information on interfacing see *Embedded Microcomputer Systems: Real Time Interfacing* by Jonathan W. Valvano.

6.1 Introduction

The `IO menu` offers the following commands, which allow you to connect external devices to the microcomputer (these commands are available only when an IO file is active):

Switch	allows you to connect up to 8 toggle switches;
LED	allows you to connect up to 8 colored light emitting diodes;
CRT	allows you to connect a CRT terminate via the SCI port;
LCD Display	allows you to connect a liquid crystal display;
Analog	allows you to connect an analog input;
Keyboard	allows you to connect scanned matrix keyboard;
DC Motor	allows you to connect a DC Motor;
IR Remote	allows you to connect an IR remote control;
HD44780	allows you to connect a HD44780-controller LCD display;
Stepper	allows you to connect stepper motors;
WhiteBackground	allows you to toggle between white and black backgrounds.

6.2 Switch Interfacing

Use the **IO->Switch. . .** command to connect toggle switches to specific input ports of the microcomputer. Figure 6.1 shows the dialog box used to interface switches to the simulated microcomputer. The specific microcomputer should first be selected before executing this command. Up to eight switches can be added. For each switch, you can specify

I/O Port Pin
Label (optional)
One of eight colors

Figure 6.1
Switch dialog.

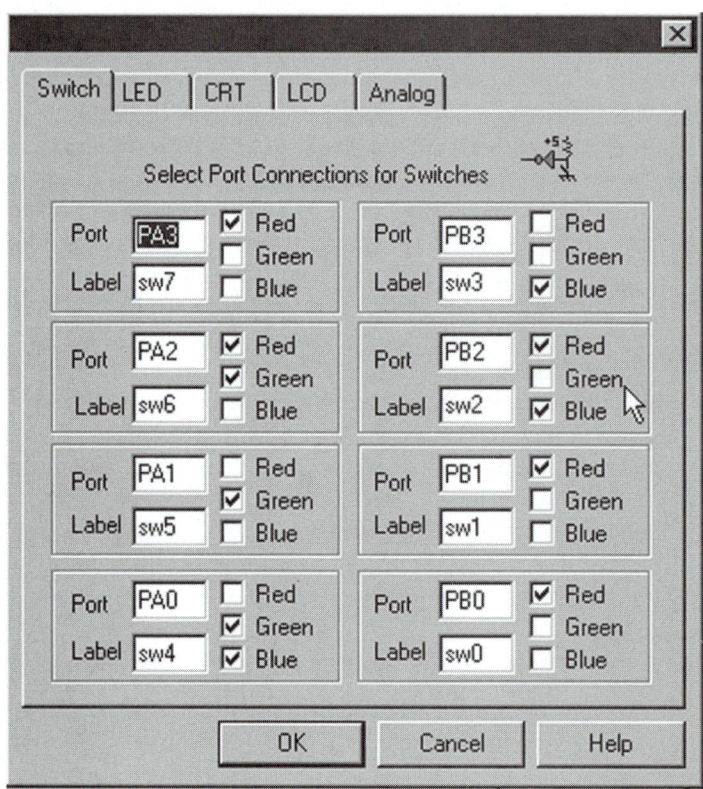

Observation: The system currently does not check for conflicts. For example, you should not connect two switches to the same pin.

Each circuit is similar to the electronics shown in Figure 6.2. The 1-kΩ resistor converts the switch resistance into a digital signal. The 33-μF capacitor removes the mechanical switch bounce that many switches have. The 22-Ω resistor prevents sparks when the switch is first touched, and the charge on the capacitor is discharged through the switch. The 74LS14 Schmitt trigger also removes noise that may exist. When the switch is pressed, its resistance is zero, V_1 is low, and the Output is high. When the switch is not pressed, its resistance is infinity, V_1 is high, and the Output is low.

Figure 6.2
Switch interface.

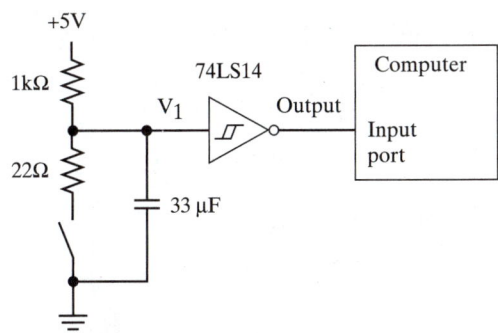

The 3-bit color scheme combines Red, Green, and Blue to make eight different colors as shown in Table 6.1.

Table 6.1
Switch colors.

Red	Green	Blue	Color
			Black
		x	Blue
	x		Green
	x	x	Cyan (greenish blue)
x			Red
x		x	Purple
x	x		Yellow
x	x	x	White

Checkpoint 6.1: *What is the purpose of the 1-kΩ resistor?*

Checkpoint 6.2: *What is the purpose of the 33-μF capacitor? In particular, would it still work if the 33 μF capacitor were removed?*

6.3 LED Interfacing

The **IO->LED. . .** command allows you to connect individual light-emitting diodes to specific output ports of the microcomputer. Figure 6.3 shows the dialog box used to interface LEDs to the simulated microcomputer. The specific microcomputer should first be selected before executing this command. Up to eight LEDs can be added. For each LED you can specify

I/O Port Pin
Label (optional)
One of eight colors (See Table 6.1)

Figure 6.3
LED dialog.

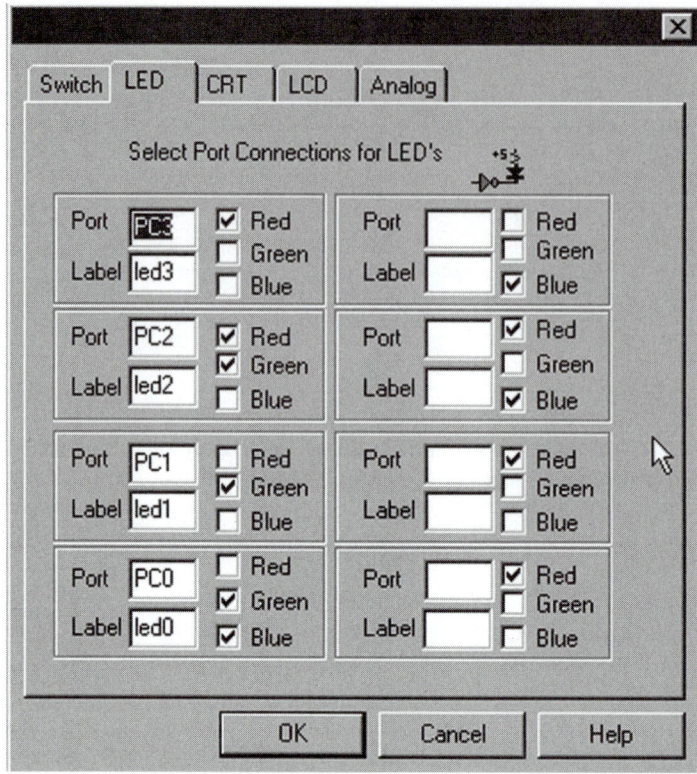

The equivalent LED interface circuit is shown in Figure 6.4. Normally, we connect a microcomputer output to the LED. In this way, the software can control the LED state. When the software writes a logic 1 to the output port, the input to the 7405 becomes high, output from the 7405 becomes low, 10 mA travels through the LED, and the LED is on. When the software writes a logic 0 to the output port, the input to the 7405 becomes low, output from the 7405 floats (neither high or low), no current travels through the LED, and the LED is dark. If the pin has another signal connected to it (like a switch or CRT serial input), then it is valid to connect LEDs to microcomputer input pins. The value of the resistor is selected to create the proper LED current. When active, the LED voltage will be about 2 V, and the power delivered to the LED will be controlled by its current. The resistor value should be

$$R = \frac{5\text{ V} - (\text{LED voltage}) - (\text{output low voltage of the 7405})}{\text{desired LED current}} = \frac{(5 - 2 - 0.5\text{ V})}{0.01\text{ A}} = 250\Omega$$

Figure 6.4
LED interface.

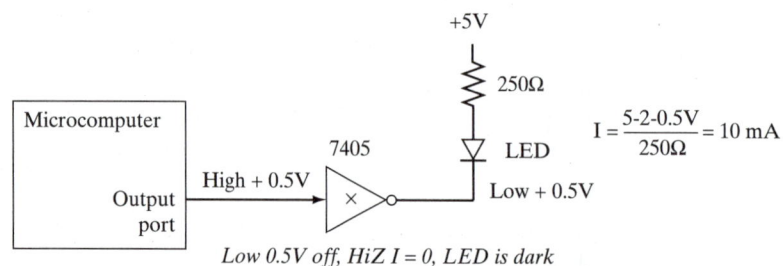

Low 0.5V off, HiZ I = 0, LED is dark

Checkpoint 6.3: *What resistor value is needed if the desired LED operating point is 1.5V and 5 mA?*

6.4 **Serial Port Interfacing**

Table 6.2 shows some of RS232 signals. The RS232 standard uses a DB25 connector that has 25 pins. The EIA-574 standard uses RS232 voltage levels and a DB9 connector that has only 9 pins. The most commonly used signals of the full RS232 standard are available with the EIA-574 protocols. Only TxD, RxD and SG are required to implement a simple bidirectional serial channel. We define the *data terminal equipment* (DTE) as the computer or a terminal and the *data communication equipment* (DCE) as the modem or printer.

Table 6.2
Pin assignments for some of the signals on the RS232 and EIA-574 protocols.

DB25 Pin	RS232 Name	DB9 Pin	EIA-574 Name	Signal	Description	True	DTE	DCE
2	BA	3	103	TxD	Transmit Data	−12V	out	in
3	BB	2	104	RxD	Receive Data	−12V	in	out
4	CA	7	105/133	RTS	Request to Send	+12V	out	in
5	CB	8	106	CTS	Clear to Send	+12V	in	out
6	CC	6	107	DSR	Data Set Ready	+12V	in	out
7	AB	5	102	SG	Signal Ground			
8	CF	1	109	DCD	Data Carrier Detect	+12V	in	out
20	CD	4	108	DTR	Data Terminal Ready	+12V	out	in
22	CE	9	125	RI	Ring Indicator	+12V	in	out

A Maxim converter chip is used to generate the +12 and −12 V RS232 voltage levels, as shown in Figure 6.5. The RS232 timing is generated automatically by the SCI. During transmission, the Maxim chip translates a digital high on microcomputer side to −12V on the RS232/EIA-574 cable, and a digital low is translated to +12V. During receiving, the Maxim chip translates negative voltages on RS232/EIA-574 cable to a digital high on the microcomputer side, and a positive voltage is translated to a digital low. The computer is classified as DTE, so its serial output is pin 3 in the EIA-574 cable, and its serial input is pin 2 in the EIA-574 cable. When connecting a DTE to another DTE, we use a cable with pins 2 and 3 crossed (i.e., pin 2 on one DTE is connected to pin 3 on the other DTE and pin

Figure 6.5
Hardware interface implementing an asynchronous RS232 channel.

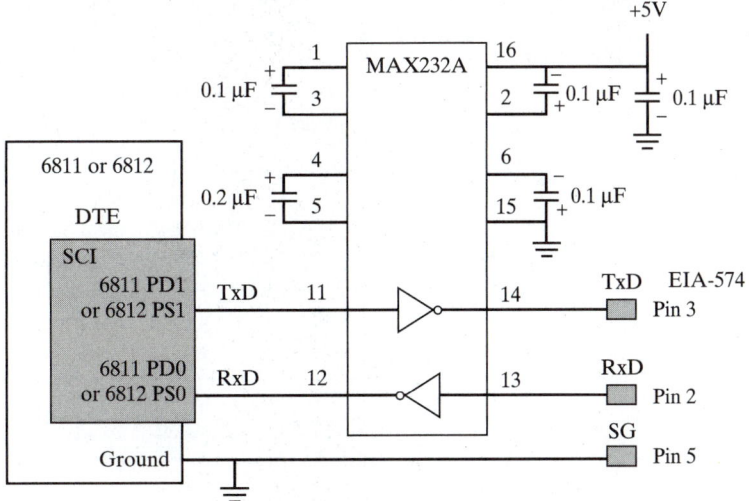

3 on one DTE is connected to pin 2 on the other DTE). When connecting a DTE to a DCE, then the cable passes the signals straight across. In all situations, the grounds are connected together using the SG wire in the cable. This channel is classified as full duplex, because transmission can occur in both directions simultaneously.

The **IO->CRT. . .** command allows you to connect a full-duplex CRT remote terminal using the SCI port of the microcomputer. Figure 6.6 shows the dialog box used to interface a CRT to the SCI serial port of the simulated microcomputer. You can connect LED's to the same pins and observe the serial transmission. The specific microcomputer should first be selected before executing this command. Only one CRT terminal can be added. You can specify one or two connections to the CRT module. In each case you give an I/O port name that represents a SCI serial pin:

TxD should be the SCI serial output (6812 PS1, or 6811 PD1)
RxD should be the SCI serial input (6812 PS0, or 6811 PD0)

The baud rate is the communication rate in bits/sec implemented by the CRT. This CRT baud rate must match within 5% of the rate created by the microcomputer's SCI port. A baud rate of 0 turns off this mode of the simulation. It is possible to implement either simplex communication from the CRT to the microcomputer (leaving the TxD field blank) or simplex communication from the microcomputer to the CRT (leaving the RxD field blank).

Figure 6.6
SCI/CRT interface dialog.

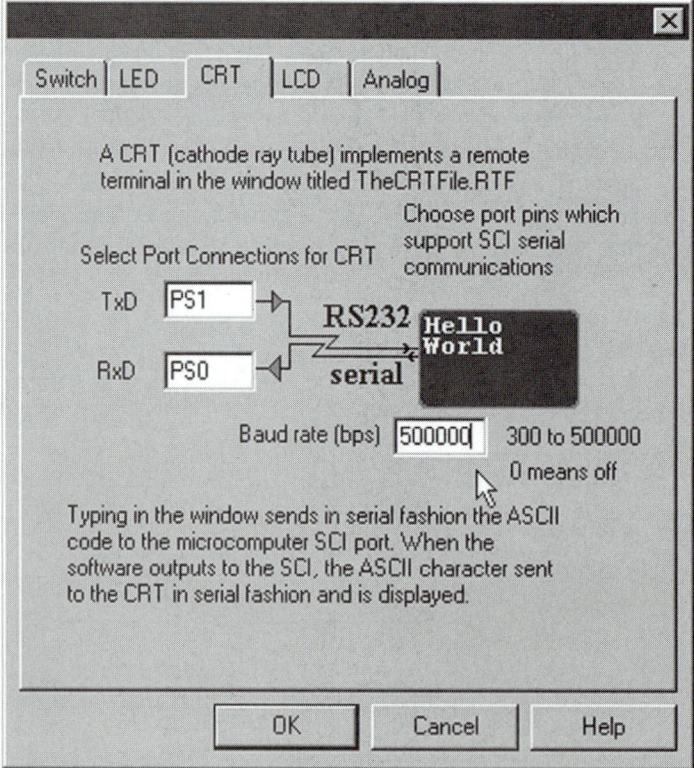

Serial transmission occurs as 10- or 11-bit frames. Figure 6.7 shows the RS232 protocol for transmitting one frame.

Figure 6.7
RS232 serial timing showing one data frame.

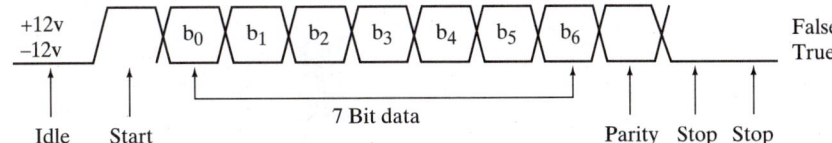

The TxData serial output is generated by a shift register, as shown in Figure 6.8.

Figure 6.8
Block diagram of the SCI transmitter.

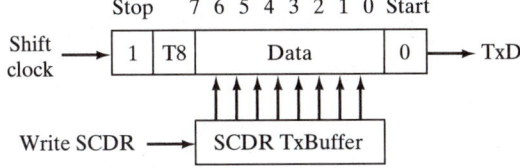

The RxData serial input is accepted by a shift register, as shown in Figure 6.9.

Figure 6.9
Block diagram of the SCI receiver.

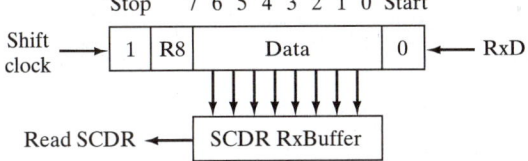

A device driver for the SCI port was presented earlier as Programs 5.7 and 5.8.

Checkpoint 6.4: Is it possible to interface a 6811 to a 6812 using their serial ports?

Checkpoint 6.5: If you are connecting two 6811/6812s together, are MAX232 interface chips required?

6.5 LCD Interfacing

The **IO->LCD. . .** command allows you to connect an integrated liquid crystal display (LCD) to specific output ports of the microcomputer. Figure 6.10 shows the dialog box used to interface a simple LCD to the simulated microcomputer. If the pin has another signal connected to it (like a switch) then it is valid to connect LCDs to microcomputer input pins. The specific microcomputer should first be selected before executing this command. Only one LCD module can be added. You can specify one, two, three, or four connections to the LCD module. In each case you give an I/O port name that represents four contiguous bits:

PA0 means PA3,PA2,PA1,PA0
PB4 means PB7,PB6,PB5,PB4

Figure 6.10
Dialog box for
interfacing a simple LCD
display.

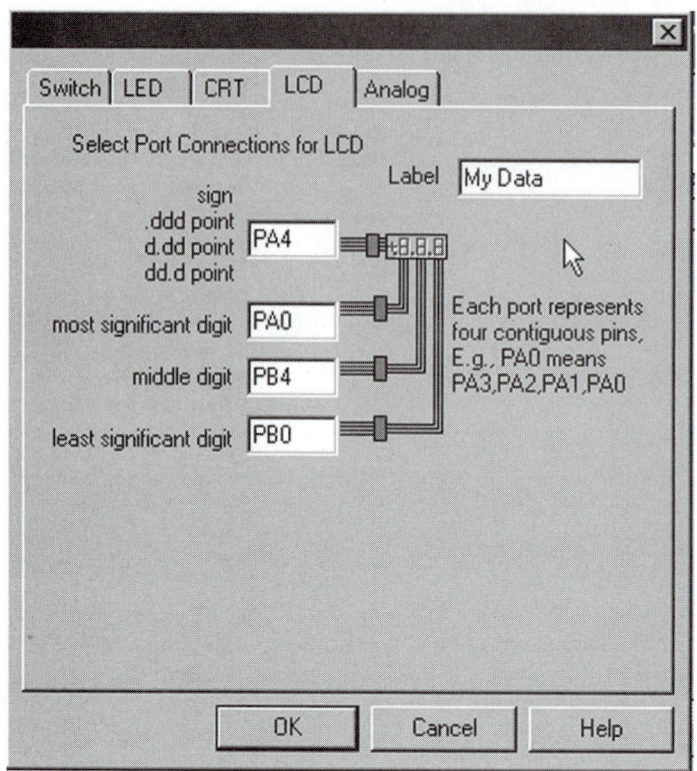

Each of the four connections has a specific function:

1) sign, ddd point field specifies 4 bits used for the

   sign   1 for minus, 0 for plus

   .ddd   1 for decimal point on in the leftmost position, 0 for this point off

   d.dd   1 for decimal point on between the left and middle digits, 0 for this point off

   dd.d   1 for decimal point on between the middle and right digits, 0 for this point off

If this field is left blank there will be no sign or decimal points

2) most signficant digit specifies 4 bits used for the leftmost digit

   0,1,2,3,4,5,6,7,8,9,A,b,C,d,E,F

   If this field is left blank there will be no most signficant digit or .ddd decimal point

3) middle digit specifies 4 bits used for the middle digit

   0,1,2,3,4,5,6,7,8,9,A,b,C,d,E,F

   If this field is left blank there will be no middle digit or d.dd decimal point

4) least significant digit specifies 4 bits used for the rightmost digit

   0,1,2,3,4,5,6,7,8,9,A,b,C,d,E,F

   If this field is left blank there will be no least significant digit or dd.d decimal point

Observation: *Some I/O ports do not have a full 8 bits (e.g., 6811 Port D, MC68HC812A4
Port F). Connecting an LCD to these incomplete ports may limit the software's ability to
control the entire display. For example, if you connect to 6811 PD4, then you will not be
able to control the top two pins.*

Figure 6.11 shows the equivalent circuit to interface one LCD digit to the microcomputer. The ABCD inputs to the MC14542 controller determine the LCD pattern. All LCD drivers use a clock to generate the low-level signals required by the LCD display itself.

Figure 6.11
Equivalent circuit for each digit of the simple LCD display.

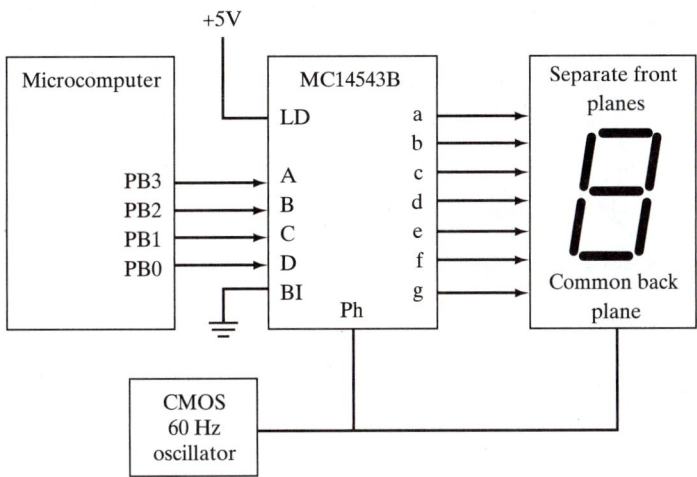

The data between the microcomputer and the LCD is packed BCD. Each LCD digit will be connected to 4 parallel port output bits from the microcomputer. The LCD digits can display decimal or hexadecimal information, as shown in Table 6.3.

Table 6.3
Conversion from 4-bit binary to LCD character.

4 bit value	LCD digit
0000	0
0001	1
0010	2
0011	3
0100	4
0101	5
0110	6
0111	7
1000	8
1001	9
1010	A
1011	b
1100	C
1101	d
1110	E
1111	F

Table 6.4 shows some typical results assuming the LCD has three digits without sign or decimal points.

Table 6.4
LCD display examples.

24-bit data	display
$123	123
$094	094
$FFE	FFE

An optional sign and decimal points can be added. Table 6.5 shows the LCD display, assuming the 24-bit data are $123.

Table 6.5
LCD display examples.

sign	display
0000	+123
0001	+12.3
0010	+1.23
0100	+.123
1000	-123
1001	-12.3
1010	-1.23
1100	-.123

Program 6.1 is a simple device driver for an LCD module connected to Port B and Port C. Digits are either in decimal format 0-9 or in hexadecimal format 0-F.

```
B7      sign (0 is plus, 1 is minus)
B6      left decimal point (0 is off, 1 is on)
B5      middle decimal point (0 is off, 1 is on)
B4      right decimal point (0 is off, 1 is on)
B3-B0   most significant digit
C7-C4   middle digit
C3-C0   least significant digit
```

Program 6.1   Device driver for the simple LCD display.

```
*****Initialize LCD****           *****Initialize LCD****
* Inputs: none                    * Inputs: none
* Outputs: none                   * Outputs: none
LCD_Init ldaa #$ff                LCD_Init movb #$ff,DDRB
         staa DDRC                         movb #$ff,DDRC
         rts                               rts
*****Output packed BCD to LCD**** *****Output packed BCD to LCD****
* Inputs: RegD is sign,decimal,   * Inputs: RegD is sign,decimal,
*              three digits       *              three digits
* Outputs: none                   * Outputs: none
LCD_Out staa PORTB sign, Msdigit  LCD_Out staa PORTB sign, MSdigit
        stab PORTC middle Lsdigit         stab PORTC middle LSdigit
        rts                               rts
```

Checkpoint 6.6: *What value should you make Register D in Program 6.1 in order to make the LCD show −90.6?*

6.6 Analog Interfacing

The **IO->Analog. . .** command allows you to connect analog signals to the ADC port of the microcomputer. Figure 6.12 shows the dialog box used to connect analog signals to the ADC of the simulated microcomputer.

Figure 6.12
Dialog box for connecting an analog signal to the built-in ADC.

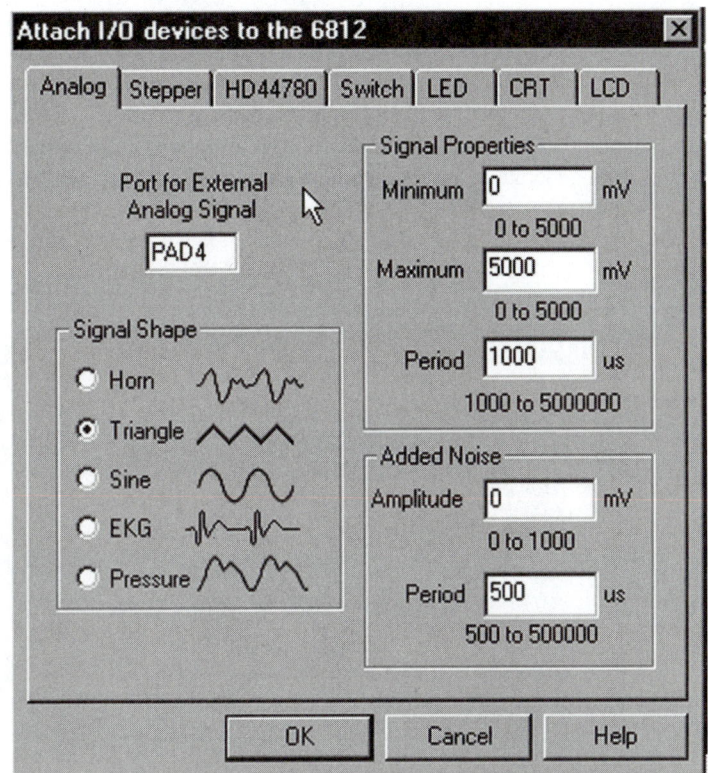

Signal choices include
    horn (trumpet sound)
    triangle
    sine
    EKG (human heart electrical wave)
    pressure (human aortic blood pressure)
You can specify
    the maximum voltage
    the minimum voltage
    the period (in each case the analog signal is periodic)
You can add noise as follows:
    Amplitude (0 means no added noise)
    Period (the noise is periodic sine wave)

A device driver for the ADC port was presented earlier in Chapter 5, as Programs 5.14 and 5.15.

Checkpoint 6.7: *What is the ADC resolution, which is the smallest change in analog input that can be detected?*

6.7 Scanned Key Pad Interfacing

The **IO->Keyboard. . .** command allows you to connect a matrix scanned keyboard to specific input/output ports of the microcomputer. Figure 6.13 shows the dialog box used to interface a matrix keyboard to the simulated microcomputer. The specific microcomputer should first be selected before executing this command. Up to sixteen keys in a four by four matrix can be added. For each row you specify an output Port Pin (shown as PD7,PD6,PD5,PD4 in the following figure). The row pins must either support open collector or have a direction pin. For each column you specify an input Port Pin (shown as PD3,PD2,PD1,PD0 in the following figure). An optional AND gate (74LS21) can be added (shown connected to PT0 in the following figure.) Later in the book, we will use this AND gate to generate an interrupt when a key is typed. For each key you specify a one- or two-letter label. A keyboard bouncing can be simulated.

Figure 6.13
Dialog box for connecting a matrix keyboard to the microcomputer.

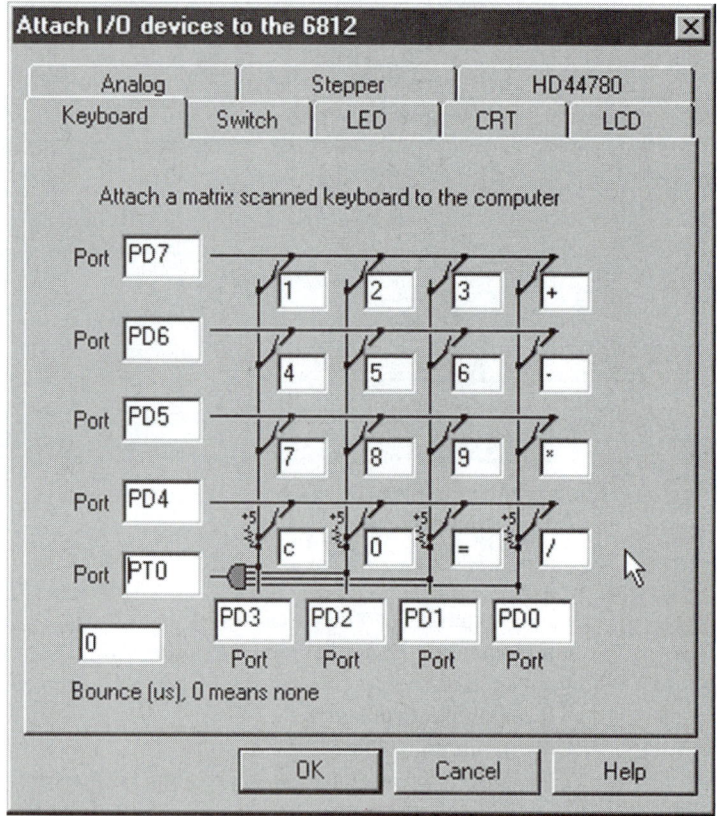

The corresponding keyboard generated by the above configuration is shown in Figure 6.14.

Figure 6.14
Example I/O window showing a matrix keyboard.

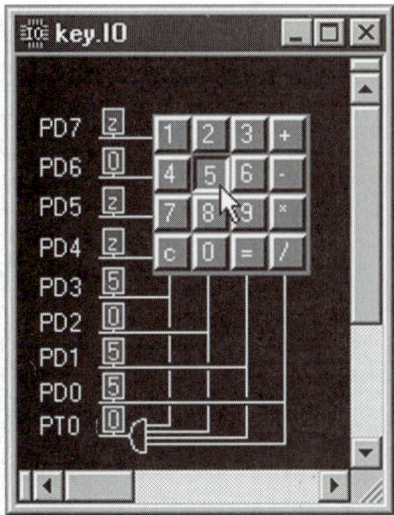

Figure 6.15 shows a corresponding interface circuit. We interface a matrix keyboard by connecting the rows to output ports and the columns to input ports. Each key has a unique (row, column) position. When a key is pressed the corresponding row is connected to the corresponding column.

Figure 6.15
Equivalent circuit for the matrix keyboard.

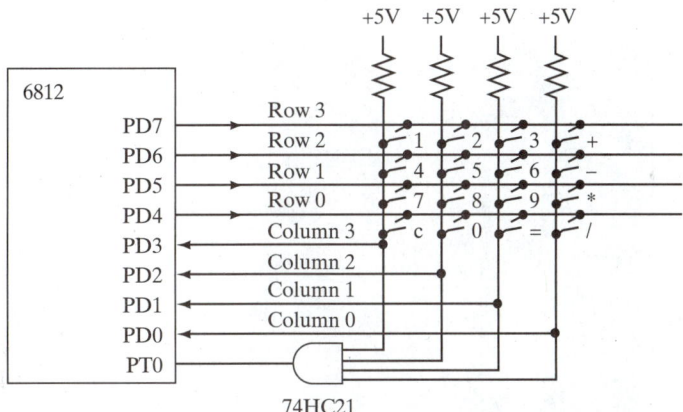

Individual rows, columns, AND gate and keys can be eliminated by removing the label from the port name or the key name. In the example shown in Figure 6.16, 10 keys are configured in a four-by-three matrix.

Figure 6.16
Example keyboard dialog
box.

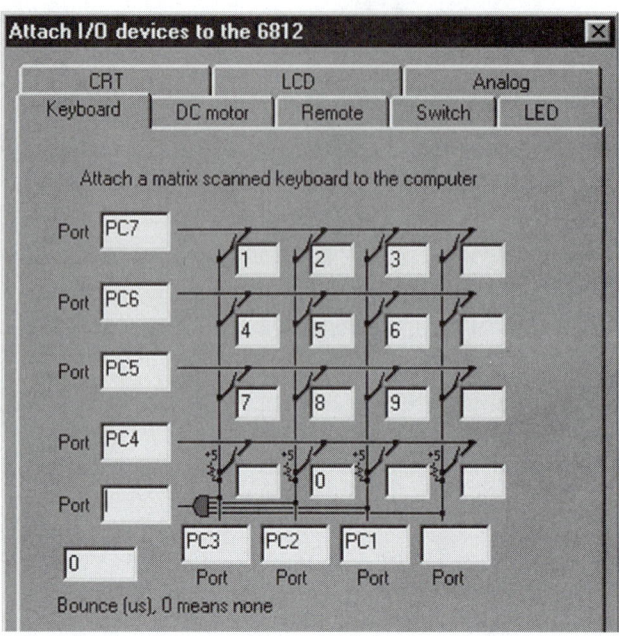

The corresponding keyboard generated by this second configuration is shown in Figure 6.17.

Figure 6.17
Example keyboard I/O
window.

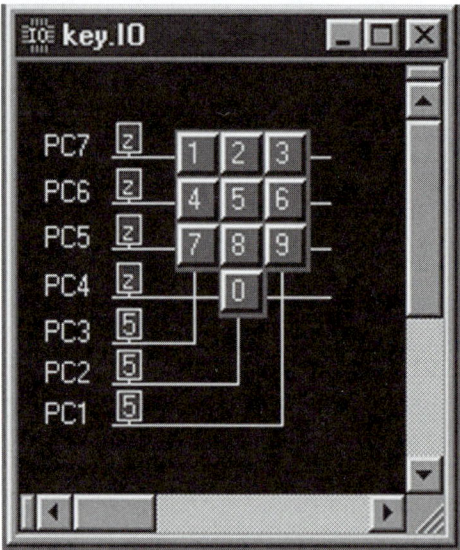

A keyboard bounce time of 0 means no bouncing. Even though real keyboards bounce for 2 to 5 ms when touched or released, because the simulator requires a lot of "human" time to simulate 2 ms, you may wish to set for educational purposes the bounce time to an unrealistic 50 μsec.

Observation: *The system does check for conflicts. For example, your software sets one row to an output high and a second row to an output low; then if two keys are touched in the same column, there is a fault (+5V is shorted to ground). Conflicts of this type will cause the simulation to halt.*

Programs 6.2, 6.3, and 6.4 constitute a device driver for the keyboard shown in Figure 6.17. Program 6.2 is the initialization ritual, Program 6.3 scans the keyboard waiting for one key to be touched, and Program 6.4 waits until all keys are released. The basic approach to software scanning is to drive exactly one row to zero, while making the other rows HiZ (floating). Next the column inputs are read, and any zeros in the result represents a key being pressed in that (row, column) position.

Program 6.2   Initialization ritual for a simple matrix keyboard.

```
***6811****Initialize Keyboard*******      **6812*****Initialize Keyboard*******
*Inputs: None                              *Inputs: None
*Outputs: None                             *Outputs: None
Key_Init ldaa #$00   ;PC3-PC1 from columns Key_Init movb #$00,DDRC
         staa DDRC   ;PC7-PC4 to rows               movb #$00,PORTC
         ldaa #$00   ;PC0 unused input              rts
         staa PORTC
         rts
```

Program 6.3   Program to read a key from a simple matrix keyboard.

```
***6811****Input from Keyboard*******       ***6812****Input from Keyboard*******
*waits for exactly one key to be pressed    *waits for exactly one key to be pressed
*Inputs: none                               *Inputs: none
*Outputs: RegA is ASCII code of key typed   *Outputs: RegA is ASCII code of key typed
Key_In ldaa #$80                            Key_In movb #$80,DDRC ; activate row0
     staa DDRC ; 0xxx activates row0              ldaa PORTC
     ldaa PORTC                                   anda #$0E
     anda #$0E                                    cmpa #$06     ;"1" pressed?
     cmpa #$06       ;"1" pressed?                bne  not1
     bne  not1                                    ldaa #'1'
     ldaa #'1'                                    bra  found
     jmp  found                             not1 cmpa #$0A     ;"2" pressed?
not1 cmpa #$0A       ;"2" pressed?               bne  not2
     bne  not2                                    ldaa #'2'
     ldaa #'2'                                    bra  found
     jmp  found                             not2 cmpa #$0C     ;"3" pressed?
not2 cmpa #$0C       ;"3" pressed?               bne  not3
     bne  not3                                    ldaa #'3'
     ldaa #'3'                                    bra  found
     jmp  found                             not3 movb #$40,DDRC ; x0xxx activates row1
not3 ldaa #$40                                    ldaa PORTC
     staa DDRC ; x0xxx activates row1             anda #$0E
     ldaa PORTC                                   cmpa #$06     ;"4" pressed?
     anda #$0E                                    bne  not4
     cmpa #$06       ;"4" pressed?                ldaa #'4'
     bne  not4                                    bra  found
     ldaa #'4'                              not4 cmpa #$0A     ;"5" pressed?
```

continued

```
        jmp   found                          bne   not5
not4 cmpa #$0A        ;"5" pressed?          ldaa  #'5'
        bne   not5                           bra   found
        ldaa  #'5'                 not5 cmpa #$0C        ;"6" pressed?
        jmp   found                          bne   not6
not5 cmpa #$0C        ;"6" pressed?          ldaa  #'6'
        bne   not6                           bra   found
        ldaa  #'6'                 not6 movb #$20,DDRC ; xx0x activates row2
        jmp   found                          ldaa  PORTC
not6 ldaa  #$20                              anda  #$0E
        staa  DDRC ; xx0x activates row2     cmpa  #$06        ;"7" pressed?
        ldaa  PORTC                          bne   not7
        anda  #$0E                           ldaa  #'7'
        cmpa  #$06        ;"7" pressed?       bra   found
        bne   not7                  not7 cmpa #$0A        ;"8" pressed?
        ldaa  #'7'                           bne   not8
        jmp   found                          ldaa  #'8'
not7 cmpa #$0A        ;"8" pressed?          bra   found
        bne   not8                  not8 cmpa #$0C        ;"9" pressed?
        ldaa  #'8'                           bne   not9
        jmp   found                          ldaa  #'9'
not8 cmpa #$0C        ;"9" pressed?          bra   found
        bne   not9                  not9 movb #$10,DDRC ; xxx0 activates row3
        ldaa  #'9'                           ldaa  PORTC
        jmp   found                          anda  #$0E
not9 ldaa  #$10                              cmpa  #$0A        ;"0" pressed?
        staa  DDRC ; xxx0 activates row3     bne   not0
        ldaa  PORTC                          ldaa  #'0'
        anda  #$0E                           bra   found
        cmpa  #$0A        ;"0" pressed?  not0 bra   Key_In ; wait for exactly one
        bne   not0                   found rts
        ldaa  #'0'
        bra   found
not0 jmp Key_In ; wait for exactly one
found rts
```

For example assume the "5" is pressed. First the DDRC is set to $80, which makes only PC7 an output. In the initialization ritual, PORTC was set to zero, so PC7-4 becomes 0zzz (where z means HiZ or floating.) Next the software reads PORTC. Since no keys are pressed in the top row, PC3-1 is 111. Second, the DDRC is set to $40, which makes PC7-4 z0zz. Next the software reads PORTC. Since the "5" is pressed in this row, PC3-1 is 101. The "0" in the column signifies a key is pressed. Third, the DDRC is set to $20, which makes PC7-4 zz0z. Next the software reads PORTC. Since no keys are pressed in the third row, PC3-1 is 111. Fourth, the DDRC is set to $10, which makes PC7-4 zzz0. Next the software reads PORTC. Since no keys are pressed in the fourth row, PC3-1 is 111.

Program 6.4 Program to wait for all keys to be released on a simple matrix keyboard.

```
***6811****wait for release        ***6812****wait for release
*inputs: none                      *inputs: none
*outputs: none                     *outputs: none
Key_Release                        Key_Release
     ldaa #$F0                         movb #$F0,DDRC ; activate all rows
     staa DDRC ; activate all rows wait ldaa PORTC
wait ldaa PORTC                         anda #$0E
     anda #$0E                          cmpa #$0E      ; none pressed?
     cmpa #$0E      ; none pressed?     bne  wait
     bne  wait                          rts
     rts
```

Observation: *An n-by-n matrix keypad has n^2 keys, but requires only 2n I/O pins. You can detect any zero, one, or two key combinations, but it has trouble when three or more are pressed.*

Checkpoint 6.8: *What happens if the three keys "a" "b" and "e" are all pressed?*

Checkpoint 6.9: *Why wouldn't you use a matrix approach when creating a music keyboard for an electric piano?*

6.8   DC Motor Interfacing

The **IO->DCMotor. . .** command allows you to connect a DC Motor to the microcomputer. The DC motor has an 8-bit DAC to deliver power to the motor. A tachometer measures motor speed, and is connected to the 8-bit ADC of the microcomputer. The DAC must be connected to an 8-bit output port. The range of power (displayed as a voltage) is from **MinP** to **MaxP**. The 8-bit DAC is straight binary, where 0 maps to **MinP** and 255 maps to **MaxP**. The range of speed (displayed as a voltage) is from **MinS** to **MaxS**. The finite difference simulation equations will maintain the speed in this range. The 8-bit ADC is also straight binary, where **MinS** maps to 0 and **MaxS** maps to 255. The I/O window displays the current power and speed (in volts). Figure 6.18 shows a block diagram of the motor control system.

Figure 6.18
Block diagram of the DC
motor interface.

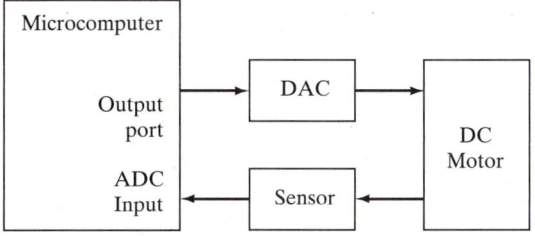

To make the DC motor active, you must specify an 8-bit port for the DAC actuator and an analog ADC input pin for the tachometer. Figure 6.19 shows the dialog box used to interface a DC motor to the simulated microcomputer. Figure 6.20 shows the corresponding I/O window. In particular, there are many parameters to this simulation:

1. The range of the power is specified with the minimum and maximum power parameters, **MinP** and **MaxP**. The units of power, when plotted on the scope and implemented with the finite difference equation, is volts. The minimum power must be less than the maximum power, and both parameters must be between -9.9 and $+9.9$ V.

2. Any pin on the 8-bit output port for the DAC actuator is specified, and the simulator will select the entire port (e.g., you type PH0 to specify all of PORTH).

3. The range of the motor speed is specified with the minimum and maximum speed parameters, **MinS** and **MaxS**. The unit of speed, when plotted on the scope and implemented with the finite difference equation, is also volts. The minimum speed must be less than the maximum speed, and both parameters must be between -9.9 and $+9.9$ V.

4. The analog ADC input pin for the tachometer is specified (e.g., you type PAD2 for channel 2 on the 6812).

5. The digital simulation time-step interval (in μsec) is specified. The smaller the time, the more accurate the simulation, but the slower the application will run. This time should be much less than the time constant of the motor.

6. Select "single-pole" mode to enter the 7) and 8) parameters, ignoring 10) parameters. With this simple mode, the correct finite difference coefficients are automatically calculated from the simulation time step, the motor gain, and the motor time constant.

7. The motor gain is the steady-state slope of the speed versus power response. A gain of 1.5 means a steady-state power of 2.0V will spin the motor with a steady state speed of 3.0V.

8. The time constant of the motor is the time to reach 0.707 of the final value after a step change in the power.

9. Select "linear system" mode to enter the 10) parameters, ignoring 7) and 8) parameters. With this complex mode, it is your responsibility to determine the correct finite difference coefficients from the simulation time step and the motor response.

10. These five parameters can be specified to implement the 5-parameter digital model relating motor speed to motor power.

Figure 6.19
Dialog box for interfacing a DC motor to the microcomputer.

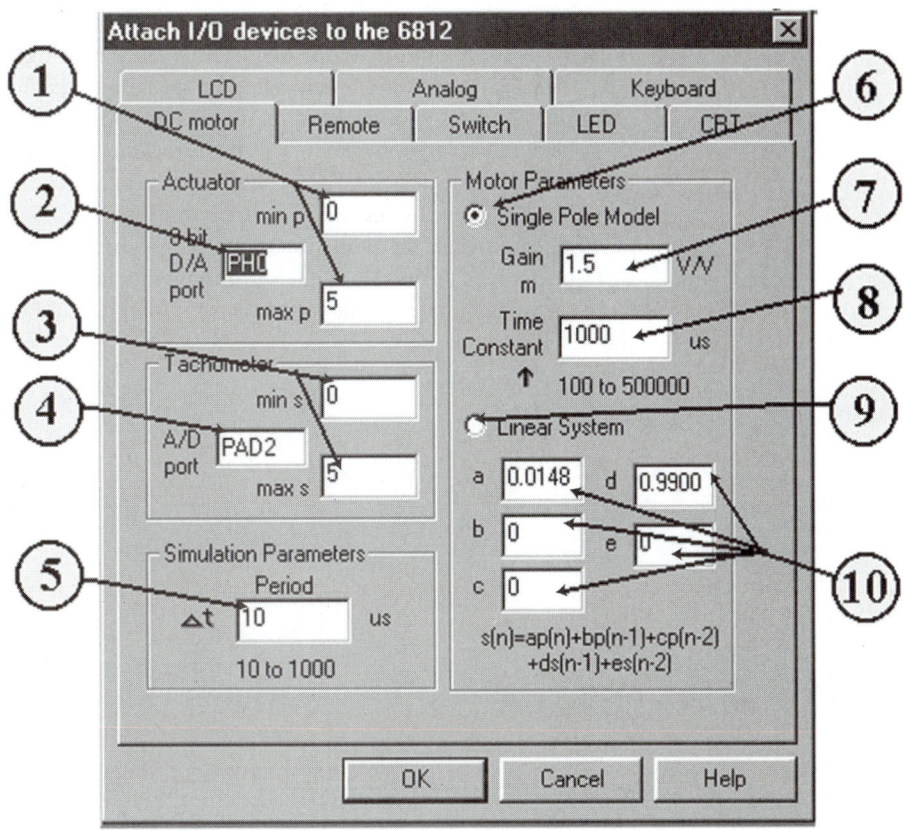

The DC motor is simulated using the five-parameter finite difference equation executed at a fixed step (Δt), shown as parameter five in the Figure 6.19. In particular, the following equation is executed every Δt:

$$s(n) = a*p(n) + b*p(n-1) + c*p(n-2) + d*s(n-1) + e*s(n-2)$$

Here, *a, b, c, d,* and *e* are constants, $s(n)$ is the new calculated speed, $s(n-1)$ is the previous calculated speed, $s(n-2)$ is the calculated speed two Δt times ago, $p(n)$ is the current applied power from the DAC, $p(n-1)$ is the previous applied power, and $p(n-2)$ is the applied power two Δt times ago.

Figure 6.20
Example I/O window showing a DC motor.

The objective of this control system is to control the speed of the motor. The `SetPoint` is the desired speed of the motor. An incremental control algorithm simply adds or subtracts a constant from the `power`, depending on the sign of the error. (See Figure 6.21 and Programs 6.5 and 6.6.) In other words, if the `speed` is too small, then `power` is incremented; if `speed` is too large, then `power` is decremented. It is important to choose the proper rate at which the incremental control software is executed. If it is executed too many times per second, then the actuator will saturate resulting in a bang-bang system. If it is not executed often enough, then the system will not respond quickly to changes in the physical plant or changes in the `SetPoint`. In this incremental controller we add or subtract 1 from the actuator, but a value larger than 1 would have a faster response at the expense of introducing oscillations.

Figure 6.21
Flowchart of a speed controller implemented using incremental control.

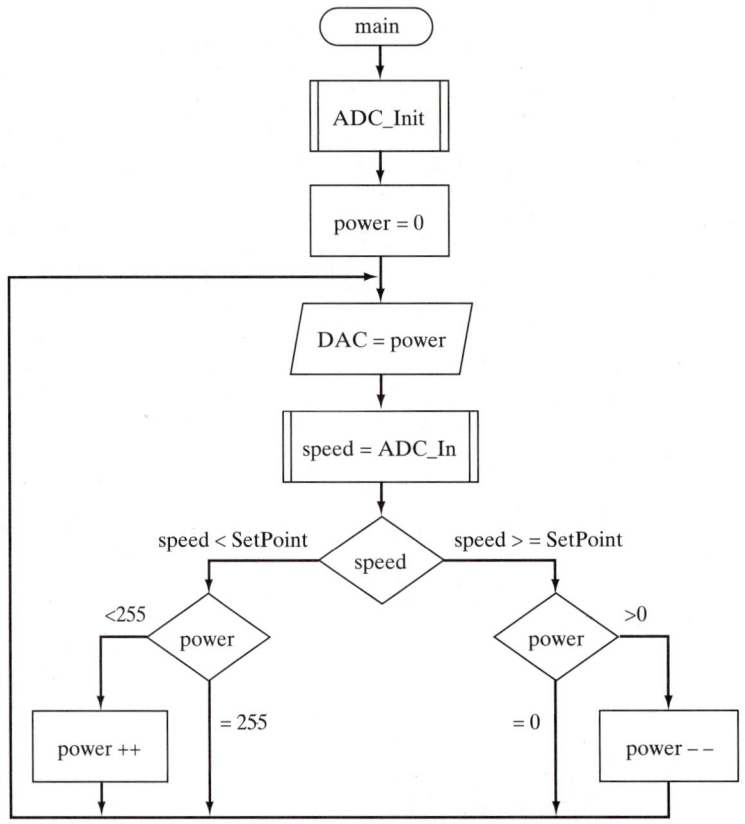

Common error: *An error will occur if the software does not check for overflow and underflow after U is changed.*

Observation: *If the incremental control algorithm is executed too frequently, then the resulting system behaves like a simple bang-bang controller.*

Observation: *Many control systems operate well when the control equations are executed about 10 times faster than the step response time of the physical plant.*

The ADC is used to measure the motor speed. The DAC is used to deliver power to the motor. Later in Chapter 11, we will study periodic interrupts. It will be both convenient and effective to run the controller in the background using interrupt synchronization. Typically, we run a digital controller from 10 to 100 times faster than the time constant of the physical plant. For example, if the motor time constant were 0.1 sec, then time in between executions of the digital controller should be from 1ms to 10ms.

Observation: *Incremental control will work moderately well (with accuracy and stability) for an extremely wide range of applications. Its only shortcoming is that the controller response time can be quite slow.*

Program 6.5   Assembly language incremental controller.

```
*6811                                    *6812
main lds  #$00FF                         main lds #$0C00
     jsr  ADC_Init                            jsr ADC_Init
     ldaa #0        ; actuator control        movb #$FF,DDRH  ; output pins
loop staa PORTB     ; output to actuator      ldaa #0         ; actuator control
     ldab #2                             loop staa PORTH      ; set DAC actuator
     jsr  ADC_In    ; read sensor             ldab #2         ; read sensor
     cmpb #SetPoint ; desired speed            jsr ADC_In
     blo  slow                                cmpb #SetPoint  ; desired speed
fast cmpa #0        ; already at min           blo slow
     beq  loop                           fast cmpa #0         ; already at min
     deca           ; decrease speed           beq loop
     bra  loop                                deca            ; decrease speed
slow cmpa #255      ; already at max           bra loop
     beq  loop                           slow cmpa #255       ; already at max
     inca           ; increase speed           beq loop
     bra  loop                                inca            ; increase speed
                                              bra loop
```

Program 6.6   C language incremental controller.

```
/* 6811 */                            // 6812
unsigned char power,speed;            unsigned char power,speed;
#define SetPoint 128                  #define SetPoint 128
void main(void){                      void main(void){
  power=0;                              TSCR=0x80;     // enable TCNT
  OPTION=0x80;  /* turn on ADC power */  power=0;
  while(1){                             DDRH=0xFF;
    PORTB=power; /* output to actuator */ ATDCTL2=0x80; // ASPU=1 enables ADC
    ADCTL=2;      /* channel 2 */        while(1){
    while((ADCTL&0x80)==0){};             PORTH=power; // output to actuator
    speed=ADR1;                           ATDCTL5=2;   // channel 2
    if(speed<SetPoint){ /* too slow */    while((ATDSTAT&0x8000)==0){};
       if(power<255) power++;             speed=ADR0H;
    }                                     if(speed<SetPoint){ // too slow
    else{ /* too fast */                     if(power<255) power++;
       if(power>1) power--;              }
    }                                     else{ // too fast
  }                                          if(power>1) power--;
}                                         }
                                        }
                                      }
```

Checkpoint 6.10: *What would happen if the incremental controller didn't check for underflow or overflow?*

Checkpoint 6.11: *What would be the effect of adding/subtracting 2 instead of 1 in an incremental controller?*

Observation: *The motor controller system can be used to simulate a simple DAC, simply by not connecting the tachometer signal.*

6.9   Infrared Remote Interfacing

The **IO->IRremote. . .** command allows you to connect an IR remote control to an input port of the microcomputer. Figure 6.22 shows the dialog box used to interface an IR remote to the simulated microcomputer. As an option, you are allowed to connect the sensor simultaneously to two inputs: the primary and the secondary port. In this way, you could use two input capture lines to separately measure the rise and fall of the waveform. The specific microcomputer should first be selected before executing this command. Up to 16 keys, each with its own waveform pattern, can be added. If you need just one connection, you can use either the primary or the secondary port. For each key, you specify a one- or two letter label as well as a sequence of 0's and 1's that will occur when that key is hit. The key will appear to release when the binary pattern is complete. The key letters are case sensitive.

Figure 6.22
Dialog box for interfacing an infrared controller to the microcomputer.

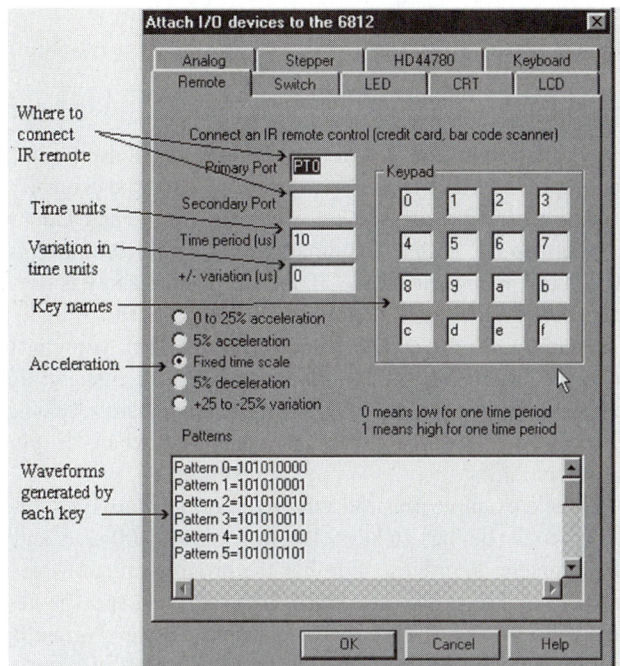

The corresponding remote control generated by the above configuration is shown in Figure 6.23.

Figure 6.23
Example I/O window showing an infrared controller.

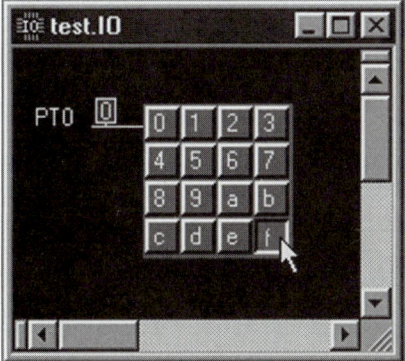

Figure 6.24 illustrates the overall behavior of the IR interface.

Figure 6.24
Block diagram of an infrared interface.

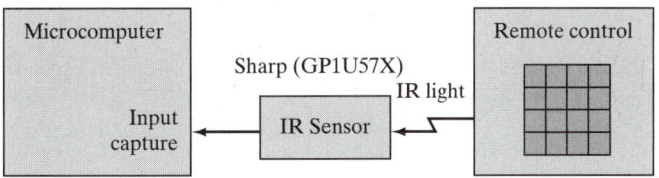

Individual keys can be eliminated by removing the label from the key name.

The Time period is the basic unit of time for the IR remote simulation. For example, if the Time period is 50 μs and the pattern is 1011101, then the resulting waveform is shown in Figure 6.25.

Figure 6.25
IR timing of Pattern 1.

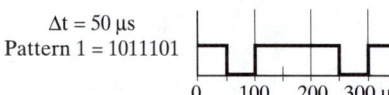

A $+/-$ variation of 0 means the scan velocity will always start at the same speed. This is the appropriate setting when simulating an IR remote control.

For credit card readers or bar code readers, you can add a variable speed, which simulates what happens when the human arm motion determines the speed. If the Time period is 50 and the $+/-$ variation is 20, then each time a key is pressed, the simulator will randomly choose a time period between 30 and 70 μsec.

In addition to the scan velocity (Time period), you can specify acceleration. If you choose "Fixed time scale," the time base does not change. This is the appropriate setting when simulating an IR remote control. For credit card readers or bar code readers, you can add acceleration, which simulates what happens when the human arm motion determines the acceleration.

The next example illustrates the waveform patterns generated by the ITZA remote control. This controller has 16 keys. The time base is 400 μsec and each pattern takes about 20 ms to complete. In order to simplify the programming of these waveforms we will create some macros. These four patterns do not represent specific keys on the remote, but rather are the building blocks of the more complicated waveforms. B and E are the common beginning and end sequences. The L is a long pulse and the S is a short pulse. These building blocks are defined with the following lines:

```
Macro B=1100000001
Macro L=0001
Macro S=001
Macro E=11111
```

These patterns are shown in Figure 6.26.

Figure 6.26
IR timing of the four macros.

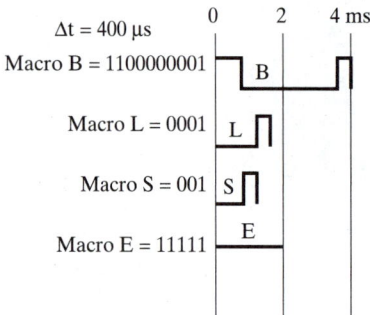

The semicolon can be used as a delimiter between the pattern and an optional comment fields; for example,

```
; ITZA remote control
Macro B=1100000001      ;beginning code
Macro L=0001            ;long pulse
Macro S=001             ;short pulse
Macro E=11111           ;d code
```

The key patterns themselves are defined using these definitions

```
Pattern 1=BSSSSSSSLSSSSE          ;"1" key
Pattern 2=BLSSSSSSLSSSSE          ;"2" key
Pattern 3=BSLSSSSSLSSSSE          ;"3" key
Pattern 4=BLLSSSSSLSSSSE          ;"4" key
Pattern 5=BSSLSSSSLSSSSE          ;"5" key
Pattern 6=BLSLSSSSLSSSSE          ;"6" key
Pattern 7=BSLLSSSSLSSSSE          ;"7" key
Pattern 8=BLLLSSSSLSSSSE          ;"8" key
Pattern 9=BSSSLSSSLSSSSE          ;"9" key
Pattern 0=BLSSLSSSLSSSSE          ;"0" key
Pattern v+=BSLSSLSSSLSSSSE        ;"v+" key, volume up
Pattern v-=BLLSSLSSSLSSSSE        ;"v-" key, volume down
Pattern c+=BSSSSLSSSLSSSSE        ;"c+" key, channel up
Pattern c-=BLSSSLSSSLSSSSE        ;"c-" key, channel down
Pattern e=BLLSLSSSLSSSSE          ;"e" key, enter
Pattern m=BSSSLSLSSSLSSSSE        ;"m" key, mute
```

The resulting patterns are shown in Figure 6.27.

Figure 6.27
IR timing of each of the patterns.

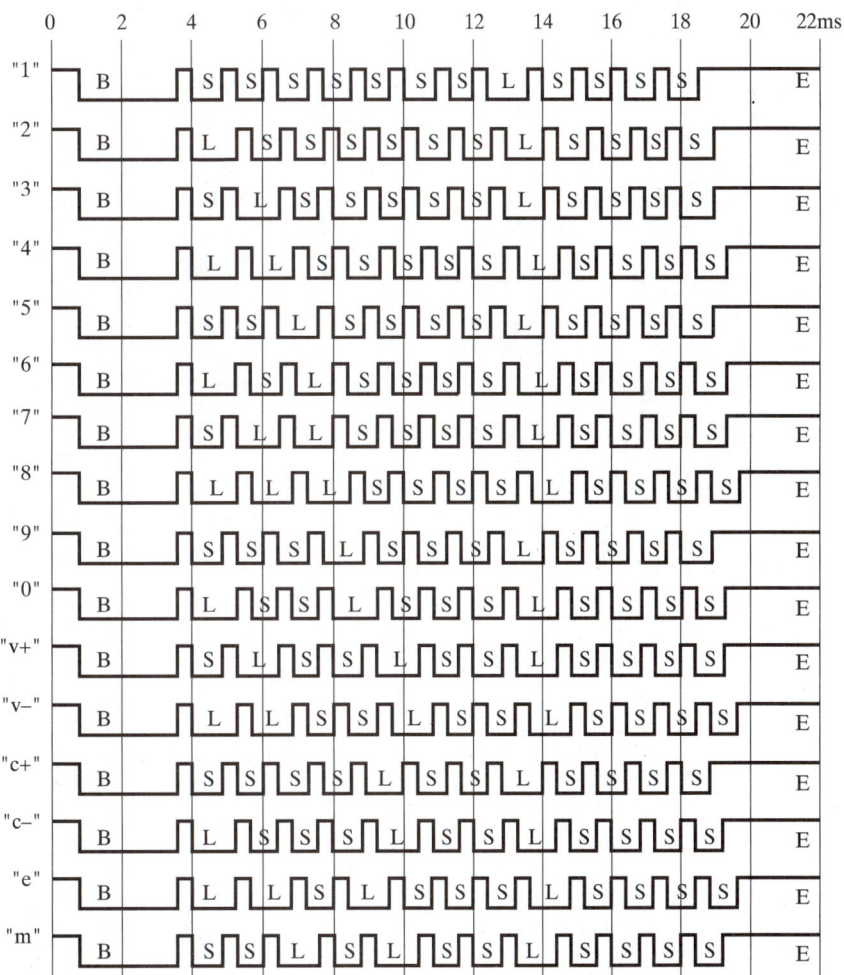

Macros must be defined only with 0's and 1's (i.e., there can be no nested macro calls), but Patterns can have 0's, 1's or macros. Lower case is different from upper case, and macros are different from patterns. For example key "e," key "E," macro "e," and macro "E" would be all different.

> **Observation:** *If you press a second button before the first is finished, the first pattern will be terminated and the second pattern started immediately.*

Notice that the information is encoded in the pulse width of the IR signal. Program 6.7 is a simple system that measures the time from one falling edge to the next rising edge from the IR remote. The proper way to develop the software to decode the IR signals is to use the input capture system. For information about using input capture to measure pulse width, see Chapter 6 of *Embedded Microcomputer Systems: Real Time Interfacing* by Jonathan W. Valvano, published by Brooks/Cole, copyright (c) 2000.

Program 6.7 Assembly code that measures the pulse width.

```
* 6811                                    * 6812
* IR remote connected to PA0              * IR remote connected to PT0
      org   $0000                               org   $0800
width rmb   2   in cycles                 width rmb   2   in cycles
fall  rmb   2   time at falling edge      fall  rmb   2   time at falling edge
      org   $F000                               org   $F000
main  ldx   #PORTA                        main  movb  #$80,TSCR ;enable TCNT
high  brset 0,x,#$01,high                       bclr  DDRT,#$01 ;PT0 input
      ldd   TCNT ; time of falling edge   high  brset PORTT,#$01,high
      std   fall                                movw  TCNT,fall ; falling edge
low   brclr 0,x,#$01,low                  low   brclr PORTT,#$01,low while low
      ldd   TCNT ; time at rising edge          ldd   TCNT  ; time at rising edge
      subd  fall                                subd  fall
      std   width ;pulse width measurement      std   width ;pulse width measurement
      bra   high                                bra   high
      org   $FFFE                               org   $FFFE
      fdb   main                                fdb   main
```

> **Checkpoint 6.12:** *What causes the acceleration and deceleration in a bar code or credit card scanner?*

6.10 Hitachi HD44780 LCD Display

The **IO->HD44780. . .** command allows you to connect a LCD display based on the Hitachi HD44780 controller. Microprocessor-controlled LCD displays are widely used, having replaced most of their LED counterparts, because of their low power and flexible display graphics. This example will illustrate how a handshaked parallel port of the microcomputer will be used to output to the LCD display. The hardware for the display uses an industry standard HD44780 controller, as shown in Figure 6.28. The low-level software initializes and outputs to the HD44780 controller.

Figure 6.28
Interface of a HD44780
LCD controller.

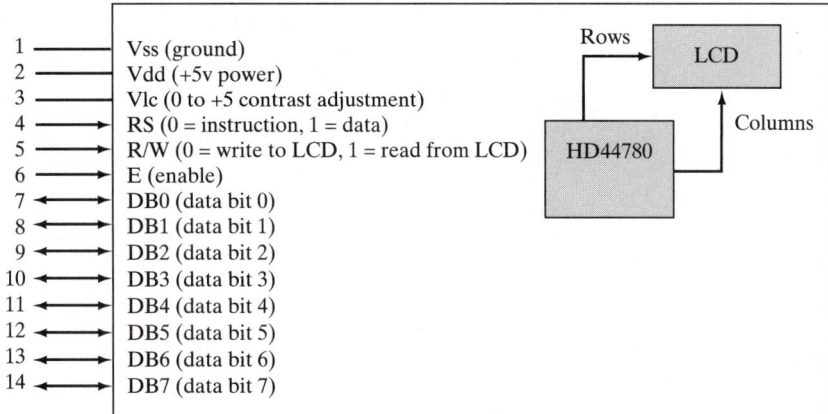

There are four types of access cycles to the HD44780 depending on RS and R/W, as illustrated in Table 6.6.

Table 6.6
Two control signals
specify the type of
access to the HD44780.

RS	R/W	Cycle
0	0	Write to Instruction Register
0	1	Read Busy Flag (bit 7)
1	0	Write data from μP to the HD44780
1	1	Read data from HD44780 to the μP

The dialog box used to interface the HD44780 display is shown in Figure 6.29. Five parameters can be selected: 1) choose the size of the LCD display; 2) decide on the timing of the LCD; 3) add an optional label; 4) interface the data signals; and 5) connect the control lines.

Figure 6.29
Dialog box for
interfacing HD44780
LCD to the
microcomputer.

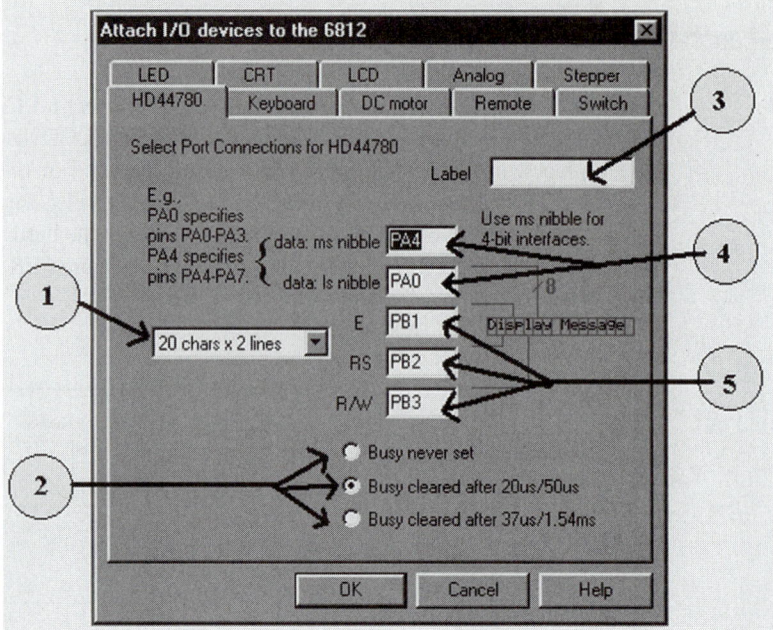

Figure 6.30 shows a typical I/O window with a 2-by-20-line HD44780 display.

Figure 6.30
An I/O window with a
HD44780 LCD.

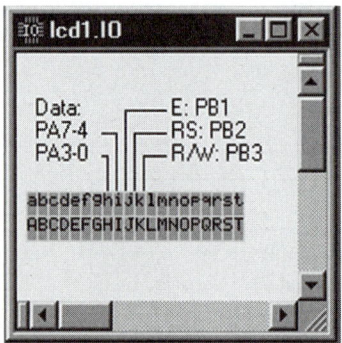

Two types of synchronization can be used, blind-cycle and busy-waiting. Most operations require 40 μs to complete while some require 1.64 ms. This implementation uses OC5 to create the blind-cycle wait. A busy-waiting interface would have provided feedback to detect a faulty interface, but has the problem of creating a software crash if the LCD never finishes. The best interface utilizes both busy-waiting and blind-cycle, so that the software can return with an error code if a display operation does not finish on time (due to a broken wire or damaged display). First we present a low-level private helper function. (See Program 6.8.) This function would not have a prototype in the LCD.H file.

Program 6.8   Private functions for an HD44780 controlled LCD display.

```
// MC68HC11A8                               // MC68HC812A4
// 1 by 16 char LCD Display (HD44780)       // 1 by 16 char LCD Display (HD44780)
//   ground = pin 1 Vss                     //   ground = pin 1 Vss
//   power  = pin 2 Vdd +5v                 //   power  = pin 2 Vdd +5v
//   10Kpot = pin 3 Vlc contrast adjust     //   10Kpot = pin 3 Vlc contrast adjust
//   PB2    = pin 6 E enable                 //   PJ2    = pin 6 E enable
//   PB1    = pin 5 R/W 1=read, 0=write      //   PJ1    = pin 5 R/W 1=read, 0=write
//   PB0    = pin 4 RS 1=data, 0=control     //   PJ0    = pin 4 RS 1=data, 0=control
//   PC0-7  = pins7-14 DB0-7 8-bit data      //   PH0-7  = pins7-14 DB0-7 8-bit data
#define LCDdata    1 // PB0=RS=1            #define LCDdata    1 // PJ0=RS=1
#define LCDcsr     0 // PB0=RS=0            #define LCDcsr     0 // PJ0=RS=0
#define LCDread    2 // PB1=R/W=1           #define LCDread    2 // PJ1=R/W=1
#define LCDwrite   0 // PB1=R/W=0           #define LCDwrite   0 // PJ1=R/W=0
#define LCDenable  4 // PB2=E=1             #define LCDenable  4 // PJ2=E=1
#define LCDdisable 0 // PB2=E=0             #define LCDdisable 0 // PJ2=E=0
void LCDcycwait(unsigned short cycles){     void LCDcycwait(unsigned short cycles){
    TOC5=TCNT+cycles; // 500ns cycles           TC5=TCNT+cycles;  // 500ns cycles to wait
    TFLG1 = 0x08;     // clear C5F               TFLG1 = 0x20;     // clear C5F
    while((TFLG1&0x08)==0){};}                   while((TFLG1&0x20)==0){};}
```

Next we show the high-level public members, see Program 6.9. These functions would have prototypes in the LCD.H file.

Program 6.9   Public functions for an HD44780 controlled LCD display.

```
// MC68HC11A8                                   // MC68HC812A4
void LCD_putchar(unsigned short letter){        void LCD_putchar(unsigned short letter){
// letter is ASCII code                         // letter is ASCII code
    PORTC=letter;                                   PORTH=letter;
    PORTB=LCDdisable+LCDwrite+LCDdata;              PORTJ=LCDdisable+LCDwrite+LCDdata;
    PORTB=LCDenable+LCDwrite+LCDdata;              PORTJ=LCDenable+LCDwrite+LCDdata;  // E=1
// E goes 0,1                                        PORTJ=LCDdisable+LCDwrite+LCDdata; // E=0
    PORTB=LCDdisable+LCDwrite+LCDdata;              LCDcycwait(80);} // 40 us wait
// E goes 1,0                                    void LCD_putcsr(unsigned short command){
    LCDcycwait(80);} // 40 us wait                   PORTH=command;
void LCD_putcsr(unsigned short command){            PORTJ=LCDdisable+LCDwrite+LCDcsr;
    PORTC=command;                                  PORTJ=LCDenable+LCDwrite+LCDcsr;   // E=1
    PORTB=LCDdisable+LCDwrite+LCDcsr;              PORTJ=LCDdisable+LCDwrite+LCDcsr;  // E=0
    PORTB=LCDenable+LCDwrite+LCDcsr;              LCDcycwait(80);}    // 40 us wait
// E goes 0,1                                    void LCD_clear(void){
    PORTB=LCDdisable+LCDwrite+LCDcsr;               LCD_putcsr(0x01);  // Clear Display
// E goes 1,0                                        LCDcycwait(3280);  // 1.64ms wait
    LCDcycwait(80);}    // 40 us wait               LCD_putcsr(0x02);  // Cursor to home
void LCD_clear(void){                               LCDcycwait(3280);} // 1.64ms wait
    LCD_putcsr(0x01);  // Clear Display         void LCD_init(void){
    LCDcycwait(3280);  // 1.64ms wait               DDRH=0xFF; DDRJ=0xFF;
    LCD_putcsr(0x02);  // Cursor to home            TIOS |= 0x20; // enable OC5
    LCDcycwait(3280);} // 1.64ms wait               TSCR |= 0x80; // enable, no fast clear
```

continued

```
void LCD_init(void){                    TMSK2=0xA2;      // 500 ns clock
   DDRC=0xFF;                            LCD_putcsr(0x06);
   LCD_putcsr(0x06);                // I/D=1 Increment, S=0 nodisplayshift
// I/D=1 Increment, S=0 nodisplayshift    LCDputcsr(0x0C);    // D=1 displayon,
   LCD_putcsr(0x0C); // D=1 displayon,  // C=0 cursoroff, B=0 blinkoff
// C=0 cursoroff, B=0 blinkoff            LCD_putcsr(0x14);
   LCD_putcsr(0x14); /               // S/C=0 cursormove, R/L=0 shiftright
/ S/C=0 cursormove, R/L=0 shiftright      LCD_putcsr(0x30);
   LCD_putcsr(0x30);                 // DL=1 8bit, N=0 1 line, F=0 5by7dots
// DL=1 8bit, N=0 1 line, F=0 5by7dots    LCD_clear(); }    // clear display
   LCD_clear();} // clear display
```

Figure 6.31 shows a rough sketch of the E, RS, R/W and data signals as the `LCD_putchar` function is executed.

Figure 6.31
Timing diagram of the LCD signals as data is sent to the HD44780 display.

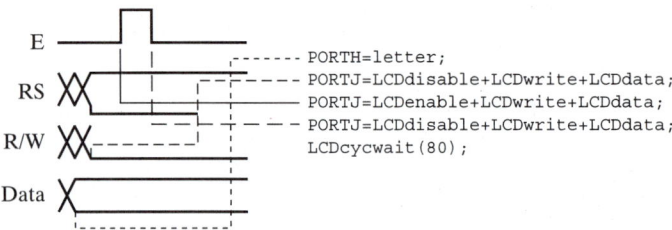

Checkpoint 6.13: Draw a rough sketch of the E, RS, R/W and data signals as the `LCD_putcsr` function is executed.

6.11 Stepper Motor

The **IO->Stepper. . .** command allows you to connect stepper motors to the microcomputer. Figure 4.9 shows a typical stepper motor interface. Figure 6.32 shows the dialog used to interface one or two stepper motors to the microcomputer. Each stepper has four control wires. Each output within the sequence 1010, 1001, 0101, 0110, . . . will cause the motor to make one clockwise step. Each output within the reverse sequence 0110, 0101, 1001, 1010, . . . will cause the motor to make one counterclockwise step. There are 200 steps per rotation. The two motors independently control the two drive wheels of a robot car. There are bumper switches on the front and back of the car. More details of the stepper motor robot can be found within the help system of **TExaS**.

Figure 6.32
Dialog box for interfacing stepper motors to the microcomputer.

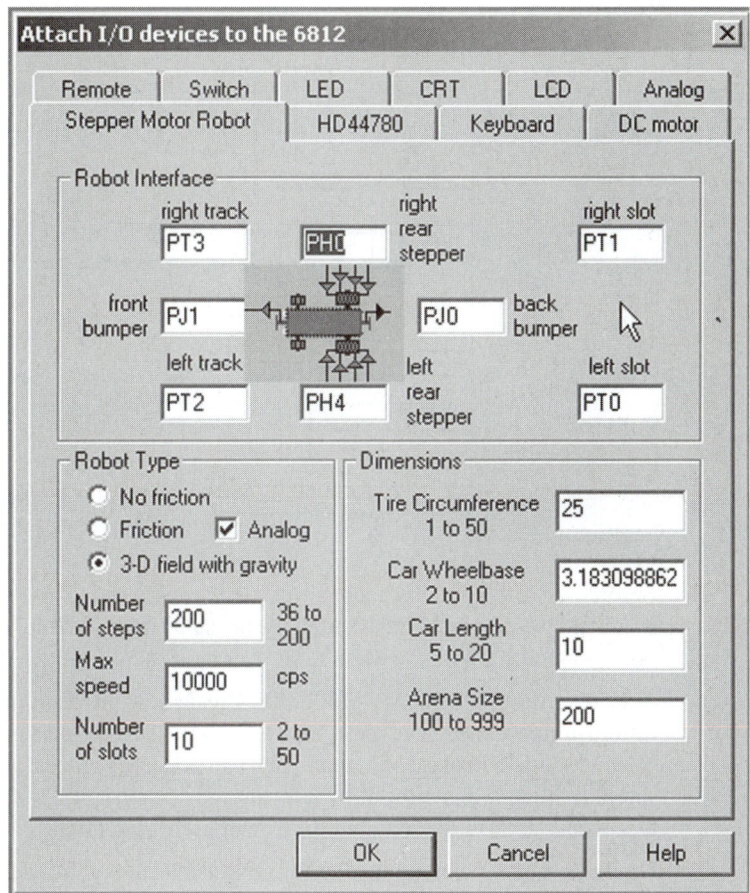

Maintenance Tip: New versions of TExaS may include new I/O devices. Check for the availability of TExaS upgrades at: www.ece.utexas.edu/~valvano.

6.12 Tutorial 6. Microcomputer-Based Lock

To illustrate the software development process, we will implement a simple digital lock. The lock system has seven toggle switches and a solenoid as shown in Figure 6.33. If the 7-bit binary pattern on Port C bits 6-0 becomes 0100011 for at least 1 ms, then the solenoid will activate. The 1-ms delay will compensate for the switch bounce. For information on switches and solenoids see Chapter 8 of *Embedded Microcomputer Systems: Real Time Interfacing* by Jonathan W. Valvano. For now, what we need to understand is that Port C bits 6-0 are input signals to the computer and Port C bit 7 is an output signal.

Figure 6.33
Hardware configuration for a microcomputer-controlled lock.

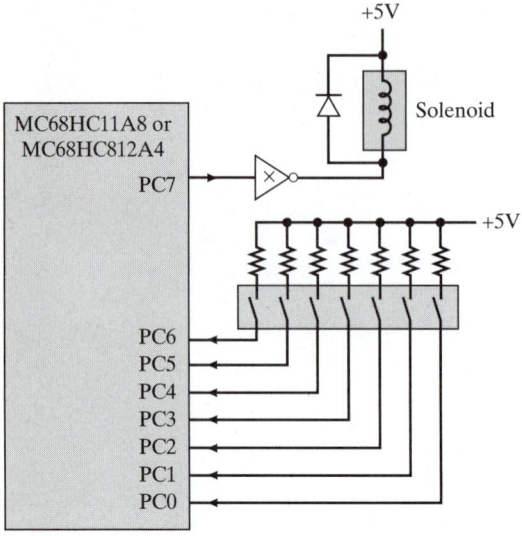

Before we write assembly code, we need to develop a software plan. Software development is an iterative process. Even though we list steps in the development process in a 1,2,3, ... order, in reality we iterate these steps over and over.

Action: Start a fresh copy of **TExaS**. Create new program microcomputer and I/O files from within **TExaS**. Save these files as `Chap6.rtf`, `Chap6.uc`, and `Chap6.io`. Execute the **Mode->Processor...** command and select the processor you wish to use. You could shortcut through this tutorial by copying the `Chap6.rtf`, `Chap6.uc`, and `Chap6.io` files from the CD instead of building them up from scratch.

Question 6.1.   What are the RAM and ROM locations for your microcomputer?

Action: Type the following assembly code into the `chap6.rtf` file, replacing RAM with the first RAM address, STCK with the last RAM address, and ROM with the first ROM address. The 6811 STCK value is the last actual RAM address, while for the 6812 you set STCK to the last RAM address plus 1.

```
        org   RAM
; global variables will go here
        org   ROM
main    lds   #STCK
; program will go here
        stop
; constant data will go here
        org   $FFFE
        fdb   main
```

Action: We begin with a list of the inputs and outputs. We specify the range of values and their significance. In this example, we will use PORTC. Bits 6-0 will be inputs. The seven input signals represent an unsigned integer from 0 to 127. Port C bit 7 will be an output. If PC7 is 1, then the solenoid will activate and the door will be unlocked. Click on the `Chap6.io` window and add seven switches to PC6-PC0, and one LED to PC7. The LED will simulate the solenoid. Figure 6.34 shows the resulting IO window.

Action: For the 6811, open the `Port11.rtf` file, and for the 6812, open the `Port12.rtf` file. Copy the `PORTC` and `DDRC` lines, and paste them into your `Chap6.rtf` file.

Figure 6.34
I/O window for the microcomputer-controlled lock.

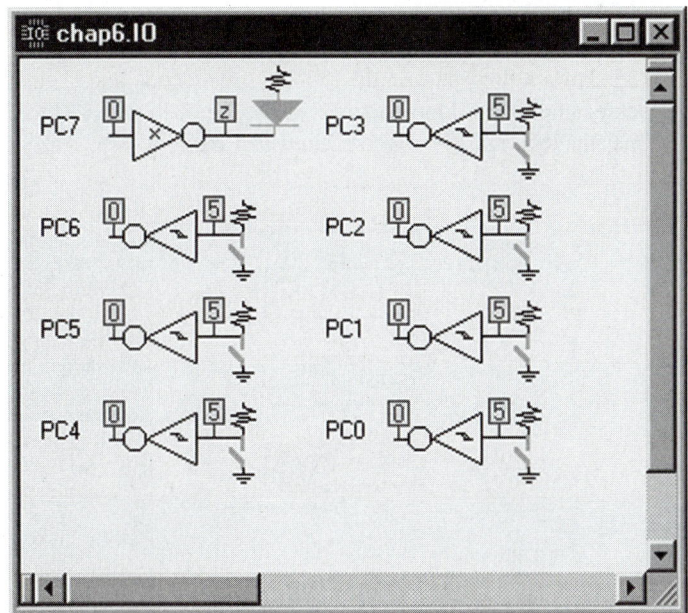

Question 6.2.   What are the addresses of `PORTC` and `DDRC`?

Action: Next we make a list of the required data structures. Data structures are used to save information. If the data need to be permanent, then they are allocated in global space. If the software will change its value then it is allocated in RAM. In this example we need a 16-bit unsigned counter. Add this code to the global variable section. The `rmb` pseudo-op will reserve multiple bytes.

```
cnt    rmb    2      ; 16-bit  counter
```

If data structure can be defined at assembly time and will remain fixed, then it can be allocated in EEPROM. In this example, we will define an 8-bit fixed constant to hold the key code, which the operator needs to set to unlock the door. We will place these lines directly after the program so that they will be defined in ROM or EEPROM memory. The `fcb` pseudo-op defines an 8-bit constant. Add this code to the constant data section (after the `stop` and before the `org $FFFE`). This line also assigns the symbolic name `key` to the corresponding address of the information:

```
key    fcb    %00100011   ; key  code
```

It is not really clear at this point exactly where in EEPROM this constant will be, but luckily for us, the assembler will calculate the exact address automatically. After the program is assembled, we can look at the line in the listing file or in the symbol table to see where in memory each structure is allocated.

Action: Next we develop the software algorithm, which is a sequence of operations we wish to execute. There are many approaches to describing the plan. Experienced programmers can develop the algorithm directly in assembly language. On the other hand, most of us need an abstractive method to document the desired sequence of actions. Flowcharts, pseudocode, and high-level language code are three common descriptive formats. The **TExaS** application is unique in that if you draw the flowchart on the computer, you can paste it directly into the program as a comment. There are no formal rules regarding pseudocode; rather it is a shorthand for describing what to do and when to do it. We can place our pseudocode as documentation into the comment fields of our program. Figure 6.35 shows a flowchart on the left and pseudocode and C code on the right for our digital lock example. The loop counter (74 for the 6811 and 400 for the 6812) is the number of times the loop must be executed to wait 1 ms.

Figure 6.35
Flowchart for a microcomputer-controlled lock.

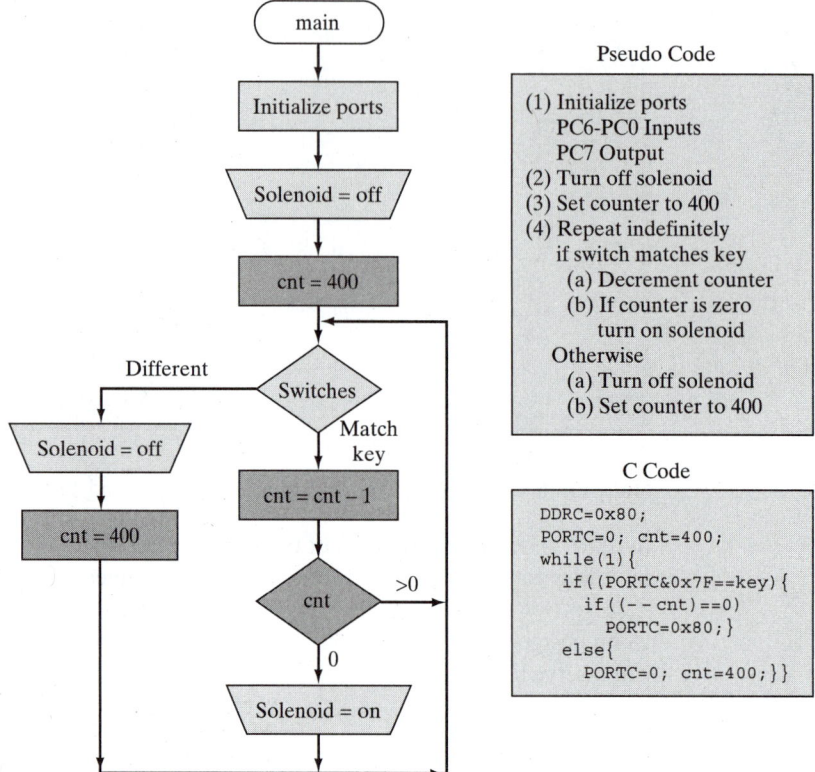

Pseudo Code

(1) Initialize ports
    PC6-PC0 Inputs
    PC7 Output
(2) Turn off solenoid
(3) Set counter to 400
(4) Repeat indefinitely
    if switch matches key
        (a) Decrement counter
        (b) If counter is zero
            turn on solenoid
    Otherwise
        (a) Turn off solenoid
        (b) Set counter to 400

C Code

```
DDRC=0x80;
PORTC=0; cnt=400;
while(1){
    if((PORTC&0x7F==key){
        if((--cnt)==0)
            PORTC=0x80;}
    else{
        PORTC=0; cnt=400;}}
```

Next we write assembly code to implement the algorithm as illustrated in the above flowchart and pseudocode. In step 1, we initialize Port C so that PC7 is an output and PC6-PC0 are inputs.

```
ldaa  #$80
staa  DDRC    ; PC6-PC0 input, PC7 output
```

In step 2, we turn off the solenoid. Remember, writing to an input pin has no effect, so this operation only changes bit 7.

```
clr PORTC    ; disable solenoid lock
```

In step 3, we initialize the counter to 400, which is the number of loops required to wait 1 ms. The 6811 requires 27 cycles to execute the loop. You should replace the 400 with a 74 when running a 2 MHz 6811. The 6812 requires 20 cycles to execute the loop.

```
ldx  #400
stx  cnt ; 1,000,000ns/(125*20)
```

In step 4, we implement the indefinite loop. We place an assembly label at the program locations to which we wish to branch. The bra instruction is an unconditional branch.

```
loop
      bra   loop
```

Inside the indefinite loop we test to see if the switch pattern matches the key code. In this implementation we branch to off if the switches do not match the key code. If they do match, we will execute the instruction immediately after the bne off.

```
loop    ldaa    PORTC   ; [3]  input from 7 switches
        anda    #$7F    ; [1]
        cmpa    key     ; [3]  match key code?
        bne     off     ; [3]
```

If the switches match the key code, then the 16-bit counter is decremented.

```
ldx    cnt  ; [3]
dex         ; [1]
stx    cnt  ; [3]
```

If the counter becomes zero, then the door is unlocked. The bne instruction will go to loop if cnt is not equal to zero.

```
bne     loop    ; [2]=20 cycles/loop
ldaa    #$80
staa    PORTA   ; enable solenoid lock
bra     loop
```

If the switches do not match the key code, then the solenoid is turned off and the cnt set back to 400.

```
off     ldx    #400
        stx    cnt     ; 1,000,000ns/(125*20)
        clr    PORTA   ; disable solenoid lock
        bra    loop
```

We put the foregoing pieces together to create the source code, as shown in Program 6.10. The order of the instructions is very important because it determines the sequence of execution. The last two lines will define where the computer will start execution after a reset.

Program 6.10

Lock program for Tutorial 6.

```
; activate solenoid (PC7=1) if switches match key code
PORTC   equ    $0004   ; PC6-PC0 switches, PC7 solenoid lock
DDRC    equ    $0006   ; specifies input or output
        org    $0800   ; RAM
cnt     rmb    2       ; 16-bit counter
        org    $F000   ; EEPROM
main    lds    #$0C00
        ldaa   #$80
        staa   DDRC    ; PC6-PC0 input, PC7 output
        clr    PORTC   ; disable solenoid lock
        ldx    #400
        stx    cnt     ; 1,000,000ns/(125*20)
loop    ldaa   PORTC   ; [3] input from 7 switches
        anda   #$7F    ; [1]
        cmpa   key     ; [3] match key code?
        bne    off     ; [3]
        ldx    cnt     ; [3]
        dex            ; [1]
        stx    cnt     ; [3]
        bne    loop    ; [2]=20 cycles/loop
```

continued

```
; 7 switches match key code for more than 10 ms
        ldaa   #$80
        staa   PORTC   ; enable solenoid lock
        bra    loop
off     ldx    #400
        stx    cnt     ; 1,000,000ns/(125*20)
        clr    PORTC   ; disable solenoid lock
        bra    loop
key     fcb    %00100011  ; key code
        org    $FFFE
        fdb    main
```

Action: The last stage is debugging. You should run the system to verify its proper behavior.

Question 6.3. What switch pattern activates the solenoid (turns on the LED)?

Question 6.4. How do you change the program so the key is Sw6,5,4=off, Sw3,2,1,0=on?

6.13 Homework Problems

Homework 6.1: You are given a 16-position rotary switch, which has 17 wires, as shown in Figure 6.36. There is one wire called common, and the other 16 wires are labeled S0 through S15. The common wire is connected to exactly one of other 16 wires. You are to design an interface the creates a 4-bit digital signal representing the switch position. These signals are to be connected to Port C bits 3,2,1,0. Write an initialization ritual. Write an input subroutine that reads Port C and returns in Register A the current switch position 0 to 15.

Figure 6.36
16-position rotary switch.

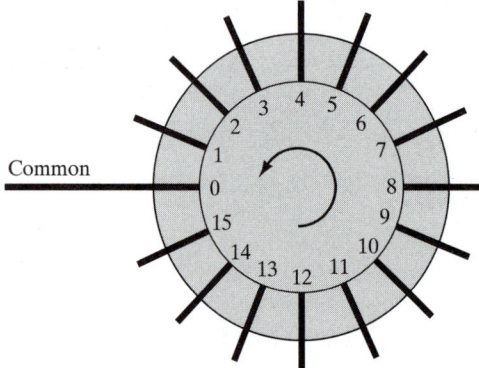

Homework 6.2: A solid-state relay has an LED as its controlling element, as shown in Figure 6.37. In particular, if there is 10 mA of current through the LED, then the relay will be on, and the 120 VAC power will be delivered to the load. If no current follows through the LED, then the relay will be off, and the load will not receive power. Assume the LED voltage is 2.2V. Show the interface between the microcomputer Port C bit 0 and the relay. Write an initialization ritual. Write two output subroutines that write Port C, one routine to turn the relay on and another to shut it off.

Figure 6.37
Solid-state relay.

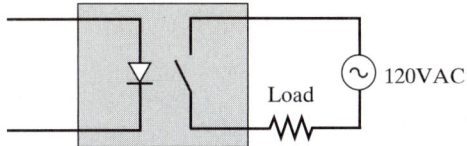

Homework 6.3: Rewrite Program 6.1 so that the input parameter to the function LCD_Out is an integer ranging from -999 to 999. For example, if Register D is -25, -025 should be displayed on the LCD.

Homework 6.4: Rewrite Program 6.1 so that the input parameter to the function LCD_Out is a fixed-point integer with a resolution of 0.01, ranging from -9.99 to 9.99. For example, if Register D is -125, -1.25 should be displayed on the LCD.

Homework 6.5: Assume an AC waveform is connected to analog channel 0. Write an initialization ritual. Write a subroutine that samples the analog input 256 times and returns the DC amplitude (average) in Register A, and the AC amplitude (maximum-minimum) in Register B.

Homework 6.6: Design an interface for a 64-key keyboard that is configured with eight rows and eight columns. Show the hardware interface to Ports B and C. Show the initialization ritual. Assume there are either no keys or one key pressed. Write an input subroutine that returns the key number 0 to 63 if a key is pressed or -1 if no key is pressed. Assume the keys do not bounce.

Homework 6.7: Assume an analog input signal is connected to the ADC channel 2 on computer 1. Assume the transmit serial output of computer 1 is connected to the receive serial input of computer 2. A MAX550A is connected to computer 2, like the interface shown in the previous chapter. Write the dedicated software in both systems, so that the analog input is sampled by the first computer, transmitted via the serial link, and converted back to analog form by the DAC connected to the second computer.

6.14   Laboratory Assignments

Lab 6.1
Calculator
Purpose

The objectives of this lab are to

- interface a matrix keyboard and HD44780 LCD display to the microcomputer,
- write device drivers for the keyboard and HD44780 LCD display,
- implement a four-function integer calculator.

Description

In this lab you will design a four-function 8-bit unsigned integer calculator. The matrix keypad will include the numbers $0 - 9$, and the letters '+', '$-$', '*', '/', '=', and 'C'. The HD44780 LCD display will show both an 8-bit global accumulator, and an 8-bit temporary register. You are free to design the calculator functionality in any way you wish, but you must be able to 1) clear the accumulator and temporary; 2) type numbers in using the matrix keyboard; 3) add, subtract, multiply, and divide; 4) display the results on the HD44780 LCD display. Recall that a device driver is a set of software functions that facilitate the use of an I/O port.

Part A. Create new program, microcomputer, and I/O files. Attach a 16-key matrix keyboard and HD44780 display. You can assume the matrix keyboard does not bounce. During the initial debugging stages of the lab, you may disable the HD44780 busy flag, but your final demonstration will have to include the realistic timing for the LCD.

Part B. Write a device driver for the HD44780. You should be able to 1) initialize the interface; 2) clear the display; 3) output a character; 4) output an 8-bit integer; and 5) output a

string. The names of all the public driver subroutines should start with the letters "LCD_".
Draw flowcharts of these subroutines.

Part C. Write a device driver for the matrix keyboard. You should design subroutines as
needed. All software that directly accesses the I/O ports connected to the keyboard must be
included in this driver. The names of all the public driver subroutines should start with the
letters "Key_". Draw flowcharts of these subroutines.

Part D. Write the main program that implements the calculator functionality. Include a
"call-graph" of the system.

Lab 6.2
AC/DC Voltmeter

The objectives of this lab are to

Purpose

- interface three-digit LCD display to the microcomputer,
- write device drivers for a switch, an ADC and a LCD,
- implement functions for addition, multiplication, division, and square root,
- implement AC/DC voltmeter.

Description

In this lab you will design an AC/DC voltmeter. The ADC will be used to sample an analog input. The DC amplitude is the simple average of multiple samples. Let v[n] be 256
voltage-sampled voltage values:

$$DC = (v[0]+v[1]+\ldots+v[255])/256)$$

The AC amplitude is calculated as a root-mean-squared value:

$$AC = sqrt(((v[0]\text{-}DC)^2+(v[1]\text{-}DC)^2+\ldots+(v[255]\text{-}DC)^2)/256)$$

The LCD output will be in decimal fixed point with a Δ of 0.01 V. You will find it much
simpler to perform the AC/DC calculations on the raw integer sample, then convert integer
AC/DC results to fixed-point voltages. A toggle switch will allow the operator to select either AC or DC mode. The three-digit LCD display will show either the calculated DC or
AC value as a fixed-point number with Δ equal to 0.01 V. A device driver is a set of software functions that facilitate the use of an I/O port.

Part A. Create new program, microcomputer and I/O files. Attach a toggle switch, an analog signal to the ADC and simple three-digit LCD display. You can assume the toggle
switch does not bounce.

Part B. Write a device driver for the three-digit LCD. You should be able to initialize the interface and output a fixed-point number. The names of all the public driver subroutines
should start with the letters "LCD_". Draw flowcharts of these subroutines.

Part C. Write a device driver for the ADC interface. You should design subroutines as
needed. The names of all the public driver subroutines should start with the letters "ADC_".
Draw flowcharts of these subroutines.

Part D. Write a device driver for the switch interface. You should design subroutines as
needed. All software that directly accesses the I/O ports connected to the switch must be included in this driver. The names of all the public driver subroutines should start with the letters "Switch_". Draw flowcharts of these subroutines.

Part E. Write the main program that implements the voltmeter functionality. Sample the
ADC as fast as possible, and use the TCNT timer to estimate the sampling rate. Calculate

the DC and AC results independent of the switch position, so that the sampling rate will be approximately constant. Include a "call-graph" of the system.

Part F. Evaluate the accuracy of the meter, which is the difference between the true signal and the results measured with the system. Use a signal period that is about 16 times larger than the ADC sample period. In this way there will be about 16 ADC samples per wave, and about 16 waves per block of 256 ADC samples. The first two tests will be performed on the pure sine wave. Determine the mathematical relationship between the peak-to-peak sine wave amplitude and its RMS value. First, keeping the DC value fixed, evaluate the accuracy of the system, using the five different AC values. Second, keeping the AC value fixed, evaluate the accuracy of the system, using the five different DC values. Last, keeping the minimum and maximum of the signal constant, test the voltmeter with each of the signal shapes. Explain the differences in the results.

Lab 6.3
Stepper Motor
Controller
Purpose

The objectives of this lab are to

- interface a matrix keyboard, a LCD display and stepper motor to the microcomputer,
- write device drivers for the keyboard, LCD display and stepper motor,
- implement a stepper motor controller.

Description

In this lab you will design a simple stepper motor controller. The matrix keypad will include the numbers $0 - 9$ and the letters 'c' and 'g'. To move the motor, the operator types in the desired angle (0 to 359), then hits the 'g' key. As the operator enters the numbers, the digits are displayed on the three-digit LCD. If the operator types 'c', the command is cleared, and no motion occurs. The system should move clockwise or counterclockwise, whichever is fewer steps. While the motor is moving, the three-digit LCD display will show the current angle of the stepper motor (0 to 359). Recall that a device driver is a set of software functions that facilitate the use of an I/O port.

Part A. Create new program, microcomputer and I/O files. Attach a 10-key matrix keyboard, a three-digit LCD display and one stepper motor. You can assume the matrix keyboard does not bounce.

Part B. Write a device driver for the three-digit LCD. You should be able to initialize the interface and output an angle as a number from 0 to 359. The names of all the public driver subroutines should start with the letters `"LCD_"`. Draw flowcharts of these subroutines.

Part C. Write a device driver for the matrix keyboard. You should design subroutines as needed. All software that directly accesses the I/O ports connected to the keyboard must be included in this driver. The names of all the public driver subroutines should start with the letters `"Key_"`. Draw flowcharts of these subroutines.

Part D. Write a device driver for the stepper interface. You should design subroutines as needed. All software that directly accesses the I/O ports connected to the stepper motor must be included in this driver. The names of all the public driver subroutines should start with the letters `"Step_"`. Draw flowcharts of these subroutines.

Part E. Write the main program that implements the controller functionality. Include a "call-graph" of the system.

7 Debugging

Chapter 7 objectives are to

❏ describe the basic approach to verifying the correctness of programs,
❏ discuss the special features of the **TExaS** debugger,
❏ develop various methods for observing program execution.

Every programmer is faced with the need to debug and verify the correctness of his or her software. In this chapter, we will study hardware probes like the logic analyzer and in-circuit-emulator (ICE); software tools like simulators, monitors, and profilers; and manual tools like inspection and print statements. The specific approaches for debugging software using the **TExaS** application will be presented. **Nonintrusiveness** is the characteristic or quality of a debugger that allows the software/hardware system to operate normally as if the debugger did not exist. On the other hand, intrusiveness is a measure of the degree of perturbation caused in program performance by the debugging itself. For example, a `printf` statement added to your source code is very intrusive because it significantly affects the real-time interaction of the hardware and software. A debugging instrument is classified as **minimally intrusive** if it has a negligible effect on the system being debugged. In a real microcomputer system, breakpoints and single-stepping are also intrusive, because the real hardware continues to change while the software has stopped. When a program interacts with real-time events, the performance can be significantly altered when using intrusive debugging tools. On the other hand, dumps, dumps with filters and monitors (e.g., output strategic information on LEDs) are much less intrusive. A logic analyzer that passively monitors the activity of the software is completely non-intrusive. An in-circuit emulator is also noninterusive because the software input/output relationships will be the same with and without the debugging tool. Similarly, breakpoints and single-stepping on a simulator like **TExaS** are nonintrusive, because the simulated hardware and the software are affected together.

> **Checkpoint 7.1:** *What does it mean for a debugging instrument to be minimally intrusive? Give both a general answer and a specific criterion.*

7.1 Debugging Theory

Research in the area of program monitoring and debugging has not kept pace with developments in other areas of computer programming. This area is comparatively deficient in the structure and unity now common to programming languages. The area is compartmentalized because of specialized tools. For example, run-time profile generators and execution monitors are tools that have many commands and functions in common with "debuggers," but in practice, because of tool specialization, a user needs to use two tools: a monitor to examine time behavior and a debugger for functional behavior. This area is also fragmented because of a multiplicity of terms. Terms such as program testing, diagnostics, performance debugging, functional debugging, tracing, profiling instrumentation, visualization, optimization, verification, performance measurement, and execution measurement have specialized meanings, but they are also used interchangeably, and they often describe overlapping functions. For example, the terms profiling, tracing, performance measurement, or execution measurement may be used to describe the process of examining a program from a time viewpoint. But "tracing" is also a term that may be used to describe the process of monitoring a program state or history for functional errors, or to describe the process of stepping through a program with a debugger. Usage of these terms among researchers and users varies.

Furthermore, the meaning and scope of the term "debugging" itself is not clear. In this book, the goal of debugging is to maintain and improve software, and the role of a debugger is to support this endeavor. The debugging process is defined as testing, stabilizing, localizing, and correcting errors. Although testing, stabilizing, and localizing errors are important and essential to debugging, they are auxiliary processes: the primary goal of debugging is to remedy faults or to correct errors in a program.

Although a wide variety of program monitoring and debugging tools are available today, in practice it is found that an overwhelming majority of users either still prefer or rely mainly upon "rough and ready" manual methods for locating and correcting program errors. These methods include desk-checking, dumps, and print statements, with print statements being one of the most popular manual methods. Manual methods are useful because they are readily available, and they are relatively simple to use. But the usefulness of manual methods is limited: they tend to be highly intrusive, and they do not provide adequate control over repeatability, event selection, and event isolation. A real-time system, where software execution timing is critical, usually can not be debugged with simple print statements, because the print statement itself will require too much time to execute.

A debugging **instrument** is defined as software code that is added to the program for the purpose of debugging. A print statement is a common example of an instrument. Using the editor, one adds print statements to the code that either verify proper operation or illustrate the programming errors. A key to writing good debugging instruments is to provide for a mechanism to reliably and efficiently remove all them when the debugging is done. Consider using one of these mechanisms as you develop your own debugging style.

- Place all print statements in a unique column, so that the only code that exists in this column must be a debugging instrument.
- Define all debugging instruments as functions that all have a specific pattern in their names. In this way, the find/replace mechanism of the editor can be used to find all the calls to the instruments.
- Define the instruments so that they test a run-time global flag. When this flag is turned off, the instruments perform no function. Notice that this method leaves a permanent copy of the debugging code in the final system, causing it to suffer a run-time overhead, but the debugging code can be activated dynamically without recompiling. Many commercial software applications utilize this method because it simplifies "on-site" customer support.
- Use conditional compilation (or conditional assembly) to turn on and off the instruments when the software is compiled. When the assembler or compiler supports this feature, it can provide both performance and effectiveness.

The emergence of concurrent languages and the increasing use of embedded real-time systems place further demands on debuggers. The complexities introduced by the interaction of multiple events or time-dependent processes are much more difficult to debug than errors associated with sequential programs. The behavior of non-real-time sequential programs is reproducible: for a given set of inputs their outputs remain the same. In the case of concurrent or real-time programs this does not hold true. Control over repeatability, event selection, and event isolation is even more important for concurrent or real-time environments.

Checkpoint 7.2: Consider the difference between a run-time flag that activates a debugging command versus an assembly/compile-time flag. In both cases it is easy to activate/deactivate the debugging statements. For each method, list one factor for which that method is superior to the other.

Checkpoint 7.3: What is the advantage of making an input pin activate/deactivate a debugging statement?

7.2   *Hardware Debugging Tools

Microcomputer-related problems often require the use of specialized equipment to debug the system hardware and software. Two very useful tools are the Logic Analyzer and In-circuit Emulator (ICE). A logic analyzer is essentially a multiple-channel digital storage scope with many ways to trigger. As a troubleshooting aid, it allows the experimenter to observe numerous digital signals at various points in time and thus make decisions based upon such observations. Typically, the logic analyzer is attached to the address/data bus in a passive manner, which allows the user to view the real-time execution of the software (see Figure 7.1). One problem with logic analyzers is the massive amount of information that they generate. To use an analyzer effectively, one must learn proper triggering mechanisms to capture data at appropriate times, eliminating the need to sift through volumes of output. With less expensive logic analyzers, the user must interpret the R/W, address, data information by hand. This can be quite tedious, especially when the original program is written in a high-level language. More expensive logic analyzers can disassemble the output and show the assembly-level instructions. This type of output is also difficult to interpret because the comments, labels, and program structure are not integrated into the assembly instructions. With the advent of today's microprocessor technology (in particular the cache, multiple instruction queues, branch predictions and internal buses) it is very difficult and sometimes impossible to interpret software activity simply by observing read/write cycles to external memory. Even though the Motorola 8-bit microcomputer architectures are quite simple (having none of the aforementioned fancy features and therefore being good candidates for a logic analyzer), this discussion will focus on debugging techniques that can be applied to most computer architectures. The 6812 does have an instruction queue so the address/data bus activity precedes program execution.

Figure 7.1
A logic analyzer and example output.

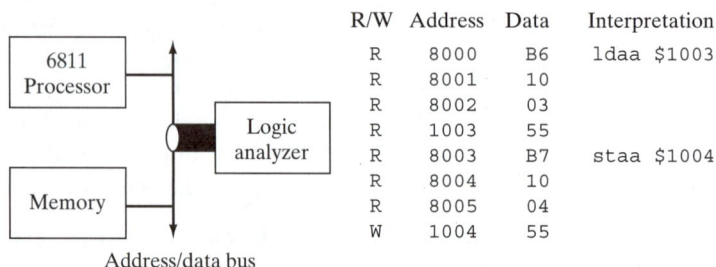

R/W	Address	Data	Interpretation
R	8000	B6	ldaa $1003
R	8001	10	
R	8002	03	
R	1003	55	
R	8003	B7	staa $1004
R	8004	10	
R	8005	04	
W	1004	55	

Checkpoint 7.4: The logic analyzer can measure the sequence of address, R/W, and data values that occur on the memory bus as the program executes. How can this information be used to determine the software state?

An in-circuit emulator is a hardware debugging tool that recreates the input/output signals of the processor chip. To use an ICE, we remove the processor chip. One side of the cable is inserted into the vacated processor chip socket, and the other side is connected to the ICE. Figure 7.2 shows the microcomputer system with and without the ICE. Notice the cable between the debugging instrument (ICE) and the microcomputer socket on the target board. In most cases, the emulator/computer system operates at full speed. The emulator allows the programmer to observe and modify internal registers of the processor. Emulators are often integrated into a personal computer, so that its editor, hard drive, and printer are available for the debugging process.

Figure 7.2 In-circuit emulator and example output.

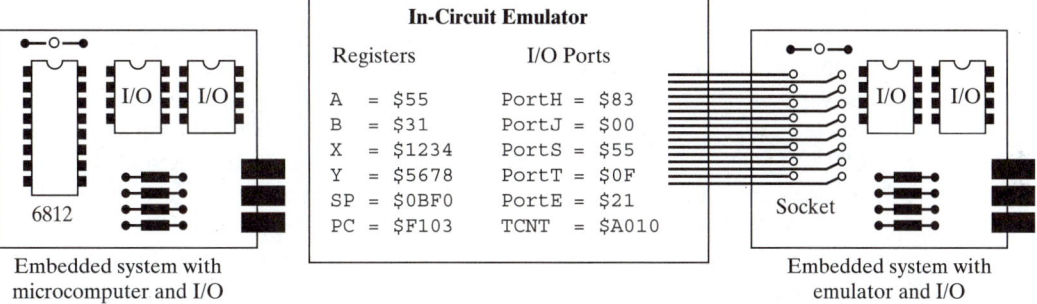

Embedded system with
microcomputer and I/O

Embedded system with
emulator and I/O

Observation: Many target microcomputer systems have the microcomputer chip soldered onto the circuit board, and thus can not be removed.

Most software-based debuggers implement a breakpoint by replacing the existing instruction with a software trap (6811/6812 instruction `swi`). This procedure can not be performed when the software is programmed in ROM. To debug this type of system, we can use another class of emulator called the ROM emulator (see Figure 7.3) This debugging tool replaces the ROM with cable connecting to a dual-port RAM within the emulator. While the software is running, it fetches information from the emulator RAM just as though it were the ROM. While the software is halted, you can modify its contents.

Figure 7.3
In-circuit ROM emulator
and example output.

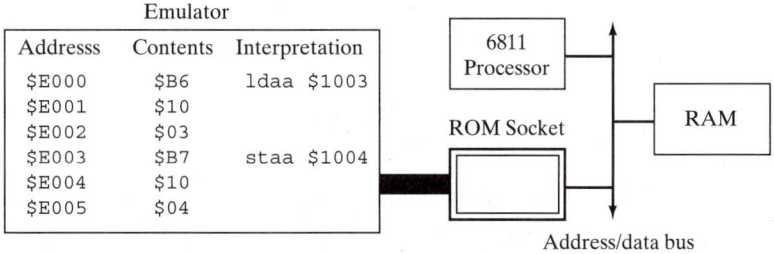

Observation: An in-circuit ROM emulator can be used only in a microcomputer system that stores the program into an external ROM chip.

The only disadvantage of the in-circuit emulator is its cost. To provide some of the benefits of this high-priced debugging equipment, the MC68HC812A4 and MC68HC912B32 both have a *background debug module* (BDM). The BDM hardware exists on the microcomputer chip itself and communicates with the debugging computer via a dedicated two- or three-wire serial interface. Although not as flexible as an ICE, the BDM can provide the ability to observe software execution in real-time, the ability to set breakpoints, the ability to stop the computer and the ability to read and write registers, I/O ports, and memory. Although both the MC68HC812A4 and the MC68HC912B32 both have background debug modules, only the 'B32 version supports hardware breakpoints. The registers can be observed only when the computer is halted, but the memory and I/O ports are accessible while the program is executing.

7.3   Texas Viewbox

You use the **ViewBox** to specify which registers, memory and I/O devices you wish to view. Usually the data at this address will be displayed. When you specify the **v** or V format, however, the value of the address itself is displayed. The **ViewBox** displays numbers in decimal constants (e.g., 0 100 10000), hexadecimal (e.g., $00 $1000 $F000) or binary (e.g., %100 %111000 %1111111111). You can observe the CPU registers (e.g., RegA RegX, RegSP) The register names are not case sensitive (e.g., regA rega REGA) and multiple spellings are allowed (e.g., A RegA PC CC CCR RegCC). The IR and EAR registers are observable. Your program symbols are case sensitive, and your program must be assembled first because the **ViewBox** needs the symbol table. I/O port names are available as symbols if your program includes the appropriate header file. The files HC11.RTF and HC12.RTF contain all the I/O ports. The files Port11.RTF and Port12.RTF contain the specific I/O ports supported by this simulator. Any expression which evaluates to a constant is also allowed; for example,

```
$F000+5*(1+2)
data+1
```

assuming `data` is a valid symbol.

The **V** format can be used to implement a calculator. First, specify one of the value formats for the result in the Format Field (v V +v or +V). Next, type the desired expression in the Address Field, then hit Enter. For example, to calculate `Main+50` in hexadecimal (where `Main` is a symbol in your program), you could

1. Type `Main+50` in the address field,
2. Type V in the format field to specify 16-bit unsigned hex,
3. Click the Enter button.

To calculate `Program2-Program1` in decimal (where `Program1` and `Program2` are two symbols in your program), you could

1. Type `Program2-Program1` in the address field,
2. Type +V in the format field to specify 16-bit signed decimal,
3. Click the Enter button.

For the CCR register, the data field can be a regular number like above or be specified as individual CCR bits like the following. Note that the upper-case letter will set the CCR bit and a lower-case letter will clear the CCR bit.

```
C Z V N I H S X    to set bits in the 6811/6812 CCR
c z v n i h s x    to clear bits in the 6811/6812 CCR
```

For example, if you wished to clear the zero bit and set the carry/negative bits, you could

1. Click on the CCR entry in the **ViewBox** or type CCR in the address field,
2. Type zCN in the data field,
3. Click the Enter button.

Any expression that evaluates to a constant is also allowed. For example,

```
(5+'3')/3
$F000+5*12
$F0|($23&$36)
$F0F0%10
```

If the source code has been assembled, symbols can also be utilized. For all symbols except those generated by equ set or =, the value of a symbol is its address. For example if

```
     org $F800
data fcb 1,2,3
```

then the expression `data+1` specifies the address $F801.

Multiple data values can be entered as a list, separated by commas. All of the preceding examples (except the null-terminated ASCII string) can be concatenated together as a list—for example,

```
1,2,3,4
(3+3*2)-4,'f',4&-7,data+1
```

You use the format field to specify the format of the entry. Each address can have a separate format. If the field is blank or does not make one of the possibilities listed below, it will use the 16-bit unsigned hexadecimal format. Table 7.1 shows the **ViewBox** format field options.

Table 7.1   ViewBox formats.

format	description	range
h	8-bit unsigned hexadecimal	$00 to $FF
d	8-bit unsigned decimal	0 to 255
b	8-bit unsigned binary	%00000000 to %11111111
H	16-bit unsigned hexadecimal	$0000 to $FFFF
D	16-bit unsigned decimal	0 to 65535
B	16-bit unsigned binary	%0000000000000000 to %1111111111111111
−h or +h	8-bit signed hexadecimal	−$80 to +$7F
−d or +d	8-bit signed decimal	−128 to +127
−b or +b	8-bit signed binary	−%10000000 to +%01111111
−H or +H	16-bit signed hexadecimal	−$8000 to +$7FFF
−D or +D	16-bit signed decimal	−32768 to +32767
−B or +B	16-bit signed binary	−%1000000000000000 to +%0111111111111111
b3	3-bit binary (least significant bits)	%000 to %111
b4	4-bit binary (least significant bits)	%0000 to %1111
cc	8-bit binary showing bits in the CCR	
c or C	ASCII character	
s or S	NULL or EOT terminated ASCII string	
v	address itself unsigned decimal	0 to 65535
V	address itself unsigned hexadecimal	$0000 to $FFFF
+v or −v	address itself signed decimal	−32768 to +32767
+V or −V	address itself signed hexadecimal	−$8000 to +$7FFF

Each of these formats (except for s, S, v, and V) can be preceded by a number (2 through 9) which will create a list. In the following examples, assume the data is $313233343536373800:

```
-d shows    +49
3h shows    $31,$32,$32
4c shows    '1','2','3','4'
s shows     "12345678"
2H shows    $3132,$3334
```

7.4   Functional Debugging

Functional debugging involves the verification of input/output parameters. It is a static process where inputs are supplied, the system is run, and the outputs are compared against the expected results. Seven methods of functional debugging are presented.

7.4.1 Single Stepping

Many debuggers allow you to set the program counter to a specific address and then execute one instruction at a time. The **TExaS** simulator provides four stepping commands: **Step**, **Few**, **StepOver** and **StepOut** commands. **Action->Step** is the usual execute one assembly instruction. **Action->Few** will execute some instructions and stop (you can set how many "some" is.) **Action->StepOver** will execute one assembly instruction, unless that instruction is a subroutine call, in which case the simulator will execute the entire subroutine and stop at the instruction following the subroutine call. **Action->StepOut** assumes the execution has already entered a subroutine, and will finish execution of the subroutine and stop at the instruction following the subroutine call.

7.4.2 Breakpoints without Filtering

The first step of debugging is to *stabilize* the system with the bug. In the debugging context, we stabilize the problem by creating a test routine that fixes (or stabilizes) all the inputs. In this way, we can reproduce the exact inputs over and over again. Once inputs are stabilized, if we modify the program, we are sure that the change in our outputs is a function of the modification we made in our software and not due to a change in the input parameters. A **breakpoint** is a mechanism to tag places in our software, which when executed will cause the software to stop. A **ScanPoint** is similar to a breakpoint in that we place them at strategic places in our software. When that address is encountered, information is logged in a debugging file and the software continues to run.

7.4.3 Conditional Breakpoints

One of the problems with breakpoints is that sometimes we have to observe many breakpoints before the error occurs. One way to deal with this problem is the conditional breakpoint. Add a global variable called count and initialize it to zero in the ritual. Add the following conditional breakpoint to the appropriate location. And run the system again (you can change the 32 to match the situation that causes the error.)

```
if (++count==32)
    bkpt
```

Notice that the breakpoint occurs only on the 32nd time the break is encountered. Any appropriate condition can be substituted.

**7.4.4
Instrumentation:
Print Statements**

The use of print statements is a popular and effective means for functional debugging. One difficulty with print statements in embedded systems is that a standard "printer" may not be available. Another problem with printing is that most embedded systems involve time-dependent interactions with their external environment. The print statement itself may so slow, that the debugging process itself causes the system to fail. In this regard, the print statement is *intrusive*. Therefore; this section will develop other debugging methods that do not rely on the availability of a printer.

**7.4.5
Instrumentation:
Dump Into Array
without Filtering**

As mentioned earlier, one of the difficulties with print statements are that they can significantly slow down the execution speed in real-time systems. Many times the bandwidth of the print functions can not keep pace with the existing system. For example, our system may wish to call a function 1000 times a second (or every 1 ms). If we add print statements to it that require 50 ms to perform, the presence of the print statements will significantly affect the system operation. In this situation, the print statements would be considered extremely intrusive. Another problem with print statements occurs when the system is using the same output hardware for its normal operation, as is required to perform the print function. In this situation, debugger output and normal system output are intertwined.

To solve both these situations, we can add a debugger instrument that dumps strategic information into an array at run time. We can then observe the contents of the array at a later time. One of the advantages of dumping is that the 6812 BDM module allows you to visualize memory even when the program is running. So this technique will be quite useful in systems with a background debug module. Assume `happy` and `sad` are strategic 8-bit variables. The first step when instrumenting a dump is to define a buffer in RAM to save the debugging measurements.

```
size equ 20
buffer rmb size*2
cnt rmb 1
```

The `cnt` will be used to index into the buffers. `cnt` must be initialized to zero, before the debugging begins. The debugging instrument, shown in Program 7.1, saves the strategic variables into the buffer.

Program 7.1
Instrumentation dump.

```
Save   pshb
       pshx      ;save
       ldab cnt
       cmpb #size*2 ; full?
       beq  done
       ldx  #buffer
       abx       ; place to put next
       ldab happy
       stab 0,x ; save happy
       ldab sad
       stab 1,x ; save sad
       inc  cnt
       inc  cnt
done   pulx
       pulb
       rts
```

Next, you add `jsr Save` statements at strategic places within the system. You can either use the debugger to display the results, or add software that prints the results after the program has run and stopped.

> *Observation:* You should save registers at the beginning and restore them at the end, so the debugging instrument itself doesn't cause the software to crash.

7.4.6 Instrumentation: Dump into Array with Filtering

One problem with dumps is that they can generate a tremendous amount of information. If you suspect a certain situation is causing the error, you can add a filter to the instrument. A filter is a software/hardware condition that must be true in order to place data into the array. In this situation, if you suspect the error occurs when another variable gets large, you could add a filter that saves in the array only when the variable is above a certain value. In the example shown in Program 7.2, the instrument saves the strategic variables into the buffer only when sad is greater than 100.

Program 7.2
Instrumentation dump with filter.

```
Save  pshb
      pshx      ;save
      ldab  sad
      cmpb  #100
      ble   done ; save only when sad >100
      ldab  cnt
      cmpb  #size*2 ; full?
      beq   done
      ldx   #buffer
      abx        ; place to put next
      ldab  happy
      stab  0,x ; save happy
      ldab  sad
      stab  1,x ; save sad
      inc   cnt
      inc   cnt
done  pulx
      pulb
      rts
```

7.4.7 Monitor Using the LED Display

Another tool that works well for real-time applications is the monitor. A monitor is an independent output process, somewhat similar to the print statement, but one that executes much faster and thus is much less intrusive. The LCD display can be an effective monitor for small amounts of information. Small LCDs can display up to 16 characters, while the larger ones can hold 4 lines by 40 characters. The hardware/software interface for such a display was presented in Section 5.10. You can place one or more LEDs on individual otherwise unused output bits. Software toggles these LEDs to let you know what parts of the program are running. An LED is an example of a BOOLEAN monitor. Assume an LED is attached to Port B bit 6. Program 7.3 will toggle the LED.

Program 7.3
An LED monitor.

```
Toggle psha
       ldaa PORTB
       eora #$40
       staa PORTB
       pula
       rts
```

Next, you add `jsr Toggle` statements at strategic places within the system. On the 6812, the `DDRB` must be initialized so that bit 6 is an output, before the debugging begins. You can either observe the LED directly, or look at the LED control signals with a high speed oscilloscope.

> **Observation:** *When using LED monitors, it is better to modify just the one bit, leaving the other seven as is. In this way, you can use additional LED monitors.*

7.5 Performance Debugging

Performance debugging involves the verification of timing behavior of our system. It is a dynamic process during which the system is run, and the dynamic behavior of the input/outputs is compared against the expected results. Two methods of performance debugging are presented; then the techniques are applied to measure execution speed.

7.5.1 Instrumentation Measuring with an Independent Counter, TCNT

A 16-bit counter called TCNT is incremented every E clock. There is a prescaler that can be placed between the E clock and the TCNT counter. It automatically rolls over when it gets to $FFFF. If we are sure the execution speed of our function is less than (65535 counts), we can use this timer to collect timing information with only a modest amount of intrusiveness.

7.5.2 Instrumentation Output Port

Another method to observe time-dependent behavior of our software involves an output port and an oscilloscope. Assume an oscilloscope is attached to Port B bit 6. This bit is used just for debugging. Program 7.4 can be used to set and clear the bit.

Program 7.4
Instrumentation output port.

```
; 6811                        ; 6812
Set psha                       Set bset PORTB, #$40
    ldaa PORTB                     rts
    oraa #$40                  Clr bclr PORTB, #$40
    staa PORTB                     rts
    pula
    rts
Clr psha
    ldaa PORTB
    anda #$BF
    staa PORTB
    pula
    rts
```

Next you add `jsr Set` and `jsr Clr` statements at strategic places within the system. On the 6812, the DDRB must be initialized so that bit 6 is an output, before the debugging begins. You can observe the signal with a high-speed oscilloscope.

7.5.3 Measurement of Dynamic Efficiency

There are three ways to measure dynamic efficiency of our software. To illustrate these three methods, we will consider measuring the execution time of the **sqrt** function. The first method is to count bus cycles using the assembly listing. This approach is appropriate only for very short programs, and becomes difficult for long programs with many conditional branch instructions. Often this is a very tedious process, but luckily the **TExaS** assembler will look up and keep a running count of the number of cycles. The assembly pseudo-op `org *` will reset the cycle counter, shown between the parentheses. A portion of the assembly output is presented in Program 7.5. Notice that the total cycle count for a 6812 implementation is 71 cycles. At 8 MHz, 71 cycles is 8.875 μs. Because the loop (between `next` and `bne next`) is executed exactly four times, the actual time will be 71+3*41 cycles or 24.25 μs. For most programs, it is actually very difficult to get an accurate time measurement using this technique.

Program 7.5   Assembly listing from **TExaS** of the sqrt subroutine.

```
$F000                                      org  *
$F000 36          [ 2] (  0){Os       }sqrt psha
$F001 C7          [ 2] (  2){O        }     clrb
$F002 97          [ 2] (  3){O        }     tsta
$F003 272E        [ 2] (  4){PPP/P    }     beq  done  ; test for Input==0
$F005 C610        [ 2] (  7){P        }     ldab #16
$F007 12          [ 2] (  8){ffO      }     mul
$F008 7C0802      [ 2] ( 11){WOP      }     std  s16   ; 16*input
$F00B 8620        [ 2] ( 14){P        }     ldaa #32
$F00D 7A0800      [ 2] ( 15){wOP      }     staa t     ; t=2.0, initial
F010 8604         [ 2] ( 18){P        }     ldaa #4
$F012 7A0801      [ 2] ( 19){wOP      }     staa cnt
$F015 F60800      [ 2] ( 22){rOP      }next ldab t     ; RegA=t
$F018 87          [ 2] ( 25){O        }     clra
$F019 B7C5        [ 2] ( 26){P        }     xgdx       ; RegX=t
$F01B B60800      [ 2] ( 27){rOP      }     ldaa t
$F01E 180E        [ 2] ( 30){OO       }     tab        ; RegB=t
$F020 12          [ 2] ( 32){ffO      }     mul        ; RegD=t*t
$F021 F30802      [ 2] ( 35){ROP      }     addd s16   ; RegD=t*t+16*s
$F024 1810        [12] ( 38){Offffffff+}    idiv       ; RegX=(t*t+16*s)/t
F026 B7C5         [ 2] ( 50){P        }     xgdx       ; RegD=(t*t+16*s)/t
$F028 49          [ 2] ( 51){O        }     lsrd       ; RegB=((t*t+16*s)/t)/2
$F029 C900        [ 2] ( 52){P        }     adcb #0    ; round up?
$F02B 7B0800      [ 2] ( 53){wOP      }     stab t     ; t=((t*t+16*s)/t)/2
$F02E 730801      [ 2] ( 56){rOPw     }     dec  cnt
$F031 26E2        [ 2] ( 60){PPP/P    }     bne  next
$F033 32          [ 2] ( 63){ufO      }done pula
$F034 3D          [ 2] ( 66){UfPPP    }     rts        ;   RegB=sqrt(s)
$F035 CF0900      [ 2] ( 71){OP       }main lds  #$0900
```

The second method uses an internal timer called TCNT, as shown in Program 7.6. Both the 6811 and the 6812 have a 16-bit unsigned counter that is automatically incremented at a regular rate. If the function completes in a time less than 65535 clock counts, then the internal timer can be used to measure execution speed empirically. The assembly language call to the function is modified so that TCNT is read before and after the subroutine call. The elapsed time is the difference. Since the execution speed may be dependent on the input data, it is often wise to measure the execution speed for a wide range of input parameters. There is a slight overhead in the measurement process itself. To be more accurate, you could measure this overhead and subtract it from your measurements.

Program 7.6
Empirical measurement of dynamic efficiency in assembly.

```
before  rmb 2        ; TCNT value before the call
elasped rmb 2        ; number of cycles required to execute sqrt
     movw TCNT,before
     jsr  sqrt   ; subroutine call to the module "sqrt"
     ldd  TCNT   ; TCNT value after the call
     subd before
     std  elasped ; execute time in cycles
```

Common error: *Debugging code should not alter the program operation. In particular, the debugging code in Program 7.6 destroys the result parameter returned in Register B.*

This same technique can also be used in C language programs. (See Program 7.7.)

Program 7.7
Empirical measurement of dynamic efficiency in C.

```
unsigned short before,elasped;
void main(void){
   ss=100;
   before=TCNT;
   tt=sqrt(ss);
   elasped=TCNT-before;
}
```

The third technique can be used in situations where TCNT is unavailable or where the execution time might be larger than 65535 counts. In this empirical technique, we attach an unused output pin to an oscilloscope or to a logic analyzer. We will set the pin high before the call to the function and set the pin low after the function call. In this way a pulse is created on the digital output with a duration equal to the execution time of the function. Assume Port B is available, and bit 7 is connected to the scope. By placing the function call in a loop, the scope can be triggered. With a storage scope or logic analyzer, the function need be called only once. The assembly-language debugging code is shown in Program 7.8. The input is stablized, which means fixed.

Program 7.8
Another empirical measurement of dynamic efficiency in assembly.

```
     movb #$FF,DDRB  ; make Port B an output
loop bset PORTB,#$80 ; set PB7 high
     ldaa #100       ; typical input
     jsr  sqrt       ; subroutine call to the module "sqrt"
     bclr PORTB,#$80 ; clear PB7 low
     bra  loop
```

Program 7.9 shows this approach in C.

Program 7.9
Another empirical measurement of dynamic efficiency in C.

```
void main(void){
  DDRB=0xFF;  // PB7 is connected to a scope
  ss=100;
  while(1){
    PORTB |= 0x80;  // set PB7 high
    tt=sqrt(ss);
    PORTB &= ~0x80; // clear PB7 low
  }
}
```

7.6   Profiling

Profiling is similar to performance debugging because both involve dynamic behavior. Profiling is a debugging process that collects the time history of strategic variables. For example, if we could collect the time-dependent behavior of the program counter, then we could see the execute patterns of our software. We can profile the execution of a multiple-thread software system to detect reentrant activity.

7.6.1
Profiling Using a
Software Dump to
Study Execution
Pattern

In this section, we will discuss software instruments that study the execution pattern of our software. In order to collect information concerning execution, we will add a debugging instrument that saves the time and location in an array (like a dump), as shown in Program 7.10. By observing this data we can determine both a time profile (when) and an execution profile (where) of the software execution.

Program 7.10
A time/position profile dumping into a data array.

```
unsigned short time[100];
unsigned short place[100];
unsigned short n;
void profile(unsigned short p){
  time[n]=TCNT; // record current time
  place[n]=p;
  n++;
}
unsigned short sqrt(unsigned short s){ unsigned short t,oldt;
profile(0);
  t=0;         // based on the secant method
  if(s>0) {
profile(1);
    t=32;    // initial guess 2.0
    do{
profile(2);
      oldt=t;  // calculation from the last iteration
      t=((t*t+16*s)/t)/2;} // t is closer to the answer
    while(t!=oldt);}    // converges in 4 or 5 iterations
profile(3);
  return t;}
```

7.6.2
Profiling Using an
Output Port

In this section, we will discuss a hardware/software combination to visualize program activity. Our debugging instrument will set output port bits. We will place these instruments at strategic places in the software. If we are using a regular oscilloscope, then we must stabilize the system so that the function is called over and over. We connect the output pins to a scope or logic analyzer and observe the program activity. Program 7.11 uses an output port to profile.

Program 7.11
A time/position profile using two output bits.

```c
unsigned int sqrt(unsigned int s){ unsigned int t,oldt;
PORTB=0;
  t=0;          // based on the secant method
  if(s>0) {
PORTB=1;
    t=32;    // initial guess 2.0
    do{
PORTB=2;
      oldt=t;  // calculation from the last iteration
      t=((t*t+16*s)/t)/2;} // t is closer to the answer
    while(t!=oldt);}    // converges in 4 or 5 iterations
PORTB=3;
  return t;}
```

Checkpoint 7.5: Write two friendly debugging instruments, one that sets Port B bit 3 high, and the other that sets it low.

7.6.3
*Thread Profile

When more than one program (multiple threads) is running, you could use the previous technique to visualize the thread that is currently active (the one running). For each thread, we assign an output pin. The debugging instrument would set the corresponding bit high when the thread starts and clear the bit when the thread stops. We would then connect the output pins to a multiple channel scope to visualize in real time the thread that is currently running. For an example of this type of profile, run one of the `thread.*` examples included with the **TExaS** simulator, and observe the logic analyzer.

7.6.4
*Reentrant Behavior

When more than one program (multiple threads) is running, it is possible for two threads to be executing the same program. To detect if a subroutine has been reentered (one thread starts and a second thread starts before the first one is finished), we need two output pins. We increment the port at the start and decrement it at the end. The thread has been reentered if the value goes above 1, as shown in Program 7.12. In this example, Port B is not part of the original code, but rather used just for the purpose of debugging.

Program 7.12
Detection of re-entrant behavior using two output bits.

```
; PB0=1 when one thread executes the function
; PB1=1 when two threads are executing the function
; subroutine to be tested
func: inc   PORTB ; used for debugging
      ; the function
      dec   PORTB
      rts
```

7.7 Tutorial 7. Debugging Techniques

The objective of this tutorial is to illustrate some debugging techniques. In particular, we will use **TExaS** to visualize stack overflow, stack underflow, and profiling.

Action: Copy the `Chap7.rtf` and `Chap7.uc` files from the CD onto your hard drive. Start a fresh copy of **TExaS** and open the `Chap7.rtf` program file from within **TExaS**, thereby opening the corresponding microcomputer window. This program contains a binary fixed-point subroutine ($\Delta = 2^{-4}$) that calculates the square root. A bug in it causes a stack overflow. The purpose of this main program is to exhaustively test this function by giving it all possible input patterns, and manually checking the validity of all outputs. Being able to evaluate a subroutine with a known and repeatable sequence of inputs is called **stabilization**. Once a system is stabilized (the inputs are fixed and known), changes to the subroutine can be made on the basis that changes in the output are the result of software modification and not the result of changes in the input.

Question 7.1. This is a very easy bug to spot, but it represents a typical programming error. By visual inspection of the main program, identify the programming error that causes the stack overflow, but don't fix it.

Question 7.2. What's the difference between a breakpoint and a ScanPoint?

Action: Assemble the program. Add Registers A and B with unsigned 8-bit decimal format to the **ViewBox**. Add a breakpoint at the location in the subroutine labeled `check`. You can add breakpoints in two ways. The first way is to left-click the line in the listing file, then right-click executing `BreakAtCursor`. The second way is to type the address (you should use the symbolic address `check` rather than its numerical value) into the **Break/ScanPoints** box and click the **add** button. You could have used its absolute address, but absolute addresses must be recalculated each time the software is modified. The double red arrow ($\leftrightarrow$) points in the listing file to the breakpoint. Make `check` a ScanPoint by toggling the **Mode->BreakMode** command until the check mark is removed. Figure 7.4 shows the resulting configuration.

Figure 7.4 A ScanPoint is added to the 6811 tutorial system.

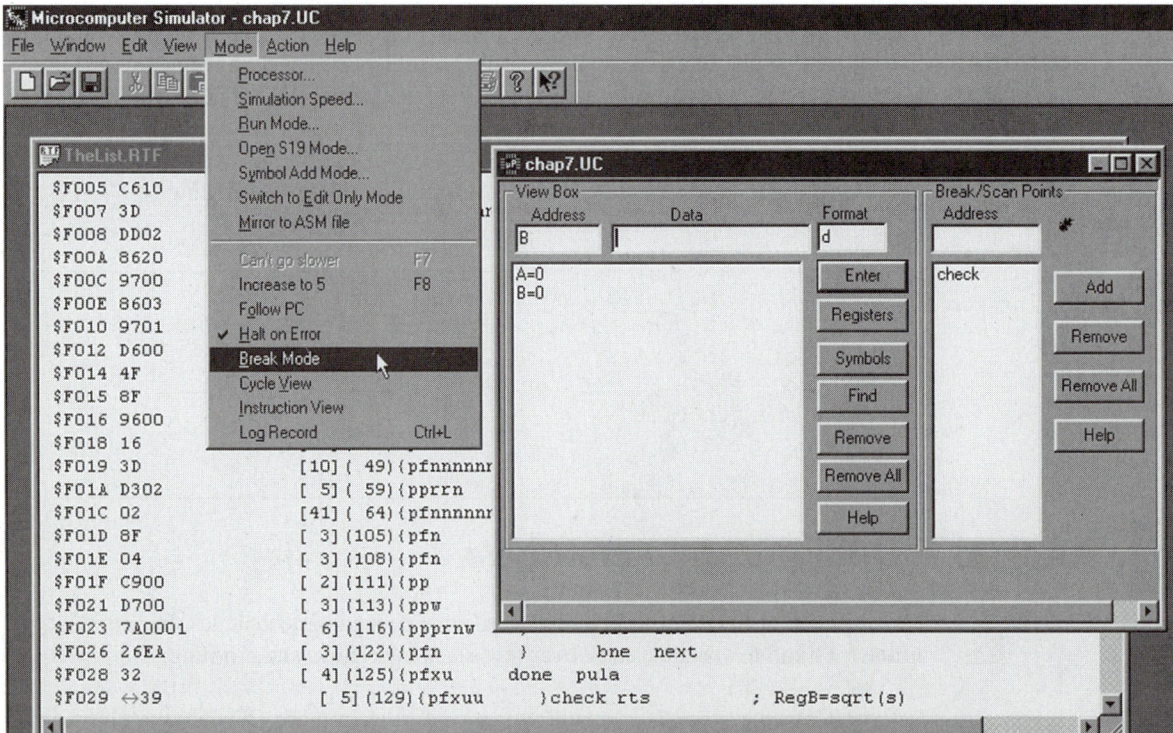

Action: Run the system until the first 10 outputs are calculated, and then stop the simulation with an **F12**. You should see the following results in `TheLog.rtf` file:

```
A=0   B=0
A=1   B=4
A=2   B=6
A=3   B=7
A=4   B=8
A=5   B=9
A=6   B=10
A=7   B=11
A=8   B=11
A=9   B=12
```

These results are correct.

Question 7.3. Explain how these first ten results are correct. In particular, verify how the values in Register B are the square root of the values in Register A.

Action: Run the system until **TExaS** gives the "**Write to ROM address 0xFFFF**" error.

Question 7.4. Look in `TheList.rtf` file and identify which instruction caused the error. The cursor arrow ($\rightarrow$) will point to the instruction after the one that caused the error.

Action: When a stack instruction causes a bug, observing the stack pointer makes sense. Add the SP to the **ViewBox**, hit reset, and run it again. The last few outputs are shown in Table 7.2.

Table 7.2
`TheLog.rtf` results showing the stack overflow.

6811	6812
A=59 B=31 SP=$0011	A=59 B=31 SP=$0812
A=60 B=31 SP=$000D	A=60 B=31 SP=$080E
A=61 B=31 SP=$0009	A=61 B=31 SP=$080A
A=224 B=32 SP=$0005	A=224 B=32 SP=$0806
Write to ROM address 0xFFFF.	**Write to undefined address 0x07FF.**

Question 7.5. Stack errors can cause weird behavior. Why did Register A skip from 61 to 224, when it should have been 62?

Action: Fix the bug (change the second `pshx` to a `pulx`), assemble and run the debugged system.

Action: Sometimes a stack error results in program branching to a location that is not part of your program. Remove the `psha` instruction from the `sqrt` subroutine. This stack underflow will cause an error. Assemble the software with this new bug, run the system. You should get a `Read from undefined address` error.

Question 7.6. You won't be able to find the cursor arrow (→) in `TheList.rtf` file. Add the PC to the **ViewBox** and check its value.

Question 7.7. There are two ways to find this bug. The first way is to execute **Action->BackDump**. What are the last five instructions to be executed just before the error?

Question 7.8. The second way to visualize the error is to activate **Mode->FollowPC**. Click this option, reset the computer and run it again. The `rts` instruction is highlighted, showing you the last instruction to execute. What does the purple color on the `pula` instruction mean?

7.8   Homework Problems

Homework 7.1: Write a debugging instrument (a subroutine) that first checks the value of Port A bit 0. If PA0 is 1, then it displays the value of Registers D, X, and Y. If PA0 is 0, the instrument returns without performing any output. You may assume the SCI port is not used for the target system, and you can call any of the routines defined in `tut2.rtf`. Save and restore any registers that you modify including the CCR. The subroutine will be added to the original software using an editor; then the combination will be assembled and downloaded to the target.

Homework 7.2: Write a debugging instrument (a subroutine) that displays the value of the PC from which the subroutine was called. You may assume the SCI port is not used for the target system, and you can call any of the routines defined in `tut2.rtf`. Save and restore any registers that you modify including the CCR. The subroutine will be added to the original software using an editor; then the combination will be assembled and downloaded to the target.

Homework 7.3: The memory cycles shown in Table 7.3 were recorded on a 6811 using a logic analyzer. Determine the sequence of executed instructions.

Homework 7.4: The memory cycles shown in Table 7.3 were recorded on a 6811 using a logic analyzer. Determine the sequence of executed instructions.

Homework 7.5: The memory cycles shown in Table 7.3 were recorded on a 6812 using a logic analyzer. Determine the sequence of executed instructions.

Table 7.3
Logic analyzer results for Problems 7.3, 7.4, and 7.5.

*Problem 7.3	*Problem 7.4	*Problem 7.5
R 0xD000 0x8E	R 0xD020 0x7F	R $F000 $8600
R 0xD001 0x00	R 0xD021 0x10	R $F002 $5A06
R 0xD002 0xFF	R 0xD022 0x07	R $F004 $86FF
R 0xD003 0x36	R 0x1007 0x00	R $F006 $5A03
R 0xD004 0x30	R 0xFFFF 0x20	W $0006 $00
W 0x00FF 0x00	W 0x1007 0x00	R $F008 $9604
R 0xD004 0x30	R 0xD023 0xFE	R $F00A $7A08
R 0xD005 0x86	R 0xD024 0xD0	W $0003 $FF
R 0x00FE 0x00	R 0xD025 0x00	R $0004 $0A
R 0xD005 0x86	R 0xD000 0xD0	R $FFFE $F000
R 0xD006 0x00	R 0xD001 0x02	R $F00C $005A
R 0xD007 0xB7	R 0xD026 0xE6	W $0800 $0A
R 0xD008 0x10	R 0xD027 0x00	R $FFFE $F000
R 0xD009 0x09	R 0xFFFF 0x20	R $F00E $0102
W 0x1009 0x00	R 0xD002 0x01	R $F010 $F700
R 0xD00A 0x86	R 0xD028 0xF7	W $0001 $0A
R 0xD00B 0x00	R 0xD029 0x10	R $F012 $0000
R 0xD00C 0xB7	R 0xD02A 0x04	R $F008 $9604
R 0xD00D 0x10	W 0x1004 0x01	R $F00A $7A08
R 0xD00E 0x07	R 0xD02B 0xF6	
W 0x1007 0x00	R 0xD02C 0x10	
R 0xD00F 0xB6	R 0xD02D 0x03	
R 0xD010 0x10	R 0x1003 0x00	
R 0xD011 0x08	R 0xD02E 0xC4	
R 0x1008 0x00	R 0xD02F 0x03	
R 0xD012 0xA7	R 0xD030 0x58	
R 0xD013 0x00	R 0xD031 0x3A	
R 0xFFFF 0x00	R 0xD031 0x3A	
W 0x00FF 0x00	R 0xD032 0xEE	
R 0xD014 0xF6	R 0xFFFF 0x20	
R 0xD015 0x10	R 0xD032 0xEE	
R 0xD016 0x03	R 0xD033 0x02	
R 0x1003 0x00	R 0xFFFF 0x20	
R 0xD017 0xAA	R 0xD004 0xD0	
R 0xD018 0x00	R 0xD005 0x0C	
R 0xFFFF 0x00	R 0xD034 0x20	
R 0x00FF 0x00	R 0xD035 0xF0	
R 0xD019 0xB7	R 0xFFFF 0x20	
R 0xD01A 0x10		
R 0xD01B 0x04		
W 0x1004 0x00		
R 0xD01C 0x20		
R 0xD01D 0xF1		
R 0xFFFF 0x00		

Homework 7.6: Write three debugging subroutines that implement a debugging array dump. Assume there are two global 16-bit variables, AA and BB, that are strategic to the system under test. The first subroutine initializes your system. The second subroutine saves AA, BB, and TCNT in the array. Your system should be able to support up to ten measurements. You may assume the SCI port is not used for the target system, and you can call any of the routines defined in tut2.rtf. The last subroutine will display the collected data. These three subroutines will be added to the original system, with the first being called at the beginning, the second placed at strategic places within the program under test, and the last one called at the end. Estimate the level of intrusiveness of this debugging process—in particular, determine the length of time it takes to call the second subroutine. These subroutines will be added to the original software using an editor; then the combination will be assembled and downloaded to the target.

Homework 7.7: Assuming the object code is running in RAM, write three debugging subroutines that implement a ScanPoint system. The first subroutine initializes your system. The second subroutine adds a ScanPoint at the address passed into it in Register D. You may assume that the ScanPoint address is the first byte of an op code. When the target program executes that scanned instruction, the values of the registers are displayed, the original instruction is executed, and the program continues execution. Your system should be able to support up to ten ScanPoints. You may assume the SCI port is not used for the target system, and you can call any of the routines defined in tut2.rtf. The last subroutine removes a ScanPoint at the address passed into it in Register D. For simplicity, you may assume the ScanPoints are placed only at single-byte instructions.

Homework 7.8: Assuming the object code is running in RAM, write debugging subroutines that implement single stepping. In particular, write a subroutine that executes the target software at the address passed into it in Register D. You may assume that the starting address is the first byte of an op code. Your system should execute the target program one instruction at a time, showing the values of the registers, and pausing for SCI input after each instruction. You may assume the SCI port is not used for the target system, and you can call any of the routines defined in tut2.rtf. If the operator types 'q', then the debugging halts and control is returned to the program that called your subroutine. For any other input, you should execute the next instruction. This is an advanced topic and will require output compare interrupts to solve.

7.9   Laboratory Assignments

Lab 7.1 Profiling

Purpose The **TExaS** simulator provides a rich set of debugging tools, but eventually we will be asked to run programs on an actual microcomputer. The objective of this lab is to develop profiling tools that do not depend on the simulator. Even though we will still be using the simulator for this lab, these techniques can be used when debugging software on an actual microcomputer.

Procedure

Part A.

Write three debugging subroutines that implement profiling. The first subroutine (`De-bug_Init`) initializes your system. The second subroutine (`Debug_Capture`) saves a profile-point (time, data, PC position) in an array. The time parameter is the current TCNT value, the data parameter is the hexadecimal value in Register D, and the PC position information can be obtained by reading the return address off the stack. You may assume the SCI port is not used for the target system, and you can call any of the routines defined in `tut2.rtf`. The last subroutine (`Debug_Display`) displays the profile on the SCI/CRT interface. Be careful to save and restore registers so the original subroutine will execute. Program 7.13 shows an example application of these debugging functions. The original program was debugged using **TExaS** as part of the tutorial for this chapter. To run the system on the 6811, you'll have to change the RAM, ROM and stack pointer initialization. Measure the execution time of the `Debug_Capture` subroutine. This time will be a quantitative measure of the intrusiveness of the debugging instrument.

Program 7.13 Profiling added to the square-root program.

```
        org   $0800
t       rmb   1      transformed to sqrt(s)
cnt     rmb   1      loop counter
s16     rmb   2      16*input
        org   $F000
* binary fixed point squareroot, 2**-4
* Input:  Reg A is s (0 to 15.9375)
* Output: Reg B is t=sqrt(s) 0 to 4.00
sqrt    psha
        clrb
        tsta
        beq   done     ; test for Input==0
        ldab  #16
        mul
        std   s16       ; 16*input
        ldaa  #32
        staa  t         ; t=2.0, initial
        ldaa  #4
        staa  cnt
next    ldab  t         ; RegA=t
        clra
        xgdx            ; RegX=t
        ldaa  t
        tab             ; RegB=t
        mul             ; RegD=t*t
        addd  s16       ; RegD=t*t+16*s
        idiv            ; RegX=(t*t+16*s)/t
        xgdx            ; RegD=(t*t+16*s)/t
        lsrd            ; RegB=((t*t+16*s)/t)/2
        adcb  #0        ; round up?
        stab  t         ; t=((t*t+16*s)/t)/2
        dec   cnt
        bne   next
```

```
* with debugging added
        org   $0800
t       rmb   1      transformed to sqrt(s)
cnt     rmb   1      loop counter
s16     rmb   2      16*input
        org   $F000
* binary fixed point squareroot, 2**-4
* Input:  Reg A is s (0 to 15.9375)
* Output: Reg B is t=sqrt(s) 0 to 4.00
sqrt    jsr   Debug_Capture
        psha
        clrb
        tsta
        beq   done     ; test for Input==0
        ldab  #16
        mul
        std   s16       ; 16*input
        ldaa  #32
        staa  t         ; t=2.0, initial
        ldaa  #4
        staa  cnt
next    ldab  t         ; RegA=t
        clra
        xgdx            ; RegX=t
        ldaa  t
        tab             ; RegB=t
        mul             ; RegD=t*t
        addd  s16       ; RegD=t*t+16*s
        idiv            ; RegX=(t*t+16*s)/t
        xgdx            ; RegD=(t*t+16*s)/t
        lsrd            ; RegB=((t*t+16*s)/t)/2
        adcb  #0        ; round up?
        stab  t         ; t=((t*t+16*s)/t)/2
```

continued

```
done    pula                                        jsr     Debug_Capture
        rts              ; RegB=sqrt(s)             dec     cnt
main    lds    #$0900                               bne     next
        clra                                done    pula
loop    pshx                                        rts              ; RegB=sqrt(s)
        bsr    sqrt                         main    lds     #$0900
check   nop                                         clra
        inca                                loop    pshx
        pulx                                        jsr     Debug_Init
        bne    loop                                 bsr     sqrt
        stop                                        jsr     Debug_Display
        org    $FFFE                        check   nop
        fdb    main                                 inca
                                                    pulx
                                                    bne     loop
                                                    stop
                                                    org     $FFFE
                                                    fdb     main
```

Part B.

In this part you will instrument the original program with debugging code that outputs to a parallel port. The purpose of this debugging is to count the number of times sqrt is called. Modify the main program so sqrt is called exactly 15 times. Connect unused parallel port bits to an external device that will assist in the visualization (LED, LCD, etc.) Run your instrumented system that visualizes the program is called 15 times. Measure the execution times of your debugging instruments. These times will be a quantitative measure of their intrusiveness.

Part C.

Again, you will instrument the original program with debugging code that outputs to a parallel port. The purpose of this debugging is to visualize the execution pattern within sqrt. Modify the main program so sqrt is called once with an input of Reg A = 100. Connect the parallel port bits to a logic analyzer. The debugging instruments should set and clear individual bits of the parallel port. Place these instruments at strategic places in the original program. Run your instrumented system that visualizes the execution pattern. In particular, you should see the subroutine start, visualize how many times it loops, and see it finish. Measure the execution times of your debugging instruments. These times will be a quantitative measure of their intrusiveness.

8 Program Structures

Chapter 8 objectives are to

❏ explain how to define local variables on the stack,
❏ show how various C compilers implement local variables and pass parameters,
❏ compare and contrast call by value versus call by reference parameter passing,
❏ define the complex assembly language blocking blocks,
❏ develop extended precision mathematical calculations,
❏ use linked structures to implement finite state machines,
❏ implement system calls using software interrupts,
❏ introduce paged memory on the MC68HC812A4.

The overall theme of this chapter is advanced program development. A program structure involves both the operations it performs and the organization of those operations. As the embedded system gets more complex, ad hoc programming methods produce code that is hard to debug, difficult to understand, and impossible to change. In this chapter, you will be presented with a wide range of programming techniques that will allow you to develop larger more complex systems. This chapter introduces advanced programming techniques, while the next chapter tackles the more difficult problem of software style.

8.1 Local Variables

8.1.1 Introduction

Because their contents are allowed to change, all variables must be allocated in RAM and not ROM. A **local variable** is temporary information used by only one software module. Local variables are typically allocated, used, then deallocated. The information stored in a local variable is not permanent. This means that if we store a value into a local variable during one execution of the module, the next time that module is executed the previous value is not available. Examples include loop counters and temporary sums. We use a local variable to store data that are temporary in nature. We can implement a local variable using the stack or registers. Reasons why we place local variables on the stack include the following:

- dynamic allocation/release allows for reuse of memory,
- limited scope of access provides for data protection,
- only the program that created the local variable can access it,
- since an interrupt will save registers and create its own stack frame, the code is reentrant,
- since absolute addressing is not used, the code is relocatable,
- the number of variables is limited only by the size of the stack allocation, more than registers.

Checkpoint 8.1: How do you create a local variable in C?

A **global variable** is information shared by more than one program module (e.g., we use globals to pass data between the main (or foreground) process and an interrupt (or background) process). Global variables may also be used to store information that is permanent in nature. Global variables are not deallocated. The information they store is permanent. Examples include time of day, date, user name, temperature, pointers to shared data. On the 6811 and 6812, we use absolute addressing (direct or extended) to access their information.

Observation: Sometimes we store temporary information in global variables out of laziness. This practice is to be discouraged because it wastes memory and may cause the module to not be reentrant.

Checkpoint 8.2: How do you create a global variable in C?

Checkpoint 8.3: How does the `static` modifier affect locals, globals, and functions in C?

Checkpoint 8.4: How does the `const` modifier affect a global variable in C?

A LIFO stack is implemented in hardware by most computers. It can be used for local variables (temporary storage), saving return addresses during subroutine calls, passing parameters to subroutines, and to save registers during the processing of an interrupt. The first advantage of placing local variables on the stack is that the storage can be dynamically allocated before usage and deallocated after usage. The second advantage is the facilitation of reentrant software. The 6811 stack operates differently from most other computers. On the other hand, the 6812 stack operates like most computers.

On the 6811, the stack pointer (SP) points to the free space that the next PUSH will store into, as shown in Figure 8.1. To PUSH a byte on the stack, first the stack pointer (SP) is decremented, then the byte is stored at the location pointed to by SP+1. To PULL a byte from the stack, first the byte is read from memory pointed to by SP+1, then SP is incremented.

On the 6812, the stack pointer (SP) points to the top entry of the stack, also shown in Figure 8.1. To PUSH a byte on the stack, first the stack pointer (SP) is decremented, then the byte is stored at the location pointed to by SP. To PULL a byte from the stack, first the byte is read from memory pointed to by SP, then SP is incremented.

Figure 8.1
6811 and 6812 stack.

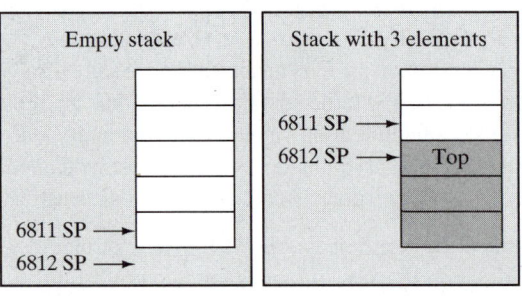

To access local variables on the 6811 stack, the stack pointer must first be transferred into Reg X (or Y). The instruction `tsx` will move a copy of the stack pointer into Register X. Although the `tsx` and `tsy` instructions work a little different on the 6811 versus the 6812, in both cases the instruction causes the index register to point to the top element of the stack, as shown in Figure 8.2. The `tsx` and `tsy` instructions do not modify the stack pointer.

Figure 8.2
The tsx instruction creates a stack frame pointer.

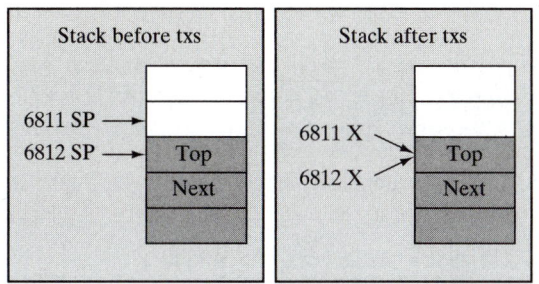

Then index mode reads and writes are possible. For example to read the next to the top byte,

```
tsx       Reg X points to the top byte of the stack
ldaa 1,X  Reg A = the next to the top byte
```

The LIFO stack has a few rules:

1. Program segments should have an equal number of pushes and pulls;
2. Stack accesses (PUSH or PULL) should not be performed outside the allocated area;
3. Stack reads and writes should not be performed within the *free area*;
   Stack PUSH should first decrement SP, then store the data,
   Stack PULL should first read the data, then increment SP.

Programs that violate rule number 1 will probably crash when a `rts` instruction pulls an illegal address from the stack at the end of a subroutine. The **TExaS** simulator will usually recognize this error as an illegal memory access when the processor tries to fetch an op code at this incorrect address. The backdump command will be useful to retrace the steps leading up to the crash.

Violations of rule number 2 can be the result of a stack underflow or overflow. Stack underflow is caused when there are more pulls than pushes, and is always the result of a software bug. The **TExaS** simulator will recognize this error as an illegal memory access when the processor tries to pull data from an address that doesn't exist. A stack overflow can be caused by two conditions. If the software mistakenly pushes more than it pulls, then the stack pointer will eventually overflow its bounds. Even when there is exactly one pull for each push, a stack overflow can occur if the stack is not allocated enough space. Stack overflow is a very difficult bug to recognize, because the first consequence occurs when the computer pushes data onto the stack and overwrites data stored in a global variable. At this point the local variables and global variables exist at overlapping addresses. Setting a breakpoint at the first address of the allocated stack area allows you to detect a stack overflow situation.

Checkpoint 8.5: *How do you specify the size of the stack*

The following 6812 assembly code violates rule 3, and will not work if interrupts are active. The objective is to save register A onto the stack. When an interrupt occurs, registers will automatically be pushed onto the stack, destroying the data.

```
staa -1,SP  Store zero onto the stack (***illegal***)
```

To use the stack, one first allocates, then saves. The following assembly code also violates rule 3, because it first stores it onto the stack, then allocates space. The objective is to push a zero onto the stack.

```
tsx         Reg X points to the top of the stack
dex
clr 0,X     Store zero onto the stack (***illegal***)
des         Make space for the zero
```

If an interrupt were to occur between the `clr` and `des` instructions, the zero will be destroyed when all the registers are pushed on the stack by the interrupt process. The proper technique is to allocate first,

```
des         Allocate stack space first
tsx         Reg X points to the top of the stack
clr ,X  Store zero onto the stack
```

The 6811 push instructions (e.g., `psha pshx`) do store data on the stack first, then decrement the stack pointer. These 6811 operations do not violate rule 3, because the store and decrement operations are atomic. An atomic operation is a sequence that once started will always finish, and can not be interrupted. Most instructions on the 6811 and 6812 are atomic. The exceptions are `wai rev` and `revw`, which can be suspended to process an interrupt.

8.1.2
Implementation

Stack implementation of local variables has four stages; binding, allocation, access, and deallocation.

1. Binding is the assignment of the address (not value) to a symbolic name. This address will be the actual memory location to store the local variable. The assembler binds the symbolic name to a stack index. The computer calculates the actual location during execution. For example,

```
sum    set    0   16-bit  local variable, stored on the stack
```

Checkpoint 8.6: Why is `set` better than `equ` for binding?

2. Allocation is the generation of memory storage for the local variable. The computer allocates space during execution by decrementing the SP. In this first example, the software allocates the local variable by pushing a register on the stack. An 8-bit push (e.g., `psha`) creates an uninitialized 8-bit local variable, and a 16-bit push (e.g., `pshx`) creates an uninitialized 16-bit local variable. The value in the register is irrelevant; the instruction is used because it decrements the SP.

```
pshx    allocate sum
```

In this next example, the software allocates the local variable by decrementing the stack pointer. This local variable is also uninitialized.

```
des     allocate sum
des
```

If you wished to allocate the 16-bit local and initialize it to zero, you could execute.

```
; 6811 or  6812              ; 6812 only
    ldx    #0                    movw #0,2,-sp allocate sum=0
    pshx       allocate sum=0
```

In this last example, the technique provides a mechanism for allocating large amounts of stack space. This example allocates 20 bytes for the structure `big[20]`. Local variables are so important that the 6812 has special instructions to simplify the implementation of local variables.

```
; 6811 or 6812                        ; 6812 only
     tsx        allocate big[20]           leas  -20,sp  allocate big[20]
     xgdx
     subd   #20
     xgdx
     txs
```

3. The **access** to a local variable is a read or write operation that occurs during execution. In the next code fragment, the local variable `sum` is set to 0.

```
; 6811 or 6812                          ; 6812 only
     tsx        X points to locals             movw #0,sum,sp   sum=0
     ldd    #0
     std    sum,x sum=0
```

In the next code fragment, the local variable `sum` is incremented.

```
; 6811 or  6812                      ; 6812 only
     tsx                                  ldd   sum,sp
     ldd    sum,x                         addd  #1
     addd   #1                            std   sum,sp    sum=sum+1
     std    sum,x     sum=sum+1
```

4. Deallocation is the release of memory storage for the location variable. The computer deallocates space during execution by incrementing SP. In this first example, the software deallocates the local variable by pulling a register from the stack.

```
pulx deallocate sum
```

> **Observation:** When the software uses the "push-register" technique to allocate and the "pull-register" technique to deallocate, it looks as though it is saving and restoring the register. Because most applications of local variables involve storing into the local, the value pulled will NOT match the value pushed.

In this next example, the software deallocates the local variable by incrementing the stack pointer.

```
ins
ins    deallocate sum
```

In this last example, the technique provides a mechanism for allocating large amounts of stack space.

```
; 6811 or 6812                        | ; 6812 only
    tsx       deallocate big[20]      |    leas 20,sp deallocate big[20]
    ldab  #20                         |
    abx                               |
    txs                               |
```

Checkpoint 8.7: Write a 6811/6812 subroutine that allocates, then deallocates, three 8-bit locals.

8.1.3
Stack Frames

The 6812 provides a negative-offset index-addressing mode, whereas on the 6811 only positive index values can be used. With either computer, it is possible to establish a stack frame pointer using either register X or Y. It is important in this implementation that once the stack frame pointer is established (e.g., the `tsx` instruction), that the stack frame register (X) not be modified. The term **frame** refers to the fact that its value is fixed. Because the stack frame pointer should not be modified, every subroutine will save the old stack frame pointer of the function that called the subroutine (e.g., `pshx` at the top) and restore it before returning (e.g., `pulx` at the bottom.) The stack frame will allow you to use the `txs` to deallocate the local variables. Local variable access uses indexed addressing mode. The difference between the two versions is the position of the stack frame pointer. As always, the 6811 example will function properly on the 6812. This example will be extended to include subroutine parameters later in the chapter. Notice the subroutine deallocates the locals by moving the stack frame pointer back into SP with the `txs` instruction.

Observation: One advantage of using a stack frame is that the `tsx` instruction needs to be executed only once at the beginning of the function.

Observation: Another advantage of using a stack frame is that you can push and pull within the body of the function, and still be able to access local variables using their symbolic name.

Observation: One disadvantage of using a stack frame is that a register is dedicated as the frame pointer, and thus it is unavailable for general use.

This example calculates the sum of the first 100 numbers. Program 8.2 shows the assembly code that implements the C function given in Program 8.1. The result will be returned by value in Register D.

Program 8.1
A simple function with two local 16-bit variables.

```
unsigned short calc(void){ unsigned short sum,n;
   sum = 0;
   for(n=100;n>0;n--){
      sum=sum+n;
   }
   return sum;
}
```

Program 8.2 Assembly language implementation of a function with two local 16-bit variables.

```
; 6811 or 6812                              ; 6812 only
; *****binding phase***************         ; *****binding phase**************
sum  set  0  16-bit number                 sum  set  -4  16-bit number
n    set  2  16-bit number                 n    set  -2  16-bit number
; ******allocation phase *********          ; ******allocation phase *********
calc pshx        save old Reg X            calc pshx        save old Reg X
     pshx        allocate n                     tsx         stack frame pointer
     pshx        allocate sum                   leas -4,sp  allocate n,sum
     tsx         stack frame pointer       ; ********access phase ***********
; *******access phase ***********               movw #0,sum,x   sum=0
     ldd  #0                                     movb #100,n,x   n=100
     std  sum,x  sum=0                     loop ldd  n,x     RegD=I
     ldd  #100                                  addd sum,x   RegD=sum+n
     std  n,x    n=100                          std  sum,x   sum=sum+n
loop ldd  n,x    RegD=n                         ldd  n,x     n=n-1
     addd sum,x  RegD=sum+n                      subd #1
     std  sum,x  sum=sum+n                       std  n,x
     ldd  n,x    n=n-1                           bne  loop
     subd #1
     std  n,x
     bne  loop
; *******deallocation phase *****           ; ********deallocation phase *****
     txs         deallocation                   txs         deallocation
     pulx        restore old X                  pulx        restore old X
     rts         RegD=sum                       rts         RegD=sum
```

8.2 Parameter Passing

The **input parameters** (or arguments) are pieces of data passed from the calling routine into the module during execution. The **output parameter** (or argument) is information returned from the module to the calling routine after the module has completed its task. There are two methods to pass parameters: **call by reference** and **call by value**. With call by reference, a pointer to the object is passed. In this way, the subroutine and the module that calls the subroutine have access to the exact same object. Call by reference can be used to pass a large quantity of data, and can be used to implement a parameter that is both an input and an output parameter. With call by value, a copy of the data itself is passed.

8.2.1
Parameter Passing in C

The call-by-reference method passes a pointer to the object. In other words, references (pointers) to the actual arguments are passed, instead of copies of the actual arguments themselves. In this scheme, assignment statements have implied side effects on the actual arguments; that is, variables passed to a function are affected by changes to the formal arguments. Sometimes side effects are beneficial, and sometimes they are not. As an example, consider a stepper motor program shown in Program 8.3. The read and write accesses to the parameter affect the original variable.

Program 8.3
An input/output parameter is implemented using call by reference.

```
void next(unsigned short *pt){
  (*pt)++;                   /* next angle */
  if((*pt) == 200){
    (*pt) = 0;               /* one complete rotation */
  }
}
void main(void){
unsigned short angle=0;   /* stepper motor angle 0 to 199 */
  Stepper_Init();
  while(1){
    Stepper_Step();
    next(&angle);
  }
}
```

Since C supports only one formal output parameter, we can implement additional output parameters using call by reference. The calling program passes pointers to empty objects, and the function fills the objects with data. Program 8.4 shows a function that returns two parameters using call by reference.

Program 8.4
Multiple output parameters are implemented using call by reference.

```
short xx,yy;               /* position */
void where(short *Xpt, short *Ypt){
  (*Xpt) = xx;             /* return xx */
  (*Ypt) = yy;             /* return yy */
}
short myX,myY;
void main(void){
  where(&myX,&myY);
}
```

When we use the call by value scheme, the values, not references, are passed to functions. With call by value, copies are made of the parameters. Within a called function, references to formal arguments see copied values on the stack, instead of the original objects from which they were taken. At the time when the computer is executing within next, as shown in Program 8.5, there will be two separate and distinct copies of the angle data.

Program 8.5
Parameters are implemented using call by value.

```
unsigned short next(unsigned short theAngle){
  theAngle++;                /* next angle */
  if(theAngle == 200){
    theAngle = 0;            /* one complete rotation */
  }
  return(theAngle);
}
void main(void){
unsigned short angle;   /* stepper motor angle 0 to 199 */
  Stepper_Init();
  while(1){
    Stepper_Step();
    angle = next(angle);
  }
}
```

An important point to remember about passing arguments by value in C is that there is no connection between an actual argument and its source. Changes to the arguments made within a function have no effect whatsoever on the objects that have supplied their values. They can be changed and the original values will not be affected. This removes a burden of concern from the programmer since he may use arguments as local variables without side effects. It also avoids the need to define temporary variables just to prevent side effects.

It is precisely because C uses call by value that we can pass expressions, not just variables, as arguments. The value of an expression can be copied, but it cannot be referenced since it has no existence in global memory. Therefore, call by value adds important generality to the language. Since expressions may include assignment, increment, and decrement operators, it is possible for argument expressions to affect the values of arguments lying to their right. Consider, for example,

```
func(y=x+1, 2*y);
```

where the first argument has the value `x+1` and the second argument has the value `2* (x+1)`. The value of the second argument depends on whether the arguments are evaluated right-to-left or left-to-right. This kind of situation should be avoided, since the C language does not guarantee the order of argument evaluation. The safe way to write this expression is

```
y=x+1;
func(y, 2*y);
```

The value of the expression is calculated at the time of the call, and that value is passed into the subroutine.

8.2.2 Parameter Passing in Assembly Language

In contrast to C, it is easy to return multiple parameters in assembly language. If just a few parameters need to be returned we can use the registers. In Program 8.6, the values of ports A, B, C, and D are to be returned. Notice that it packs two 8-bit parameters into the 16-bit Register X.

Program 8.6   Multiple return parameters implemented with registers.

```
; Reg A = Port A, Reg B= Port B          ********calling sequence******
; Reg X = Ports C and D                        jsr  GetPorts
GetPorts ldaa PORTC                      * Reg A,B,X have four results
         ldab PORTD                            staa first
         xgdx                                  stab second
         ldaa PORTA                            xgdx
         ldab PORTB                            staa third
         rts                                   stab fourth
```

If many parameters are needed, then the stack can be used. Space for the output parameters is allocated by the calling routine, and `GetPorts` stores the results into those stack locations. Program 8.7 also returns the values of ports A, B, C, and D.

Program 8.7   Multiple return parameters passed on the stack.

```
dataA    set 2                          dataA     set 2
dataB    set 3                          dataB     set 3
dataC    set 4                          dataC     set 4
dataD    set 5                          dataD     set 5
GetPorts tsx                            GetPorts  movb PORTA,dataA,sp
         ldaa PORTA                               movb PORTB,dataB,sp
         staa dataA,x                             movb PORTC,dataC,sp
         ldaa PORTB                               movb PORTD,dataD,sp
         staa dataB,x                             rts
         ldaa PORTC
         staa dataC,x
         ldaa PORTD
         staa dataD,x
         rts
********calling sequence******          ********calling sequence******
    pshx                                      leas -4,sp       ;allocate
    pshx            ;allocate                 jsr  GetPorts
    jsr  GetPorts                             pula             ;first
    pula            ;first                    staa first
    staa first                                pula             ;second
    pula            ;second                   staa second
    staa second                               pula             ;third
    pula            ;third                     staa third
    staa third                                pula             ;fourth
    pula            ;fourth                   staa fourth
    staa fourth
```

An input parameter is information passed from the calling program into the subroutine before the subroutine is executed. An output parameter is information passed out of the subroutine back to the calling program after the subroutine is executed. A parameter can be both an input and an output. The purpose of the next set of examples is to illustrate parameter passing. The subroutine `Add8` adds M=M+N, and sets the flag P if there is an unsigned overflow. M is a 16-bit input/output parameter, N is an 8-bit input parameter, and P is a 1-bit output parameter.

The simplest and fastest method is to pass parameters using registers. In this method, the information is contained in the registers. Because concurrent programs have "separate" registers and stack areas, the subroutine is reentrant (more about reentrancy will be presented in Chapter 11). Program 8.8 shows the addition module.

Program 8.8   Addition function that passes parameters call by value in registers.

```
* Subroutine Calling Sequence
* place information in A,X
     bsr  Add8
* use information in CC,X
* Subroutine Definition
* N is an input parameter, an unsigned 8-bit byte, passed in Reg A
* M is an input/output, a 16-bit number, passed/returned in Reg X
* P is an output parameter, a Boolean flag,
*          returned in Reg CC carry bit
```

```
Add8    psha        Put N on the stack
        tsy         Reg Y points to N
        xgdx        Place M in Reg D
        addb ,Y     Add N to the LSByte of M
        adca #0     Reg D=M+N, CC(carry bit) = P
        xgdx        Return result in Reg X
        ins         Deallocate local variable
        rts
```

A simple but not very elegant method is to pass parameters using global variables. In this method the information is contained in global memory variables. Because of the writes to global memory M and P, the subroutine, shown in Program 8.9, is not reentrant. Many embedded systems use this approach because the processor has limited or no facilities with handling data on the stack.

Program 8.9 Addition function that passes parameters call by value in global variables.

```
* These three variables can be anywhere in RAM memory
N       rmb   1     N is an input parameter, an unsigned 8-bit number
M       rmb   2     M is an input/output parameter, 16 bits
P       rmb   1     P is an output parameter, a Boolean flag,
*                     0 means no overflow, -1 means overflow
* Subroutine Calling Sequence
* place information in N,M
        bsr   Add8
* use information in M,P
* Subroutine Definition
Add8    clr   P         Assume no overflow, P=0
        ldd   M         Place M in Reg D
        addb  N         Add N to the LSByte of M
        adca  #0        Reg D=M+N, CC(carry bit) = P
        bcc   POK       Skip if P should remain zero
        com   P         Overflow, P=-1
POK     std   M         Return result in M
        rts
```

A flexible and elegant method is to pass parameters using the stack. In this method the information is placed on the system or user stack. Most high-level language compilers (Pascal, C, and FORTH) use the stack to pass parameters. So far in this book, we have not activated the interrupt system. Without interrupts, the computer executes one main program. There may be data-dependent branching and subroutine calls, but the execution sequence is simple and predictable. When interrupts are enabled, it is possible have multiple programs active at the same time. There is still only one processor, so exactly one program is actually running at a time, but we define concurrent programming as the state where multiple programs are "ready to run" at the same time. The interrupt hardware, which will be presented in Chapter 11, provides the mechanism to switch from one program to the next. Because concurrent programs have "separate" registers and stack areas, software that uses the stack will operate properly in a concurrent environment. Conversely, extreme care is required when using global variables (including the I/O ports) in a concurrent environment. As we will see later in Section 8.2.3, most C compilers pass multiple input parameters on the stack, but return the single output parameter in a register. The other advantage of using the stack is that memory space is used temporarily, then deallocated. Program 8.10 passes both input and output parameters on the stack.

Program 8.10
Addition function that passes parameters call by value on the stack.

```
* Subroutine Calling Sequence
        des               Make room on the stack for P
* push M (16 bits) onto the stack
* push N (8 bits) onto the stack
        bsr   Add8
        ins               Discard input only parameter, N
* pop M (16 bits) off the stack
* pop P (8 bits) off the stack
* Subroutine Definition
* N is an input parameter, a unsigned 8-bit number,
*       passed on the top of the stack
* M is an input/output , a 16-bit number,
*       passed/returned on top-1, top-2
* P  is an output parameter, a Boolean flag, returned on top-3
*               Access       Contents
*               0,Y          16-bit return address
N       set   2   N,Y        8-bit N
M       set   3   M,Y        16 nit M
P       set   5   P,Y        8-bit P
Add8    tsy.          Reg Y points to stack
        clr   P,Y     Assume no overflow, P=0
        ldd   M,Y     Place M in Reg D
        addb  N,Y     Add N to the LSByte of M
        adca  #0      Reg D=M+N, CC(carry bit) = P
        bcc   POK     Skip if P should remain zero
        com   P,Y     Overflow, P=-1
POK     std   M,Y     Return result in M
        rts           Return
```

Another flexible method is to pass parameters using the In-Line Argument List. In this method, the parameter values or address pointers are stored in a list following the `jsr` instruction. This method of parameter passing is used by many FORTRAN compilers. In the example shown in Program 8.11, the information is stored after the JSR instruction. The subroutines are reentrant, but the calling program is not. This approach will not work for embedded systems, where the program is burned into ROM or EEPROM. This approach also does not work in a concurrent environment, because the multiple programs would share the common parameter space, causing the parameters generated by one program to be confused with the parameters of another program.

Program 8.11
Addition function that passes parameters call by value using an argument list.

```
* Subroutine Calling Sequence
* store M (16-bit) into the position M1
* store N (8-bits) into the position N1
        jsr   Add8
N1      rmb   1           Place to pass N            0,X
```

```
M1      rmb   2              Place to pass/return M    1,X
P1      rmb   1              Place to return P         3,X
* read M (16 bits) from M1                             4,X
* read P (8 bits) from P1
* Subroutine Definition
*                     Access        Contents
N       equ   0       N,X           8-bit N
M       equ   1       M,X           16-bit M
P       equ   3       P,X           8-bit P
*                     4,X           16-bit return address
Add8    pulx                 Reg X points to N1,M1,P1
        clr   P,X            Assume no overflow, P=0
        ldd   M,X            Place M in Reg D
        addb  N,X            Add N to the LSByte of M
        adca  #0             Reg D=M+N, CC(carry bit) = P
        bcc   POK            Skip if P should remain zero
        com   P,X            Overflow, P=-1
POK     std   M,X            Return result in M
        jmp   4,X            Return
```

In the example shown in Program 8.12, the pointers to the information are stored after the JSR instruction. This approach will work for embedded systems, because the addresses of the variables are fixed. Just as in the previous example, this approach should not be used in a concurrent programming environment.

Program 8.12
Addition function that passes parameters call by reference using an argument list.

```
* Subroutine that uses an In-line Argument List
* Pointers to the data are stored after the jsr instruction.
*Subroutine Calling Sequence
* M,N,P are memory variables
        jsr   Add8
        fdb   N              Pointer to  N      0,X
        fdb   M              Pointer to  M      2,X
        fdb   P              Pointer to  P      4,X
* Return place                                 6,X
* Subroutine Definition
Add8    pulx                 Reg X points to list of addresses for N, M, P
        ldy   4,X            Pointer to P
        clr   ,Y             Assume no overflow, P=0
        ldy   2,X            Pointer to M
        ldd   ,Y             Place M in Reg D
        ldy   ,X             Pointer to N
        addb  ,Y             Add N to the LSByte of M
        adca  #0             Reg D=M+N, CC(carry bit) = P
        bcc   POK            Skip if P should remain zero
        ldy   4,X            Pointer to P
        com   ,Y             Overflow, P=-1
POK     ldy   2,X            Pointer to M
        std   ,Y             Return result in M
        jmp   6,X            Return
```

Checkpoint 8.8: Write 6811/6812 assembly code that implements Program 8.3.

Checkpoint 8.9: Write 6811/6812 assembly code that implements Program 8.4.

Checkpoint 8.10: Write 6811/6812 assembly code that implements Program 8.5.

8.2.3
C Compiler
Implementation
of Local and
Global Variables

In order to understand both the machine architecture and the C compiler, we can look at the assembly code generated. Program 8.13 shows a simple C program with a global variable x, two local variables both called y and a function parameter z.

Program 8.13
An example used to illustrate the C compiler's use of the stack.

```
short x;        /* definition of a global variable */
void main(void){
short y;        /* definition of a local variable */
   x = 5;       /* access global variable */
   y = 6;       /* access local variable */
   x = sub(y);  /* call function, pass parameter */
   return(0);
}
short sub(short z){ short y;
   y = z+1;
   return(y);
}
```

The first compiler we will study is Symantec Think C version 7 for the 68K Macintosh. The disassembled output, shown as Program 8.14, has been edited to clarify its operation. The loader will allocate three segmented memory areas: code pointed to by the PC; global pointed to by A5; and local pointed to by the stack pointer A7. The global symbol x will be assigned or bound by the loader. "Binding" means establishing its address. The compiler can bind the local variables and subroutine parameters. The link instruction establishes a stack frame pointer, A6, and allocates local variables. The actual ThinkC compiler optimized the subroutine by placing the local variable, y, in register D7, but in this example, the code was "unoptimized" to illustrate the use of local variables. The addressing mode -(A7) first decrements the stack pointer, A7, then stores the value at that place. The unlink instruction deallocates the local variables, and restores the register to its original value.

Program 8.14
Assembly code generated for the 68000 by the ThinkC compiler.

```
y      equ      -2               local variable binding A6 relative
main:  LINK     A6,#-2           allocate y for main
       MOVE.W   #5,x(A5)         x=5;
       MOVE.W   #6,y(A6)         y=6;
       MOVE.W   y(A6),-(A7)      call by value
       JSR      sub
       MOVE.W   D0,x(A5)         x=result of sub
       MOVEQ    #0,D0
       UNLK     A6
       RTS
y      equ      -2               local variable binding A6 relative
z      equ      8                Parameter binding A6 relative
```

```
sub:    LINK    A6,#-2          allocate y for sub
        MOVEQ.W #1,y(A6)
        ADD.W   z(A6),y(A6)     y=z+1;
        MOVE.W  y(A6),D0        D0 is the return parameter
        UNLK    A6              deallocate y
        RTS
```

The stack frame at the time of the `ADD.W z(A6),y(A6)` instruction is shown in Figure 8.3. Within the subroutine the local variables of main are not accessible.

Figure 8.3
The 68000 stack contains local variables, parameters and the return address.

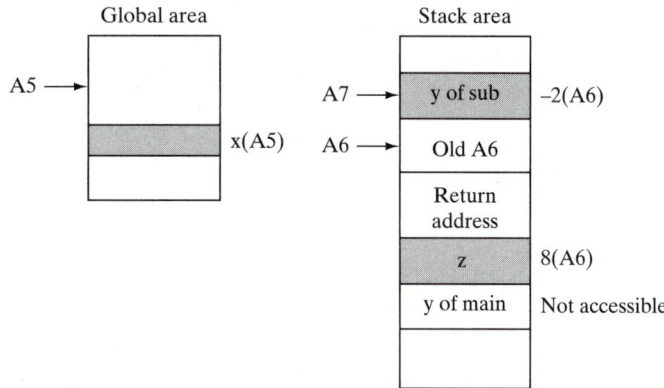

The next compiler we will study is ImageCraft ICC11 for the Motorola 6811. Again, the disassembled output has been edited to clarify its operation, and shown as Program 8.15. The linker/loader also allocates three segmented memory areas: code pointed to by the PC; global accessed with absolute addressing; and locals pointed to by the stack pointer SP. The global symbol, _x, will be assigned or bound by the linker/loader. The `pshx` instruction allocates the local variable, and the `tsx` instruction establishes a stack frame pointer, X. This compiler passes the first input parameter into the subroutine by placing it in register D. The remaining parameters (none in this example) would have been pushed onto the stack.

Program 8.15
Assembly code generated for the 6811 by the ICC11 compiler.

```
y       set     0       local variable binding X relative
main:   PSHX            allocate y for main
        TSX             establish stack frame pointer
        LDD     #5
        STD     _x      x=5;
        LDD     #6
        STD     y,X     y=6;
        JSR     sub
        STD     _x      x=result of sub
        CLRA
        CLRB
        PULX            deallocate y
        RTS
y       equ     0       local variable binding X relative
z       equ     2       Parameter binding X relative
```

continued

```
sub:    PSHB                    put parameter on stack
        PSHA
        PSHX                    allocate y for sub
        TSX                     establish stack frame pointer
        LDD     z,X
        ADDD    #1
        STD     y,X     y=z+1;
        PULX            deallocate y
        PULX            discard z
        RTS             RegD is the return parameter
```

The stack frame at the time of the `ADDD #1` instruction is shown in Figure 8.4. Within the subroutine the local variables of main are not accessible.

Figure 8.4
The 6811 stack contains local variables, parameters and the return address.

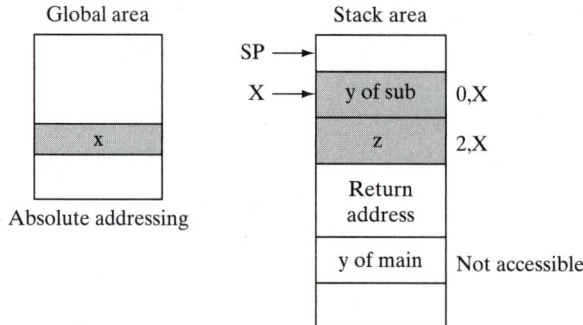

The third compiler we will study is ImageCraft ICC12 for the Motorola 6812. Again, the disassembled output has been edited to clarify its operation (see Program 8.16). Like the 6811, the linker/loader also allocates three segmented memory areas: code pointed to by the PC; global accessed with absolute addressing; and locals pointed to by the stack pointer SP. The `leas -2,sp` instruction allocates the local variable, and the `tfr s,x` instruction establishes a stack frame pointer, X. ImageCraft ICC12 compiler passes the first input parameter into the subroutine by placing it in register D. The remaining parameters (none in this example) would have been pushed onto the stack.

Program 8.16
Assembly code generated for the 6812 by the ICC12 compiler.

```
y       set     -2          ;local variable binding X relative
main:   pshx                ; main()
        tfr     s,x         ; X is the stack frame pointer
        leas    -2,sp       ; allocate y short y;
        movw    #5,_x       ; x=5;
        movw    #6,-2,x     ; y=6;
        ldd     -2,x        ; parameter in RegD
        jsr     sub         ; x=sub(y);
        std     _x          ; store return in global x
        ldd     #0          ; return(0);}
        tfr     x,s
        pulx
        rts
```

```
Y      set    -4         ; local variable binding X relative
Z      set    -2         ; Parameter binding X relative
sub:   pshx              ; short sub(short z){
       tfr    s,x
       pshd
       leas   -2,sp      ; allocate y short y;
       ldd    Z,x        ; y=z+1;
       addd   #1
       std    Y,x
       ldd    Y,x        ; return(y);}
       tfr    x,s
       pulx
       rts
```

The stack frame at the time of the ADDD #1 instruction is shown in Figure 8.5. Within the subroutine the local variables of main are not accessible.

Figure 8.5
The 6812 stack contains local variables, parameters and the return address.

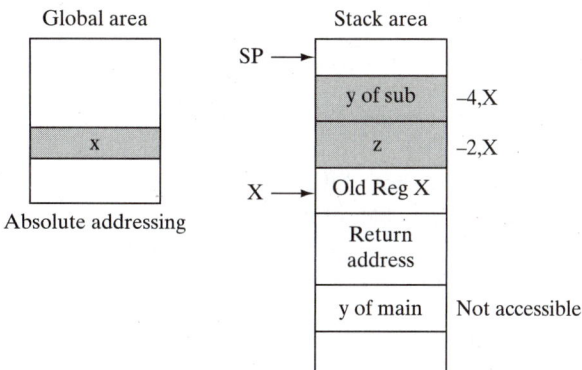

The fourth compiler we will study is Borland C version 3 for the Intel 386 IBM-PC. The disassembled output, shown as Program 8.17, has been edited to clarify its operation. The small memory model was used. With the large memory models, the short call would be replaced with a far callf. Similar to the MC68000, the loader will allocate three segmented memory areas: code pointed to by the CS:IP; global accessed with data segment addressing DS:offset; and local pointed to by the stack pointer SP. The offset of global symbol x can be bound by the compiler. The loader will establish the position in memory and set the segment register, DS. The compiler can bind the local variables and subroutine parameters. The SUB SP,2 instruction allocates the local variable, and the PUSH BP MOV BP,SP instructions establishes a stack frame pointer, SS:BP. In later versions of the compiler, these three instructions are replaced by the single instruction ENTER. The MOV SP,BP POP BP instructions are replaced by LEAVE.

Program 8.17
Assembly code generated for the x86 by the Borland compiler.

```
ymain: PUSH  BP
       MOV   BP,SP                        establish stack frame
       SUB   SP,2                         allocate y for main
       MOV   word ptr[x],5                x=5;
       MOV   word ptr[BP+y],6             y=6;                    continued
```

```
            PUSH   word ptr[BP+y]        call by value
            CALL   sub
            POP    CX                    discard parameter
            MOV    [x],AX                x=result of sub
            XOR    AX,AX
            MOV    SP,BP
            POP    BP
            RET
    y       equ    -2                    local binding BP relative
    z       equ    4                     Parameter binding BP relative
    sub:    PUSH   BP
            MOV    BP,SP                 establish stack frame
            SUB    SP,2                  allocate y for sub
            MOV    AX,[BP+z]
            INC    AX
            MOV    [BP+y],AX             y=z+1;
            MOV    AX,[BP+y]             AX is the return parameter
            MOV    SP,BP
            POP    BP
            RET
```

The stack frame at the time of the `INC AX` instruction is shown in Figure 8.6. Within the subroutine the local variables of main are not accessible.

Figure 8.6
The Intel x86 stack contains local variables, parameters and the return address.

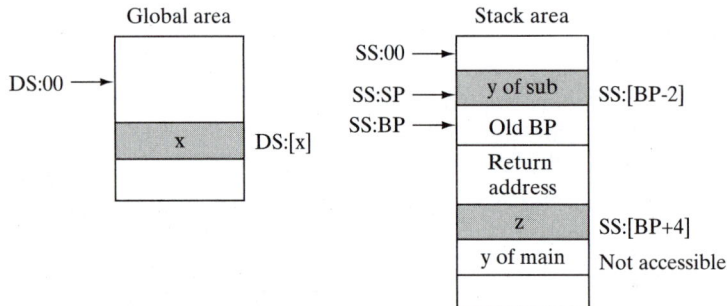

Observation: *Although the local variables of the main program are on the stack, and it IS possible to access them while executing the subroutine, the compiler will NOT allow the subroutine to access them. In C, there is a clear distinction between the parameters pushed onto the stack that are supposed to be accessed by the subroutine and the local variables of the calling program, which are not supposed to be accessed.*

Common error: *It would be a grievous programming error to access the local variables of the main program. Therefore, in assembly language, it is essential to make the distinction between local variables and data passed onto the stack to the subroutine.*

8.2.4
*Recursion

A **recursive** program is one that calls itself. Each time the subroutine is started, a new instantiation occurs. There is a unique set of parameters and local variables for each instantiation. The stack is a convenient way to separate the parameters and variables of one instantiation from another. In order for the recursive function to finish, there must be a situation where a direct result is generated, which is called the **end condition**. For example, the factorial has two possibilities

$$Fact(1) = 1 \qquad\qquad \text{end condition}$$
$$Fact(n) = n*Fact(n-1) \text{ if } n>1 \qquad \text{recursion}$$

Program 8.18 shows two assembly language implementations of factorial. The one on the left uses iteration, and the one on the right uses recursion. It is usually the case that a recursive algorithm can be rewritten in iterative form. Nevertheless, sometimes it is more convenient to implement the algorithm in recursive form.

Program 8.18   Iterative and recursive implementations of factorial.

```
; iterative implementation          ; recursive implementation
; Input: RegA is n                  ; Input: RegA is n
; Output: RegA is Fact(n)           ; Output: RegA is Fact(n)
n       set  0      ; local         Fact   cmpa #1    ; end condition
Fact    psha        ; save n               beq  done
        ldab #1     ; r=1                   psha       ; save n
        tsx                                 deca       ; n-1
loop    ldaa n,x    ; RegA=n                bsr  Fact  ; RegA=Fact(n-1)
        cmpa #1     ; end condition         pulb       ; RegB=n
        beq  done                           mul
        mul         ; r=r*n                 tba        ; RegA=n*Fact(n-1)
        dec  n,x    ; n=n-1          done   rts
        bra  loop
done    tba         ; RegA=Fact(n)
        ins         ; deallocate
        rts
```

Table 8.1 shows the execution time in cycles for these two implementations. Notice that the recursive implementation is shorter, but the execution speeds are similar.

Table 8.1
Execution times for
Program 8.18.

Input	6811 iterative	6811 factorial	6812 iterative	6812 factorial
1	33	16	22	11
2	61	53	36	31
3	89	90	50	51
4	117	127	64	71
5	145	164	78	91

Checkpoint 8.11: How many stack bytes are required for each instantiation of Fact? How much stack space is required to execute Fact(5)?

A recursive implementation, OutUDec, written in C, of a 16-bit unsigned decimal output was presented earlier as Program 2.12. Program 8.19 shows two assembly language implementations of this 16-bit output decimal function. The one on the left uses iteration, and the one on the right uses recursion. There is no fundamental rule that states whether iteration or recursion is better. A good programmer has both in her toolbox, and uses whichever is easier to understand and easier to debug.

Program 8.19   Iterative and recursive implementations of output decimal.

```
; Reg D is input                          ; Reg D is input,n
OutUDec  ldy  #0     Number of char        OutUDec  cpd  #10   end condition
ODloop   ldx  #10    Reg D = input                  blo  end
         idiv        RegB is digit                  ldx  #10    RegD = n
         pshb        Save for later                 idiv        RegX = n/10
         iny                                        pshb        RegB = n%10
         xgdx                                       xgdx        RegD = n/10
         cpd  #0     Continue until                 bsr  OutUDec
         bne  ODloop                                pulb        RegB = n%10
ODout    pula        next digit            end      tba
         adda #'0'   convert ASCII                  adda #'0'   convert ASCII
         jsr  OutCh                                 jsr  OutCh
         dey                               out      rts
         bne  ODout
         rts
```

To illustrate the execution of the recursive implementation of `OutUDec`, we can place a ScanPoints on the first line and observe the stack and Register D, see Table 8.2. In this way we can observe each instantiation. Let the initial input be 12345.

Table 8.2
ScanPoint results for recursive version of Program 8.19.

RegD	Stack
12345	Return
1234	Return,5,Return
123	Return,5,Return,4,Return
12	Return,5,Return,4,Return,3,Return
1	Return,5,Return,4,Return,3,Return,2,Return

Observation: In general, recursive algorithms are shorter to write, but require additional stack space.

8.3   Control Structures

Control structures are the building blocks of a structured program. The control structures in C involve the `if`, `if-else`, `while`, `do-while`, `for`, and `switch` statements. Two local variables are used in the examples of this section. For most situations, local variables are more appropriate than globals. Assume there are two bytes allocated on the stack pointed to by the stack frame pointer Register X, and the following local variable binding.

```
M  set  0  8-bit local
N  set  1  8-bit local
```

8.3.1
Conditional if-then-else Statements

Simple if-then structures were presented in Chapter 3. We can use the unconditional branch to add an `else` clause to any of the previous `if-then` structures. A simple example of an unsigned conditional is illustrated in the Figure 8.7 and presented in Program 8.20.

Figure 8.7
Flowchart of an if-then-else structure.

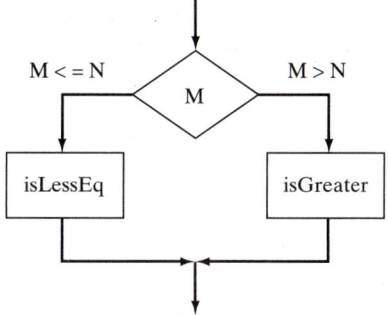

Program 8.20
A simple if-then-else structure.

```
        ldaa   M,x
        cmpa   N,x
        bhi    high       ; branch if M>N
low     jsr    isLessEq   ; M<=N
        bra    next
high    jsr    isGreater  ; M>N
next
```

The label `low` in Program 8.20 exists even though there are no references to it. Its presence serves two purposes. Acting like a comment, it makes the code easier to understand. Second, it simplifies debugging by allowing you to use the symbolic name as a breakpoint, rather than using its absolute address.

Using these basic building blocks, you can implement more complex tests. The conditional or (| |), is implemented as a sequential test. Program 8.21 implements the following C structure:

```
if((M==100) || (N>50)) True(); else False();
```

Notice that Program 8.21 first tests `M==100`, then tests `N>50`. If either is true, it will jump to T and execute `True`. If neither is true, then `False` is executed.

Program 8.21
A conditional or structure.

```
        ldaa   M,x
        cmpa   #100
        beq    T       ; true if M==100
        ldaa   N,x
        cmpa   #50
        bhi    T       ; true if N>50
F       jsr    False   ; (M!=100)&&(N<=50)
        bra    next
T       jsr    True    ; (M==100)||(N>50)
next
```

The conditional and (&&) can be implemented as a nested if. It can also be implemented using the conditional or. The following C statements, implemented as Program 8.22, are equivalent:

```
if((M==100)&&(N>50)) True(); else False();
if((M!=100)||(N>=50)) False(); else True();
if(M==100){if(N>50) True(); else False();} else False();
```

Checkpoint 8.12: *If the above three lines are the same, which is best?*

Notice that Program 8.22 first tests M==100, then tests N>50. If both are true, it will jump to T and execute True. If either is false, then False is executed.

Program 8.22
A conditional and
structure.

```
        ldaa    M,x
        cmpa    #100
        bne     F     ; false if M!=100
        ldaa    N,x
        cmpa    #50
        bls     F     ; false if N<=50
T       jsr     True  ; (M==100)&&(N>50)
        bra     next
F       jsr     False ; (M!=100)||(N<=50)
next
```

If the body of the if-then structure is a long assembly code sequence, rather than the simple subroutine call, then a *branch out of range* error might occur. The 8-bit PC relative branching can be converted to 16-bit addressing to accommodate larger program segments. Table 8.3 illustrates various methods to solve the *branch out of range* error. The "simple assembly" column shows the original code that has the branch out of range. Since, the 6812 has long condition branches, you fix this error simply by converting the ble to lble instruction as shown in the "6812 assembly" column. The last column shows an implementation that will work for either the 6811 or 6812. When using jmp, the conditional is changed to its logical complement, e.g., ble to bgt.

Table 8.3
Conditional structures
with a large program
segment.

C code	simple assembly	6812 assembly	using jmp

if(M>N){
//lots of code
}
	ldaa M,x
	cmpa N,x
	ble next
	;lots of code
	next
		ldaa M,x
		cmpa N,x
		lble next
		;lots of code
		next
			ldaa M,x
			cmpa N,x
			bgt ok
			jmp next
			ok ;lots of code
			next

Observation: *Branch out of range errors may indicate the program logic is too complicated or the program structure is not very modular. Before fixing a branch out of range error, consider simplifying the logic or creating a new subroutine to perform some of the tasks.*

8.3.2

Switch
Statements

Switch statements provide a noniterative choice between any number of paths based on specified conditions. They compare an expression to a set of constant values. Selected statements are then executed depending on which value, if any, matches the expression. Switch statements have the form

```
switch ( ExpressionList ) { Statement?...}
```

where `ExpressionList` is a list of one or more expressions. `Statement?...` represents the statements to be selected for execution. They are selected by means of `case` and `default` prefixes, which are special labels that are used only within switch statements. These prefixes locate points to which control jumps depending on the value of `ExpressionList`. They are to the `switch` statement what ordinary labels are to the `goto` statement. They may occur only within the braces that delimit the body of a switch statement. The `case` prefix has the form

```
case ConstantExpression :
```

and the default prefix has the form

```
default:
```

Any expression involving only numeric and character constants and operators is valid in the case prefix. After evaluating `ExpressionList`, a search is made for the first matching `case` prefix. Control then goes directly to that point and proceeds normally from there. Other `case` prefixes and the `default` prefix have no effect once a `case` has been selected; control flows through them just as though they were not even there. If no matching `case` is found, control goes to the `default` prefix, if there is one. In the absence of a `default` prefix, the entire compound statement is ignored and control resumes with whatever follows the `switch` statement. Only one `default` prefix may be used with each `switch`.

If it is not desirable to have control proceed from the selected prefix all the way to the end of the `switch` block, `break` statements may be used to exit the block. Break statements have the form

```
break;
```

Some examples may help clarify these ideas (See Program 8.23). Assume that Port B is specified as an output, and bits 3,2,1,0 are connected to a stepper motor. The `switch` statement will first read Port B, then *and* the data with 0x0F (PORTB&0x0F). If the result is 5, then Port B is set to 6 and control is passed to the end of the `switch` (because of the break). The other three possibilities are treated similarly. On the 6812, one could replace the ldab stab sequence with a movb instruction. Notice that the `break` statement is translated into the `bra stend`.

Program 8.23   A stepper motor example using the `switch` statement.

```
step  ldaa PORTB   ;present state        /* turn stepper motor one step
      anda #$0F                              the sequence is 5,6,10,9,5 */
case5 cmpa #$05                          void step(void){
      bne  case6                           switch(PORTB&0x0F){
      ldab #$06                              case 0x05:
      stab PORTB                               PORTB=0x06;
      bra  stend   ;break                      break;
```
continued

```
case6 cmpa    #$06                          case 0x06:
      bne     caseA                           PORTB=0x0A;
      ldab    #$0A                            break;
      stab    PORTB                         case 0x0A:
      bra     stend   ;break                  PORTB=0x09;
caseA cmpa    #$0A                            break;
      bne     case9                         case 0x09:
      ldab    #$09                            PORTB=0x05;
      stab    PORTB                           break;
      bra     stend   ;break                default:
case9 cmpa    #$09                            PORTB=0x05;
      bne     caseD                     }
      ldab    #$05                    }
      stab    PORTB
      bra     stend   ;break
caseD ldab    #$05
      stab    PORTB
stend
```

Program 8.24 shows that the multiple tests can be performed for the same condition.

Program 8.24
A switch statement used to convert ASCII to decimal.

```
// ASCII to decimal digit conversion
unsigned char convert(unsigned char letter){
unsigned char digit;
  switch(letter){
    case 'A':
    case 'B':
    case 'C':
    case 'D':
    case 'E':
    case 'F':
      digit=letter+10-'A';
      break;
    case 'a':
    case 'b':
    case 'c':
    case 'd':
    case 'e':
    case 'f':
      digit=letter+10-'a';
      break;
    default:
      digit=letter-'0';
  }
  return digit;
}
```

The body of the `switch` is not a normal compound statement since local declarations are not allowed in it or in subordinate blocks. This restriction enforces the C rule that a block containing declarations must be entered through its leading brace.

8.3.3
For-loops

A `for-loop` control structure is a convenient way to perform repetitive tasks as illustrated in Figure 8.8. The first field is the starting task (e.g., `i=0`). The next field specifies the conditions with which to continue execution (e.g., `i<100`), and the last field is the operation to perform after each interaction (e.g., `i++`). For loops can iterate up or down.

```
for(i=0;i<100;i++) process(); for(i=100;i!=0;i--) process();
```

Figure 8.8
Two flowcharts of a
`for-loop` structure.

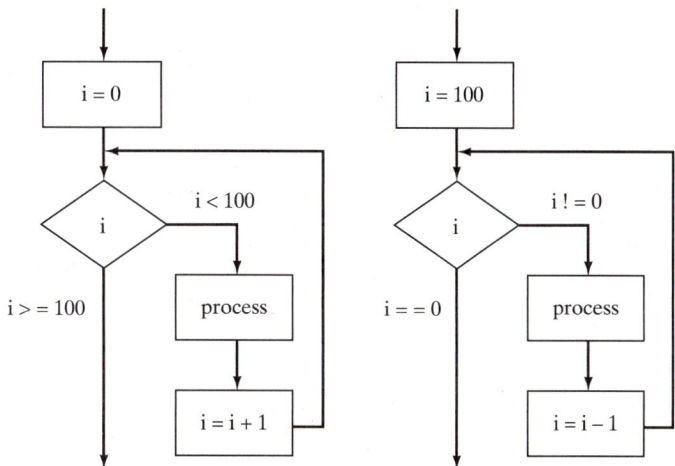

The first implementation, which works for the 6811 and 6812, places the loop counter in the Register B, as shown in Program 8.25.

Program 8.25
A simple `for-loop`.

```
;  for(i=0;i<100;i++) process();
        ldab    #0          ;  i=0
loop    cmpb    #100
        bhs     done
        jsr     process
        incb                ;  i=i+1
        bra     loop
done
```

If you don't really need the index to go up, it is more efficient to count down, as shown in Program 8.26.

Program 8.26
A `for-loop` that
counts down.

```
;  for(i=100;i!=0;i--) process();
        ldab    #100        ;  i=100
loop    jsr     process
        decb                ;  i=i-1
        bne     loop
```

The 6812 has special op codes to facilitate the implementation of `for-loops`, as shown Program 8.27.

Program 8.27
The 6812 `dbne` instruction optimizes `for-loop` implementation.

```
; for(i=100;i!=0;i—) process();
        ldab   #100     ; i=100
L1      jsr    process
        dbne   B,L1     ; i=i-1
```

Checkpoint 8.13: How do you modify Programs 8.25, 8.26, and 8.27 to handle a 16-bit counter?

If a register is not available, the `for-loop` can be implemented by placing the counter in a variable, as shown in Program 8.28. In general, temporary variables, like `i`, should be placed in a register or as a local variable on the stack. This example implements the C expression.

```
for(i=100;i!=0;i--) process();
```

Program 8.28   A `for-loop` that uses a local variable.

```
; 6811 assembly                          ; 6812 assembly
i       set  0                           i       set  0
        ldab #100       ; i=100                  movb #100,1,-sp  ; i=100
        pshb            ; create local   loop    jsr  process
loop    jsr  process                             dec  i,sp       ; i=i-1
        tsx                                       bne  loop
        dec  i,x        ; i=i-1                   ins                   ;deallocate
        bne  loop
        ins             ;deallocate
```

8.4   Extended Precision Calculations

In this section, we will study various techniques to perform extended precision calculations. Sometimes complex calculations can be performed simply by combining simpler operations, while at other times, more sophisticated algorithms will be required. Three 32-bit local variables are used in the examples of this section. For most situations, local variables are more appropriate than globals, although using globals is often faster and easier to debug. Assume that there are 12 bytes allocated on the stack pointed to by the stack frame pointer Register X, and the following local variable binding:

```
N   set   0     32-bit  local
M   set   4     32-bit  local
P   set   8     32-bit  local
```

8.4.1
Addition and
Subtraction

Program 8.29 gives a 32-bit addition algorithm. The approach starts with the least significant byte and uses the add-with-carry operation to combine the 8-bit additions to form the 32-bit operation.

Program 8.29
A 32-bit addition
operation.

```
; 32-bit addition P=N+M
; Input:    Two 32-bit numbers N,M
; Output:   One 32-bit sum P
; Error:    C/V set for unsigned/signed overflow
add32 ldaa   N+3,x   ; start with least significant byte
      adda   M+3,x
      staa   P+3,x
      ldaa   N+2,x   ; next byte
      adca   M+2,x   ; carry from previous addition
      staa   P+2,x
      ldaa   N+1,x   ; next byte
      adca   M+1,x   ; carry from previous addition
      staa   P+1,x
      ldaa   N,x     ; last byte
      adca   M,x     ; carry from previous addition
      staa   P,x
; C bit set if unsigned overflow
; V bit set if signed overflow, Z bit is not correct
```

Checkpoint 8.14: Why isn't the Z bit correct?

Program 8.30 gives a 32-bit subtraction algorithm. Again, the approach starts with the least significant byte and uses the subtract-with-borrow operation to combine the 8-bit subtractions to form the 32-bit operation. Similarly to addition, the V and C bits are properly set, while the Z bit is incorrect.

Program 8.30
A 32-bit subtraction
operation.

```
sub32 ldaa   N+3,x   ; start with least significant byte
      suba   M+3,x
      staa   P+3,x
      ldaa   N+2,x   ; next byte
      sbca   M+2,x   ; carry from previous addition
      staa   P+2,x
      ldaa   N+1,x   ; next byte
      sbca   M+1,x   ; carry from previous addition
      staa   P+1,x
      ldaa   N,x     ; last byte
      sbca   M,x     ; carry from previous addition
      staa   P,x
; C bit set if unsigned overflow
; V bit set if signed overflow, Z bit is not correct
```

Program 8.31 presents functions that add and subtract two unsigned 8-bit values, using promotion to detect for errors. The assembly language version implements the 16-bit local `result` in Register D. This C program was previously presented as Program 2.2.

Program 8.31   Using promotion to detect and compensate for unsigned overflow errors.

```
add  ldab _A                        unsigned char A,B,R;
     clra     promote to 16 bits    void add(void){
     addb _B                        unsigned short result;
     adca #0   A+B (16 bits)          result = A+B;    /* promote */
     cpd  #255                        if(result>255){  /* overflow ?*/
     bls  aOK                           result = 255;  /* yes */
     ldd  #255 ceiling                }
aOK  stab _R   demote                 R = result;      /* demote */
     rts                            }
sub  ldab _A                        void sub(void){
     clra     promote to 16 bits    short result;
     subb _B                          result = A-B;    /* promote */
     sbca #0   A-B (16 bits)          if(result<0){    /* underflow? */
     cpd  #0                            result = 0;    /* yes */
     bge  sOK                         }
     ldd  #0    floor                 R = result;      /* demote */
sOK  stab _R   demote                }
     rts
```

Program 8.32 presents functions that add and subtract two signed 8-bit values, using promotion to detect errors. It is much easier to promote signed numbers on the 6812. The C version was previously presented as Program 2.3. A 16-bit local variable holds the temporary result.

Program 8.32   Using promotion to detect and compensate for signed overflow errors.

```
; 6811 or 6812 assembly              ; 6812 assembly
temp set  0                          add  ldab  _A
add  pshx      allocate temp              sex   b,d    promote to 16 bits
     tsx       stack frame pointer        pshd
     clra                                 ldab  _B
     ldab _A                              sex   b,d    promote to 16 bits
     bpl  aOK1                            addd  2,s+   add A+B
     coma      promote to 16 bits         cpd   #127
aOK1 std  temp,x                          ble   aOK1
     clra                                 ldd   #127   ceiling
     ldab _B                         aOK1 cpd   #-128
     bpl  aOK2                            bge   aOK2
     coma      promote to 16 bits         ldd   #-128 floor
aOK2 addd temp,x  add _A+_B          aOK2 stab  _R     demote
     cpd  #127                            rts
     ble  aOK3
     ldd  #127   ceiling
aOK3 cpd  #-128
     bge  aOK4
     ldd  #-128 floor
aOK4 stab _R      demote
     pulx         deallocate
     rts
```

```
sub   pshx       allocate temp
      tsx        stack frame pointer
      clra
      ldab  _B
      bpl   sOK1
      coma       promote to 16 bits
sOK1 std   temp,x
      clra
      ldab  _A
      bpl   sOK2
      coma       promote to 16 bits
sOK2 subd temp,x subtract AA-BB
      cpd   #127
      ble   sOK3
      ldd   #127   ceiling
sOK3 cpd   #-128
      bge   sOK4
      ldd   #-128 floor
sOK4 stab _R       demote
      pulx         deallocate
      rts
```

```
sub   ldab  _B
      sex   b,d   promote to 16 bits
      pshd
      ldab  _A
      sex   b,d   promote to 16 bits
      subd  2,s+  subtract A-B
      cpd   #127
      ble   sOK1
      ldd   #127   ceiling
sOK1 cpd   #-128
      bge   sOK2
      ldd   #-128 floor
sOK2 stab  _R      demote
      rts
```

Checkpoint 8.15: How do you force the C compiler to promote an intermediate calculation to signed 32-bits?

8.4.2
Shift Operations

The 32-bit shift left operation is described in Figure 8.9, and presented in Program 8.33. In particular, $N = N<<1$. When shifting left, you start from the least significant byte and proceed to the most significant byte. The operation can be used for signed or unsigned numbers. The C bit will be set if there was an unsigned overflow. The V bit will be set if there was a signed overflow.

Figure 8.9 A 32-bit shift left.

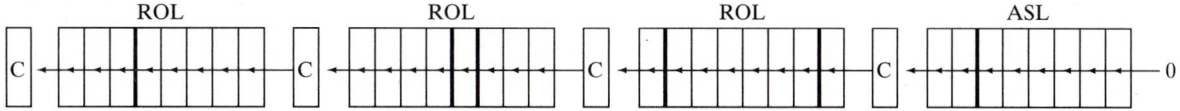

Program 8.33
A 32-bit shift left operation.

```
asl32  asl  N+3,x  ; start with least significant byte
       rol  N+2,x  ; next byte
       rol  N+1,x  ; next byte
       rol  N,x    ; last byte
; C bit set if unsigned overflow
; V bit set if signed overflow, the Z bit is incorrect
```

Observation: In assembly language it is critical to keep track of whether a number is signed or unsigned.

The signed 32-bit shift right operation is described in Figure 8.10, and presented in Program 8.34. In particular, $N = N>>1$. When shifting right, you start from the most significant byte and proceed to the least significant byte. The signed shift maintains the most significant bit, and overflow can not occur. When you shift right, the least significant bit is lost. The second shift program will round the result up if the input was originally odd. For example, in the first program 5>>1 is 2, and in the second program 5>>1 is 3.

Figure 8.10   A 32-bit signed shift right.

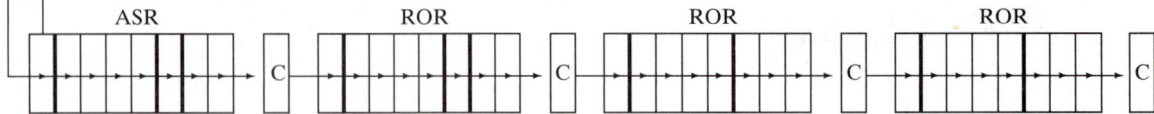

Program 8.34   Two 32-bit signed shift right operations.

```
; simple shift without rounding      ;shift right with rounding
asr32 asr  N,x      ; MSByte         asr32r asr    N,x    ; MSbyte
      ror  N+1,x    ; next byte             ror    N+1,x ; next byte
      ror  N+2,x    ; next byte             ror    N+2,x ; next byte
      ror  N+3,x    ; last byte             ror    N+3,x ; last byte
; C bit set if was odd                ; C bit set if you should round up
; V and N not correct                       bcc    Done
                                            inc    N+3,x ; round up
                                            bcc    Done
                                            inc    N+2,x ; round up
                                            bcc    Done
                                            inc    N+1,x ; round up
                                            bcc    Done
                                            inc    N,x    ; round up
                                     Done    ; no valid CCR flags
```

> **Checkpoint 8.16:** *Assuming the input is 100001 in decimal, how are the results of the two shifts in Program 8.34 different?*

The unsigned 32-bit shift right operation is described in Figure 8.11 and presented in Program 8.35. In particular, N = N>>1. The unsigned shift right will clear the most significant bit, and overflow cannot occur. The second shift program will round the result up if the input was originally odd.

Figure 8.11   A 32-bit unsigned shift right.

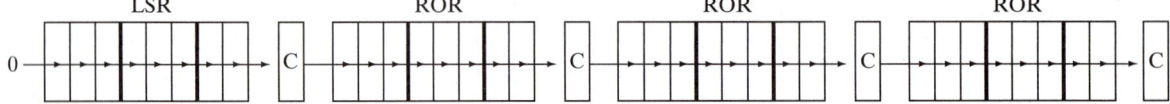

Program 8.35   Two 32-bit unsigned shift right functions.

```
lsr32 lsr  N,x    ; MSbyte           ;shift right with rounding
      ror  N+1,x ; next byte         lsr32r lsr    N,x    ; MSbyte
      ror  N+2,x ; next byte                ror    N+1,x ; next byte
      ror  N+3,x ; last byte                ror    N+2,x ; next byte
; C bit set if was odd                      ror    N+3,x ; last byte
```

```
; C bit set if you should round up
        bcc   Done
        inc   N+3,x  ; round up
        bcc   Done
        inc   N+2,x  ; round up
        bcc   Done
        inc   N+1,x  ; round up
        bcc   Done
        inc   N,x    ; round up
Done  ; no valid CCR flags
```

8.4.3 Mathematical Instructions on the 6812

When designing the 6812, Motorola added a few instructions not available on the 6811. These instructions are quite useful when implementing mathematical calculations. For 6811 systems, these operations are so convenient that we will need subroutines for the same calculations. Many of these subroutines can be found in the `math.rtf` file in the MC6811 folder as part of the **TExaS** simulator.

The `emul` instruction performs a 16-bit by 16-bit unsigned multiply, RegY:D= RegY*RegD, as shown in Figure 8.12. The `emuls` instruction is a 16-bit by 16-bit signed multiply, using the same registers and generating the same condition code bits.

Figure 8.12
The `emul` and `emuls` instructions take two 16-bit inputs and generate a 32-bit product.

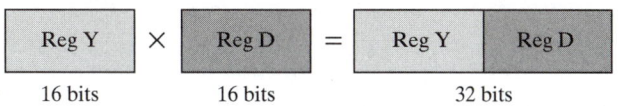

Condition Code Bits after R=Y*D

    N: result is negative, N=R31
    Z: result is zero, Z=not(R31)*not(R30)*. . .*not(R2)*not(R1)*not(R0)
    C: R15, bit 15 of the result

The `ediv` instruction performs a 32-bit by 16-bit unsigned divide, RegY=(Y:D)/ RegX; RegD is remainder, as shown in Figure 8.13. The `edivs` instruction is a 32-bit by 16-bit signed divide, using the same registers. The overflow bit calculation is different, but the other three condition code bits are the same.

Figure 8.13
The `ediv` and `edivs` instructions perform extended precision division.

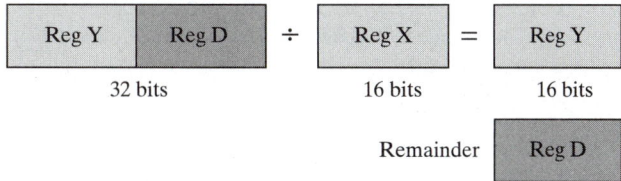

Condition Code Bits after R=(Y:D)/X

    N: result is negative (undefined after an overflow or a divide by zero), N=R15
    Z: result is zero (undefined after an overflow or a divide by zero)
        Z=not(R15) • not(R14) •. . .• not(R2) • not(R1) • not(R0)
    V: overflow (undefined after a divide by zero),
        `ediv` **result** >$FFFF
        `edivs` **result** >$7FFF or less than -$8000
    C: divide by zero, C=not(X15) • not(X14) •. . .• not(X2) • not(X1) • not(X0)

The `emacs` instruction performs a 16-bit by 16-bit signed multiply, followed by a 32-bit signed addition. It uses indexed addressing to access the two 16-bit inputs and extended addressing to access the 32-bit sum. Recall that {X} and {Y} represent the 16-bit contents pointed to by Registers X and Y respectively. If we define <U> as the 32-bit contents of memory location U, then `emacs` U calculates

$$<U>=<U>+\{X\}*\{Y\}$$

Condition Code Bits, first P=X*Y, then R=M+P (M,P,R are 32 bits)

N: result is negative, N=R31
Z: result is zero, Z=not(R31) • not(R30) •. . .• not(R2) • not(R1) • not(R0)
V: signed overflow (after addition), V=P31 • M31 • not(R31)+not(P31) • not(M31) • R31
C: unsigned overflow (after addition), C=P31 • M31+M31 • not(R31)+not(R31) • P31

This instruction is quite useful for calculating fixed point equations, as illustrated in Program 8.36. We place the input variables xx, yy, and zz consecutively in RAM, and the constants 902, −1810, and 45 consecutively in ROM. Registers X and Y are not automatically incremented, so the program performs that task explicitly.

Program 8.36
Fixed-point calculation using the 6812 `emac` instruction.

```
        org    $0800
; rr=0.902*xx-1.81*yy+0.045*zz
; first we can convert this equation to fixed point
; without introducing any error:
; rr=(902*xx-1810*yy+45*zz)/1000
xx      ds     2
yy      ds     2
zz      ds     2
rr      ds     2        ; result
acc     ds     4        ; temporary 32-bit result
        org    $F000
cc      dc.w   902,-1810,45
Calc ldx       #xx      ; pointer to data
        ldy    #cc      ; pointer to coeficients
        movw   #0,acc   ; initially clear temporary result
        movw   #0,acc+2
        ldaa   #3       ; number of terms
loop emacs acc          ;acc=acc+{X}*{Y}
        leax   2,x
        leay   2,y
        dbne   A,loop
        ldy    acc
        ldd    acc+2    ;Y:D=902*xx-1810*yy+45*zz
        ldx    #1000
        edivs
        sty    rr
        rts
```

The design of digital filters is beyond the scope of this book, so we simply present the following equation, which implements a high-Q high-pass digital filter:

$$y_n = 0.9025x_n - 1.805x_{n-1} + 0.9025x_{n-2} + 1.8y_{n-1} - 0.81y_{n-2}$$

We show the implementation of this equation using the emacs instruction on the 6812 in Program 8.37. First though, we convert the filter to fixed-point without introducing any error:

$$y_n = (9025x_n - 18050x_{n-1} + 9025x_{n-2} + 18000y_{n-1} - 8100y_{n-2})/10000$$

Program 8.37
A fixed-point digital filter implemented using the 6812 emac instruction.

```
         org    $0800
XN       ds     2     ; x(n)   current 16-bit data point
XN1      ds     2     ; x(n-1)   previous data point
XN2      ds     2     ; x(n-2)   data point two samples ago
YN1      ds     2     ; y(n-1)   previous filter output
YN2      ds     2     ; y(n-2)   filter output two samples ago
YN       ds     2     ; y(n)   current 16-bit filter output
acc      ds     4     ; temporary 32-bit result
         org    $F000
COEF     dc.w   9025,-18050,9025,18000,-8100
DF       ldx    #XN     ; data
         ldy    #COEF   ; coef
         ldd    #0
         std    acc     ; initially clear temporary result
         std    acc+2
         ldaa   #5      ; number of terms
loop     emacs  acc     ;acc=acc+{X}*{Y}
         leax   2,x
         leay   2,y
         dbne   A,loop
         ldy    acc
         ldd    acc+2
;Y:D=9025x(n)-18050x(n-1)+9025x(n-2)+18000y(n-1)-8100y(n-2)
         ldx    #10000
         edivs
         sty    YN
         rts
```

8.4.4 Multiplication and Division

Program 8.38 presents assembly programs that multiply two 32-bit unsigned numbers (N,M), yielding a 64-bit product (P=N*M). The approach first considers each input as having two 16-bit components (i.e., $N=2^{16}*msN+lsN$ and $M=2^{16}*msM+lsM$). The next step is to perform four executions of the 16-by-16-bit emul instruction. The 6811 version requires a call to a emul subroutine, which can be found as part of the math.rtf file in the MC6811 directory of the **TExaS** application. The final 64-bit product is the sum of the four partial products.

$$\begin{aligned} P = N*M &= (2^{16}*msN+lsN)*(2^{16}*msM+lsM) \\ &= (2^{32}*msN*msM) + (2^{16}*lsN*msM) + (2^{16}*msN*lsM) + (lsN*lsM) \end{aligned}$$

Program 8.38 A 32-bit multiply function.

```
; 6811                                    ; 6812
; Input:Two 32-bit numbers N(SP+2),M(SP+4) ; Input: Two 32-bit numbers N(SP+2),M(SP+4)
; Output:One 64-bit product P(SP+6)        ; Output: One 64-bit product P(SP+6)
; modifies Reg A,B,X,Y                     ; modifies Reg A,B,X,Y
mult32   ; P=(2**32)*msN*msM               mult32
        tsx                                ; P=(2**32)*msN*msM
        ldx  2,x   ; pointer to N                  ldx  2,sp   ; pointer to N
        ldy  0,x   ; msw of N                      ldy  0,x    ; msw of N
        tsx                                        ldx  4,sp   ; pointer to M
        ldx  4,x   ; pointer to M                  ldd  0,x    ; msw of M
        ldd  0,x   ; msw of M                      emul        ; Y:D=msN*msM
        jsr  emul  ; Y:D=msN*msM                   ldx  6,sp   ; pointer to P
        tsx                                        sty  0,x
        ldx  6,x   ; pointer to P                  std  2,x    ; set high 32 bits of P
        sty  0,x
        std  2,x   ; set high 32 bits of P
; P=P+lsN*lsM                              ; P=P+lsN*lsM
        tsx                                        ldx  2,sp   ; pointer to N
        ldx  2,x   ; pointer to N                  ldy  2,x    ; lsw of N
        ldy  2,x   ; lsw of N                      ldx  4,sp   ; pointer to M
        tsx                                        ldd  2,x    ; lsw of M
        ldx  4,x   ; pointer to M                  emul        ; Y:D=lsN*lsM
        ldd  2,x   ; lsw of M                      ldx  6,sp   ; pointer to P
        jsr  emul  ; Y:D=lsN*lsM                   sty  4,x
        tsx                                        std  6,x    ; set low 32 bits of P
        ldx  6,x   ; pointer to P
        sty  4,x
        std  6,x   ; set low 32 bits of P
; P=P+65536*lsN*msM                        ; P=P+65536*lsN*msM
        tsx                                        ldx  2,sp   ; pointer to N
        ldx  2,x   ; pointer to N                  ldy  2,x    ; lsw of N
        ldy  2,x   ; lsw of N                      ldx  4,sp   ; pointer to M
        tsx                                        ldd  0,x    ; msw of M
        ldx  4,x   ; pointer to M                  emul        ; Y:D=lsN*msM
        ldd  0,x   ; msw of M                      ldx  6,sp   ; pointer to P
        jsr  emul  ; Y:D=lsN*msM                   addd 4,x
        tsx                                        std  4,x    ; add to P
        ldx  6,x   ; pointer to P                  tfr  y,d
        addd 4,x                                   adcb 3,x
        std  4,x   ; add to P                      adca 2,x
        xgdy                                       std  2,x    ; add to P
        adcb 3,x                                   ldd  0,x
        adca 2,x                                   adcb #0
        std  2,x   ; add to P                      adca #0
        ldd  0,x                                   std  0,x
        adcb #0
        adca #0
        std  0,x
```

```
; P=P+65536*msN*lsM
      tsx
      ldx   2,x    ; pointer to N
      ldy   0,x    ; msw of N
      tsx
      ldx   4,x    ; pointer to M
      ldd   2,x    ; lsw of M
      jsr   emul   ; Y:D=msN*lsM
      tsx
      ldx   6,x    ; pointer to P
      addd  4,x
      std   4,x    ; add to P
      xgdy
      adcb  3,x
      adca  2,x
      std   2,x    ; add to P
      ldd   0,x
      adcb  #0
      adca  #0
      std   0,x
      rts
```

```
; P=P+65536*msN*lsM
      ldx   2,sp   ; pointer to N
      ldy   0,x    ; msw of N
      ldx   4,sp   ; pointer to M
      ldd   2,x    ; lsw of M
      emul         ; Y:D=msN*lsM
      ldx   6,sp   ; pointer to P
      addd  4,x
      std   4,x    ; add to P
      tfr   y,d
      adcb  3,x
      adca  2,x
      std   2,x    ; add to P
      ldd   0,x
      adcb  #0
      adca  #0
      std   0,x
      rts
```

Program 8.39 presents assembly programs that divide a 64-bit unsigned dividend (N) by a 32-bit divisor (M), yielding a 32-bit quotient (Q=N/M) and a 32-bit remainder (R). The following RAM-based global variables are used to pass data into and out of the function. The implementations of the subroutines can be found in the `math.rtf` file installed as part of **TExaS**.

Program 8.39
A 64-bit by 32-bit divide function.

```
N     rmb   4    ; 64-bit dividend
R     rmb   4    ; low part of dividend becomes remainder
M     rmb   4    ; 32-bit divisor
      rmb   4    ; pad
Q     rmb   4    ; quotient
; 64-bit by 32-bit into 32-bit divide Q=N/M, R=N%M
; Input:     64-bit dividend, global N (replaced with R)
;            32-bit divisor, global M (destroyed)
; Output:    32-bit quotient, global Q
;            32-bit remainder, global R
;            V bit set on overflow (R>=M)
;            C,V bits set on divide by zero (M=0)
; modifies Reg A,B,X,Y
i     set   0    ; loop counter
div32 leas  -1,s ; allocate i
      ldd   M
      bne   d32A ; divisor not zero
      ldd   M+2
      bne   d32A ; divisor not zero
      sev        ; divide by zero
      sec
      bra   d32E
```

continued

```
d32A  movw  #0,M+4
      movw  #0,M+6  ; divisor is 64 bits, right justified
      movw  #0,Q
      movw  #0,Q+2  ; quotient=0
      movb  #32,i,s ; i=0
d32B  ldx   #M
      jsr   lsr64   ; M=M>>1
      ldx   #Q
      jsr   lsl32   ; Q=Q<<1
      ldx   #N
      ldy   #M
      jsr   cmp64   ; N-M
      blo   d32C    ; skip if N<M
      ldx   #Q
      jsr   inc32
      ldx   #N
      ldy   #M
      jsr   sub64
d32C  dec   i,s
      bne   d32B
      ldx   #R
      ldy   #M+4    ; no overflow if R<M
      jsr   cmp32   ; R-M
      blo   d32D    ; skip if R<M
      clc
      sev           ; overflow, R>=M
      bra   d32E
d32D  clc
      clv           ; no error
d32E  leas  1,s
      rts
```

8.4.5 Table Lookup and Interpolation

Many applications require the representation of complex waveforms in digital form. A typical application is a calibration curve that describes the input/output behavior of the system (see Figure 8.14 and Program 8.40). The actual hardware takes the measurand (y is position, pressure, temperature etc.) as input and has the ADC conversion (x is 0 to 255) as output. In this situation, the software algorithm is asked to reverse the process, taking as input (x) the ADC measurement and giving as output (y) the measurand. One of the most efficient, yet simple, techniques for describing nonlinear equations is to provide a small table of (x,y) points then use linear interpolation between the points. In this way, the response is piece-wise linear. This technique works only for any single-valued data set (a unique y for each x). There is a clear tradeoff between accuracy and software efficiency (static and dynamic) (e.g., you can add more points to improve accuracy, but it requires more memory and runs slower).

Figure 8.14
A table contains specific points and the software will use linear interpolation to fill in the gaps.

Xtable	0	10	100	150	255
Ytable	20	0	200	100	150

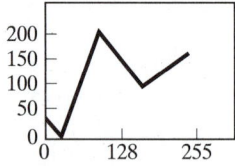

Program 8.40
Interpolation using the
6812 `tbl` instruction.

```
        org    $800
in      ds     1      ; 8-bit input
out     ds     1      ; 8-bit output
        org    $F000
; The table consists of multiple unsigned (x,y)
; which define a piece-wise linear function
; first X entry must be less than or equal to minimum input
; last X entry must be bigger than maximum input
Xtable  dc.b 0,10,100,150,255 ; must be monotonic
Ytable  dc.b 20,0,200,100,150
Main    lds    #$0C00
        clra
loop    staa   in        ; ranges from 0 to 254
        bsr    Lookup
        staa   out
here    ldaa   in
        inca
        cmpa   #255
        bne    loop
        stop
;**********Lookup*******************
;Inputs: RegA is 0 to 254 Xdata point, xL
; RegA input must be greater than or equal to first Xdata point
; RegA input must be less than last Xdata point
;Output: RegA is 0 to 255 Ydata point
;Registers destroyed: X,Y,B,CCR
Lookup  ldx    #Xtable    ; first find x1<=xL<x2
        ldy    #Ytable
lookx1  cmpa   1,x        ; check xL<x2
        blo    found      ; stops when X points to x1
        inx
        iny
        bra    lookx1
found   suba   0,x    ; xL-x1
        clrb          ; D=256*(xL-x1)
        pshd
        ldab   1,x    ; x2
        subb   0,x    ; x2-x1
        clra          ; D=x2-x1
        tfr    D,X    ; X=x2-x1
        puld          ; D=256*(xL-x1)
        idiv          ; X=(256*(xL-x1))/(x2-x1)
        tfr    X,D
; B=(256*(xL-x1))/(x2-x1)
; Y points to y1,y2
        tbl 0,y
; A=Y1+B*(Y2-Y1)
        rts
        org    $FFFE
        fdb    Main
```

8.5 Expression Evaluation

In this section, we will develop methods to implement complex mathematical and logical operations. When the expression is simple, no formal approach is required, and registers can be used to store intermediate results. For example, Program 8.41 shows the assembly code to calculate P=3*M+N, where P, M, and N are 8-bit unsigned global variables.

Program 8.41
A simple expression evaluation.

```
    ldaa #3
    ldab M
    mul        ; Reg D=3*M
    addb N     ; Reg B=3*M+N
    stab P
```

To increase the precision of the expression, we replace instructions with subroutine calls as needed. Program 8.42 shows the assembly code to calculate P=3*M+N, where P, M, and N are 16-bit unsigned global variables.

Program 8.42
16-bit expression evaluation.

```
;6811                                ;6812
    ldd  #3                              ldd  #3
    ldy  M                               ldy  M
    jsr  emul ; Reg D=3*M                emul       ; Reg D=3*M
    addd N    ; Reg D=3*M+N              addd N     ; Reg D=3*M+N
    std  P                               std  P
```

As the complexity of the expression increases, we need a place to save intermediate results. The stack is a natural place to store temporary results. Polish mathematician Jan Lucasiewicz developed a method to write expressions without parentheses. His technique is called **Polish Notation** or **Reverse Polish Notation**. Table 8.4 illustrates the notation in which the parameters come first. Any expression that is written in regular format with parentheses can be rewritten in an equivalent Polish Notation. Compilers possess the ability to perform this translation automatically, but assembly language programmers can translate their expressions by hand. The following rules are used when evaluating expressions in Polish Notation:

- numbers are pushed onto the stack,
- values of the variables are pushed onto the stack,
- for unary operations the parameter is popped and the result is pushed,
- for binary operations two parameters are popped and the result is pushed.

Table 8.4
Examples of Polish Notation.

Regular expression	Polish Notation
3*M+N	3 M * N +
~(M|(N&P))	N P & M | ~
M*(5+P)-N/10	5 P + M * N 10 / -
w−x+y+z−4	w x − y + z + 4 −

Program 8.43 shows the assembly code to calculate P=M*(5+P)-N/10, where P, M, and N are 8-bit unsigned global variables. Let `buffer` be a 10-byte RAM area, which will be used as a data stack. Register Y points to this data stack, which is used in a manner similar to Register SP and the regular stack. The subroutine `Add` pops two elements off the data stack, adds them, and pushes the result onto the data stack. The subroutine `Mult` pops two elements off the data stack, multiplies them, and pushes the result onto the data stack. Notice how easy it is to implement a data stack with the 6812.

Program 8.43   16-bit expression evaluation of P=M*(5+P)-N/10.

```
; 6811                              ; 6812
        ldy  #buffer+10  ; empty            ldy  #buffer+10  ; empty
        ldaa #5                             movb #5, -1,y ; push 5
        dey                                 movb P,-1,y    ; push P
        staa 0,y  ; push 5                  jsr  Add
        ldaa P                              movb M,-1,y    ; push M
        dey                                 jsr  Mult
        staa 0,y  ; push P                  movb N,-1,y    ; push N
        jsr  Add                            movb #10,-1,y  ; push 10
        ldaa M                              jsr  Div
        dey                                 jsr  Sub
        staa 0,y  ; push M                  movb 1,y+,P    ; pop
        jsr  Mult                           stop
        ldaa N
        dey                         Add     ldaa 1,y+ ;add top two,pop
        staa 0,y  ; push N                  adda 0,y
        ldaa #10                            staa 0,y
        dey                                 rts
        staa 0,y  ; push 10
        jsr  Div                    Mult    ldaa 1,y+ ;add top two,pop
        jsr  Sub                            ldab 0,y
        ldaa 0,y                            mul
        iny       ; pop                     stab 0,y
        staa P                              rts
        stop
Add     ldaa 0,y  ; add top two
        iny       ; pop
        adda 0,y
        staa 0,y
        rts
Mult    ldaa 0,y  ; add top two
        iny       ; pop
        ldab 0,y
        mul
        stab 0,y
        rts
```

Checkpoint 8.17: *Write an assembly subroutine to implement* `Sub`, *which pops two elements off the data stack, subtracts them, and pushes the result onto the data stack. Write an assembly subroutine to implement* `Div`, *which pops two elements off the data stack, divides them, and pushes the result onto the data stack.*

8.6   OS Calls Using Software Interrupts

A trap is a software interrupt. The 6811 and 6812 have one software interrupt,

```
swi
```

The 680x0 family of processors has 16 independent software interrupts,

```
TRAP #n
```

where n ranges from 0 to 15. The Intel x86 family of processors has 256 independent software interrupts,

```
INT #n
```

where n ranges from 0 to 255. A trap is similar to a subroutine. A program executes a trap when it needs to perform a software function. The difference between a trap and a subroutine is that the subroutine address binding occurs during assembly or compilation time. In other words, the address of the subroutine is specified in the object code of the instruction. For a `bsr`, the subroutine address is specified as a PC relative offset from the current PC location. In this case, the calling program (the software with the `bsr`) and the subroutine are assembled together, and linked by the relative offset between them. For a `jsr`, the subroutine address is specified as an absolute 16-bit address. In this case, the calling program (the software with the `jsr`) and the subroutine are assembled together, and linked by the absolute address of the subroutine.

> **Checkpoint 8.18:** *Explain how to use the unimplemented instruction trap on the 6811/6812 to provide for additional software interrupts.*

The most common use of a trap is calling operating system functions. The user application executes the trap and the OS software performs the desired function in the trap handler. The difference with a trap is that the calling program (e.g., user application) need not know where in memory the trap handler (e.g., OS) exists either at assembly or at run time. The programmer who writes the trap handler places a pointer to it in a trap vector table. The trap vector table contains pointers to all the trap handlers. Since the 6811 and 6812 have only one trap, the trap vector table is one 16-bit entry. When the `swi` is executed, the 6811/6812 pushes all the registers on the stack (including the PC) and places the 16-bit contents of memory locations $FFF6 and $FFF7 into the PC. The trap handler performs the desired function and returns results in the registers, or on the stack. The control is passed back to the calling program by executing an `rti` at the end of the trap handler, which pops all the registers off the stack (including the PC). The programmer who writes the SWI handler places a pointer to it at $FFF6 and $FFF7. In this way the program which calls the trap (the software with the `swi`) need not be assembled with the trap handler. This separation provides protection for both programs. In other words, neither user nor OS needs to know details of each other's programs like address in memory, object code or source code.

Traps are used to define operating systems. The OS can be implemented as a long list of useful software functions which are defined as traps. In this way the OS can be assembled and loaded into memory. The OS also sets the addresses in the trap vector table. Then the application program (like an editor or database) can be independently assembled (without knowledge of the OS source code). When the application requires an OS function (like InCh) it simply executes a trap.

Another application of traps is breakpoints. To place a breakpoint, the debugger replaces the existing instruction with a `swi`. If the program arrives at the breakpoint, it stops executing the user program and enters the trap handler. To remove the breakpoint, the debugger replaces the `swi` with the previously operative instruction. This use of `swi` can not be used to set breakpoints if the program is in PROM or ROM.

Program 8.44 places the vectors in ROM. These vectors can not be changed at run time.

Program 8.44
The software interrupt vector points to the `SWI` handler.

```
        org   $F000    ; $D000 for the MC68HC711E9
main    lds   #$00FF   ; initialize stack, $01FF for the MC68HC711E9
        bsr   ritual   ; initialization of hardware, data structures
loop
;* main program
        bra   loop     ;repeat forever
;***called by software using the SWI instruction ********
SWIhan
        ;SWI handler
        rti
        org   $FFF6
        fdb   SWIhan ; SWI vector
        org   $FFFE
        fdb   main    ; reset vector
```

Program 8.45 places indirect `jmp` instructions in RAM so that the interrupt vectors can be changed at run time.

Program 8.45   A software interrupt vector that can be redirected.

```
        org   $0800    ; RAM locations, 0 for the MC68HC711E9
SWIpt   rmb   3        ; contains a jump to the actual SWI interrupt handler
        org   $F000    ; $D000 for the MC68HC711E9
; Set the SWI interrupt vector dynamically
; input RegX points to SWIhandler
SetSWI  ldaa  #$06
        staa  SWIpt    ; install jmp opcode (extended)
        stx   SWIpt+1  ; operand field of the jmp instruction
        rts
main    lds   #$0C00   ; initialize stack, $01FF for the MC68HC711E9
        ldx   #SWIhand
        bsr   SetSWI   ; dynamically install vector
        bsr   ritual
loop    bsr   func1
        bsr   func2
        bra   loop     ;repeat forever
SWIhan
        ;... handler
        rti
        org   $FFF6
        fdb   SWIpt    ; redirection
        org   $FFFE
        fdb   main     ; reset vector
```

Figure 8.15 shows the stack before and after the `swi` instruction is executed on the 6812. The `swi` instruction first pushes the CC, then sets the Ibit. A hardware interrupt will also push registers onto the stack, and set the Ibit.

Figure 8.15
A 6812 `swi` instruction can be used to call OS functions. On the 6811, the SP points to just above the top of the stack.

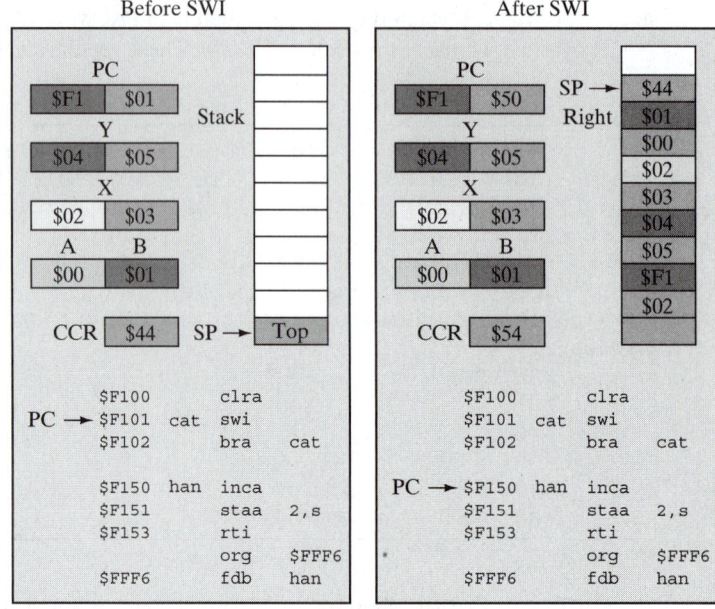

The `rti` instruction is used to return from a hardware or a software interrupt on the 6812. (See Figure 8.16.) The `rti` instruction restores the Ibit back to its original value, the one that existed before the `swi`.

Figure 8.16
A 6812 `rti` instruction returns from an OS function. On the 6811, the SP points to just above the top of the stack.

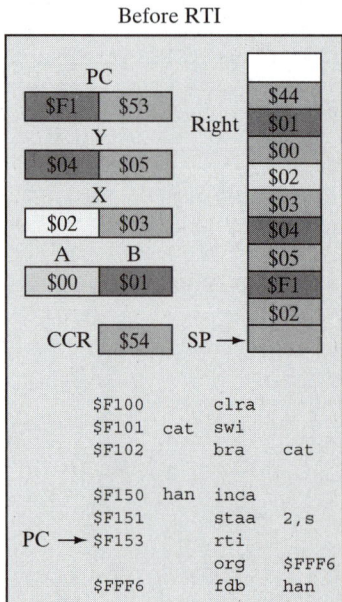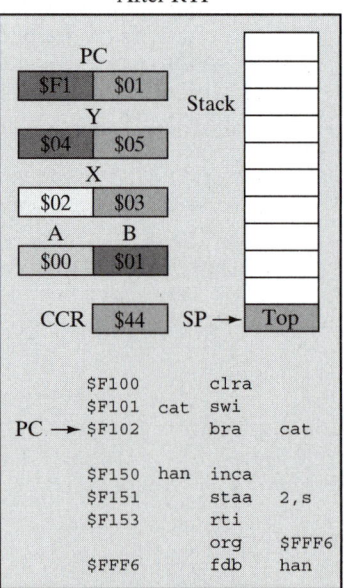

Programs 8.46 and 8.47 show serial port functions implemented as a software interrupt. Notice that to return a parameter from a `swi`, you have to store the result into the stack posi-

tion from which one of the registers will be pulled (e.g., the `staa 4,sp` within Program 8.46, and the `staa 2,sp` within Program 8.47).

Program 8.46   6812 C program showing the use of `SWI` to implement an OS function.

```
#define RDRF 0x20    // Receive Data Register Full Bit
// Wait for new serial port input, return ASCII code for key typed
char InChar(void){
  while ((SC0SR1 & RDRF) == 0){};
  return(SC0DRL);
}
#define TDRE 0x80    // Transmit Data Register Empty Bit
// Wait for buffer to be empty, output ASCII to serial port
void OutChar(char data){
  while ((SC0SR1 & TDRE) == 0){};
  SC0DRL = data;
}
#pragma interrupt_handler SWIHan()
void SWIHan(void){ char command,data;
// command is placed in Reg B, 0 for InChar 1 for OutChar
// data placed in and out using Reg A
asm(" staa %data");
asm(" stab %command");
  if(command==0)
    data=InChar();
  else if(command==1)
    OutChar(data);
asm(" ldaa %data");
asm(" staa 4,sp");  // place on the stack to return Reg A
}
```

Program 8.47
6812 assembly program showing the use of `SWI` to implement an OS function.

```
In       equ  0
Out      equ  1
         org  $F000
SWIhan cmpb #In
         bne  notIC
         bsr  InChar
         staa 2,sp        ; position on the stack for Reg A
         bra  done
notIC  cmpb #Out
         bne  done
         bsr  OutChar
done   rti
; Read 8 bits from serial port
; Return 8-bit byte in RegA
RDRF     equ  $20
InChar ldaa SC0SR1   ; Reg A has the serial status
         bita #RDRF       ; new input available?
         beq  InChar      ; wait for RDRF to be set
         ldaa SC0DRL      ; read ASCII character
         rts
```

continued

```
; Write 8 bits to serial port
; Input 8-bit byte in RegA
TDRE    equ  $80
OutChar ldab SC0SR1  ; Reg B has the serial status
        andb #TDRE   ; output channel ready for transmission?
        beq  OutChar ; wait for TDRE to be set
        staa SC0DRL  ; write ASCII character
        rts
main    lds  #$0C00
        ldaa #13     ; CR
        ldab #Out
        swi          ; OutChar(13)
loop    ldab #In
        swi          ; RegA=InChar()
        ldab #Out
        swi          ; OutChar(RegA)
        bra  loop
        org  $FFF6
        fdb  SWIhan  ; $FFF6 is SWI vector
        org  $FFFE
        fdb  main    ; reset vector
```

8.7 Finite State Machines with Statically Allocated Linked Structures

8.7.1
Abstraction

Software abstraction is the definition of a complex problem with a set of basic abstract principles. If we can construct our software system using these abstract building blocks, then we have a better understanding of both the problem and its solution. This is because we can separate what we are doing (policies) from the details of how we are getting it done (mechanisms.) This separation also makes it easier to optimize. Abstraction provides for a proof of correct function, and simplifies both extensions and customization. The abstraction presented in this section is the **Finite State Machine** (FSM.) The abstract principles of FSM development are the inputs, outputs, states, and state transitions. The FSM state graph defines the time-dependent relationship between its inputs and outputs. If we can take a complex problem and map it into an FSM model, then we can solve it with a simple FSM software tool. Our FSM software implementation will be easy to understand, debug, and modify. Other examples of software abstraction include **Proportional Integral Derivative** digital controllers, fuzzy logic digital controllers, neural networks, and linear systems of differential equations. In each case, the problem is mapped into a well-defined model with a set of abstract yet powerful rules. Then the software solution is a matter of implementing the rules of the model. In our case, once we prove that our software correctly solves one FSM, then we can make changes to this FSM and be confident that our software solution correctly implements the new FSM.

The FSM controller employs a well-defined model or framework to solve our problem. The state graph will be specified using a linked data structure. The three advantages of this abstraction are (1) it can be faster to develop because many of the building blocks preexist; (2) it is easier to debug (prove correct) because it separates conceptual issues from implementation; and (3) it is easier to change. In this chapter, we will implement Moore FSMs, where the output values depend only on the current state and the inputs affect the state transitions. On the other hand, the outputs of a Mealy FSM depend both on the current state and the inputs.

A linked structure consists of multiple identically structured nodes. One or more of the entries in the node is a pointer (or link) to other nodes. In an embedded system, we usually use statically allocated fixed-size linked structures, which are defined at compile time and exist throughout the life of the software. In a simple embedded system, the state graph is fixed, so we can store the linked data structure in nonvolatile memory. For complex systems where the control functions change dynamically (e.g., the state graph itself varies over time), we could implement dynamically allocated linked structures, which are constructed at run time and number of nodes can grow and shrink in time. A linked structure will be used to define the FSM. An important factor when implementing FSMs using linked structures is that there should be a clear and one-to-one mapping between the FSM state graph and the linked structure (i.e., there should be one structure for each state). A Moore FSM has the outputs a function of only the current state. In constrast, the outputs are a function of both the input and the current state in a Mealy FSM.

8.7.2 Stepper Motor Controller

Figure 8.17 shows a circular linked graph containing the output commands to control a stepper motor. This simple FSM has no inputs, four output bits and four states. There is one state for each output pattern in the usual stepper sequence 5,6,10,9,. . . The circular FSM is used to spin the motor in a clockwise direction. Notice the one-to-one correspondence between the state graph in Figure 8.17 and the `fsm[4]` data structure in Program 8.48.

Figure 8.17
This stepper motor FSM has four states. The 4-bit outputs are given in binary.

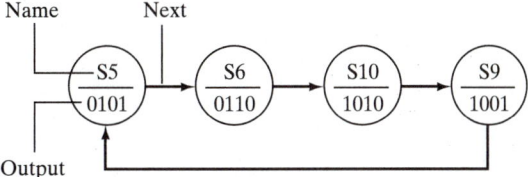

We connect four high-current drivers to Port B of the Motorola MC68HC11, as shown in Figure 8.18. The output low current of the driver must be large enough to energize the stepper coils. The ULN2074 data sheet states that its maximum I_{CE} is 1.25A. But since the ULN2074 is a high-current Darlington switch, its I_{CE} will also be limited by its input base current, which comes from the I_{OH} of the 6811. In this case, the I_{OH} of the MC68HC11A8 is 0.8 mA, so this ULN2074 circuit can sink up to 500 mA.

Figure 8.18
A unipolar stepper motor interfaced to a Motorola MC68HC11.

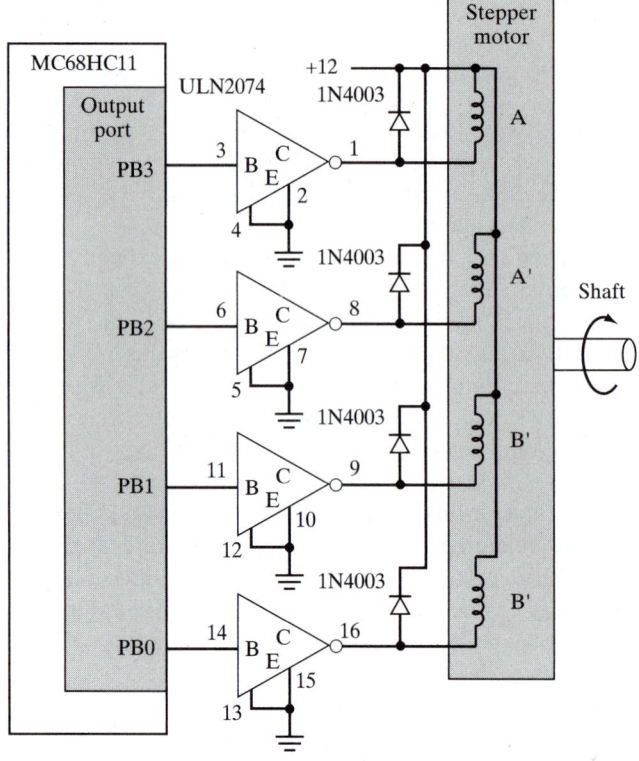

The main program, Program 8.48, begins by initializing the Port B output and the state pointer. To implement this example on the 6812, you would have to set DDRB to $FF. Every 5 ms the program outputs a new stepper command. The function `Wait()` uses the built-in timer to generate an appropriate delay between outputs to the stepper. To illustrate how easy it is to make changes to this implementation, let's consider these two modifications. To make it spin in the other direction, we simply change pointers to sequence in the other direction. To implement an eight-step sequence (e.g., 5, 4, 6, 2, 10, 8, 9, 1. . .), we add the four new states and link all eight states in the desired sequence. Both of these changes can be made without modification to the main program.

Program 8.48
Stepper motor controller.

```
/* Port B bits 3-0 are outputs to the stepper */
const struct State {
  unsigned char Out;           /* stepper command */
  const struct State *next;}; /* clockwise */
typedef const struct State StateType;
#define S5 &fsm[0]
#define S6 &fsm[1]
#define S10 &fsm[2]
#define S9 &fsm[3]
StateType fsm[4]={
  { 5, S6},   /* Out=0101, Next=S6 */
  { 6,S10},   /* Out=0110, Next=S10 */
  {10, S9},   /* Out=1010, Next=S9 */
  { 9, S5}}; /* Out=1001, Next=S5 */
```

```
/* delay time is given 500ns units */
void Wait(unsigned short delay){ short Endt;
  Endt=TCNT+delay;  /* TCNT at the end of delay */
  while((Endt-(short)TCNT)>0);
}
void main(void){ StateType *Pt;
  PORTB=5;   /* initial output */
  Pt=S5;     /* initial state */
  while(1){  /* embedded systems never quit */
    Wait(10000);    /* 5ms wait */
    Pt=Pt->next;    /* Clockwise step */
    PORTB=Pt->Out;  /* stepper drivers */
  }
}
```

8.7.3
Traffic Light
Controller

Controlling traffic is a good illustrative example because we all know what is supposed to happen at the intersection of two busy one-way streets. We place car sensors on the North and East roads, which we can simulate with two switches connected to Port C, see Figure 8.19. If the digital input is high (1), we will consider that there is a car on that road. To simulate the traffic light, we will interface six colored LEDs to Port B.

Figure 8.19

A simulated traffic intersection interfaced to a Motorola MC68HC11.

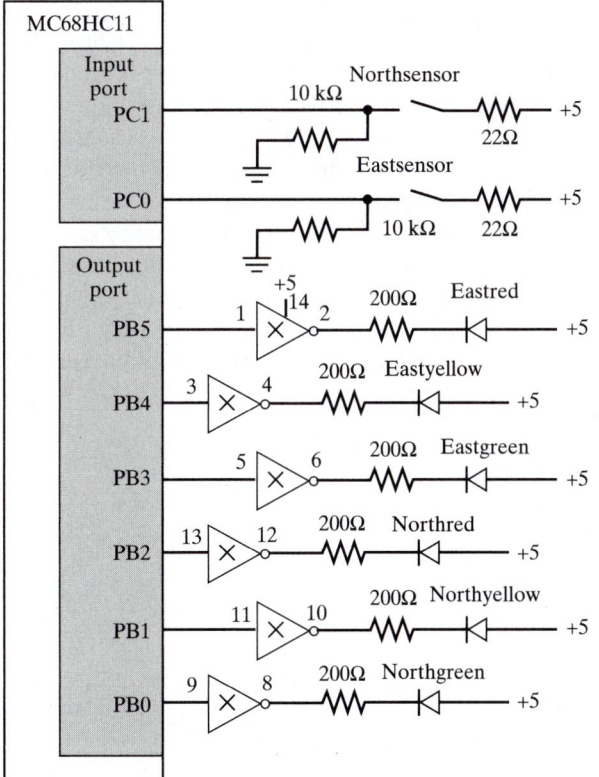

Figure 8.20 shows the simple Moore FSM that controls traffic at the intersection. The goal is to maximize traffic flow, minimize waiting time at a red light, and avoid accidents. For example, if we are in state goN, we will set the Port B output to 100001_2 (green light on North and red light on East), and then we wait 30 seconds. Next, we read the sensor inputs. If the sensor value is 01_2 (meaning that a car has entered the intersection on the East road, but no car exists on the North road), then we will go to state waitN (yellow light on North and red light on East). After showing the yellow light for five seconds on the North road, our controller will switch to state goE (red light on North and green light on East), regardless of the input.

Figure 8.20
This Moore FSM controls traffic in our intersection.

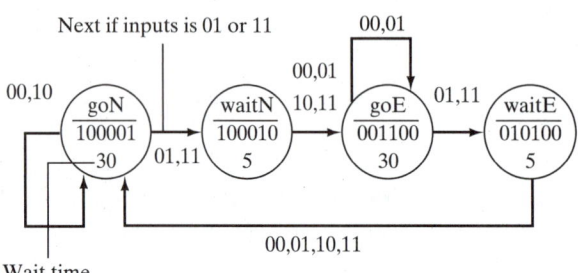

The main program, Program 8.49, begins by specifying the Port C bits 1 and 0 to be inputs. To implement this example on the 6812, you would have to set DDRB to $FF. With CMOS microcomputers, one must define unused I/O pins either as outputs or specify them as inputs and tie the pin high (or low) in hardware. In this example, we define the unused pins PC7-PC2 as outputs. Because of the extremely high impedance of CMOS inputs, an unconnected input pin may oscillate, dissipating power unnecessarily. The initial state is defined as goN. Our controller software first outputs the desired light pattern to the six LEDs, waits for the specified amount of time, reads the sensor inputs from Port C, then switches to the next state, depending on the input data. The function Wait1sec() calls the Wait() function defined in Program 8.49. In order to make it easier to understand, thereby simplifying verification and modification, we have made a one-to-one correspondence between the state graph in Figure 8.20 and the fsm[4] data structure in Program 8.49. Notice also how this implementation separates the civil engineering policies (the FSM specifies what we do) from the computer engineering mechanisms (the C program specifies how it is done). Once we have proven the C program operational, we can modify the policies and be confident that the mechanisms will still work. When an accident occurs, we can blame the civil engineer who designed the state graph.

Program 8.49 Traffic light controller.

```
/* Port C bits 1,0 are sensor inputs,
   Port B bits 5-0 are LED outputs */
const struct State {
  unsigned char Out;                /* Output to Port B */
  unsigned short Time;              /* Time in sec to wait */
const struct State *Next[4];};      /* Next if input=00,01,10,11*/
typedef const struct State StateType;
#define goN &fsm[0]
```

```
#define waitN &fsm[1]
define goE &fsm[2]
#define waitE &fsm[3]
StateType fsm[4]={
  {0x21, 30,{goN,waitN,goN,waitN}}, /* goN state */
  {0x22,  5,{goE,goE,goE,goE}},     /* waitN state */
  {0x0C, 30,{goE,goE,waitE,waitE}}, /* goE state */
  {0x14,  5,{goN,goN,goN,goN}}};    /* waitE state */
void Wait1sec(unsigned short delay){ unsigned short i;
  for(i=0;i<delay;i++)
    Wait(1000);    /* 1 second wait */
}
void main(void){ StateType *Pt;  /* Current State */
  unsigned char Input;
  DDRC = 0xFC;   /* Port C bits 1,0 are inputs from the sensors */
  Pt = goN;      /* Initial State */
  while(1){
    PORTB = Pt->Out;       /* Perform output for this state */
    Wait1sec(Pt->Time);    /* Time to wait in this state */
    Input = PORTC&0x03;    /* Input=00, 01, 10, or 11 */
    Pt = Pt->Next[Input];  /* Move to next state */
  }
}
```

Checkpoint 8.19: Why is it good to use labels for the states (e.g., `goN` is better than `&fsm[0]`)?

Again, the FSM approach makes it easy to change. To change the wait time for a state, we simply change the value in the data structure. To add more complexity (e.g., put a red/red state after each yellow state, which will reduce accidents caused by bad drivers), we simply increase the size of the `fsm[]` structure and define the `Out`, `Time`, and `Next` pointers for these new states.

To add two more output signals (e.g., walk and left turn lights), we simply use all 8 bits of the `Out` field. To add even more output bits we could increase the precision of the `Out` field. To add two more input lines (e.g., wait button, left turn car sensor), we increase the size of the pointer field to `Next[16]`. Because now there are four input lines, there are 16 possible combinations, where each input possibility requires a `Next` state pointer specifying where to go if this combination occurs. In this simple scheme, the size of the `Next[]` field will be 2 raised to the power of the number of input signals.

8.7.4
A Mealy FSM

In a Mealy FSM, the outputs are a function of both the input and the current state. To illustrate the Mealy machine, a simple finite state machine will be implemented, as shown in Figure 8.21 and Programs 8.50 and 8.51. For example, if we are in State C, we will wait 15 time units, and if the input is 1, then we will make the output 0 and go to State D. The `org` pseudo-ops are used to place the FSM data structure in EEPROM and the assembly language program in ROM of a single chip 6811.

Figure 8.21
A simple Mealy FSM.

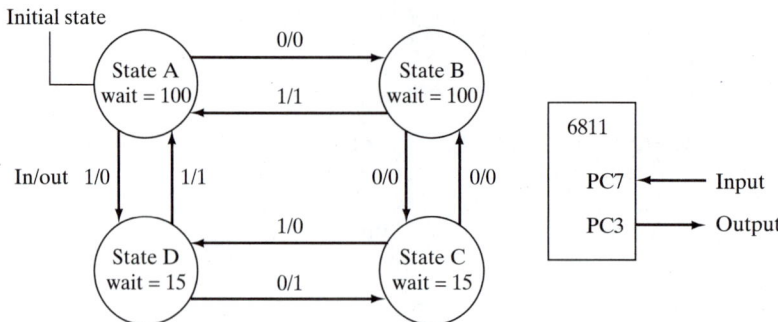

Program 8.50
Assembly language
Mealy FSM.

```
        org   $B600   Put in EEPROM so it can be changed
* Finite State Machine
Time  equ   0       Index for time to wait in this state
Out0  equ   1       Index for output pattern if input=0
Out1  equ   2       Index for output pattern if input=1
Next0 equ   3       Index for next state if input=0
Next1 equ   5       Index for next state if input=1
IS    fdb   SA      Initial state
SA    fcb   100     Time to wait
      fcb   0,0     Outputs for inputs 0,1
      fdb   SB      Next state if Input=0
      fdb   SD      Next state if Input=1
SB    fcb   100     Time to wait
      fcb   0,8     Outputs for inputs 0,1
      fdb   SC      Next state if Input=0
      fdb   SA      Next state if Input=1
SC    fcb   15      Time to wait
      fcb   0,0     Outputs for inputs 0,1
      fdb   SB      Next state if Input=0
      fdb   SD      Next state if Input=1
SD    fcb   15      Time to wait
      fcb   8,8     Outputs for inputs 0,1
      fdb   SC      Next state if Input=0
      fdb   SA      Next state if Input=1
      org   $E000      Place assembly program in ROM
* 6811 program
* Initialization of 6811 and linked list
GO    ldaa  #$08    PC3=output, rest are input
      staa  $1007   Set DDRC
      ldx   IS      Reg X => current state
* Linked list interpreter
LL    ldaa  Time,X    Time to wait
      bsr   WAIT      Reg A is call by value
      ldaa  $1003
      bita  #$80
      bpl   is0       Go to is0 if Input=0
is1   ldaa  Out1,X    Get desired output from linked list
      staa  $1003     Set PC3=output
      ldx   Next1,X   Input is 1
      bra   LL
```

```
is0    ldaa  Out0,X     Get desired output from linked list
       staa  $1003      Set PC3=output
       ldx   Next0,X    Input is 0
       bra   LL         Infinite loop
       org   $FFFE
       fdb   GO         reset vector
```

Program 8.51   C language Mealy FSM.

```
struct State{
  unsigned char Time;    /* Time to wait in each state */
  unsigned char Out[2];  /* Output if input=0,1 */
  struct State *Next[2]; /* Next state if input=0,1 */
};
typedef struct State StateType;
typedef StateType * StatePtr;
StateType fsm[4]={
  {100,{0,0},{&fsm[1],&fsm[3]}},
  {100,{0,8},{&fsm[2],&fsm[0]}},
  { 15,{0,0},{&fsm[1],&fsm[3]}},
  { 15,{8,8},{&fsm[2],&fsm[0]}}
};
void main(void){ StatePtr Pt; /* Current State */
unsigned char Input;
  Pt = &fsm[0];              /* Initial State */
  DDRC = 0x08;               /* PortC bit3 is output */
  while(1){
    Wait(Pt->Time);        /* Time to wait in this state */
    Input = PORTC>>7;      /* Input=0 or 1 */
    PORTC = Pt->Out[Input];
    Pt = Pt->Next[Input];
  }
}
```

8.7.5
FSM
Implementation
of a Vending
Machine

This vending machine example illustrates additional flexibility that we can build into our FSM implementations. In particular, rather than simple digital outputs, we will implement general functions for each state. We could have solved this particular vending machine using the approach in the previous example, but this approach provides an alternative mechanism when the output operations become complex. Our simple vending machine has two coin sensors, one for dimes and one for nickels, which we will again simulate with two switches connected to Port C of our 6811 (see Figure 8.22). If the digital input is high (1), we will consider there to be a coin in the slot. When a coin is inserted into the machine, the sensor goes high, then low. Because of the nature of vending machines, we will assume there can not be both a nickel and a dime at the same time. To simulate the soda and change dispensers, we will interface two solenoids to Port B. The coil current of the solenoids is less than 40 mA, so we can use the 7406 open collector driver.

Figure 8.22
A simulated vending
machine interfaced to a
Motorola MC68HC11.

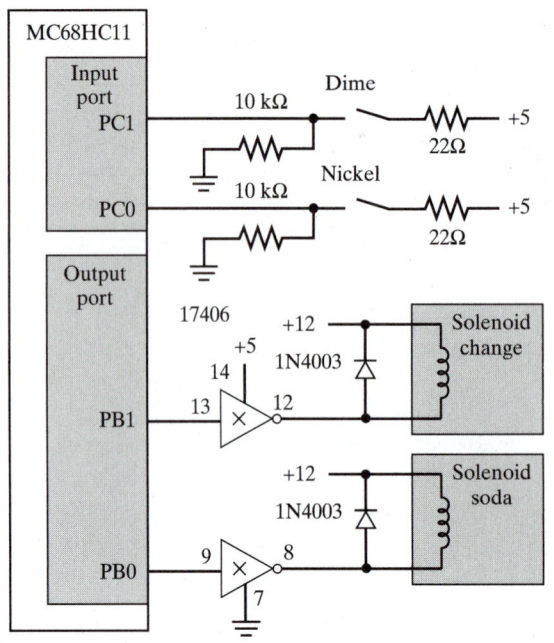

Figure 8.23 shows the Moore FSM that implements the vending machine. A soda costs 15 cents, and the machine accepts nickels and dimes. We have an input sensor to detect nickels (bit 0) and an input sensor to detect dimes (bit 1). We choose the wait time in each state to be 20ms, which is smaller than the time it takes the coin to pass by the sensor. Waiting in each state will debounce the sensor, preventing multiple counting of a single event. Notice that we wait in all states, because the sensor may bounce both on touch and release. Each state also has a function to execute. The function `soda` will trigger the Port B output so that a soda is dispensed. Similarly, the function `change` will trigger the Port B output so that a nickel is returned. The M states refer to the amount of collected money. When we are in a W state, we have collected that much money, but we're still waiting for the last coin to pass the sensor. For example, we start with no money in state `M0`. If we insert a dime, the input will go 10(in binary), and our state machine will jump to state `W10`. We will stay in state `W10` until the dime passes by the coin sensor. In particular, when the input goes to 00, we go to state `M10`. If we insert a second dime, the input will go 10, and our state machine will jump to state `W20`. Again, we will stay in state `W20` until this dime passes. When the input goes to 00, then we go to state `M20`. Now we call the function `change` and jump to state `M15`. Lastly, we call the function `soda` and jump back to state `M0`.

Figure 8.23
This Moore FSM implements a vending machine.

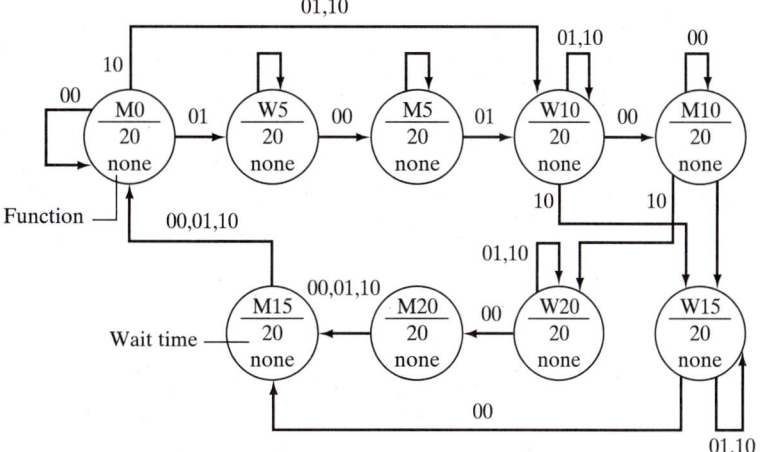

The main program, Program 8.52, begins by specifying the Port C bits 1 and 0 to be inputs. To implement this example on the 6812, you would have to set DDRB to $FF. The initial state is defined as MO. Our controller software first calls the function for this state, waits for the specified amount of time, reads the sensor inputs from PORTC, then switches to the next state depending on the input data. The Wait() function is defined previously in Program 8.48. Notice again the one-to-one correspondence between the state graph in Figure 8.23 and the fsm[9] data structure in Program 8.52.

Program 8.52
Vending machine controller.

```
/* Port C bits 1,0 are coin sensor inputs,
   Port B bits 1-0 are solenoid outputs */
void none(void){};
void soda(void){
  PORTB = 1;     /* activate soda solenoid */
  Wait(20000);   /* 10 msec */
  PORTB = 0;     /* deactivate solenoid */
}
void change(void){
  PORTB = 2;     /* activate change solenoid */
  Wait(20000);   /* 10 msec */
  PORTB = 0;     /* deactivate solenoid */
}
const struct State {
  void (*CmdPt)(void);           /* function to execute */
  unsigned short Time;           /* Time in msec to wait */
  const struct State *Next[3];}; /* Next if input=00,01,10*/
typedef const struct State StateType;
#define M0  &fsm[0]
#define W5  &fsm[1]
#define M5  &fsm[2]
#define W10 &fsm[3]
#define M10 &fsm[4]
```

continued

```
#define W15 &fsm[5]
#define M15 &fsm[6]
#define W20 &fsm[7]
#define M20 &fsm[8]
StateType fsm[9]={
  {&none,  40000U,{M0,W5,W10}},      /* M0 state */
  {&none,  40000U,{M5,W5,W5}},       /* W5 state */
  {&none,  40000U,{M5,W10,W15}},     /* M5 state */
  {&none,  40000U,{M10,W10,W10}},    /* W10 state */
  {&none,  40000U,{M10,W15,W20}},    /* M10 state */
  {&none,  40000U,{M15,W15,W15}},    /* W15 state */
  {&soda,  40000U,{M0,M0,M0}},       /* M15 state */
  {&none,  40000U,{M20,W20,W20}},    /* W20 state */
  {&change,40000U,{M15,M15,M15}}};   /* M20 state */
void main(void){ StateType *Pt; /* Current State */
  unsigned char Input;
  DDRC = 0xFC;  /* Port C bits 1,0 are inputs from sensors */
  Pt = M0;        /* Initial State */
  while(1){
    (*Pt->CmdPt)();         /* execute function */
    Wait(Pt->Time);         /* Time to wait in this state */
    Input = PORTC&0x03;   /* Input=00, 01, or 10 */
    Pt = Pt->Next[Input]; /* Move to next state */
  }
}
```

8.8 *6812 Paged Memory

The MC68HC812A4 and MC68HC912DG128 support paged memory. The extended address contains 22 bits and can access up to 4 Mbytes of memory. The top 8 bits are retrieved from the PPAGE register, and the bottom 14 bits come from the regular 16-bit address. Each of the 256 program pages is a fixed size of 16 Kbytes. To access a subroutine located in the paged memory, the `call` instruction is used. Figure 8.24 shows the stack before and after the **call** instruction is executed on the MC68HC812A4.

Figure 8.24
The `call` instruction is used to call a subroutine in paged memory.

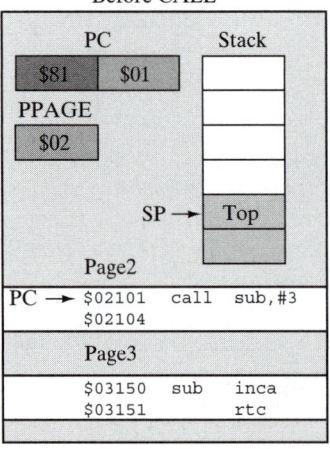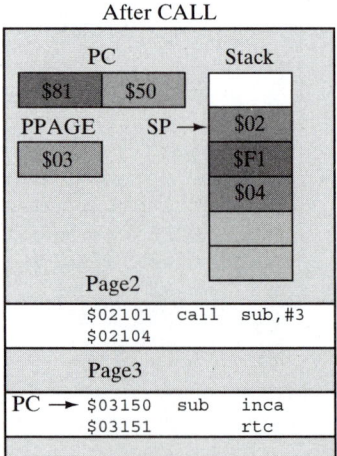

To activate the PPAGE system, you set the PWEN bit (bit 6) in the WINDEF register. If activated, any access within the $8000 to $BFFF will invoke the PPAGE hardware. When any data is fetched from the $8000 to $BFFF window, the 8-bit PPAGE is combined with the lower 14 bits of the PC to form a 22-bit address. The translation occurs automatically in hardware. Consider the case where the PPAGE register equals $07, and the PC is $8101.

```
PPAGE = $07 = 00000111
PC = $8101 = 1000000100000001
PPAGE + Lower 14 bits of PC = 00000111+00000100000001 = $01C101
```

The `rtc` instruction will return to the program that called the subroutine. Figure 8.25 shows the stack before and after execution of the `rtc` instruction.

Figure 8.25
The `rtc` instruction is used to return from a subroutine in paged memory.

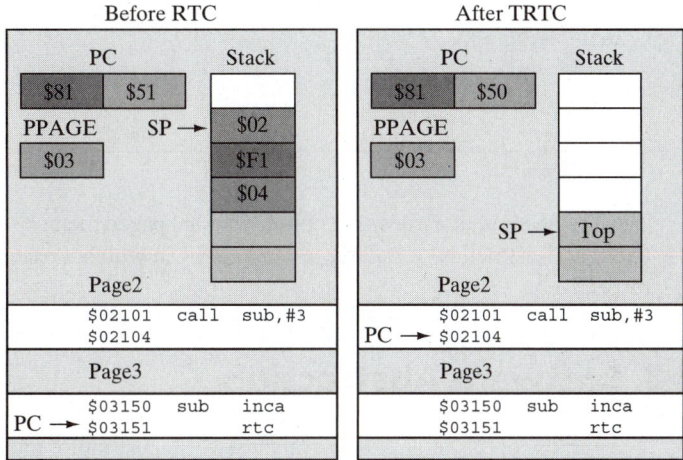

Programs 8.53, 8.54 and 8.55 illustrate the use of `call` and `rtc` to create a paged memory system on the 6812. Program 8.53 will be programmed into main EEPROM.

Program 8.53
Main memory programs for this paged memory system.

```
func1 equ  0          ; relative offset in paged memory
func2 equ  3          ; relative offset in paged memory
      org  $F000      ; main EEPROM memory
main  lds  #$0C00     ; stack in main RAM
      clra
loop  call func1,#1   ; call function 1 in page 1 (add 1)
      call func1,#2   ; call function 1 in page 2 (add 2)
      call func2,#1   ; call function 2 in page 1 (add 3)
      call func2,#2   ; call function 2 in page 2 (add 4)
      bra  loop
```

Program 8.54 will be programmed into external page 1.

Program 8.54
Page 1 programs for this
paged memory system.

```
        org   $0000        ; page 1 external memory
        lbra  fun1         ; link to actual function
        lbra  fun2         ; link to actual function
fun1    adda  #1
        rtc
fun2    adda  #2
        rtc
```

Program 8.55 will be programmed into external page 2.

Program 8.55
Page 2 programs for this
paged memory system.

```
        org   $0000        ; page 2 external memory
        lbra  fun1         ; link to actual function
        lbra  fun2         ; link to actual function
fun1    adda  #3
        rtc
fun2    adda  #4
        rtc
```

The **TExaS** simulator does not support external paged memory, but it will execute the `call` and `rtc` instructions similarly to the manner in which it executes `jsr rts` instructions respectively.

8.9 Tutorial 8. Software Abstraction

The purpose of this tutorial is to evaluate two stepper motor interfaces. `Chap8a.rtf` spins a stepper motor using the switch statement, similar to Program 8.23. `Chap8b.rtf` spins a stepper motor using a linked structure, similar to Program 8.48. You will first be asked to calculate the execution speed for each example. Then you will study its ease of modification by adding additional states to the system.

Action: Open and assemble the `switch` statement program `Chap8a.rtf`.

Question 8.1. What is the static efficiency of the `step` subroutine in the `Chap8a.rtf` system in ROM bytes?

Action: Run the `Chap8a.rtf` system and observe the stepper motor signals.

Question 8.2. Put a ScanPoint somewhere in the loop. Run the system and measure the minimum and maximum time (in cycles) to step the motor.

Question 8.3. Add four more output values to implement half-stepping. The new sequence should be `0x05,0x04,0x06,0x02,0x0A,0x08,0x09,0x01`.

Question 8.4. What is the static efficiency of the new system? Also, measure the minimum and maximum time (in cycles) to step the motor.

Action: Open and assemble the linked-structure program `Chap8b.rtf`.

Question 8.5. What is the static efficiency of the linked structure and the `step` subroutine in the `Chap8b.rtf` system in ROM bytes?

Action: Run the `Chap8b.rtf` system and observe the stepper motor signals.

Question 8.6. Put a ScanPoint somewhere in the loop. Run the system and measure the minimum and maximum time (in cycles) to step the motor.

Question 8.7. Add four more output values to implement half-stepping. The new sequence should be `0x05,0x04,0x06,0x02,0x0A,0x08,0x09,0x01`.

Question 8.8. What is the static efficiency of the new system? Also, measure the minimum and maximum time (in cycles) to step the motor.

8.10   Homework Problems

Homework 8.1: Write assembly code that finds the average value of a 10-element array. The two parameters are passed by reference on the stack. Local variables must be allocated on the stack.

```
void average(unsigned short *pt, unsigned short *ave){
unsigned short sum,n;
  sum = 0;
  for(n=0;n<10;n++){
    sum = sum+(*pt);
    pt = pt+1;
  }
  (*ave) = sum/10;
}
```

A typical calling sequence is

```
ldx   #mydata   ; pointer to 10-element structure
pshx
ldx   #myave    ; pointer to result
pshx
jsr   average
pulx            ; balance stack
pulx
```

Homework 8.2: Write assembly code that calculates the average of three numbers. The three parameters are passed by value on the stack. The return parameter is passed back in Register D. Local variables must be allocated on the stack.

```
short average(short data1, short data2, short data3){ short sum;
  sum = data1+data2+data3;
  return = sum/3;
}
```

A typical calling sequence is

```
ldx   var1   ; first parameter
pshx
ldx   var2   ; second parameter
pshx
ldx   var3   ; third parameter
pshx
jsr   average
std   var4   ; var4=(var1+var2+var3)/3
pulx         ; balance stack
pulx
pulx
```

Homework 8.3: Write assembly code that finds the maximum value of a 10-element array. The two parameters are passed by reference on the stack. Local variables must be allocated on the stack.

```
void max(unsigned short *pt, unsigned short *result){
unsigned short n;
  (*result) = 0;
  for(n=0;n<10;n++){
    if((*result)<(*pt))
      (*result) = (*pt);
    pt = pt+1;
  }
}
```

A typical calling sequence is

```
ldx   #mydata      ; pointer to 10-element structure
pshx
ldx   #myresult    ; pointer to result
pshx
jsr   max
pulx               ; balance stack
pulx
```

Homework 8.4: Write assembly code that implements a median filter. The three parameters are passed by value on the stack. The return parameter is passed back in Register D. Local variables must be allocated on the stack.

```
short median(short data1, short data2, short data3){ short temp;
  if(data1 > data2){
    temp = data1; data1 = data2; data2 = temp; // switch
  }
  if(data1 > data3){
    temp = data1; data1 = data3; data3 = temp; // switch
  }                    // data1 is now the smallest
  if(data2 < data3)
    return data2; // return the middle value
  else
    return data3
}
```

A typical calling sequence is

```
ldx   var1   ; first parameter
pshx
ldx   var2   ; second parameter
pshx
ldx   var3   ; third parameter
pshx
jsr   median
std   var4   ; var4=median(var1,var2,var3)
pulx         ; balance stack
pulx
pulx
```

Homework 8.5: Write assembly code that converts temperature in Fahrenheit to temperature in Centigrade. The input parameter is passed by value on the stack. The return parameter is passed back in Register D.

```
short FtoC(short tempF){
   return = (5*(tempF-32))/9;
}
```

A typical calling sequence is

```
ldx   myTempF  ; temperature in F
pshx
jsr   FtoC
std   myTempC  ; temperature in C
pulx           ; balance stack
```

Homework 8.6: Write assembly code that converts temperature in Centigrade to temperature in Fahrenheit. The input parameter is passed by value on the stack. The return parameter is passed back in Register D.

```
short CtoF(short tempC){
   return = (9*tempF)/5+32;
}
```

A typical calling sequence is

```
ldx   myTempC  ; temperature in C
pshx
jsr   CtoF
std   myTempF  ; temperature in F
pulx           ; balance stack
```

Homework 8.7: Write assembly code that finds the median of value of a five-element array. The input parameter is passed by reference on the stack. The return parameter is passed back in Register D. Local variables must be allocated on the stack. A typical calling sequence is

```
ldx   #mydata  ; pointer to 5-element structure
pshx
jsr   median
std   myMedian
pulx           ; balance stack
```

Homework 8.8: Write assembly code that calculates random numbers. The return parameter is passed back in Register D. Show all the code, including subroutines.

```
unsigned short Next;  // 16-bit
unsigned short random(void) {
  Next = Next*0x4E6D + 12345;
  return(Next);        // ignore overflow
}
```

Homework 8.9: Write assembly code that calculates random numbers. The return parameter is passed back in Register D. You may call any of the subroutines presented in this chapter.

```
unsigned long Next;  // 32-bit
unsigned short random(void) {
  Next = Next*0x1A504E6D + 123456789L;
  return(Next);   // return lower 16 bits
}
```

Homework 8.10: In this question, choose whether the *stack activity* is

- **MUST** must be performed because not doing so leads to an illegal stack operation
- **SHOULD** should be followed because it leads to clearer and easier-to-debug code
- **ILLEGAL** may not be performed because doing so leads to an illegal stack operation

Part A. Reading RAM memory at an address less than the SP.

Part B. Performing all the stack operations which decrement the stack pointer at the beginning of a subroutine and performing all the stack operations that increment the stack pointer at the end.

Part C. Writing to RAM memory at an address greater than the SP.

Part D. Using `set` or `equ` to specify the stack relative position of local variables.

Part E. Having the same number of pulls as you have pushes.

Homework 8.11: Consider the concepts of local variables and promotion.

Part A. List the four stages of a local variable.

Part B. Write the explicit assembly code for the following C function. Place the corresponding C code into the comment field of the assembly program. Also identify the four stages of the local variables. For each line of C code, show the corresponding assembly, without optimization.

```
void fun(void){
  char small;      \\ creates an 8-bit signed local variable
  int large;       \\ creates a 16-bit signed local variable
  small = -100;
  large = small;   \\ 8-bit signed is promoted to 16-bit signed
}
```

Homework 8.12: Write a subroutine, called FUZZY, which performs the following input/output function. `In` and `Th` are inputs and `Out` is the result. All parameters are 8-bit unsigned integers.

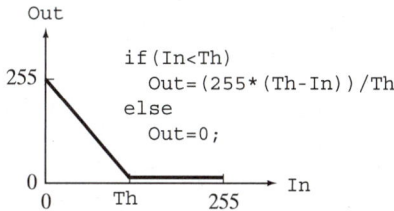

A typical calling shows the two inputs, `In` and `Th`, passed on the stack and the return parameter, `Out`, returned in Reg B.

```
ldaa #150    value for Th
psha         Th pushed on the stack
ldaa #90     value for In
psha         In on the stack
jsr  FUZZY   your function
ins
ins          pop off Th and In
* Reg B = (255*(150-90))/150 = 102
```

Homework 8.13: In this problem you will write an assembly language subroutine that outputs data to the following printer using a *busy-waiting handshake* protocol.

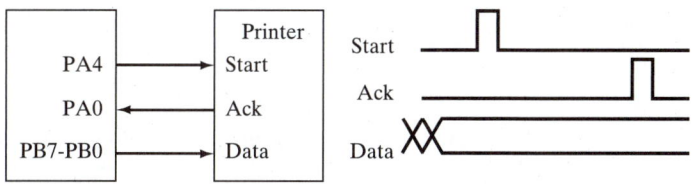

The following sequence will print one ASCII character:

1. The microcomputer puts the 8-bit ASCII on the Data lines
2. The microcomputer issues a Start pulse (does not matter how wide)
3. The microcomputer waits for the Ack pulse (Printer is done)

Part A. Show the subroutine that outputs a character. You may assume that the Ack pulse is larger than 10 μs. The 8-bit ASCII data to print is **passed by value on the stack**. An example calling sequence is

```
ldab #'V
pshb         The ASCII 'V' is pushed on the stack
jsr  Output
ins          The 'V' is discarded
```

Your solution should include the appropriate use of the `bset`, `bclr`, `brset`, and `brclr` instructions.

Part B. How long is your Start pulse? Explain your calculation.

Homework 8.14: Normally, we use the microcomputer hardware stack (RAM pointed to by RegS) to

1. Hold return addresses when we call subroutines, and software interrupts;
2. Store temporary data like our local variables; and
3. Pass data into functions.

In this problem, we will develop the concept of a separate *data stack* (LIFO) to pass parameters between our functions. We will still use the microcomputer hardware stack for the return addresses and local variables. We begin with a fixed allocation of the *data stack* area. In RAM we define:

```
Size    equ  10         Data stack size in bytes
Stack   rmb  Size
```

If we dedicate index RegY as the *data stack* pointer (rather than defining a global memory pointer), then the program execution speed will be improved at the expense of the inconvenience of dedicating RegY to be used for only this one purpose throughout all our programs.

The application we will solve involves the calculation of logic equations. We will define True to be the 8-bit value $FF and False to be 0. Logic values (true or false) will be stored on the *data stack*. For example, if False,False,True is pushed on the *data stack* (with True on top):

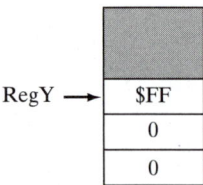

RegY always points to the top *data stack* entry. If another False is pushed on the *data stack*:

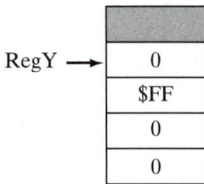

In all of the subroutines of this problem, you do not need to save/restore registers. Also, you do not need to worry about *data stack* underflow or overflow. Assume that we have global variables in RAM which represent logic variables—for example,

```
P    rmb  1     logic variable      True=$FF and False=0
Q    rmb  1
R    rmb  1
S    rmb  1
```

Part A. Show the assembly subroutine that initializes the *data stack*.

Part B. Write a subroutine Qpush, which pushes the value of the logic variable Q onto the *data stack.* Assume (but do not show) that there are similar routines Ppush, Rpush,Spush etc.

Part C. Write a subroutine popQ, which pops the value off the top of the *data stack* and stores it in the logic variable Q. Assume (again do not show) there are similar routines popP, popR,popS etc.

Part D. Write the subroutine NOT, which performs the *logical not* of the top *data stack* entry. This routine simply changes the value of the top of the *data stack* without performing any *data stack* pushes or pops.

Part E. Write the subroutine AND, which performs the *logical and* of the top two *data stack* entries. This routine pops two values off the top of the *data stack,* performs the function, and pushes the result back. Assume (again do not show) there are routines OR, NOR, NAND, XOR with similar parameter passing which perform their respective logic functions.

Part F. Now comes the fun part. In order to develop a really powerful software solution, we need a one-to-one correlation between the mathematical expression and the software implementation. To do this, we will develop the concept of a command sequence. A command sequence is simply a list of subroutines to execute in order. The programming of logic expressions involves two steps. The first step is to rewrite the logic expressions in **Reverse Polish Notion**. For example,

$$R = (P+Q) \bullet S'$$
$$S = R \oplus Q$$

can be written as

P Q or S not and StoreR R Q XOR StoreS

This is indeed a one-to-one translation and can be performed for any logic equation. The next step is specify the command sequence as an assembly language data structure. This variable-length vector contains 16-bit subroutine pointers and is terminated with a 0. Each subroutine passes data using the *data stack* (e.g., `CmdSeq fdb Ppush,Qpush,OR,Spush, NOT,AND,popR,Rpush,Qpush,XOR,popS,0`).

Write the assembly code that executes the logic expressions defined by the `CmdSeq` data structure. Full credit will be given to the proper usage of local variables.

Homework 8.15: Consider the reasons why one chooses a given technique to create a variable.

Part A. List three reasons why one would implement a variable using a register.

Part B. List three reasons why one would implement a variable on the stack and access it using RegX indexed mode addressing.

Part C. List three reasons why one would implement a variable in RAM and access it using direct or extended mode addressing.

Homework 8.16: Consider reasons for implementing "call by value" versus "call by reference."

Part A. List two reasons for implementing "call by value."

Part B. List two reasons for implementing "call by reference."

Homework 8.17: Convert the two subroutines shown in Programs 6.2 and 6.3 to software interrupts. Show the software interrupt handler and the code to initialize the software interrupt vector. The calling sequence for `Key_Init` would be

```
ldab   #0    means Key_Init
swi
```

The calling sequence for `Key_In` would be

```
ldab   #1    means Key_In
swi          Reg A returned with next input
```

Homework 8.18: Convert the two subroutines shown in Program 5.14 to software interrupts. Show the software interrupt handler and the code to initialize the software interrupt vector. The calling sequence for `ADC_Init` would be

```
ldaa   #0    means ADC_Init
swi
```

The calling sequence for `ADC_In` would be

```
     ldaa  #1    means ADC_In
 *  Reg B contains ADC channel to convert 0 to 7
     swi         Reg A returned with next ADC sample
```

Homework 8.19: Write C code that implements the operations shown in Program 8.46 on the 6811.

Homework 8.20: Write assembly code that implements the operations shown in Program 8.47 on the 6811.

Homework 8.21: Write assembly code that implements the stepper motor controller shown in Program 8.48.

Homework 8.22: Write assembly code that implements the traffic controller shown in Program 8.49.

Homework 8.23: Write assembly code that implements the vending machine shown in Program 8.52.

Homework 8.24: Assume we have 10-dimensional vectors, stored as 10-element arrays. For example, let the vector **X** equal $(x_0, x_1, \ldots x_9)$. Each value is a signed 8-bit integer. Write assembly code that finds the dot-product of two 10-element vectors.

$$\mathbf{X} \cdot \mathbf{Y} = x_0{*}y_0 + x_1{*}y_1 + \ldots + x_9{*}y_9$$

The two parameters are passed by reference using registers. The result is to be returned as a 16-bit signed value in Register D. Local variables must be allocated on the stack. A typical calling sequence is

```
ldx   #vector1   ; pointer to the first vector
ldy   #vector2   ; pointer to the second vector
jsr   DotProduct
```

Homework 8.25: Let P be the 16-bit unsigned period of a squarewave in cycles. Each cycle is 500 ns. Calculate the equivalent frequency, f, in Hz. In particular,

$$f = 2000000/P$$

The input is passed by value in Register D, and the result is also returned by value in Register D. One way to implement this on the 6811 using a full 32-bit divide is to divide 2000000 into 30*65536+33920. Plugging this value into the equation yields

$$f = (30*65536)/P + 33920/P$$

The first divide can be performed using the `fdiv` instruction, and the second can use the `idiv` instruction.

Homework 8.26: Using recursion, write a subroutine that calculates the Fibonacci function. In particular,

$$f ib(0) = 1$$
$$f ib(1) = 1$$
$$f ib(n) = fib(n-1)+fib(n-2) \text{ for } n>1$$

The input is passed by value in Register D, and the result is also returned by value in Register D.

8.11   Laboratory Problems

Lab 8.1
Analyzing the Code Generated by a C Cross-Compiler

Purpose   This lab has these major objectives:

- to evaluate the static and dynamic efficiency of software;
- to learn how to pass subroutine parameters on the stack
  By value pushing the value onto the stack
  By reference pushing a pointer onto the stack;
- to implement local variables on the stack or in registers;
- to understand how and why we promote numbers during arithmetic calculations;
- to understand how C programs are converted into assembly.

Description   In preparation for this assignment, study the source files `lab8_1.c` and `lab8_1.rtf` on the computer. **Fuzzy logic** is used to control complex systems that can not be described by linear differential equations. One part of a fuzzy logic controller is the **Input Membership** function. The input (called **Crisp Input**) comes from the physical device being controlled. It is a measured parameter like speed, position, temperature, pressure, force, or acceleration that describes the current state of the system. The Fuzzy Membership Value tells us if a condition (like too fast) is true or false. Unlike digital logic, where booleans are either true or false, fuzzy logic allows for partial truth, or shades of gray. Figure 8.26 shows a typical function that converts crisp input to a fuzzy logic variable. A fuzzy logic variable that describes an input is called an input membership set.

Figure 8.26
Fuzzification is the process of converting crisp input to input membership.

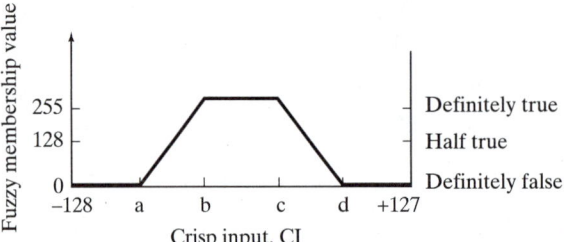

In this laboratory, there are four such membership functions, as shown in Figure 8.27. In each case the crisp input is temperature that can range from $-128°C$ to $+127°C$. The output of each function is a fuzzy membership value that specifies the level of truth for the four variables **Cold**, **Perfect**, **Warm**, and **Hot**.

Figure 8.27
The particular functions used in Lab 8.1.

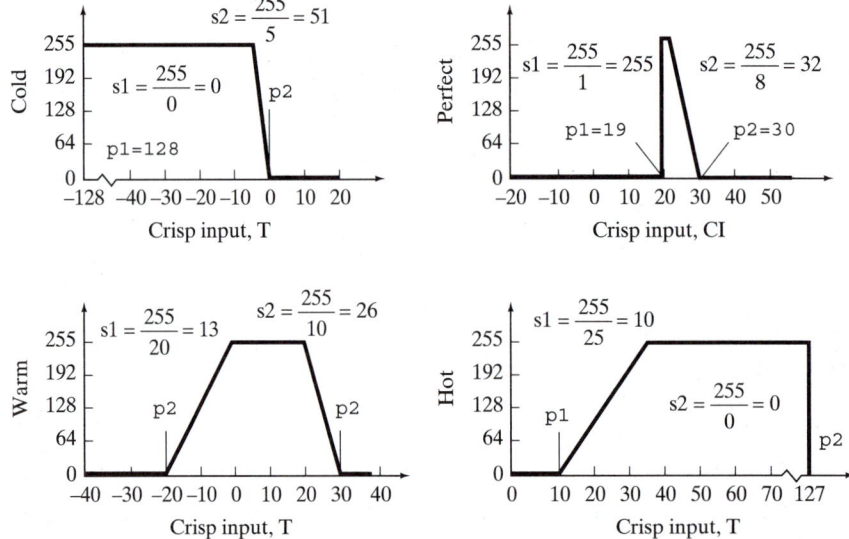

Most fuzzy logic membership functions can be represented by two points (p1 and p2) and two slopes (s1 and s2). A slope of zero specifies a vertical line.

1) Assemble and Run the `lab8_1.rtf` program using the simulator. The simulator `TheLog.rtf` window should display the input and output values for the many calls to the subroutine `InputMember`. TCNT is a built-in unsigned 16-bit counter giving the number of cycles (500 ns each on the 6811, and 125 ns each on the 6812) since the hardware reset. The main program reads TCNT before and after calling `InputMember`. The difference is the dynamic efficiency (e.g., 1000 cycles is 500 μsec on the 6811 and 125 μsec on the 6812). The execution speed will depend on the input parameter.

2) Analyze `lab8_1.c` and `lab8_1.rtf`

> How are parameters passed?
> How are for-loops implemented?
> How are numbers promoted?
> How are local variables allocated, accessed and deallocated?

The `InputMember` subroutine that you are analyzing and optimizing in this lab is an important part of the fuzzy logic controller. (See Program 8.56.) It is not necessary at this point to understand fuzzy logic. Just think of the subroutine simply as a numerical calculation with five inputs and one output. The `InputMember` subroutine has five inputs (CI, p1, p2, s1, and s2), and one output. The p1, p2 inputs are signed 8-bit numbers that range from −128 to +127; the s1, s2 inputs are unsigned 8-bit numbers that range from 0 to 255, and the output is an unsigned 8-bit number that ranges from 0 to +255. The C code to implement and test the function, shown below, can be found in `lab8_1.c`. The ImageCraft ICC11/ICC12 compiler was used to convert the C into the assembly found in the file `lab8_1.rtf`.

Program 8.56   Fuzzy Logic C program for Lab 8.1.

```c
unsigned char Hot,Warm,Cold,Perfect;   // Fuzzy membership variables
const struct threasholds{
  char p1,p2;
  unsigned char s1,s2; }; // fuzzy logic membership parameters
typedef const struct threasholds threasholdType;
threasholdType FuzzyCold=    { -128,    0,    0, 51 };    // Cold
threasholdType FuzzyPerfect={   19,   30,  255, 32 };    // Perfect
threasholdType FuzzyWarm=    {  -20,   30,   13, 26 };    // Warm
threasholdType FuzzyHot=     {   10,  127,   10,  0 };    // Hot
unsigned char InputMembership(char , threasholdType * );
unsigned int First,Elasped;
#define TCNT *(unsigned short volatile *)(0x84)
define TSCR *(unsigned char volatile *)(0x86)
char T;
void main(void){
  TSCR=0x80;
  for( T=-36; T<=34; T+=2 ){
    Cold    = InputMembership(T,&FuzzyCold);
    First   = TCNT;
    Warm    = InputMembership(T,&FuzzyWarm);
    Elasped = TCNT-First;
    Hot     = InputMembership(T,&FuzzyHot);
    Perfect = InputMembership(T,&FuzzyPerfect);
  }
}
unsigned char InputMembership(char CI, threasholdType * pt){
// see Motorola CPU12 reference manual Section 9.4, Figure 9.5
unsigned char grade; int delta1,delta2; unsigned int grade1,grade2;
  delta1 =   CI   - pt->p1;
  delta2 = pt->p2 -   CI;
  if((delta1<0) || (delta2<0))
    return(0);
```

continued

```
    grade1= pt->s1 * delta1;
    grade2= pt->s2 * delta2;
    if((pt->s2==0) || (grade2>0xFF))
      grade=0xFF;
    else
      grade=grade2;
    if((pt->s1!=0) && (grade1<=0xFF))
      grade=grade1;
    return(grade);
}
```

Part A. Using the assembler, determine the static number of bytes of object code of the existing `InputMember` subroutine. Run the existing main program, and record the range (minimum and maximum) of the dynamic efficiencies (number of cycles to execute) of the existing `InputMember` subroutine. Is the dynamic efficiency dependent on the input parameter CI? Write two or three sentences explaining your answer. Put this answer in the comments for the software you write for Part B.

Part B. In this part you will modify the entire assembly software (`main` and `InputMember`) so that it conforms with the programming style introduced in class and the book. First add labels and use them in the operands in order to document the existing code. In particular, use the set pseudo-op to add binding for all parameters and all local variables. Add any comments that better explain what the software is doing. Second, change the `InputMember` code so there is one entry at the top and one exit point at the bottom. Make a printout of this documented `Lab8_1.rtf` file.

Part C. Without changing the main program, rewrite the subroutine, `InputMember`, in order to improve both its static and dynamic efficiency. This means that you cannot change the way `main` and `InputMember` pass parameters. In particular, your subroutine must work for the calculations of all four fuzzy membership functions. Measure the static and dynamic efficiency of your new program. Your revised subroutine should work for all four cases. In particular, you should get an answer of ±1 from the original results shown next. You may assume without checking that p1 is less than p2. Record the static and dynamic efficiency measurements in the comments of your program.

In the following log, a **ScanPoint** was placed after the calls to the four functions (_T is the crisp input):

```
_Cold=$FF _Elasped=83  _Hot=$00 _Perfect=$00 _T=-36 _Warm=$00
_Cold=$FF _Elasped=83  _Hot=$00 _Perfect=$00 _T=-34 _Warm=$00
_Cold=$FF _Elasped=83  _Hot=$00 _Perfect=$00 _T=-32 _Warm=$00
_Cold=$FF _Elasped=83  _Hot=$00 _Perfect=$00 _T=-30 _Warm=$00
_Cold=$FF _Elasped=83  _Hot=$00 _Perfect=$00 _T=-28 _Warm=$00
_Cold=$FF _Elasped=83  _Hot=$00 _Perfect=$00 _T=-26 _Warm=$00
_Cold=$FF _Elasped=83  _Hot=$00 _Perfect=$00 _T=-24 _Warm=$00
_Cold=$FF _Elasped=83  _Hot=$00 _Perfect=$00 _T=-22 _Warm=$00
_Cold=$FF _Elasped=169 _Hot=$00 _Perfect=$00 _T=-20 _Warm=$00
_Cold=$FF _Elasped=169 _Hot=$00 _Perfect=$00 _T=-18 _Warm=$1A
_Cold=$FF _Elasped=169 _Hot=$00 _Perfect=$00 _T=-16 _Warm=$34
_Cold=$FF _Elasped=169 _Hot=$00 _Perfect=$00 _T=-14 _Warm=$4E
```

```
_Cold=$FF _Elasped=169 _Hot=$00 _Perfect=$00 _T=-12 _Warm=$68
_Cold=$FF _Elasped=169 _Hot=$00 _Perfect=$00 _T=-10 _Warm=$82
_Cold=$FF _Elasped=169 _Hot=$00 _Perfect=$00 _T=-8 _Warm=$9C
_Cold=$FF _Elasped=169 _Hot=$00 _Perfect=$00 _T=-6 _Warm=$B6
_Cold=$CC _Elasped=169 _Hot=$00 _Perfect=$00 _T=-4 _Warm=$D0
_Cold=$66 _Elasped=169 _Hot=$00 _Perfect=$00 _T=-2 _Warm=$EA
_Cold=$00 _Elasped=166 _Hot=$00 _Perfect=$00 _T=+0 _Warm=$FF
_Cold=$00 _Elasped=166 _Hot=$00 _Perfect=$00 _T=+2 _Warm=$FF
_Cold=$00 _Elasped=166 _Hot=$00 _Perfect=$00 _T=+4 _Warm=$FF
_Cold=$00 _Elasped=166 _Hot=$00 _Perfect=$00 _T=+6 _Warm=$FF
_Cold=$00 _Elasped=166 _Hot=$00 _Perfect=$00 _T=+8 _Warm=$FF
_Cold=$00 _Elasped=166 _Hot=$00 _Perfect=$00 _T=+10 _Warm=$FF
_Cold=$00 _Elasped=166 _Hot=$14 _Perfect=$00 _T=+12 _Warm=$FF
_Cold=$00 _Elasped=166 _Hot=$28 _Perfect=$00 _T=+14 _Warm=$FF
_Cold=$00 _Elasped=166 _Hot=$3C _Perfect=$00 _T=+16 _Warm=$FF
_Cold=$00 _Elasped=166 _Hot=$50 _Perfect=$00 _T=+18 _Warm=$FF
_Cold=$00 _Elasped=166 _Hot=$64 _Perfect=$FF _T=+20 _Warm=$FF
_Cold=$00 _Elasped=166 _Hot=$78 _Perfect=$FF _T=+22 _Warm=$D0
_Cold=$00 _Elasped=166 _Hot=$8C _Perfect=$C0 _T=+24 _Warm=$9C
_Cold=$00 _Elasped=166 _Hot=$A0 _Perfect=$80 _T=+26 _Warm=$68
_Cold=$00 _Elasped=166 _Hot=$B4 _Perfect=$40 _T=+28 _Warm=$34
_Cold=$00 _Elasped=166 _Hot=$C8 _Perfect=$00 _T=+30 _Warm=$00
_Cold=$00 _Elasped=87 _Hot=$DC _Perfect=$00 _T=+32 _Warm=$00
_Cold=$00 _Elasped=87 _Hot=$F0 _Perfect=$00 _T=+34 _Warm=$00
```

The 6811 version will yield the same data but have a larger `_Elasped` time.

Lab 8.2
Traffic Light
Controller

This lab has these major objectives:

- The usage of linked list data structures;
- Creation of a segmented software system;
- Real-time synchronization by design of an input-directed traffic light controller.

In preparation for this assignment, review finite state machines, linked lists, and memory allocation. You should also run and analyze the linked list controllers found in example files `moore.rtf` and `mealy.rtf`.

Description You will create a segmented software system putting global variables into RAM, local variables into RAM, constants and changeable data structures into EEPROM, and program object code into ROM. For computers with no ROM, you will pretend some of the EEPROM is ROM, as shown in Table 8.5. The reset vector $FFFE-$FFFF will also be in ROM.

Table 8.5
Memory allocation for various microcomputers

	MC68HC11A8	MC68HC711E9	MC68HC812A4	MC68HC912B32
globals	$0000-$007F	$0000-$00FF	$0800-$0AFF	$0800-$0AFF
stack	$0080-$00FF	$0100-$01FF	$0B00-$0C00	$0B00-$0C00
constants	$B600-$B7FF	$B600-$B7FF	$FE00-$FEFF	$0D00-$0FFF
programs	$E000-$FF00	$D000-$FF00	$F000-$FF00	$8000-$FF00
vectors	$FFD0-$FFFF	$FFD0-$FFFF	$FFD0-$FFFF	$FFD0-$FFFF

As you have experienced, the simulator requires more actual time to simulate one cycle of the microcomputer. On the other hand, the correct simulation time is maintained in the

TCNT register, which is incremented every cycle of simulation time. The simulator speed depends on the amount of information it needs to update into the windows. Unfortunately, even with the least amount of window updates, it would take too long for the simulator to process the typical three minutes it might take for a "real" car to pass through a "real" traffic intersection. Consequently, the cars in this traffic intersection travel much faster than "real" cars.

Both the 6811 and 6812 has a rich set of timer functions. The *input capture* feature can be used to measure the period, frequency, pulse width or phase of external signals. The *output compare* feature can be used to create squarewaves, pulses, or precise time delays. Remember to set TSCR equal to $80 on the 6812 to activate the TCNT operation. For this lab, you will design a subroutine that executes in 1 ms of simulated time. A *for-loop* may not be used; rather, you will read the value in TCNT, add a constant, then wait until TCNT goes beyond it, as in the following code:

```
*This method assumes the delay in cycles is less than 32767
delay equ  8000       8000 cycle delay
      ldd  TCNT        Current time
      addd #delay      EndT=TCNT value at the end of the delay
Wait  cpd  TCNT        EndT-TCNT<0 when EndT<Tcnt
      bpl  Wait
```

In general, cycle-counting (simple for loops) has the problem of conditional branches and data-dependent execution times. If an interrupt were to occur during a cycle counting delay, then the delay would be inaccurate using the cycle-counting method. In the preceding method, however, the timing will be very accurate, even if an interrupt were to occur while the microcomputer was waiting. In more sophisticated systems, other timer modes provide even more flexible mechanisms for microcomputer synchronization. These techniques will be presented in Chapter 11. A linked list solution may not run the fastest, or occupy the fewest memory bytes, but it is a structured technique that is easy to understand, easy to implement, easy to debug, and easy to upgrade.

Consider a typical four-corner intersection as shown in Figure 8.28. There are two one-way streets labeled South (cars travel North) and West (cars travel East). There are three inputs to your 6812: Two are car sensors, and one is a walk button. The *South* sensor will be true (1) if one or more cars are near the South intersection. Similarly, the *West* sensor will be true (1) if one or more cars are near the West intersection. The *Walk* button will be true (1) if a pedestrian wishes to cross in any direction. There are eight outputs from your microcomputer that control the two Red/Yellow/Green traffic lights, and the two walk/don't lights. The simulator allows you to attach binary switches to simulate the three inputs and LED lights to simulate the eight outputs.

Figure 8.28
Traffic Light Intersection.

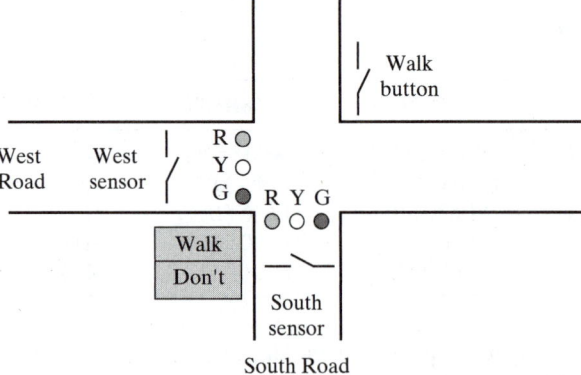

Traffic should not be allowed to crash (i.e., there should not be a green or yellow on South at the same time there is a green or yellow on West). You should exercise common sense when assigning the length of time that the traffic light will spend in each state, so that the simulated system changes at a speed convenient for the teaching assistant (TA). (Stuff changes fast enough so that the TA doesn't get bored, but not so fast that the TA can't see what is happening.) Cars should not be allowed to hit the pedestrians. The walk sequence should be realistic (*walk*, flashing *don't*, continuous *don't*). Your system should minimize both the average and worst case waiting time.

Part A. Build an I/O system with the appropriate names and colors on the lights and switches.

Part B. Design a finite state machine that implements a good traffic light system. Include a graphical picture of your finite state machine showing the various states, inputs, outputs, wait times and transitions.

Part C. Write the assembly code that implements the traffic light control system. There is no single, "best" way to implement your traffic light. However, your scheme must be segmented into RAM/EEPROM/ROM and you must use a linked- data structure. There should be a 1-1 mapping from the FSM states and the linked elements. A "good" solution has about 10 to 20 states in the finite state machine, and provides for input dependence. Your software will be graded on the Number of Accidents, Maximum Wait Time, and Average Wait Time. For example, if there are no cars currently on the roads and a new car approaches a red light, then the lights should change quickly to allow this car to proceed. On the other hand, if there are many cars going North/South and one car approaches East/West, it may not be efficient to quickly change the lights.

Typically in real applications using an embedded system, we put the executable instructions into the ROM. We then ask Motorola to make us thousands of microcomputers with our executable program in ROM. We then program the finite state machine linked-list data structure into the nonvolatile EEPROM. A good implementation will allow minor changes to the finite machine (adding states, modifying times, removing states, moving transition arrows, changing the initial state) simply by changing the linked datastructure (easy to change EEPROM in a real microcomputer), without changing the executable instructions (impossible to change the ROM of a real microcomputer). Obviously, if we add another input sensor or output light, it may be necessary to update the executable part of the software and re-assemble. *Hint: can you change the initial state of your FSM without modifying the ROM?*

Lab 8.3 Bubble Sort

Purpose This lab has these major objectives:

- to evaluate the static and dynamic efficiency of software;
- to learn how to pass subroutine parameters on the stack
  By value pushing the value onto the stack
  By reference pushing a pointer onto the stack;
- to implement local variables on the stack;
- to study the Bubble Sort algorithm.

Description

Part A. Write an assembly subroutine that implements the Bubble Sort. The input parameters are passed on the stack. Local variables must be allocated on the stack. The buffer size (Count) is 1 to 255.

```
void Bubble(char *Buffer, unsigned char Count){
// Count is the size of the byte array Buffer[i]
unsigned char i,j;    /* Indexes into Buffer */
char temp;            /* Used for exchange */
  for(i=1;i<Count;i++){
    for(j=Count-1;j>=i;j--){
      if(Buffer[j-1]>Buffer[j]){
        temp = Buffer[j-1];    /* Exchange */
        Buffer[j-1] = Buffer[j];
        Buffer[j] = temp;
      }
    }
  }
}
```

A typical calling sequence is

```
ldx   #mydata  ; pointer to 20-byte structure (call by reference)
pshx
ldaa #20       ; Count (call by value)
psha
jsr  Bubble
pula           ; balance stack
pulx
```

Part B. Use this simple assembly code to debug your Bubble Sort algorithm:

```
       org  $0800   ;for the 6811, change to 0
mydata rmb  5
main   lds  #$0C00  ;for 6811, change to $00FF
       ldaa #$35
       staa mydata  ; initialize
       ldd  #$3433
       std  mydata+1
       ldd  #$3231
       std  mydata+3 ; mydata[]={'5','4','3','2','1'}
       ldx  #mydata  ; pointer to 5-byte structure (call by reference)
       pshx
       ldab #5       ; Count (call by value)
       pshb
       jsr  Bubble
       ins           ; balance stack
       pulx
       stop
```

Part C. Write assembly code that tests the Bubble Sort algorithm. Copy and paste the SCI device driver software from `tut2.rtf`. This main program will input an ASCII string from a SCI-CRT interface (call `SCI_InString`), calculate its length, call the bubble sort subroutine, and output the sorted string on the SCI-CRT (call `SCI_OutString`).

Part D. Add debugging code to the test software in Part C that measures the elapsed execution time for the sort subroutine. Plot the execution time versus buffer size (using worst case initial data), for buffer sizes 10, 20, 30, and 40 bytes. Fit this data to a quadratic equation to derive a general solution for all sizes.

Lab 8.4
Heap Sort

Purpose This lab has these major objectives:

- to evaluate the static and dynamic efficiency of software;
- to learn how to pass subroutine parameters on the stack
  By value pushing the value onto the stack
  By reference pushing a pointer onto the stack;
- to implement local variables on the stack;
- to study the Heap Sort algorithm.

Description

Part A. Write assembly code that implements the Heap Sort. The input parameters are passed on the stack. Local variables must be allocated on the stack. The buffer size (Count) is 1 to 255.

```
void HeapSort(char *Buffer, unsigned char Count){
// Count is the size of the byte array Buffer[i]
  unsigned char i,j;   // used when sifting
  unsigned char ir;
  unsigned char m;     // used in the hiring phase
  char z;              // temporary, used to sort
  m = (Count>>1)+1;    // initial value Count/2+1
  ir = Count;
  for(;;){
    if(m > 1){
      --m;
      z = Buffer[m];        // still hiring
    }
    else{                          // in retirement and promotion
      z = Buffer[ir];              // clear space at end
      Buffer[ir] = Buffer[1]; // Retire top of heap into it
      if(--ir == 1){              // Done with last promotion?
        Buffer[1] = z;            // least competent worker of all
        break;
      }
    }
    i = m;      // whether in the hiring or promotion phase
    j = m+m;    //  we set up to sift down element z to
    while(j <= ir ){    // its proper level
      if((j<ir) && (Buffer[j]<Buffer[j+1])){
        j++;   // compare to underling
      }
      if(z<Buffer[j]){    // demote z
        Buffer[i] = Buffer[j];
        i = j;
        j <<= 1;
      }
```

```
      else{
         j = ir+1;    // this is z's level, set j to terminate
      }
   }
   Buffer[i] =z ;  // out z into its slot
  }
}
```

A typical calling sequence is

```
ldx   #mydata   ; pointer to 25-byte structure (call by reference)
pshx
ldaa  #25       ; Count (call by value)
psha
jsr   HeapSort
pula            ; balance stack
pulx
```

Part B. Use this simple assembly code to debug your Heap Sort algorithm.

```
        org   $0800    ;for the 6811, change to 0
mydata  rmb   5
main    lds   #$0C00   ;for the 6811, change to $00FF
        ldaa  #$35
        staa  mydata   ; initialize
        ldd   #$3433
        std   mydata+1
        ldd   #$3231
        std   mydata+3 ; mydata[]={'5','4','3','2','1'}
        ldx   #mydata  ; pointer to 5-byte structure (call by
reference)
        pshx
        ldab  #5       ; Count (call by value)
        pshb
        jsr   HeapSort
        ins            ; balance stack
        pulx
        stop
```

Part C. Write assembly code that tests the Heap Sort algorithm. Copy and paste the SCI device driver software from `tut2.rtf`. This main program will input an ASCII string from a SCI-CRT interface (call `SCI_InString`), calculate its length, call the heap sort subroutine, and output the sorted string on the SCI-CRT (call `SCI_OutString`).

Part D. Add debugging code to the test software in Part C that measures the elasped execution time for the sort subroutine. Plot the execution time versus buffer size (using worst case initial data), for buffer sizes 10, 20, 30, and 40 bytes. Fit this data to a quadratic equation to derive a general solution for all sizes.

9 Advanced Programming Techniques

Chapter 9 objectives are to

❑ develop techniques for developing modular software,
❑ introduce layered software design,
❑ present software style guidelines,
❑ define relocatable code.

In the previous chapters, basic building blocks were presented. In this chapter, we will discuss how to organize these blocks in an effective manner. The ultimate success of an embedded system project depends both on its software and hardware. Computer scientists pride themselves in their ability to develop quality software. Similarly, electrical engineers are well trained in the processes to design both digital and analog electronics. Manufacturers, in an attempt to get designers to use their products, provide application notes for their hardware devices. The main objective of this book is to combine effective design processes together with practical software techniques in order to develop quality embedded systems. As the size and especially the complexity of the software increase, the software development changes from simple "coding" to "software engineering," and the required skills also vary along this spectrum. These software skills include modular design, layered architecture, abstraction, and verification. Real-time embedded systems are usually on the small end of the size scale, but nevertheless these systems can be quite complex. Therefore, the aforementioned skills are essential for developing embedded systems. Writing good software is an art that must be developed, and can not be added on at the end of a project. Just like any other discipline (e.g., music, art, science, and religion), expertise comes from a combination of study and practice. The watchful eye of a good mentor can be invaluable, so take the risk and show your software to others, inviting praise and criticism. Good software, combined with average hardware, will always outperform average software on good hardware. In this chapter we will outline various techniques for developing quality software.

9.1 Modular Design

In this section we introduce the concept of modular programming and demonstrate that it is an effective way to organize our software projects. There are three reasons for forming modules. Functional abstraction allows us to reuse a software module from multiple locations. Complexity abstraction allows us to divide a highly complex system into smaller, less complicated components. The third reason is portability. If we create modules for I/O devices, then we can isolate the rest of the system from hardware details. This approach will be presented later, in the section on layered software.

The key to completing any complex task is to break it down into manageable subtasks. **Modular programming** is a style of software development that divides the software problem into distinct modules. The parts are as small as possible, yet relatively independent. Complex systems designed in a modular fashion are easier to debug because each module can be tested separately. Industry experts estimate that 50 to 90% of software development cost is spent in maintenance. All five aspects of software maintenance—

- correcting mistakes,
- adding new features,
- optimizing for execution speed or program size,
- porting to new computers or operating systems, and
- reconfiguring the software to solve a similar related program

—are simplified by organizing the software system into modules. The approach is particularly useful when a task is large enough to require several programmers.

A **program module** is a self-contained software task with clear entry and exit points. There is a distinct difference between a module and the assembly language subroutine or C language function. A module can be a collection of subroutines or functions that in its entirety performs a well-defined set of tasks. A collection of 32-bit math operations is an example of a module. The device driver for the SCI port is another example of a module. Modular programming involves both the specification of the individual modules and the connection scheme whereby the modules are connected together to form the software system. While the module may be called from many locations throughout the software, there should be a well-defined **entry point**. In this book, we define public and private in a manner similar to C++. A **public** function is one that can be called from another module. A **private** function is one that can only be called from functions within the same module. We make the same public/private distinction for data structures as well. In C, the entry point of a module is defined in the header file, and is specified by a list of function prototypes for the public functions. Similarly in assembly, the entry point of a module is also a list of public subroutines that can be called.

> **Common error:** In many situations, the input parameters have a restricted range. It would be inefficient for the module and the calling routine both to check for valid input. On the other hand, an error may occur if neither checks for valid input.

An **exit point** is the ending point of a program module. The exit point of a subroutine is used to return to the calling routine. We need to be careful about exit points. It is important that the stack be properly balanced at all exit points. Similarly if the subroutine returns parameters, then all exit points should return parameters in an acceptable format. If the main program has an exit point, it either stops the program or returns to the debugger.

> **Common error:** It is an error if all the exit points of an assembly subroutine do not balance the stack and return parameters in the same way.

From a formal perspective, I/O devices are considered as global. This is because I/O devices reside permanently at fixed addresses. From a syntactic viewpoint, any module has access to any I/O device. In order to reduce the complexity of the system, we will restrict the number of modules that actually do access the I/O device. It will be important to clarify which modules have access to I/O devices and when they are allowed to access it. When more than one module accesses an I/O device, then it is important to develop ways to arbitrate (which module goes first if two or more want to access simultaneously) or synchronize (make a second module wait until the first is finished). These arbitration issues are presented in detail in Chapters 4 and 5 of *Embedded Microcomputer Systems: Real-Time Inferfacing*, by Jonathan Valvano, Brooks Cole Publishing, 2000.

Information hiding is similar to minimizing coupling. It is better to separate the mechanisms of software from its policies. We should separate what the function does (the relationship between its inputs and outputs) from how it does it. It is good to hide certain inner workings of a module, and simply interface with the other modules through the well-defined input/output parameters. For example, we could implement a variable-size buffer by maintaining the current byte count in a global variable, CNT. A good module will hide how CNT is implemented from its users. If the user wants to know how many bytes are in the buffer, it calls one of the buffer routines that returns the count. A badly written module will not hide CNT from its users. The user simply accesses the global variable CNT. If we update the buffer routines, making them faster or better, we might also have to update all the programs that access CNT. Object-oriented programming environments provide well-defined mechanisms to support information hiding. This separation of policies from mechanisms is discussed further in the section on layered software.

The *Keep It Simple Stupid* approach tries to generalize the problem so that it fits an abstract model. Unfortunately, the person who defines the software specifications may not understand the implications and alternatives. As a software developer, we always ask ourselves these questions:

"How important is this feature?"

"What if it worked this different way?"

Sometimes we can restate the problem to allow for a simpler (and possibly more powerful) solution.

9.1.2 Dividing a Software Task into Modules

The overall goal of modular programming is to enhance clarity. The smaller the task, the easier it will be to understand. Coupling is defined as the influence one module's behavior has on another module. In order to make modules more independent, we strive to minimize coupling. Obvious and appropriate examples of coupling are the input/output parameters explicitly passed from one module to another. On the other hand, information stored in shared global variables can be quite difficult to track. In a similar way, shared accesses to I/O ports can also introduce unnecessary complexity. Global variables cause coupling between modules that complicate the debugging process because now the modules may not be able to be separately tested. On the other hand, we must use global variables to pass information into and out of an interrupt service routine, and from one call to an interrupt service routine to the next call. Another problem specific to embedded systems is the need for fast execution, coupled with the limited support for local variables. On the 6811 it is possible but inefficient to implement local variables on the stack. Consequently, many 6811 programmers opt for the less elegant yet faster approach of global variables. When passing information through global variables is required, it is better to use a well-defined abstract technique like a FIFO queue. Assign a logically complete task to each module. The module is logically complete when it can be separated from the rest of the system and placed into another application. The interfaces are extremely important. The interfaces determine the policies of our modules. In other words, the interfaces define the operations of our software system. The interfaces also represent the coupling between modules. In general we wish to minimize the amount of information passing between the modules yet maximize the number of modules. Of the following three objectives when dividing a software project into subtasks, it is really only the first one that matters:

- make the software project easier to understand,
- increase the number of modules,
- decrease the interdependency (minimize coupling).

Checkpoint 9.1: List some examples of coupling.

9.1.3 Visualizing the System Using Call-Graphs

A **software module** is a collection of public functions, private functions and private global variables that together perform a complete task. A **call-graph** is a graphical representation of the organizational structure of the modules pieced together to construct a software system. In a call-graph, we specify software modules as ovals. Inside the oval live the functions and variables of that module. I/O devices are essential in all computers, but particularly relevant when developing software for an embedded system. It is appropriate to group I/O ports into **hardware modules**, which together perform a complete I/O task. Assembly and C implementations of the tut3 example are included on the CD. Figure 9.1 shows the high-level call-graph of this example.

Figure 9.1
A call-graph of the tut3 software.

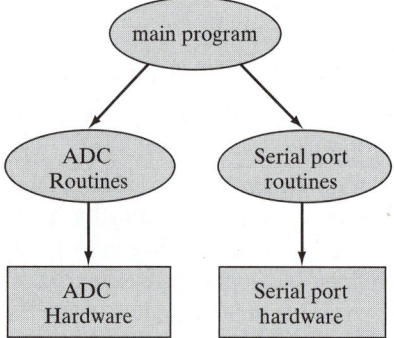

We can develop and connect modules in a hierarchical manner. Construct new modules by combining existing modules. The modules of the software project are organized using a call-graph. An arrow points from the calling routine to the module it calls. The I/O ports are organized into groups (e.g., all the Port C registers are in one group). This graph allows us to see the organization of the project. Figure 9.2 shows a call-graph for the simple example of outputting 8-bit data to Port C. `COpen` will initialize the interface. `CNib` transmits a 4-bit nibble. `COut` accepts an 8-bit unsigned byte and calls `CNib` twice to transmit the data. Figure 9.2 is an example of a detailed call-graph.

Observation: Recursive functions do not fall into this hierarchical calling structure.

Observation: If module A calls module B, and module B calls module A, then you have created a special situation that must account for these mutual calls.

Figure 9.2
A simple call-graph.

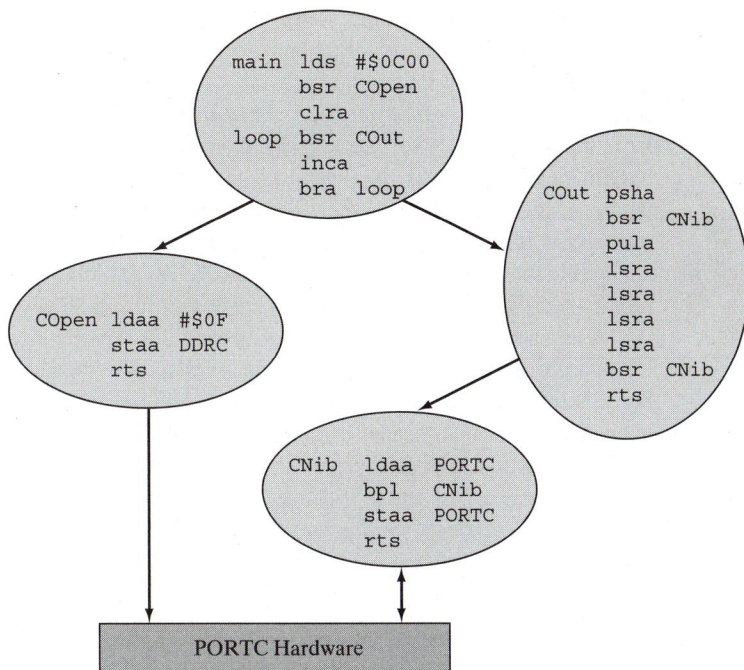

To make simpler call-graphs on large projects, we can combine multiple related sub-routines into a single module. The main program is at the top and the I/O ports are at the bottom. In a hierarchical system, the modules are organized both in a horizontal fashion (grouped together by function) and in a vertical fashion (overall policy decisions at the top and implementation details at the bottom). Since one of the advantages of breaking a large software project into subtasks is concurrent development, it makes sense to consider concurrency when dividing the tasks. In other words, the modules should be partitioned in such a way that multiple programmers can develop the subtasks as independently as possible. On the other hand, careful and constant supervision is required as modules are connected together and tested.

Figure 9.3 shows the call-graph for the **TExaS** tutorial `tut4`. This software system which contains four software modules and two hardware modules, includes the interrupting implementation of the serial I/O. There are assembly and C versions for both the 6811 and 6812. The SCI registers and the SCI interrupt vector together implement an asynchronous serial port on the 6811/6812. We specify hardware modules as rectangles. We should avoid public global variables, but, if you choose to use them, public global variables should be considered in a manner similar to I/O ports. Consequently, we will collect public global variables into groups and represent each group as a rectangle. Recall that private global variables are not accessible from other modules, so we hide them in the module and do not have to explicitly consider them in the overall structure of the software system.

A heavy arrow represents a software linkage, i.e., one software module calling another. We draw the tail of the arrow in the software module that initiates the call and we point the head of the arrow at the software module it calls. Including a header file in the implementation file of a module defines an arrow in the call-graph. The exception to this rule is including a header file that just contains constants and has no corresponding implementation file, e.g., `HC12.h`. For example, if we place an `#include "fileA.h"` statement in our `fileB.c` code, we create a call-graph arrow from module B to module A, because software in module B can call the public functions of module A. In a large complex system, we will add call-graph arrows for situations where it **can** call rather than where it **does** call. It is easier in a larger system to draw the *can-call* arrows than the *does-call* arrows, because we just have to look at the header files each code file includes. This approach will also simplify maintaining a call-graph during project phases where the software is being written, debugged, or upgraded. Changes to the list of header files included by a module are much less frequent than changes to the list of functions actually called. On the other hand, for an assembly language project every subroutine can call every other subroutine, so we will draw a does-call graph.

A thin arrow represents a hardware access; that is, the software reads from or write to an I/O port. Reading and writing public globals will also be specified using thin arrows. A double-headed arrow represents a software module that both reads and writes the hardware module or public global. A single-headed arrow from an oval to a rectangle signifies a write-only access to the hardware module or public global. A single-headed arrow from a rectangle to an oval signifies a read-only access to the hardware module or public global. Defining the specific access types allows us to identify problems such as conflict (two modules writing to the same I/O configuration registers), data consistency (e.g., the reader–writer problem), and reentrancy.

Figure 9.3
Simple call-graph of the `tut4` system.

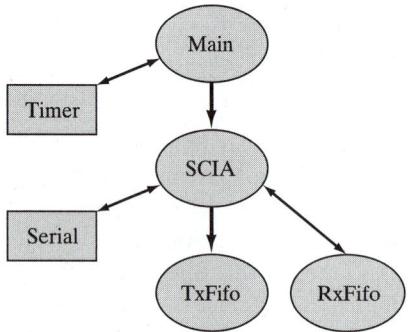

Checkpoint 9.2: Why are I/O devices considered as global?

Checkpoint 9.3: How can you implement a system that considers I/O devices as local?

9.1.4 Hierarchical Programming

Hierarchical programs have tree-structured call-graphs, like the system in Figure 9.3. Layered systems have call-graphs that group the modules into layers, such that the linkage arrows only go from a high level to a lower level or within the same level. A lower level module is not allowed to call a higher level. If at all possible, we should avoid cyclic graphs. A cycle in the call-graph will make testing difficult. Recall that we design top-down and test bottom-up. When there is a cycle in the call-graph, there is no place to start debugging.

There are two approaches to hierarchical programming. The top-down approach starts with a general overview, like an outline of a paper, and builds refinement into subsequent layers. A top-down programmer was once quoted as saying,

"Write no software until every detail is specified."

This guideline provides a better global approach to the problem. Managers like top-down because it gives them tighter control over their workers. The top-down approach works well when an existing operational system is being upgraded or rewritten. On the other hand, the bottom-up approach starts with the smallest detail, builds up the system "one brick at a time." The bottom-up approach provides a realistic appreciation of the problem because we often can not appreciate the difficulty or the simplicity of a problem until we have tried it. It allows programmers to start immediately coding, and gives programmers more input into the design. For example, a low-level programmer may be able to point out features that are not possible and suggest other features that are better. Some software projects are flawed from their conception. With bottom-up design, the obvious flaws surface early in the development cycle.

Bottom-up may be better when designing a complex system and specifications are open-ended. On the other hand, top-down is better when you have a very clear understanding of the problem specifications and the constraints of your computer system.

> ***Observation:*** *The TExaS simulator was actually written twice. The first pass was programmed bottom up and served only to provide a clear understanding of the problem and the features and limitations of my hardware. I literally threw all the source code in the trash, and programmed the second pass in a top down manner.*

9.1.5 Inappropriate I/O and Portability

One of the biggest mistakes beginning programmers make is the inappropriate usage of I/O calls (e.g., screen output and keyboard input). An explanation for their foolish behavior is that they haven't had the experience yet of trying to reuse software they have written for one project in another project. For example, assume you wrote and tested a function that found the median of three numbers. The goal of this Program 9.1 was to display the result, so you solved it with a function like the following:

Program 9.1   A median function that is not very portable.

```
unsigned int Median(unsigned int u1, unsigned int u2, unsigned int u3){
unsigned int result;
   printf("The inputs are %d, %d, %d.\n", u1, u2, u3);
   if(u1>u2)
     if(u2>u3) result=u2;      // u1>u2,u2>u3        u1>u2>u3
       else
          if(u1>u3) result=u3; // u1>u2,u3>u2,u1>u3 u1>u3>u2
          else      result=u1; // u1>u2,u3>u2,u3>u1 u3>u1>u2
   else
     if(u3>u2) result=u2;      // u2>u1,u3>u2        u3>u2>u1
       else
          if(u1>u3) result=u3; // u2>u1,u2>u3,u1>u3 u2>u1>u3
          else      result=u1; // u2>u1,u2>u3,u3>u1 u2>u3>u1
   printf("The median is %d.\n",result);
   return(result);;}
```

Software portability of this function is diminished because it is littered with `printf`. To use this function in another situation, you will almost certainly have to remove the `printf` statements. In general we avoid interactive I/O at the lowest levels of the hierarchy, rather return data and flags and let the higher level program do the interactive I/O. Often we add keyboard input and screen output calls when testing our software. It is important to remove I/O that is not directly necessary as part of the module function. You can then reuse these functions in situations where screen output is not available or appropriate. Obviously screen output is allowed if that is the purpose of the routine.

Common Error: Performing unnecessary I/O in a subroutine makes it harder to reuse at a later time.

Checkpoint 9.4: Rewrite Program 9.1 with a debugging switch that can add/remove the printf statements.

9.2   Layered Software Systems

As the size and complexity of our software systems increase, we learn to anticipate the changes that our software must undergo in the future. In particular, we can expect to re-design our system to run on new and more powerful hardware platforms. A similar expec-tation is that better algorithms may become available. The objective of this section is to use a layered software approach to facilitate these types of changes.

We can use the call-graph to visualize software layers. A module in a layer can call a module within the same layer or a module in a layer below it. Some layered systems restrict the calls only to modules within the same layer or module in the adjacent layer below. If we place all the modules that access the I/O hardware in the bottommost layer, we can call this layer a **hardware abstraction layer**. Each middle layer of modules only calls modules at the same or lower level, but not modules at a higher level. Usually the top layer consists of the main program. In a multi-threaded environment (e.g., Unix, Windows NT) there can be multiple main programs at the topmost level, but for now assume there is only one main program. The arrows in Figure 9.4 point from the calling module to the module it calls.

Figure 9.4
A layered approach to interfacing an IEEE488 parallel port printer.

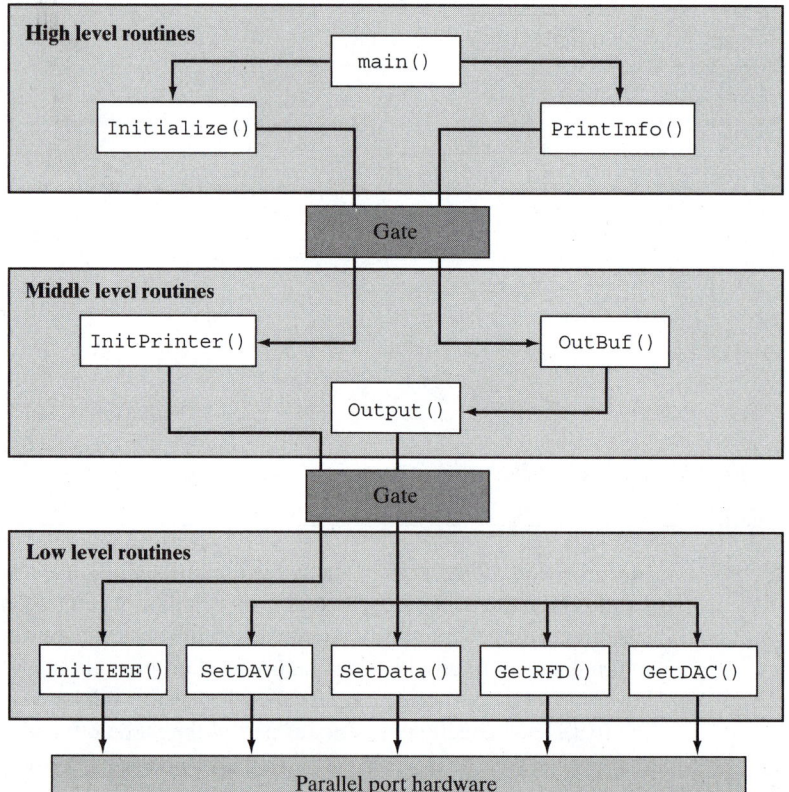

To develop a layered software system, we begin with a modular system. The main advantage of layered software is the ability to separate the modules into groups or layers such that one layer may be replaced without affecting the other layers. For example, you could change the ports to which the printer is connected by modifying the low level without any changes to the middle or high levels. Figure 9.4 depicts a layered implementation of a printer interface. In a similar way, you could replace the IEEE488 printer with a serial printer by replacing the bottom two layers. If we were to employ buffering and/or data compression to enhance communication bandwidth, then these algorithms would be added to the middle level. A layered system should allow you to change the implementation of one layer without requiring redesign of the other layers.

A *gate* is used to call from a higher to lower level routine. Another name for this gate is Application Program Interface or API. Unfortunately the 6805, 6806, 6811 and 6812 each have only one `swi` instruction (the Intel x86 and Motorola 680x0 both have multiple software interrupts.) If you wished to simulate multiple software interrupts, you could perform the trap instruction as illustrated in the `trap.rtf` example included with **TExaS**. The following rules apply to layered software systems:

1. A module may make a simple call to other modules in the same layer,
2. A module may make a call to a lower level module only by using the gate,
3. A module must use the gate only to access any function or variable in another layer,
4. A module may not call a higher level routine,
5. A module may not modify the vector address of another level's handler(s),
6. (optional) A module may not call farther down than the immediately adjacent lower level,
7. (optional) All I/O hardware access is grouped in the lowest level,
8. (optional) All user interface I/O is grouped in the highest level unless it is the purpose of the module itself to do such I/O.

The purpose of rule 6 is to prevent modifications at the low layer from affecting operation at the highest layer. On the other hand, for efficiency reasons you may wish to allow module calls further down than the immediately adjacent lower layer. To get the full advantage of layered software, it is critical to design functionally complete interfaces between the layers. The interface should support all current functions as well as provide for future expansions.

9.3 Naming Conventions

Choosing names for variables and functions involves creative thought, and is intimately connected to how we feel about ourselves as programmers. Of the policies presented in this section, naming conventions may be the hardest habit for us to break. The difficulty is that there are many conventions that satisfy the "easy to understand" objective. Good names reduce the need for documentation. Poor names promote confusion, ambiguity, and mistakes. Poor names can occur because code has been copied from a different situation and inserted into our system without proper integration (i.e., changing the names to be consistent with the new situation). They can also occur in the cluttered mind of a second-rate programmer, who hurries to deliver software before it is finished.

Names should have meaning. If we observe a name out of the context of the program in which it exists, the meaning of the object should be obvious. The object `TxFifo` is clearly the transmit first-in-first-out circular queue. The function `LCD_OutString` will output a string to the LCD display.

Avoid ambiguities. Don't use variable names in our system that are vague or have more than one meaning. For example, it is vague to use `temp`, because there are many possibilities for temporary data; in fact, it might even mean temperature. Don't use two names that look similar, but have different meanings.

Give hints about the type. We can further clarify the meaning of a variable by including phrases in the variable name that specify its type. For example, `dataPt`, `timePt`, and `putPt` are pointers. Similarly, `voltageBuf`, `timeBuf`, and `pressureBuf` are data buffers. Other good phrases include `Flag`, `Mode`, `U`, `L`, `Index`, and `Cnt`, which refer to boolean flag, system state, unsigned 16-bit, signed 32-bit, index into an array, and a counter respectively.

Use the same name to refer to the same type of object. For example, everywhere we need a local variable to store an ASCII character we could use the name `letter`. Another common example is to use the names `i`, `j`, and `k` for indices into arrays. The names `V1` and `R1` might refer to a voltage and a resistance. The exact correspondence is not part of the policies presented in this section, just the fact that a correspondence should exist. Once another programmer learns which names we use for which types of object, understanding our code becomes easier.

Use a prefix to identify public objects. An underline character will separate the module name from the function name. As an exception to this rule, we can use the underline to delimit words in all upper-case name (e.g., `#define MIN_PRESSURE 10`). Functions that can be accessed outside the scope of a module will begin with a prefix specifying the module to which it belongs. It is poor style to create public variables, but if they need to exist, they too would begin with the module prefix. The prefix matches the file name containing the object. For example, if we see a function call, `LCD_OutString("Hello world")`; we know the public function belongs to the LCD module, where the policies are defined in `LCD.h` and the implementation in `LCD.c`. Notice the similarity between this syntax (e.g., `LCD_init()`) and the corresponding syntax we would use if programming the module as a class in C++ (e.g., `LCD.init()`). Using this convention, we can easily distinguish between public and private objects.

Use upper and lower case to specify the scope of an object. We will define I/O ports and constants using no lower-case letters, like typing with caps-lock on. In other words, names without lower-case letters refer to objects with fixed values. `TRUE`, `FALSE`, and `NULL` are good examples of fixed-valued objects. As mentioned earlier, constant names formed from multiple words will use an underline character to delimit the individual words (e.g., `MAX_VOLTAGE UPPER_BOUND FIFO_SIZE`). Global objects will begin with a capital letter, but include some lower-case letters. Local variables will begin with a lower-case letter, and may or may not include upper case letters. Since all functions are global, we can start function names with either an upper-case or lower-case letter. Using this convention, we can distinguish constants, globals and locals.

> *An object's properties (public/private, local/global, constant/variable) are always perfectly clear at the place where the object is defined. The importance of the naming policy is to extend that clarity also to the places where the object is used.*

Use capitalization to delimit words. Names that combine multiple words into one should be defined using a capital letter to signify the first letter of each combined word. Recall that the case of the first letter specifies whether is local or global. Some programmers use the underline as a word-delimiter, but except for constants, we will reserve underline to separate the module name from the variable name. Table 9.1 overviews the naming convention presented in this section.

Table 9.1 Examples of names.

type	examples
constants	CR SAFE_TO_RUN PORTA STACK_SIZE START_OF_RAM
local variables	maxTemperature lastCharTyped errorCnt
private global variable	MaxTemperature LastCharTyped ErrorCnt
public global variable	DAC_MaxTemperature Key_LastCharTyped Network_ErrorCnt
private function	ClearTime wrapPointer InChar
public function	Timer_ClearTime RxFifo_Put Key_InChar

Checkpoint 9.5: How can you tell if a function is private or pubic?

Checkpoint 9.6: How can you tell if a variable is local or global?

9.4 Assembly Language Style Guidelines

The objective of this section is to present style rules for developing assembly language. This set of rules is meant to guide, not control. In other words, they serve as general guidelines rather than fundamental law.

The Single Entry Point is at the Top   In assembly language, we place a single entry point of a subroutine at the first line of the code. This guarantees that registers will be saved and local variables will be properly allocated on the stack. By default, C functions have a single entry point. Placing the entry point at the top provides a visual marker for the beginning of the subroutine.

The Single Exit Point is at the Bottom   Most programmers prefer to use a single exit point as the last line of the subroutine. Some programmers employ multiple exit points for efficiency reasons. In general, we must guarantee that the registers, stack, and return parameters are at a similar and consistent state for each exit point. In particular we must deallocate local variables properly. If you do employ multiple exit points, then you should develop a means to visually delineate where one subroutine ends and the next one starts. You could use one line of comments to signify the start of a subroutine and a different line of comments to show the end of it. Program 9.2 employs distinct visual markers to see the beginning and end of the subroutine.

Program 9.2
An assembly subroutine that uses comments to delineate its beginning and end.

```
;***************Abs***********************
;   Input: RegA is signed 8-bit
;   Output: Reg A is absolute value 0 to 127
Abs: tsta       ; already positive?
     bpl ok
     nega
ok rts
* -----------end of Abs ----------------
```

Observation: Using the first and last lines of a module as the entry and exit points makes it easier to debug, because it will be easy to place debugging instruments like breakpoints.

Common error: If you place a debugging breakpoint on the last `rts` of a subroutine with multiple exit points, then sometimes the subroutine will return without generating the break.

Write Structured Programs

A structured program is one that adheres to a strict list of program structures. When we program in C (with the exception of `goto`, which by the way you should never use), we are forced to write structured programs due to the syntax of the language. One technique for writing structured assembly language is to adhere to the same strict list of program structures available in C. In other words, restrict the assembly language branching to configurations that mimic the software behavior of `if`, `if-else`, `do-while`, `while`, `for`, and `switch`. Assembly language examples of these control structures are included with the **TExaS** simulator in the files `UIF.rtf`, `SIF.rtf`, `WHILE.rtf`, and `FOR.rtf`. Structured programs are much easier to debug, because execution proceeds only through a limited number of well-defined pathways. When we reuse existing assembly branching structures, then our debugging can focus more on the overall function and less on how the details are implemented.

The Registers Must Be Saved

On a software team it is important to establish a rule whether or not subroutines will save/restore registers. Establishing this convention is especially important when a mixture of assembly and C is being used, or if the software project remains active for long periods of time. It is safest to save and restore registers that are modified (most programmers do not save/restore the CCR) and output parameter(s) returned in a register. Exceptions to this rule can be made for those portions of the code where speed is most critical.

> **Common error:** *If the calling routine expects a subroutine to save/restore registers, and it doesn't, then information will be lost.*

> **Observation:** *If the calling routine does not expect a subroutine to save/restore registers, and it does, then the system executes a little slower and the object code is a little bigger than it need be.*

> **Common error:** *When a mixture of C and assembly language programs is integrated, an error may occur when the compiler is upgraded because there may be a change in whether registers are saved/restored or how parameters are passed.*

Use High-Level Languages Whenever Possible

It may seem odd to have a rule about high-level languages in a section about assembly language programming. It is even odder to make this statement in a book devoted to assembly language programming. In general, we should use high-level languages when memory space and execution speed are less important than portability and maintenance. When execution speed is important, you could write the first version in a high-level language, run a profiler (that will tell you which parts of your program are executed the most), then optimize the critical sections by writing them in assembly language. If a C language implementation just doesn't run fast enough, you could consider a more powerful compiler or a faster microcomputer.

> **Observation:** *High-level language programmers who are well acquainted with the underlying assembly language of the machine have a better understanding of how their machine and software work.*

Minimize Conditional Branching

Every time software makes a conditional branch, there are two possible outcomes that must be tested (branch or not branch). In the example shown in Program 9.3, assume we wish to add two 16-bit numbers: u3=u2+u1. A conditional branch could be avoided by solving the problem in another way.

Program 9.3
Sometimes we can remove a conditional branch and simplify the program.

```
; uses conditional branch
add16a ldaa u1+1    ;lsb
       adda u2+1
       staa u3+1
       ldaa u1      ;msb
       bcc noc
       inca         ;carry
noc    adda u2
       staa u3
       rts
; no conditional branch
add16b ldaa u1+1    ;lsb
       adda u2+1
       staa u3+1
       ldaa u1      ;msb
       adca u2
       staa u3
       rts
```

Observation: Software can be made easier to understand by reworking the approach in order to reduce the number of conditional branches.

9.5 *C Language Style Guidelines

9.5.1 Code File Structure, the *.c File

One of the recurring themes of this software style section is consistency. Maintaining a consistent style will help us locate and understand the different components of our software, as well as prevent us from forgetting to include a component—or worse, including it twice. The following regions should occur in this order in every code file (e.g., `file.c`).

Opening comments. The first line of every file should contain the file name, because some printers do not automatically print the name of the file. Remember that these opening comments will be duplicated in the corresponding header file (e.g., `file.h`) and are intended to be read by the client, the one who will use these programs. If major portions of this software are copied from copyrighted sources, then we must satisfy the copyright requirements of those sources. The rest of the opening comments should include

- the overall purpose of the software module,
- the names of the programmers,
- the creation (optional) and last update dates,
- the hardware/software configuration required to use the module, and
- any copyright information.

Included .h files. Next we will place the `#include` statements that add the necessary header files. Adding other code files, if necessary, will occur at the end of the file, but here at the top of the file we include just the header files. Normally the order doesn't matter, so we will list the include files in a hierarchical fashion starting with the lowest level and ending at the highest high. If the order of these statements is important, then write a comment describing both what the proper order is and why the order is important. Putting them together at the top will help us draw a call-graph, which will show us how our modules are connected. In particular, if we consider each code file to be a separate module, then the list of `#include` statements specifies which other modules can be called from this module. Of course, one header file is allowed to include other header files. In general, we should avoid having one header file include other header files. This restriction makes the organizational structure of the software system easier to observe. Be careful to include only those files that are absolutely necessary. Adding unnecessary `#include` statements will make our system seem more complex than it actually is.

extern references. After including the header files, we can declare any external variables or functions. External references will be resolved by the linker, when various modules are linked together to create a single executable application. Placing them together at the top of the file will help us see how this software system fits together (i.e., is linked to) other software systems.

#define statements. After external references, we should place the `#define` macros and `#define` constants. Since these definitions are located in the code file (e.g., `file.c`), they will be private. This means they are available within this file only. If the client does not need to use or change the macro or constant, then it should be made private by placing it here in the code file. Conversely, if we wish to create a public constant or macro, then we place it in the header file for this module.

struct union enum statements. After the define statements, we should create the necessary data structures using `struct` `union` and `enum`. Again, since these definitions are located in the code file (e.g., `file.c`), they will be private.

Global variables and constants. After the structure definitions, we should include the global variables and constants. If we specify the global as `static` then it will be private, and can be accessed only by programs in this file. If we do not specify the global as `static` then it will be public and can be accessed by any program. (That program defines it as `extern` and the linker will resolve the reference.) We put all the globals together before any function definitions to symbolize the fact that any function in this file has access to these globals. If we have a permanent variable that is only accessed by one function, then it should be defined as a static local. The **scope** of a variable includes all the software in the system that can access it. In general, we wish to minimize the scope of our data.

```
short publicGlobal; // accessible by any function via extern
static short privateGlobal; // accessible in this file only
void function(void){
static short veryPrivateGlobal; // accessible by this function only
}
```

Prototypes of private functions. After the globals, we should add any necessary prototypes. We can restrict access to private functions, like global variables, by defining them as `static`. Prototypes for the public functions will be included in the corresponding header file. In general, we will arrange the code implementations in a top-down fashion. Although it is not necessary, we will include the parameter names with all prototypes. Descriptive parameter names will help document the usage of the function. For example, which of the following prototypes is easier to understand?

```
static void plot(short, short);
static void plot(short time, short pressure);
```

Implementations of the functions. The heart of the implementation file will be, of course, the implementations. Again, private functions should be defined as static. The functions should be sequenced in a logical manner. The most typical sequence is top-down, meaning we begin with the highest level and finish with the lowest level. Another appropriate sequence mirrors the manner in which the functions will be used. For example, start with the initialization functions, followed by the operations, and end with the shutdown functions. For example:

open
input
output
close

Included .c files. At the end of the file, we will place the `#include` statements that add the necessary code files. If our compiler supports projects, then it is a good idea to take advantage of this feature. The project simplifies the management of large software systems by providing organizational structure to the software system. If we use projects, then including code files will be unnecessary, and hence should be avoided. If our compiler does not support projects, or if we are writing software for multiple compilers, then including code files allows a large software project to be constructed simply by compiling the file with the `main()` program in it. Including header files at the top of the file allows this module to access public variables and functions of the other module. On the other hand, including code files at the end of the file prevents this module from accessing private variables and functions of the other module.

If our compiler supports **assert()** functions, use them liberally. In particular, place them at the beginning of functions to test the validity of the input parameters. Place them after calculations to test the validity of the results. Place them inside loops to verify that indices and pointers are valid. There is a secondary benefit to using **assert()**. The **assert()** statements themselves provide documentation of the assumptions made by the programmer.

9.5.2
Header File
Structure, the *.h
File

Once again, maintaining a consistent style facilitates understanding and helps to avoid errors of omission. Definitions made in the header file will be public, i.e., accessible by all modules. As stated earlier, it is better to make global variables private rather than placing them in the header file. Similarly, we should avoid placing actual code in a header file.

There are two types of header files. The first type of header file has no corresponding code file. In other words, there is a `file.h`, but no `file.c`. In this type of header, we can list global constants and helper macros. Examples of global constants are I/O port addresses (e.g., `HC12.h`) and calibration coefficients. Debugging macros could be grouped together and placed in a `debug.h` file. We will not consider software in these types of header files as belonging to a particular module.

The second type of header file does have a corresponding code file. The two files (e.g., `file.h` and `file.c`) form a software module. In this type of header, we define the prototypes for the public functions of the module. The `file.h` file contains the policies (behavior or what it does) and the `file.c` file contains the mechanisms (functions or how the module works.) The following regions should occur in this order in every header file (e.g., `file.h`).

Opening comments. The first line of every file should contain the file name, because some printers do not automatically print the name of the file. Remember that these opening comments should be duplicated in the corresponding implementation file (e.g., `file.c`) and are intended to be read by the client, the one who will use these programs. We should repeat copyright information as appropriate. The rest of the opening comments should include

- the overall purpose of the software module,
- the names of the programmers,
- the creation (optional) and last update dates,
- the hardware/software configuration required to use the module, and
- any copyright information.

Including .h files. Nested includes in the header file should be avoided. As stated earlier, nested includes obscure the manner in which the modules are interconnected. On the other hand, an implementation file can include other header and implementation files.

`#define` *statements.* Public constants and macros are next. Special care is required to determine whether a definition should be made private or public. One approach to this question is to begin with everything defined as private, and then shift definitions into the public category only when deemed necessary for the client to access in order to use the module. If the parameter relates to what the module does or how to use the module, then it should probably be public. On the other hand, if it relates to how the module works or how it is implemented, it should probably be private.

`struct union enum` *statements.* The definitions of public structures allow the client software to create data structures specific for this module.

Global variables and constants. If at all possible, public global variables should be avoided. Public constants follow the same rules as public definitions. If the client must have access to a constant to use the module, then it could be placed in the header file.

Prototypes of public functions. The prototypes for the public functions are last. As we did in the implementation file, we will arrange the code implementations in a top-down fashion. Comments should be directed to the client, and these comments should clarify what the function does and how the function can be used.

9.5.3 Formatting

The rules set out in this subsection are not necessary for the program to compile or to run. Rather, the intent of the rules is to make the software easier to understand, easier to debug, and easier to change. Just like beginning an exercise program, these rules may be hard to follow at first, but the discipline will pay dividends in the future.

Make the software easy to read. I strongly object to hardcopy printouts of computer programs during the development phase of a project. At this time, there are frequent updates made by multiple members of the software development team. Because a hard-copy printout will be quickly obsolete, we should develop and debug software by observing it on the computer screen. In order to eliminate horizontal scrolling, no line of code should be more than 80 characters wide. If we do make hard-copy printouts of the software at the end of a project, this rule will result in a printout that is easy to read.

Indentation should be set at 2 spaces. When transporting code from one computer to another, the TAB settings may be different. So, what looks good on one computer may look ugly on another. For this reason, we should avoid TABs and use just spaces. Local variable definitions can go on the same line as the function definition, or in the first column on the next line.

Be consistent about where we put spaces. As in English punctuation, there should be no space before a comma or a semicolon, but there should be at least one space or a carriage return after a comma or a semicolon. There should be no space before or after open parentheses. Assignment and comparison operations should have a single space before and after the operation. One exception to the single-space rule is if there are multiple assignment statements, we can vertically line up the operators and values, for example,

```
data     = 1;
pressure = 100;
voltage  = 5;
```

Be consistent about where we put braces {}. Misplaced braces cause both syntactic and semantic errors, so it is critical to maintain a consistent style. Place the opening brace at the end of the line that opens the scope of the multistep statement. The only code that can go on the same line after an opening brace is a simple local variable declaration or a comment. Placing the open brace near the end of the line provides a visual clue that a new code block has started. Place the closing brace on a separate line to give a vertical separation showing the end of the multistep statement. For example, the horizontal placement of the close brace gives a visual clue that the following code is in a different block:

```
void main(void){ int i, j, k;
  j = 1;
  if(sub0(j)){
    for(i = 0; i < 6; i++){
      sub1(i);
    }
    k = sub2(i, j);
  }
  else{
    k = sub3();
  }
}
```

Use braces after all `if`, `else`, `for`, `do`, `while`, `case`, and `switch` commands, even if the block is a single command. This forces us to consider the scope of the block making it easier to read and easier to change. For example, assume we start the following code.

```
if(flag)
  n = 0;
```

Now we add a second statement that we want to execute also if the flag is true. The following error might occur if we just add the new statement:

```
if(flag)
  n = 0;
  c = 0;
```

If all of our blocks are enclosed with braces, we would have started with the following:

```
if(flag){
  n = 0;
}
```

Now, when we add a second statement, we get the correct software:

```
if(flag){
  n = 0;
  c = 0;
}
```

9.5.4
Code Structure

Make the presentation easy to read. We define presentation as the look and feel of our software as displayed on the screen. If at all possible, the size of our functions should be small enough so the majority of the code fits on a single computer screen. We must consider the presentation as a two-dimensional object. Consequently, we can reduce the 2-D *area* of our functions by encapsulating components and defining them as private functions, or by combining multiple statements on a single line. In the horizontal dimension, we are allowed to group multiple statements on a single line only if the collection makes sense. We should list multiple statements on a single line, if we can draw a circle around the statements and assign a simple collective explanation to the code.

Another consideration related to listing multiple statements on the same line is debugging. The compiler often places debugging information on each line of code. For example, the ICC11 and ICC12 compilers place a label specifying the starting address of the assembly code that implements the line. Breakpoints in some systems can only be placed at the beginning of a line.

Consider the next three presentations. Since the compiler generates exactly the same code in each case, the computer execution will be identical. Therefore, we will focus on the differences in style. The first example has a horrific style.

```
void testFilter(short start, short stop, short step){ short x,y;
  initFilter(); SCI_OutString("x(n) y(n)"); SCI_OutChar(CR);
  for(x=start;x<=stop; x=x+step){ y=filter(x); SCI_OutUDec(x);
  SCI_OutChar(SP); SCI_OutUDec(y); SCI_OutChar(CR);} }
```

The second example places each statement on a separate line. Although written in an adequate style, it is unnecessarily vertical.

```
void testFilter(short start, short stop, short step){
short x;
short y;
  initFilter();
  SCI_OutString("x(n) y(n)");
  SCI_OutChar(CR);
  for(x = start; x <= stop; x = x+step){
    y = filter(x);
    SCI_OutUDec(x);
    SCI_OutChar(SP);
    SCI_OutUDec(y);
    SCI_OutChar(CR);
  }
}
```

The last example groups the two variable definitions together because the collection can be considered as a single object. The variables are related to each other. Obviously, x and y are the same type (16-bit signed), but in a physical sense, they would have the same units. For example, if x represents a signal in mV, then y is also a signal in mV. Similarly, the SCI output sequences cause simple well-defined operations.

```
void testFilter(short start, short stop, short step){ short x, y;
  initFilter();
  SCI_OutString("x(n) y(n)"); SCI_OutChar(CR);
  for(x = start; x <= stop; x = x+step){
    y = filter(x);
    SCI_OutUDec(x); SCI_OutChar(SP); SCI_OutUDec(y); SCI_OutChar(CR);
  }
}
```

Observation: *Most professional programmers do not create hard-copy printouts of the software. Rather, they view software on the computer screen.*

The "make the presentation easy to read" guideline sometimes comes into conflict with the "be consistent where we place braces" guideline. For example, the following example is obviously easy to read, but violates the placement of brace rule:

```
for(i = 0; i < 6; i++) dataBuf[i] = 0;
```

When in doubt, we will always be consistent where we place the braces. The correct style is also easy to read.

```
for(i = 0; i < 6; i++){
  dataBuf[i] = 0;
}
```

Employ modular programming techniques. Complex functions should be broken into simple components, so that the details of the lower-level operations are hidden from the overall algorithms at the higher levels. An interesting question arises:

"Should a subfunction be defined if it will be called from only a single place?"

The answer to this question, in fact the answer to all questions about software quality, is yes if it makes the software easier to understand, easier to debug, and easier to change.

Minimize scope. In general, we hide the implementation of our software from its usage. The scope of a variable should be consistent with its usage. In a military sense, we ask the question, *"Which software has the need to know?"* Global variables should be used only when the lifetime of the data is permanent, or when data needs to be passed from one thread to another. Otherwise, we should use local variables. When one module calls another, we should pass data using the normal parameter-passing mechanisms. As mentioned earlier, we consider I/O ports in a manner similar to global variables. There is no syntactic mechanism to prevent a module from accessing an I/O port, since the ports are at fixed and known absolute addresses. The Intel Pentium does have a complex hardware system to prevent unauthorized software from accessing I/O ports, but the details are beyond the scope of this book. For our embedded system, therefore, we must rely on the **does-access** rather than the **can-access** method when it comes to I/O ports. In other words, we must have the discipline to restrict I/O port access only in the module that is designed to access it. For similar reasons, we should consider each interrupt vector address separately, grouping it with the corresponding I/O module. In particular, rather than having one long list of interrupt vectors for the entire system, each interrupt vector should be separately defined along with the software that supports the other I/O hardware of the module. For example, the serial port interrupt vector should be specified in the same file as the serial port interrupt handler.

Use types. Using a `typedef` will clarify the format of a variable. It is another example of the separation of mechanism and policy. New data types and structures will begin with an upper case letter. The `typedef` allows use to hide the representation of the object and use an abstract concept instead. For example

```
typedef short Temperature;
void main(void){ Temperature lowT, highT;
}
```

This allows us to change the representation of temperature without having to find all the temperature variables in our software. Not every data type requires a `typedef`. We will use types for those objects of fundamental importance to our software, and for those objects for which a change in implementation is anticipated. As always, the goal is to clarify. If it doesn't make it easier to understand, easier to debug, or easier to change, don't do it.

Prototype all functions. Public functions obviously require a prototype in the header file. In the implementation file, we will organize the software in a top-down hierarchical fashion. Since the highest level functions go first, prototypes for the lower-level private functions will be required. Grouping the low-level prototypes at the top provides a summary overview of the software in this module. Include both the type and name of the input parameters. Specify the function as `void` even if it has no parameters. These prototypes are easy to understand:

```
void start(unsigned short period, void(*functionPt)(void));
short divide(short dividend, short divisor);
void SCI_Init(void);
```

These prototypes are harder to understand:

```
start(unsigned short, (*)());
short divide(short, short);
SCI_Init();
```

Declare function return types explicitly. In general, we can remove ambiguities by clarifying exactly what we want. Unless the number of parameters is large, we will place the return type, the function name, and the input parameters on a single line. If there is still room within the 80-character line limit, we can add some local variable declarations to this line. The following are good examples of the first line of several functions:

```
void main(void){ int i;
void SCI_OutUDec(unsigned short number){
unsigned short SCI_InUHex(void){
int RxFifo_Put(char data){
```

Declare data and parameters as `const` *whenever possible.* Declaring an object as const has two advantages. The compiler can produce more efficient code when dealing with parameters that don't change. The second advantage is to catch software bugs—that is, situations in which the program incorrectly attempts to modify data that it should not modify.

Avoid all goto statements. Debugging is hard enough without adding the complexity generated when using `goto`. A corollary to this rule is that when developing assembly language software, we should restrict the branching operations to the simple structures allowed in C.

`++` *and* `--` *should not appear in complex statements.* These operations should appear only as commands by themselves. Again, the compiler will generate the same code, so the issue is readability. The statement

```
*(--pt) = buffer[n++];
```

should have been written as

```
--pt;
*(pt) = buffer[n];
n++;
```

If it makes sense to group, then put them on the same line. The following code is allowed:

```
buffer[n] = 0; n++;
```

Be a parenthesis zealot. When mixing arithmetic, logical, and conditional operations, explicitly specify the order of operations. Do not rely on the order of precedence. As always, the major issue is clarity. Even if the following code were actually to perform the intended operation (which in fact it does not),

```
if( x + 1 & 0x0F == y | 0x04)
```

the programmer assigned to modify it in the future will have a better chance if we had written

```
if( ((x + 1) & 0x0F) == (y | 0x04)).
```

Use `enum` instead of `#define` or `const`. The use of `enum` allows for consistency checking during compilation, and provides for easy to read software. A good optimizing compiler will create the exact object code for the following four examples. So once again, we focus on style. In the first example, we needed comments to explain the operations.

```
int Mode; // 0 means error
void function1(void){
  Mode = 1; // no error
}
void function2(void){
  if(Mode == 0){ // error?
    SCI_OutString("error");
  }
}
```

In the second example, no comments are needed.

```
#define NOERROR 1
#define ERROR 0
int Mode;
void function1(void){
  Mode = NOERROR;
}
void function2(void){
  if(Mode == ERROR){
    SCI_OutString("error");
  }
}
```

In the third example, the compiler performs a type-match, making sure `mode`, `NOERROR`, and `ERROR` are the same type.

```
const int NOERROR = 1;
const int ERROR = 0;
int Mode;
void function1(void){
  Mode = NOERROR;
}
void function2(void){
  if(Mode == ERROR){
    SCI_OutString("error");
  }
}
```

Enumeration provides a check of both type and value. We can explicitly set the values of the enumerated types if needed.

```
enum Mode_state{ ERROR, NOERROR};
enum Mode_state mode;
void function1(void){
  mode = NOERROR;
}
void function2(void){
  if(mode == ERROR){
    SCI_outString("error");
  }
}
```

Don't use bit-shift for arithmetic operations. Microcomputer architectures and compilers used to be so limited that it made sense to perform multiply/divide by 2 using a shift operation. For example, when multiplying a number by 4, we might be tempted to write data<<2. This notation is wrong; if the operation is multiply, we should write data*4. Compiler optimization has developed to the point where the compiler can choose to implement data*4 as either a shift or multiply depending on the instruction set of the computer. When we use data*4, we have code that is easier to understand than data<<2.

Reference:
Mike Dahlin, ed., "LESS Software Engineering", coding standard for the LESS group.

9.6 *Relocatable Code

Relocatable code (or position-independent code) is object code that will execute properly regardless of the position at which the code is loaded. Relocatable code can be loaded anywhere in memory without reassembly. One can implement relocatable code by using

1. the registers or the stack to pass parameters
2. PC relative addressing to access data within the program block (difficult with the 6811)
3. PC relative branches (bpl) and subroutine calls (bsr) within the program block
4. absolute addressing to access I/O ports (ldaa PORTC)
5. absolute jumps and subroutine calls (jsr) to fixed address software like OS programs

The 6811 does not easily support PC relative data addressing. It is awkward to write relocatable 6811 software. There are two ways to implement PC relative data addressing. The first example assumes that the program and data are in the same block (will be relocated together), but we do not assume they are within ±128 bytes of each other. The second example assumes the variable is within ±128 bytes from the program.

```
* Version 1. DATA is a constant in the same program block
Start   bsr    HERE        Push a copy of the PC onto the stack
HERE    pula               Reg D=PC value at HERE
        pulb
        addd   #DATA-HERE  Reg D points to DATA,
* Note that   DATA-HERE is PC relative
        xgdx               Reg X points to DATA
        ldaa   0,X         read from in constant
        staa   $1005       write data to fixed address I/O data port
        bra    Start
DATA    fcb    12          8-bit constant (initialized)
```

```
* Version 2. DATA is a variable within the same program block
* program and variable close together in memory
Start   bsr     HERE        Push a copy of the PC onto the stack
HERE    pulx                Reg x=PC value at HERE
* Note that  DATA-HERE is PC relative,
* and must be a positive number between 0 and 255
Loop    ldaa    $1005       read data from fixed address I/O data port
        staa    DATA-HERE,X write data into variable
        bra     Loop
DATA    rmb     1           8-bit global variable (not initialized)
```

It is important to remember the context of the 6811 microcomputer. The 6811 is a single-chip microcomputer with at most 64 Kbytes of memory; hence, a small number of programmers will probably generate all the software for the system. Relocatable code is not usually necessary for this type of simple system. The programs can still be relocated by modifying the source code and reassembling.

In larger microcomputers with gigabytes of RAM memory and hundreds of gigabytes of disk space, the software will be generated by many programmers working for many different companies. Because of these large and complex software systems, it is necessary to allow programs to relocate. There are two mechanisms for relocation:

1. With nonrelocatable code, one must modify source file, assemble, and reload;
2. With relocatable code, one simply moves the object code.

Clearly the second option is faster. For speed, as well as privacy, it will be necessary to implement relocatable code. It is more difficult to pirate (steal) object code than it is to pirate source code. By selling relocatable object code, the buyer is happy because the program will execute regardless of where it is loaded, and the seller is happy because the trade-secrets within the code can not be easily observed.

The example shown in Program 9.4 shows a 6812 program assembled at $0800 and at $0900. Notice that the object codes are identical. The **TExaS** assembler uses the PC relative addressing mode to create references to data objects within the module.

Program 9.4   Relocatable code can be loaded anywhere in memory.

```
$0800                    org   $0800          $0900                    org   $0900
$0800         data  rmb   1              $0900         data  rmb   1
$0000         local set   0              $0000         local set   0
$0001         PORTB equ   $0001          $0001         PORTB equ   $0001
$0801 1B9F    funct leas  -1,s           $0901 1B9F    funct leas  -1,s
$0803 9601          ldaa  PORTB          $0903 9601          ldaa  PORTB
$0805 6A80          staa  local,s        $0905 6A80          staa  local,s
$0807 6AD7          staa  data,pcr       $0907 6AD7          staa  data,pcr
$0809 1B81          leas  1,s            $0909 1B81          leas  1,s
$080B 3D            rts                  $090B 3D            rts
$080C 07F3  program bsr   funct          $090C 07F3  program bsr   funct
$080E 20FC          bra   program        $090E 20FC          bra   program
```

Checkpoint 9.7: What 6812 mechanisms make it easier to write relocatable code?

9.7   Comments

Discussion about comments was left for last, because they are the least important aspect involved in writing quality software. It is much better to write well-organized software with simple interfaces having operations so easy to understand that comments are not necessary than it is to write sloppy code that requires detailed explanation.

The beginning of every file should include the file name, purpose, hardware connections, programmer, date, and copyright as in the following code:

```
; filename adtest.rtf
; Test of 6812 8-bit ADC
; 1 Hz sampling and output to the serial port
; Last modified 2/7/01 by Jonathan W. Valvano
; Copyright 2001 by Jonathan W. Valvano, valvano@uts.cc.utexas.edu
;    You may use, edit, run or distribute this file
;    as long as the above copyright notice remains
```

The beginning of every function should include a line delimiting the start of the function, purpose, input parameters, output parameters, and special conditions that apply. The comments at the beginning of the function explain the policies (e.g., how to use the function). These comments, which are similar to the comments for the prototypes in the header file, are intended to be read by the client. The following code is illustrative:

```
;----------------------SCI_InUDec----------------------------------
; accepts ASCII input from the SCI in unsigned decimal format
; and converts to a 16-bit unsigned number with a maximum of 65535
; If a number is above 65535, it truncates without reporting the error
; Backspace will remove last digit typed
; Inputs: none
; Outputs: Register D is the unsigned 16-bit value
```

Comments can be added to a variable or constant definition to clarify the usage. In particular, comments can specify the units of the variable or constant. For complicated situations, we can use addition lines and include examples, as in these lines of code:

```
V1         rmb  2  ; voltage at node 1 in mV, range -5000 mV to +5000 mV
Fs         rmb  2  ; sampling rate in Hz
FoundFlag  rmb  1  ; 0 if keyword not yet found, 1 if found
RunMode    rmb  1  ; 0, 1, 2, or 3 specifies system mode
; 0 means idle
; 1 means startup
; 2 means active run
; 3 means stopped
```

Comments can be used to describe complex algorithms. These types of comments are intended to be read by our coworkers. The purpose of these comments is to assist in changing the code in the future, or applying this code into a similar but slightly different application. Comments that restate the function provide no additional information and actually make the code harder to read. The following are examples of bad comments:

```
inc time ; add one to time
clr mode ; set mode to zero
```

Good comments explain why the operation is performed, and what it means:

```
inc time ; maintain elapsed time in msec
clr mode ; switch to idle mode because no more data are available
```

We can add spaces so the comment fields line up. As stated earlier, we avoid tabs because they often do not translate from one system to another. In this way, the software is on the left and the comments can be read on the right.

I taught a large programming class one semester, and being an arrogant and lazy fellow, I thought I could write a grading program that accepts the students' programming assignments and automatically generates and records their grades. (The second step would have been to write a self-study book, then I could teach the masses without ever having to show up for work.) My grading program worked OK for the functional aspects of the students' software. My program generated inputs, called the students' program and compared the results with expected behavior. Where I utterly failed was in my attempts to automatically grade their software on style. I used the following three-part "quality" statistic. First, I measured execution speed of the student's software, s_i. Smaller times represent improved dynamic efficiency. Next, I measured the number of bytes in the object code, b_i. A smaller number represents better static efficiency. Third, I used the number of ASCII characters in the source code, c_i, as a quantitative measure of documentation. For this parameter, bigger is better. In a typical statistical fashion, I used the average and standard deviation to calculate

$$\text{quality} = \frac{\bar{s} - s_i}{\sigma_s} + \frac{\bar{b} - b_i}{\sigma_b} + \frac{c_i - \bar{c}}{\sigma_c}.$$

Halfway through the semester, I happened to look at some assignments and was horrified to find the all-time worst software ever written from both a style and content basis. To improve speed and reduce size, the students cut so many corners that their code didn't really work anymore, it just appeared to work to my grading program. Then they took the ugly mess and filled it with nonsense comments, giving it the appearance of extensive documentation. To my students in that class that semester, I sincerely apologize. We should write comments for coworkers who must change our software, or clients who will use our software.

9.8 Tutorial 9. Layered Programming

The purpose of this tutorial is to evaluate a layered software system. There are three distinct modules, which are structured into a three-layer system. Even though you will assemble the three modules as a single program, the linkage will be performed using software interrupts and traps. Then you will study its ease of modification by adding additional commands to the system.

Action: Open and assemble the program `Chap9.rtf`.

Question 9.1. Draw a memory map of the system. Where in memory are the modules?

Action: Run the `Chap9.rtf` system and observe Register A.

Question 9.2. Assuming the modules don't overlap in memory, can the physical location of one module be moved without changing the object code of the other two? For example, if the `org $F400` for module 2 were changed to `org $F500`, would the object code for modules 1 and 3 stay the same?

Question 9.3. If the modules were to have global variables, how could they be implemented?

Question 9.4. How are parameters passed into a module?

Question 9.5. How are parameters returned?

Question 9.6. Add two new commands to module 1 that logical shift right and logical shift left the value in Register A. Add two commands to module 2 that left/right shift twice, invoking the Module 1 routines. Modify Module 3 to test the new module functions. That is, change Module 3 to

```
Mod3.loop trap #AddTwo    ; A=A+2
          trap #LSLTwo    ; A=2*A
          trap #LSRTwo    ; A=A/2
          trap #SubTwo    ; A=A-2
          bra  Mod3.loop
```

9.9   Homework Problems

Homework 9.1: List three factors that we can use to evaluate the "goodness" of a program.

Homework 9.2: One mechanism to expand the number of software interrupts is to add a program byte immediately after the `swi`. For example,

```
swi
fcb 20
```

where 20 is the command number. The `swi` handler must fetch the return PC value from the stack to read the command number. After the command is executed, the return value must be updated so the command is not executed as an instruction. Write an `swi` handler that adds an 8-bit constant to Register A. In particular, the above example should add 20 to Register A.

Homework 9.3: What does it mean to say a function is public versus private? Why is this distinction important?

Homework 9.4: You are given a stopwatch module with the following functions. Try and guess what each function does. `Watch_SetTimerResolution`, `Watch_StartTimer`, `Watch_StopTimer`, `Watch_DisplayTime`.

Homework 9.5: Your job is to design a device driver for a computer mouse. Assuming it is to be written in C, give the `Mouse.h` header file that lists the prototypes for the public functions. You show just the header file, not the implementation file.

Homework 9.6: Your job is to design a device driver for a black and white text-based video screen. There are 24 lines and 80 columns. Assuming it is to be written in C, give the `Video.h` header file that lists the prototypes for the public functions.

Homework 9.7: Which of the assembly language programs in Chapter 5 are relocatable?

Homework 9.8: Which of the assembly language programs in Chapter 6 are relocatable?

Homework 9.9: Which of the assembly language programs in Chapter 8 are relocatable?

9.10　Laboratory Assignments

Lab 9.1.
Layered IEEE488
Interface

Purpose

The objectives of this lab are to

- interface a simulated IEEE488 printer to the microcomputer,
- study handshaked or interlocked communication,
- write a device driver for this printer,
- implement a three-layer software system.

Description

Hewlett Packard first defined this parallel communication standard in 1972. The IEEE488 bus is sometimes referred to as the Hewlett Packard Instrumentation Bus (HPIB) or the General Purpose Instrumentation Bus (GPIB). Its purpose is to connect tabletop instruments, typically separated by less than 2 m. As is the goal of most networks, IEEE488 is intended to interconnect devices from different manufacturers. It implements an asynchronous handshake protocol; therefore, it will operate with devices of varying speeds and will automatically improve performance as modules are upgraded with faster versions. IEEE488 is typically used for communications requiring bandwidths below 10 Kbytes/sec, but it will handle up to 1 Mbytes/sec. It can handle 15 devices, but having 8 or fewer will be more convenient. There are three types of modules, or capabilities, on the IEEE488 bus. A controller (usually attached to the computer) coordinates access and arbitrates conflicts. A talker is a device that can drive data onto the bus, and a listener is a device that can accept data from the bus. The bus supports a broadcast; that is, when one talker sends information simultaneously to multiple listeners. The controller will select which device will talk and which device(s) will listen. The microcomputer usually includes all three (controller, talker, listener) capabilities, but the table-top instruments may be

talk only	can only transmit data
listen only	can only receive data
both talker and listener	can transmit or receive data

The simulated IEEE488 printer will run in listen-only mode. The microcomputer will run in talk-only mode, and there will be no controller. Figure 9.4 overviews the software system. Figure 9.5 draws the timing signals for this handshake interface. The communication is interlocked because each device will perform an operation, then wait for the other. To print one character on this IEEE488 device, the talker (your software) must

1. Wait for RFD to be one,
2. Write the ASCII code on the 8-bit data lines,
3. Make DAV low, signaling new data are available,
4. Wait for RFD to be zero,
5. Wait for DAC to be one,
6. Make DAV high, signaling the data signals are no longer valid,
7. Wait for DAC to be zero.

Figure 9.5
IEEE488 handshake
signals during the
transfer of one ASCII
character.

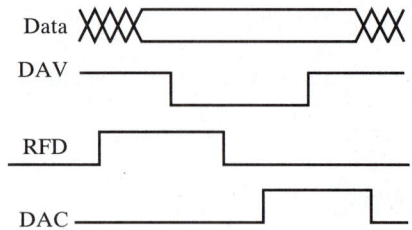

Part A. Connect a simulated printer to the microcomputer. Label each device with its IEEE488 name.

> 8 data bits connected to a simple parallel LCD display
> RFD (ready for data) input from a switch
> DAC (data accepted) input from a switch
> DAV (data available) output to a LED

Figure 9.6 shows one possible configuration. The three control lines can also be connected to a digital logic analyzer. The human operator will toggle the two switches to simulate the listen-only printer. To receive one character, the listener (human operator pretending to be a printer) must

> **1.** Make RFD high, signaling it is ready for data,
> **2.** Wait for DAV to be zero,
> **3.** Make RFD low, signaling it is working on this data,
> **4.** Print the ASCII character on the 8-bit data (you can copy the hex code on a piece of paper),
> **5.** Make DAC high, signaling you are done with this data,
> **6.** Wait for DAV to be one,
> **7.** Make DAC low, signaling this transfer is complete.

Figure 9.6
Simulated IEEE488
printer.

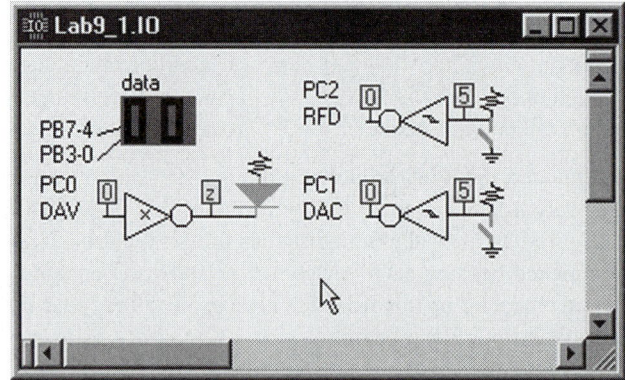

Part B. Write the lowest-level routines as a `swi` handler using an approach similar to Module 1 in Tutorial 9. The particular functions must include

> InitIEEE, which performs the low-level ritual,
> SetDAV, which can make DAV high and low,
> SetData, which sets the 8-bit data output,
> GetRFD, which returns the current status of the RFD input,
> GetDAC, which returns the current status of the DAC input.

Part C. Write the middle-level routines as a `trap` handler using an approach similar to Module 2 in Tutorial 9. You are free to add additional helper functions. The particular functions must include

> InitPrinter, which initializes system and calls InitIEEE,
> OutBuf, which prints a zero-terminated ASCII string.

Part D. Write the high-level routines as a regular main program using an approach similar to Module 3 in Tutorial 9. You are free to add additional helper functions. The particular functions must include

> Initialize, which initializes system and calls InitPrinter,
> PrintInfo, which prints a zero-terminated ASCII string by calling OutBuf.

The main loop should print "Hello World."

10 Elementary Data Structures

Chapter 10 objectives are to

❑ organize data into structures,
❑ develop software that manipulates the data,
❑ present embedded system applications.

A data structure contains two parts: the organization of information (or data) in memory and mechanisms (subroutines) to access the data. Typically the accesses involve writing, reading, or read-modify-write. Good engineers employ well-defined design processes when developing complex systems. When we work within a structured framework, it is easier to prove our system works (verification) and to modify our system in the future (maintenance.) As our software systems become more complex, it becomes increasingly important to employ well-defined software design processes. The organization of information makes a significant impact on the effectiveness of a software system. Consequently, this chapter presents many of the data structures appropriate for embedded systems.

10.1 Memory Allocation and Initialization

Statically allocated global variables are needed when sharing information between threads and when information is permanent in nature. Multiple threads will be presented in Chapter 11. If the values of variables need to change during program execution, we will place them into RAM. In C, we create global variables simply by placing their definitions not within a function. Most C compilers place the globals in a special area in the order in which they are defined. There is a fundamental difference in the way the assembler and C initialize variables. Consider the following C example:

```
char MaxVolt=5;     // 8-bit signed global, initialized (volts)
```

The C compiler will allocate this variable in RAM, but provide initialization code that will be executed at start up, which will set the variable to 5. In assembly language 8-bit initialized globals are created with the pseudo-op `fcb`, for example,

```
MaxVolt fcb 5       8-bit signed global, initialized (volts)
```

Checkpoint 10.1: What is the difference between a global and a local variable?

If we allocate this variable in RAM, then it will lose its contents when power is removed. Therefore, we must explicitly add software that sets the initial values of all variables. We should not assume any initial values of the RAM locations when power is applied to our embedded systems. We will use the concept of a ritual, which is initialization software usually executed once at the beginning to set up proper values of the globals and define the I/O hardware modes, as in Program 10.1.

Checkpoint 10.2: *ROM provides for nonvolatile storage, but can there be such a thing as nonvolatile RAM?*

Program 10.1   Initialization of a global variable.

```
; MC68HC11A8                              ; MC68HC812A4
        org   0         RAM                       org   $0800    RAM
MaxVolt rmb   1                           MaxVolt rmb   1
        org   $E000     ROM                       org   $F000    ROM
Main    lds   #$00FF    initial SP        Main    lds   #$0C00   initial SP
        bsr   Ritual                              bsr   Ritual
loop                                      loop
; ****rest of main                        ; ****rest of main
        bra   loop                                bra   loop
Ritual  ldaa  #5                          Ritual  movb  #5,MaxVolt
        staa  MaxVolt   initialize                rts
        rts                                       org   $FFFE
        org   $FFFE                               fdb   Main
        fdb   Main
```

Common Error: *If we initialize global variables at assembly time using* `fcb`, *then the program will work during the development phase because the loader will initialize all RAM when downloading, but the program will fail once the code is programmed into ROM because the globals will not be initialized.*

Observation: *When a RAM is powered up, its initial contents are undefined.*

If the values of variables do not need to change during program execution, we will place them into ROM. In C, we define constants using the `const` modifier. There is no real difference between the way the assembly and C creates constants. Consider the following C example:

```
const char MaxVolt=5;  // 8-bit signed global, initialized (volts)
```

The C compiler will allocate this constant in ROM, similar to the assembly code shown in Program 10.2.

Program 10.2   Initialization of a global constant.

```
; MC68HC11A8                              ; MC68HC812A4
        org   $E000     ROM                       org   $F000    ROM
Main    lds   #$00FF    initial SP        Main    lds   #$0C00   initial SP
        bsr   Ritual                              bsr   Ritual
loop                                      loop
; ****rest of main                        ; ****rest of main
        loop                                      bra   loop
MaxVolt fcb   5                           MaxVolt fcb   5
        org   $FFFE                               org   $FFFE
        fdb   Main                                fdb   Main
```

Maintenance Tip: It is good practice to specify whether an assembly variable is signed or unsigned in the comments. If the information has units (e.g., volts, seconds etc.) this should be included also.

Checkpoint 10.3: Why is it inefficient to leave off the `const` *modifier in C, when defining objects that are actually constants?*

Random access means one can read and write any element in any order. Random access is allowed for all indexable data structures. An indexed data structure has elements of the same size and can be accessed by the name of the structure, the size of each element, and the element number. In C, we use the syntax [] to access an indexed structure. Arrays, matrices, tables, and the multiple access circular queue are examples of indexed structures presented in this chapter.

Sequential access means one reads and writes the elements in order. Pointers are usually employed with these types of data structures. Strings, linked-lists, stacks, queues, and trees are examples of sequential structures presented in this chapter. The first-in-first-out circular queue (**FIFO**) is useful for data flow problems, and it will be presented in the next chapter.

10.2   Arrays

An **array** is made up of elements of equal precision. The **precision** is the size of each element. Typically, precision is expressed in bits or bytes. The **length** is the number of elements. The **origin** is the index of the first element. A data structure with an origin equal to zero is called **zero-origin indexing**. In C, zero-origin index is almost always used. Consider a simple byte array, shown in Figure 10.1, and defined by the following C code.

```
const char data[4]={0x05,0x06,0x0A,0x09};
```

In assembly, we define

```
data fcb $05,$06,$0A,$09
```

Figure 10.1
A byte array with four elements.

$F850	$05
$F851	$06
$F852	$0A
$F853	$09

In C, we can access an element of the array using its name and an index. Assume PORTB is an output connected to a stepper motor, and the `index` is initialized to zero.

```
unsigned char index=0;
void step(void){
  PORTB = data[index];    /* move stepper motor */
  index = 0x03&(index+1); /* next index */
}
```

In assembly, we can perform a similar function using indexed addressing. (See Program 10.3.) Assume `index` is an 8-bit variable, defined in RAM, and initialized to zero.

Program 10.3
Assembly code to access
a byte array.

```
; 6811
step    ldab index
        ldx  #data
        abx
        ldaa 0,x
        staa PORTB
        incb
        andb #$03
        stab index
        rts
```

```
; 6812
step    ldab index
        ldx  #data
        ldaa B,x
        staa PORTB
        incb
        andb #$03
        stab index
        rts
```

Consider a constant word array, shown in Figure 10.2, and defined by the following C code.

```
const short powers[5]={1,10,100,1000,10000};
```

In assembly, we define a word constant using `fdb`

```
powers fdb 1,10,100,1000,10000
```

Figure 10.2
A word array with
5 elements.

$F950	1
$F952	10
$F954	100
$F956	1000
$F958	10000

In C, the syntax for accessing all array types is independent of precision. The compiler automatically performs the correct address correction. We will assume the input is less than or equal to 4.

```
unsigned short power(unsigned char exponent){
    return powers[exponent];    /* look up answer in array */
}
```

If `I` is the index and `Base` is the base address of the array, then the address of the element at `I` is

```
Base+2*I
```

In assembly, we can access the array using indexed addressing, as shown in Program 10.4. Again, we will assume the Register B input is less than or equal to 4.

Program 10.4   Assembly code to access a word array.

```
;6811
power lslb            2*I
      ldx  #powers    Base
      abx             Base+2*I
      ldd  0,x        read access
      rts
```

```
; 6812
power lslb            2*I
      ldx  #powers    Base
      ldd  B,x        read access
      rts
```

In general, let `n` be the precision of a zero-origin indexed array in bytes. If `I` is the index and `Base` is the base address of the array, then the address of the element at `I` is

```
Base+n*I
```

The origin of an array is the index of the first element. The origin of a zero-origin indexed array is zero. In general, if `origin` is the origin of the array, then the address of the element at `I` is

```
Base+n*(I-origin)
```

> **Checkpoint 10.4:** *What are the precision, length, and total size of*
> `long data[5];`*?*

In the previous examples, the length of the array was known. Sometimes, it is desirable to allow the length to vary dynamically. There are many mechanisms that allow for a variable-length array. One simple mechanism saves the length of the array as the first element. In this way, we could add run-time checking to make sure the index bounds are not exceeded. The previous examples could be defined as

```
const char data[5]={4,0x05,0x06,0x0A,0x09};
const short powers[6]={5,1,10,100,1000,10000};
```

We could define these variable length arrays in assembly as

```
data   fcb 4,$05,$06,$0A,$09
powers fdb 5,1,10,100,1000,10000
```

Another common mechanism to handle variable length is a termination code. Typical codes are shown in Table 10.1. This method can be used only if it is not possible for the termination code to be present in the data.

Table 10.1
Typical termination codes.

ASCII	code	name
NUL	$00	null
ETX	$03	end of text
EOT	$04	end of transmission
FF	$0C	form feed
CR	$0D	carriage return
ETB	$17	end of transmission block
ESC	$1B	escape

The following array has a null-termination with extra elements added on the end:

```
char data[9]={0x05,0x06,0x0A,0x09,0,0,0,0,0};
```

In assembly, we define

```
data fcb $05,$06,$0A,$09,0,0,0,0,0,0
```

We can use the termination code to determine when to reset the index back to zero.

```
unsigned char index=0;
void step(void){
  PORTB = data[index];    /* move stepper motor */
  index++;                /* next index */
  if(data[index] == 0)    /* check for null */
    index = 0;            /* at the end, so start over */
}
```

In assembly, we can perform a similar function, as shown in Program 10.5.

Program 10.5 Assembly code to access a variable length byte array.

```
; 6811                              ; 6812
step   ldab index                  step   ldab index
       ldx  #data                         ldx  #data
       abx                                ldaa B,x
       ldaa 0,x                            staa PORTB
       staa PORTB                         incb
       incb                               tst  B,x    check for null
       ldx  #data                         bne  ok
       abx                                clrb        start over
       tst  0,x    check for null  ok     stab index
       bne  ok                            rts
       clrb        start over
ok     stab index
       rts
```

The stepper motor can be changed from full steps to half steps simply by changing the array, without modification to the program code. In particular, all we need to do is change the values to

{0x05,0x04,0x06,0x02,0x0A,0x08,0x09,0x01,0}

> *Checkpoint 10.5:* Why can't you use a termination code to signify the end of a variable-length data set where the data can be any binary value?

10.3 Strings

A **string** is a data structure with equal size elements that allows only sequential access. The bytes of the string are always read in order from the first to the last. In contrast, an **array** allows random access to any element in any order. The same mechanisms introduced for variable-length arrays will apply also to strings. In the first example, the length of the string is stored in the first byte. This approach is appropriate when the data elements can take on all possible numeric values, negating the possibility of using a termination code. Assume a DAC converter is connected to Port B. The function DAC, shown in Program 10.6, will output the string data to the DAC. The main program calls this function twice, with different data strings.

Program 10.6 A variable length string contains DAC data.

```
data1 fcb   4            ;length     unsigned const char data1[5]=
      fcb   0,50,100,50  ;data          {4,0,50,100,50};
data2 fcb   8                        unsigned const char data2[9]=
      fcb   0,25,50,75,100,75,50,25    {8,0,25,50,75,100,75,50,25};
*Reg X points to the string data    void DAC(unsigned char *pt){
DAC   ldab 0,x   ;length               unsigned int length;
```

continued

```
loop    inx         ;next element
        ldaa 0,x    ;data
        staa PORTB  ;out to DAC
        decb
        bne  loop
        rts
main    lds  #$00FF ($0C00 on 6812)
mloop   ldx  #data1  ;first string
        bsr  DAC
        ldx  #data2  ;second string
        bsr  DAC
        bra  mloop
```

```c
length = (*pt++); /* size */
do{
    PORTB = (*pt++);
}
while(--length);
}
void main(void){
    while(1){
        DAC(data1); /* first string */
        DAC(data2); /* second string */
    }
}
```

In C, ASCII strings are stored with null-termination. In C, the compiler automatically adds the zero at the end, but in assembly, the zero must be explicitly defined. The function Out-String, shown in Program 10.7, will output the string data to the serial port. The function SCI_OutChar, defined back in Chapter 5 as Program 5.7, outputs a single ASCII character to the serial port. The main program calls this function twice, with different ASCII strings.

Program 10.7   A variable length string contains ASCII data.

```
hello fcc   "Hello World"
      fcb   0
CRLF  fcb   13,10,0
*Reg X points to the string data
OutString
        ldaa 0,x      next data
        beq  done     0 means end
        jsr  SCI_OutChar
        inx
        bra  OutString
done    rts
main    lds  #$00FF ($0C00 on 6812)
        bsr  SCI_Init
mloop   ldx  #hello  ;first string
        bsr  OutString
        ldx  #CRLF   ;second string
        bsr  OutString
        bra  mloop
```

```c
unsigned const char CRLF[3]=
  {13,10,0};

void OutString(unsigned char *pt){
unsigned char letter;
  while(letter = (*pt++)){
    SCI_OutChar(letter);
  }
}

void main(void){
  SCI_Init();
  while(1){
    OutString("Hello World");
    OutString(CRLF);
  }
}
```

10.4   Matrices

A matrix is a two-dimensional data structure accessed by row and column. Each element of a matrix is the same type and precision. In C, we create matrices using two sets of brackets. Figure 10.3 shows this byte matrix with six 8-bit elements. The figure also shows two possible ways to map the two-dimensional data structure into the linear address space of memory.

```c
unsigned char M[2][3]; // byte matrix with two rows and three columns
```

Figure 10.3
A byte matrix with two
rows and three columns.

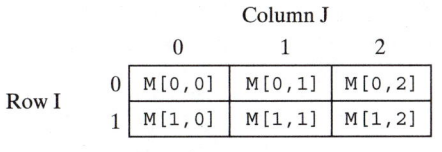

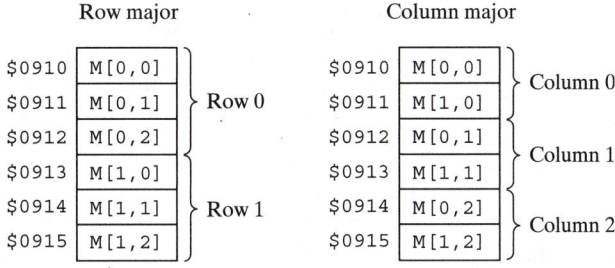

With row-major allocation, the elements of each row are stored together. Let I be the row index, J be the column index, n be the number of bytes in each row (equal to the number of columns), and Base is the base address of the byte matrix, then the address of the element at I,J is

```
Base+n*I+J
```

An assembly language program that reads elements from this row-major matrix is shown in Program 10.8. The row index (0 or 1) is passed in Register A. The column index (0, 1, or 2) is passed in Register B. The base address of the matrix is passed in Register X. The subroutine returns the value in Register A.

Program 10.8
Assembly function to
access a two-by-three
row-major matrix.

```
J     set  0       Column index
READ  pshb          Allocate and initialize
      tsy           stack frame pointer
      ldab #3       number of columns
      mul           3*I
      addb J,Y      3*I+J
      abx           Register X points to M[I,J]
      ldaa ,X       read value at M[I,J]
      ins
      rts
```

Checkpoint 10.6: Rewrite Program 10.8 using 6812 instructions, eliminating at least two instructions.

With column major allocation, the elements of each column are stored together. Let I be the row index, J be the column index, m be the number of bytes in each column (equal to the number of rows), and Base is the base address of the byte matrix, then the address of the element at I,J is

```
Base+m*J+I
```

An assembly language program that reads elements from this column-major matrix is shown in Program 10.9. The row index (0 or 1) is passed in Register A. The column index

(0, 1, or 2) is passed in Register B. The base address of the matrix is passed in Register X. The subroutine returns the value in Register A.

Program 10.9
Assembly function to access a two-by-three column-major matrix.

```
READ    aslb          Reg B = 2*J
        abx           Reg X = base + 2*J
        tab           Reg B = I
        abx           Reg X = base + 2*J + I
        ldaa ,X       Read a byte from array(I,J)
        rts
```

Checkpoint 10.7: Rewrite Program 10.9 using 6812 instructions, eliminating at least two instructions.

With a word matrix, each element requires two bytes of storage. Let I be the row index, J be the column index, n be the number of words in each row (equal to the number of columns), and Base is the base address of the word matrix, then the row major address of the element at I,J is

```
Base+2*(n*I+J)
```

An assembly language program that reads elements from a word matrix defined in row-major format is shown in Program 10.10. The number of columns is defined with an equ pseudo-op. The subroutine returns the address of the element in Register X.

Program 10.10
Assembly function to access a word matrix defined in row-major format.

```
* The matrix is m rows by n columns
* ROW MAJOR     Address = Base+2*(n*I+J)
*Zero-origin indexing  Each element is 2 bytes
n    equ  10    the number of columns (can be changed)
*Input:  Reg A is the row index(I)
*        Reg B is the column index(J)
*        Reg X is Base, points to the first element
*Output: Reg X points to the (I,J) element
* Local variables
Base    set 0          address of first element
J       set 2          column index
Access pshb           allocate and initialize J
        pshx           allocate and initialize Base
        tsx            pointer to local variables
        ldab #n        Reg B is the number of columns
        mul            n*I
        addb J,X       n*I+J
        adca #0
        lsld           2*(n*I+J)
        addd Base,X    Reg D = Base + 2*(n*I+J)
        xgdx           Reg X = Base + 2*(n*I+J)
        ins
        ins
        ins
        rts
```

Checkpoint 10.8: *Rewrite Program 10.10 using 6812 instructions, eliminating at least two instructions.*

An assembly language program that reads elements from a word matrix defined in column-major format is shown in Program 10.11. The number of rows, m, is defined with an `equ` pseudo-op. The subroutine returns the address of the element in Register X.

Program 10.11

Assembly function to access a word matrix defined in column-major format.

```
* The matrix is m rows by n columns
*COLUMN  MAJOR          Address = Base+2*(I+m*J)
*Zero-origin indexing   Each element is 2 bytes
m    equ  10   the number of rows (can be changed)
*Input:  Reg A is the row index(I)
*        Reg B is the column index(J)
*        Reg X is Base, points to the first element
*Output: Reg X points to the (I,J) element
* Local variables
Base   set 0     address of first element
I      set 2     row index
Access psha      allocate and initialize I
       pshx      allocate and initialize Base
       tsx       pointer to local variables
       ldaa #m   Reg A is the number of rows
       mul       m*J
       addb I,X  m*J+I
       adca #0
       addd Base,X  Reg D = Base + m*J+I
       xgdx         Reg X = Base + m*J+I
       ins
       ins
       ins
       rts
```

Checkpoint 10.9: *Rewrite Program 10.11 using 6812 instructions, eliminating at least two instructions.*

Bit arrays can be used to store pixel values for graphics displays. (See Figure 10.4.) Placing a 0 into a bit location will display a blank. Placing a 1 into a bit location will display that pixel. Typically the first byte bit 7 is the top left corner of the display, and the last byte bit 0 is the bottom right corner. Consider the following 32-by-12 video graphics bit array. Since there is a total of 384 bits and each byte can store 8 bits, we need 48 bytes to store the entire image. In C, we define

```
unsigned char Video[48];
```

In assembly, we define the following in global RAM:

```
Video rmb 48
```

Figure 10.4
A bit-matrix with 12
rows and 32 columns.

$0A00	$8BE4
$0A02	$2070
$0A04	$8A08
$0A06	$2088
$0A08	$FB88
$0A0A	$2088
$0A0C	$8A08
$0A0E	$2088
$0A10	$8BEF
$0A12	$BE70
$0A14	$0000
$0A16	$0000
$0A18	$0000
$0A1A	$0000
$0A1C	$89CF
$0A1E	$20F0
$0A20	$8A28
$0A22	$A088
$0A24	$AA2F
$0A26	$2088
$0A28	$522A
$0A2A	$2088
$0A2C	$51C9
$0A2E	$3EF0

```
   0123456789 ...                    31
0  10001011111010000010000001110000
1  10001010000010000010000010001000
2  11111011100010000010000010001000
3  10001010000010000010000010001000
4  10001011111011111011111001110000
5  00000000000000000000000000000000
6  00000000000000000000000000000000
7  10001001110011110010000011110000
8  10001010001010001010000010001000
9  10101010001011110010000010001000
10 01010010001010100010000010001000
11 01010001110010010011111011110000
```

Let I be the row index, where I ranges from 0 to 11. There are 4 bytes required in each row. Therefore, the starting address of row I is

```
Video + 4*I
```

Let J be the column index, where J ranges from 0 to 31. The column index specifies which byte, as well as which bit within that byte. The address of the byte containing the information at (I,J) is

```
Video + 4*I + J>>3
```

where the divide by 8 is integer math without rounding. Notice that if J is less than or equal to 31, then J divided by 8 will be less than or equal to 3. Let K be the bottom 3 bits of J.

```
K = J&0x07;
```

A mask will specify the bit location within the byte. In C, the following array can be used

```
unsigned const char Masks[8]={0x80,0x40,0x20,0x10,0x08,0x04,0x02,0x01};
```

In assembly, this array can be defined in ROM as

```
Masks fcb $80,$40,$20,$10,$08,$04,$02,$01
```

For example, if the bottom 3 bits are 010_2 then we use the bit mask of $20 to access the information

```
mask = Masks[K];
```

Program 10.12 takes the row and column index values and calculates the memory address and bit mask to access that bit in the `Video` matrix.

Program 10.12
A helper function to access a bit-matrix.

```
* ********* Access ***************
* Access the Video bit at (I,J)
*Input:    Reg A is the row index(I is 0 to 11)
*          Reg B is the column index(J is 0 to 31)
*Output:   Reg X points to the byte of interest
*          Reg A is the Mask to access that bit
Access lsla
       lsla            4*I
       pshb            save a copy of J
       lsrb
       lsrb
       lsrb            Reg B = J>>3
       aba             Reg A = 4*I + J>>3
       ldx  #Video
       tab
       abx             Reg X = Video + 4*I + J>>3
       pulb            Reg B = J again
       andb #$07       Reg B = K (bottom three bits of J)
       ldy  #Masks
       aby             Reg Y points to Masks[K]
       ldaa 0,Y        Reg A = mask = Masks[K]
       rts
```

Functions to clear, set toggle bits in the `Video` matrix are shown in Program 10.13. A function that tests the current value within the matrix is shown in Program 10.14.

Program 10.13
Functions that modify the bit-matrix.

```
* Clear the Video bit at (I,J)
*Input:    Reg A is the row index(I is 0 to 11)
*          Reg B is the column index(J is 0 to 31)
ClrBit bsr  Access
       coma        Not(mask) zero in bit location
       anda 0,x    Clear bit
       staa 0,x
       rts
* Set the Video bit at (I,J)
*Input:    Reg A is the row index(I is 0 to 11)
*          Reg B is the column index(J is 0 to 31)
SetBit bsr  Access
       oraa 0,x     Set bit
       staa 0,x
       rts
* Invert the Video bit at (I,J)
*Input:    Reg A is the row index(I is 0 to 11)
*          Reg B is the column index(J is 0 to 31)
InvBit bsr  Access
       eora 0,x     Flip bit
       staa 0,x
       rts
```

Program 10.14
A function that reads
the bit-matrix.

```
* Read the Video bit at (I,J)
*Input:   Reg A is the row index(I is 0 to 11)
*         Reg B is the column index(J is 0 to 31)
*Output:  Reg CC zero bit is the value read from the array
*         Reg A is zero or not zero depending on the bit
ReadBit bsr  Access
        anda 0,x    Z=1 if bit was zero, Z=0 if bit was one
        rts
```

10.5 Structures

A **structure** has elements with different types or precisions. In C, we use `struct` to define a structure. The `const` modifier will allocate the structure in ROM. Without the `const`, the C compiler will place the structure in RAM, allowing it to be dynamically changed. In the example shown in Figure 10.5, `Name` is a variable-length ASCII string, but as you can see, we have to specify its maximum size.

```
const struct port{
  unsigned char AndMask;      /* bits that can change */
  unsigned char OrMask;       /* bits that must stay high */
  unsigned char *Addr;        /* Port Address */
  unsigned char Name[10];     /* ASCII string */
};
typedef const struct port portType;
portType PortB={0x15,0x82,0x0001,"PORTB"};
```

Figure 10.5
A structure collects
objects of different sizes
into one object.

```
$F950  $15
$F951  $82
$F952  $0001
$F954  "PORTB",0,0,0,0,0
```

> **Checkpoint 10.10:** Most C compilers will align 16-bit elements within structures to an even address. How would Figure 10.5 have been different if the positions of OrMask and Addr had been reversed?

In Program 10.15, we can use the `equ` pseudo-op to make our software more readable.

Program 10.15
Assembly language
example of a structure.

```
AndMask   equ   0
OrMask    equ   AndMask+1
Addr      equ   OrMask+1
Name      equ   Addr+2
PortB fcb  $15,$82   AndMask,OrMask
      fdb  $0001     pointer to PORTB ($1003 on the MC68HC11)
```
continued

```
        fcc   'PORTB'  string
        fcb   0,0,0,0,0
* Reg A = data to output
* Reg X = pointer to Port structure
OutP    psha
        anda  AndMask,x   modify input with andmask
        oraa  OrMask,x    modify input with ormask
        ldy   Addr,x      get Port address
        staa  0,y         output
        ldx   Name,x      pointer to string
        jsr   OutString   print string
        pula
        rts
main    lds   #$0c00      ($00FF on the MC68HC11)
        movb  #$FF,DDRB   (skip this step on MC68HC11)
        ldaa  #$00        data
loop    ldx   #PortB      pointer to structure
        bsr   OutP
        inca
        bra   loop
```

Without the `const`, the C compiler will place the structure in RAM, allowing it to be dynamically changed. If the structure resides in RAM, then an assembly language system will have to initialize it explicitly via software execution. Again, most C compilers will implicitly initialize variable structures.

10.6   Tables

A **table** is a collection of identically sized structures. Program 10.16 and Figure 10.6 show a table containing a simple data base. Each entry in the table records the name, life span, and the year of inauguration. The names are variable length, but a fixed size will be allocated so that each table entry will be exactly 36 bytes. The C compiler will fill the unused bytes in the `Name` field with zeros.

Program 10.16
A simple data base with three entries.

```
const struct entry{
  unsigned char Name[30];      /* null-terminated string */
  unsigned short life[2];      /* birth year, year died */
  unsigned short year;         /* year of inauguration */
};
typedef const struct entry entryType;
entryType Presidents[3]={
  {"George Washington",{1732,1799},1789},
  {"John Adams",{1735,1826},1797},
  {"Thomas Jefferson",{1743,1826},1801}
};
```

Figure 10.6
A table collects structures of the same size into one object.

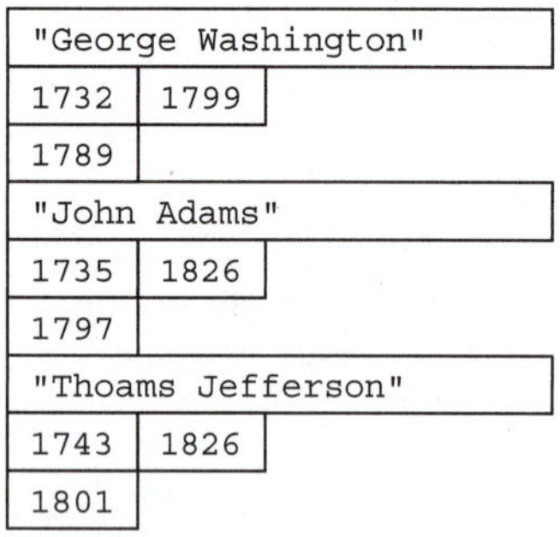

"George Washington"	
1732	1799
1789	
"John Adams"	
1735	1826
1797	
"Thoams Jefferson"	
1743	1826
1801	

Checkpoint 10.11: Why do elements of a table all have to be the same size?

Program 10.17 shows the assembly language definition of the data base. We use `equ` pseudo-ops to make the software more readable.

Program 10.17
The entries of a table written in assembly language.

```
NAME    equ    0
LIFE    equ    NAME+30
YEAR    equ    LIFE+4
SIZE    equ    YEAR+2
Presidents
        fcb   "George Washington",0
        fcb   0,0,0,0,0,0,0,0,0,0,0,0
        fdb   1732,1799
        fdb   1789
        fcb   "John Adams",0
        fcb   0,0,0,0,0,0,0,0,0,0,0,0,0,0,0,0,0,0,0
        fdb   1735,1826
        fdb   1797
        fcb   "Thomas Jefferson",0
        fcb   0,0,0,0,0,0,0,0,0,0,0,0,0
        fdb   1743,1826
        fdb   1801
```

To access the Inauguration year of the second president in C we could execute

```
theyear = Presidents[1].year;
```

This operation in assembly is

```
ldd  Presidents+SIZE+YEAR
std  theyear
```

If we wanted the year the third president died in C we could execute

```
theyear = Presidents[2].life[1];
```

This operation in assembly is

```
ldd  President+2*SIZE+LIFE+2
std  theyear
```

Program 10.18 shows an assembly language function that prints the name of the `nth` president. First it calculates the address of the nth entry (`Presidents+n*SIZE`). In general, the next step would be to add the offset (in this case `NAME` is zero). This program assumes `SIZE*n` is less than 256.

Program 10.18
A subroutine that prints the name of a president.

```
*Print the name of the nth entry
*Reg A is the index n ranging from 0 to 2
OutPresident
    ldx  #Presidents Reg X points to the table
    ldab #SIZE       36 bytes in each entry
    mul              Reg D = SIZE*n
    abx              Reg X  = base +SIZE*n
    jsr  OutString   Prints name
    rts
```

The table shown in Program 10.19 contains five identically formatted structures. Each structure (e.g., PORTA) contains five entries: an 8-bit ASCII character, two pointers, and two byte values. Again, the equ pseudo-ops clarify access to the table, which could be used to write an I/O port driver, separating the high-level software from the low-level hardware.

Program 10.19
A table containing the information about the available I/O ports.

```
PortChar equ  0       ASCII character specifying port name
DataPT   equ  2       Pointer to port address
DirPT    equ  4       Pointer to direction register
InitDir  equ  6       8-bit initial value of direction reg
InitData equ  7       8-bit initial value to output
Table
PORTA fcb 'A'      Port A
      fdb $1000    Address of PortA
      fdb $1026    Bit 7 of PACTL is DDRA7
      fcb 0,0      Initial direction, data values
PORTB fcb 'B'      Port B
      fdb $1004    Address of PortA
      fdb $FFFF    No direction, always output
      fdb $FF      No Initial direction
      fcb 0        Initial data value
PORTC fcb 'C'      Port C
      fdb $1003    Address of PortC
      fdb $1007    DDRC
      fcb 0,0      Initial direction, data values
PORTD fcb 'D'      Port D
      fdb $1008    Address of PortD
      fdb $1009    DDRD
      fcb 0,0      Initial direction, data values
PORTE fcb 'E'      Port E
      fdb $100A    Address of PortE
      fdb $0000    No direction, always input
      fcb $FF,$FF  No Initial direction, initial data value
```

10.7   Multiple-Access Circular Queues

A multiple-access circular queue (**MACQ**) is used for data flow problems. A MACQ is a fixed-length order-preserving data structure, as shown in Figure 10.7. The source process places information into the MACQ. Once initialized, the MACQ is always full. The oldest data are discarded when the newest data are entered into a MACQ. The sink process can read any of the data from the MACQ. Reading the data into the MACQ is nondestructive. This means that the MACQ is not changed by the read operation.

Figure 10.7
A multiple-access circular queue stores the most recent set of measurements.

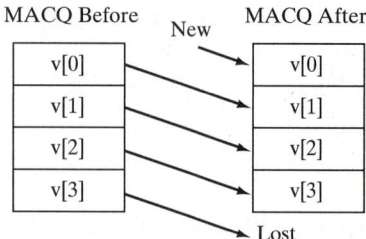

The MACQ is useful for implementing digital filters and digital controllers. The following equation illustrates a robust way to calculate the first derivative of a measured signal. Let v[0], v[1], v[2], and v[3] be the most recent data sampled at a fixed time period δt. If each v has the units of mV, and δt has the units of msec, then the derivative, d, will be in mV/msec:

$$d = \frac{v[0] + 3v[1] - 3v[2] - v[3]}{6 \cdot \delta t}$$

To measure the derivative, the following sequence of operations is executed every 1 ms:

$$v[3] = v[2]$$

$$v[2] = v[1]$$

$$v[1] = v[0]$$

$$v[0] = \text{new voltage measurement (in mV)}$$

$$d = (v[0] + 3*v[1] - 3*v[2] - v[3])/6$$

Checkpoint 10.12: Write an equation to calculate acceleration using inputs v[0], v[1], and v[2].

Checkpoint 10.13: Write a more robust equation to calculate acceleration using inputs d[0], d[1], d[2], and d[3], where d is calculated using the above equation.

10.8   First-In-First-Out Queue and Double Buffers

The **first-in-first-out** circular queue (**FIFO**) and **double buffer** are useful for data flow situations, as shown in Figure 10.8. These data structures can be used to link a source process (hardware/software that generates data) to a sink process (hardware/software that consumes data). In both cases the data are order-preserving, such that the order in which data are saved equals the order in which they are retrieved.

Figure 10.8
FIFO queues and double buffers can be used to pass data from a producer to a consumer.

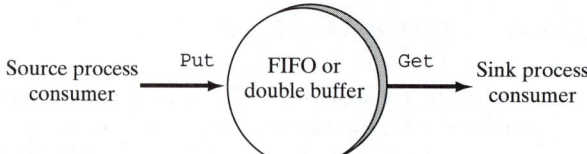

The source process puts data into the FIFO or double buffer. If there is room, the `Put` operation saves data in the structure. If the data structure is full and the user tries to put, the `Put` routine will return a full error signifying the last (newest) data was not properly saved. The sink process removes data from the FIFO or double buffer. After a `Get`, the particular information returned from the `Get` routine is no longer saved. If the structure is empty and the user tries to get, the `Get` routine will return an empty error signifying no data could be retrieved. The FIFO and double buffer are order preserving, such that the information is returned by repeated calls of `Get` in the same order as the data was saved by repeated calls of `Put`. The implementation of the FIFO will be presented in Chapter 11. A FIFO typically can store many small chunks of data, whereas a double buffer can store two large fixed-size blocks of data.

Checkpoint 10.14: What conditions might cause the FIFO to become full?

A double buffer is two buffers of fixed size. One example that uses a double buffer is a disk. Consider the situation where a large amount of data is to be read from a disk. The disk is organized into fixed size blocks. The size of each of the two buffers will match the block size of the disk. In the situation shown in Figure 10.9, the hardware is reading data from the disk filling `Buf1`. The hardware is configured to read an entire block. During this time the software is processing the data previously stored in `Buf2`. The double buffer will preserve order. This means the order in which the characters are input from the floppy disk is the same as the order in which they are processed by the software. The differences between a FIFO queue and a double buffer are data size and queue length. The data size of a FIFO is typically one or two bytes. This means that one puts and gets single bytes into and out of the FIFO queue. The data size of the double buffer is typically large (e.g., 80, 256, 1024 bytes). This means that one always saves and removes big blocks into and out of the double buffer. The FIFO queue length is large (typically ranging from 16 to 60,000 bytes). The double buffer has exactly two buffers. When the software finishes processing `Buf2` and the hardware finishes filling `Buf1`, the buffers are switched. (Hardware fills `Buf2` and the software processes `Buf1`). This means that if the hardware finishes first, then the disk hardware will have to be paused. Maximum disk efficiency occurs only if the disk can continuously read data as the blocks pass under the read head.

Figure 10.9
A double buffer allows you to store data into one buffer at the same time as retrieving data from the other buffer.

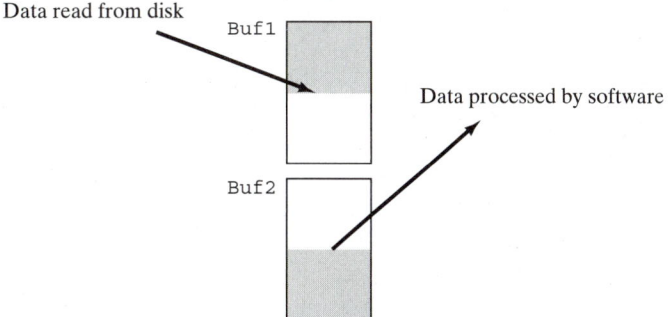

There are two events required to read data from a disk. First, the disk must be positioned so the desired data are under the read head. This operation is called **seek**, and involves specifying the head, track, cylinder, and sector number, then waiting for the physical motion to occur. The second phase of the process is the actual data transfer. If multiple blocks are to be read, it is more efficient to perform one seek followed by multiple transfers. Figure 10.10

illustrates how a double buffer can be used to read a sequence block from a disk. The system begins in state 0 by commanding the hardware to seek, then to fill `Buf1`. When `Buf1` is full, we go to state 1, where the system simultaneously fills `Buf2` and empties `Buf1`. Notice that when the system changes from state 0 to state 1, the hardware does not stop; therefore, a second seek is not required. If the software is fast enough the system progresses through states 1,2,3,4,1,2,3, . . . at the maximum rate allowed by the hardware. On the other hand, if the hardware finishes first, it must go into an idle state (e.g., state 5 or state 6) because there is no place to put more data. Once in an idle state, the system must perform another seek to restart the I/O transfer. In a similar way, a double buffer can be used to write data to a disk. The hardware I/O transfer will be performed using direct memory access or DMA. You can find more about DMA in Chapter 10 of *Embedded Microcomputer Systems: Real-time Interfacing* by Jonathan W. Valvano, Brooks Cole Publishing, 2000.

Figure 10.10
As data is read from a disk, the double buffer can be in one of seven states.

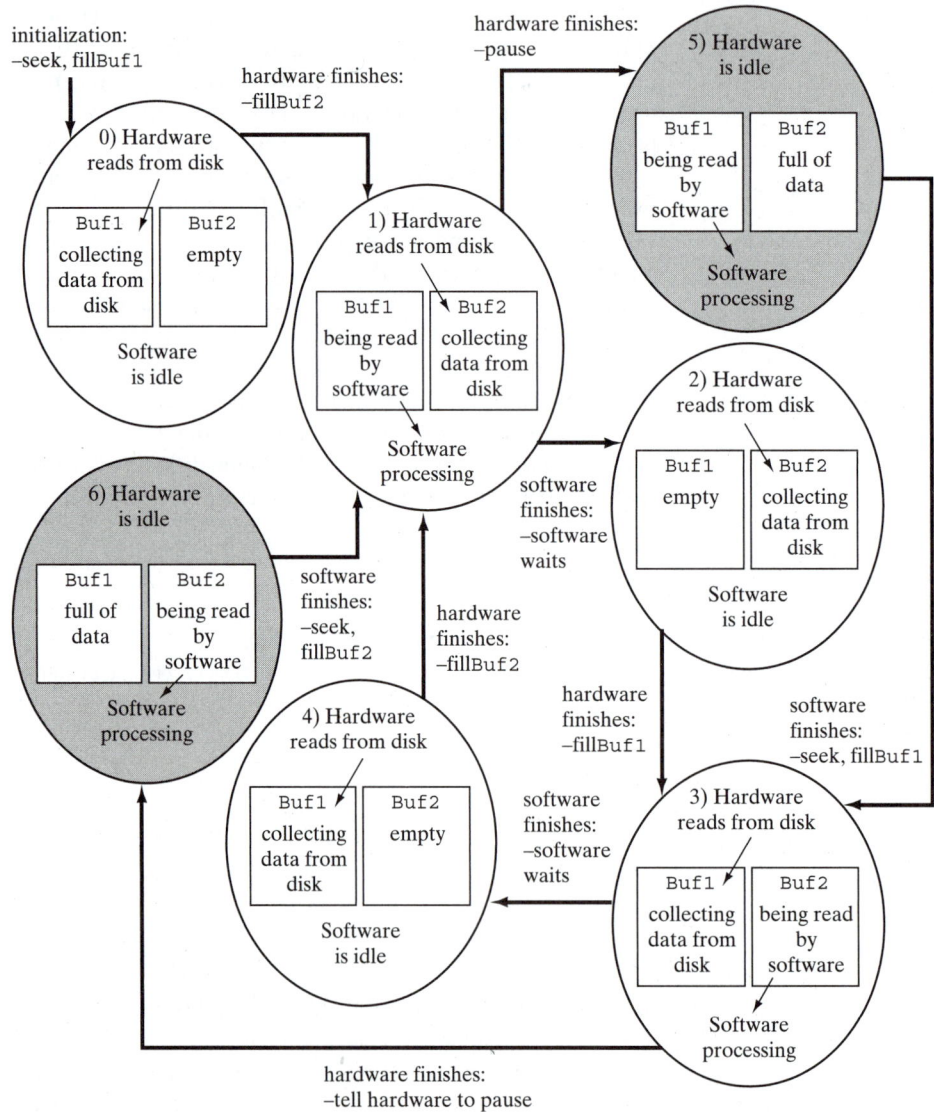

I/O devices which manipulate data in fixed size blocks are candidates for using double buffer data structures. Other examples of such devices include: graphics displays, card readers, card punches, UPC readers, credit card readers, and line printers. A graphics display uses two buffers called a front buffer and a back buffer. The graphics hardware uses the front buffer to create the visual image on the display; that is, the front buffer contains the data that you see. The software uses the back buffer to create a new image; that is, the back buffer contains the data that you see next. When the new image is ready, and the time is right, the two buffers are switched. (The front becomes the back and the back becomes the front). In this way, the user never sees a partially drawn image.

10.9   **Trees**

A graph is a general linked structure without limitations. (See Figure 10.11.) An acyclic graph is a linked structure without loops. Although there may be multiple pathways to access a node in an acyclic graph, all paths have a finite length. A tree is an acyclic graph with a single root node from which a unique path to each node can be traced.

Figure 10.11
Graphs and trees have nodes and are linked with pointers.

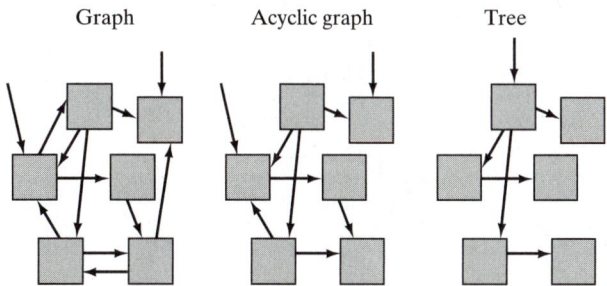

Figure 10.12 shows that an arbitrary tree can have a variable number of leaves, while a binary tree consists of a node with exactly two pointers (i.e., links, or branches).

Figure 10.12
A tree can be constructed with only down arrows, and there is a unique path to each node.

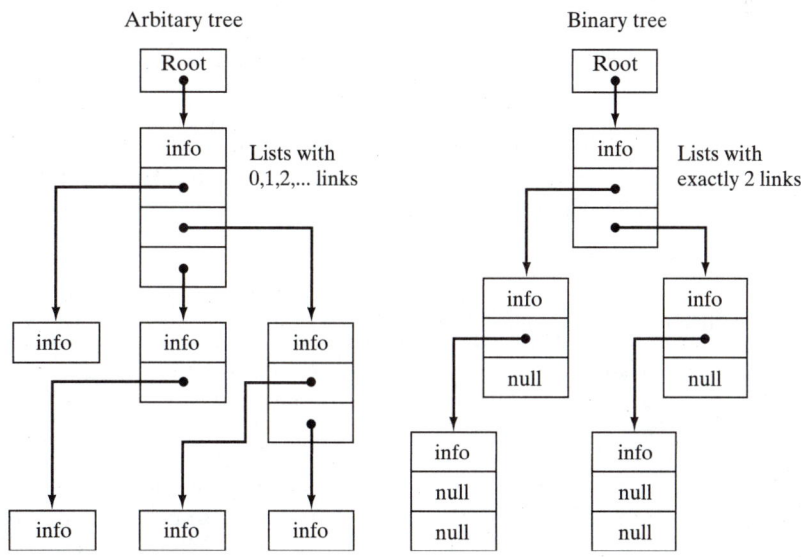

Checkpoint 10.15: Neglecting shortcuts and the StartMenu for now, what type of organization best describes the file structure on the Windows OS?

Checkpoint 10.16: Shortcuts and the StartMenu on the Windows OS allow for files and programs to be accessed in multiple ways. Observe the properties of a shortcut on your computer. Does Windows OS implement an acyclic graph?

Checkpoint 10.17: If you made an electronic dictionary where each word in the definition portion of an entry was linked to its definition, what type of structure would you have?

A null pointer signifies the end or leaf of the tree. Since each node of a tree has exactly one pointer to it, there is a unique path from the root to each node. One application of tree structure is dictionary storage, as shown in Figure 10.13. Each word is stored as a node in the tree. The position of each word in the tree is determined from its alphabetical order. In this simple dictionary the word is a single character with a single 8-bit number as its definition.

Figure 10.13
A binary tree is constructed so that earlier elements are to the left and later ones to the right.

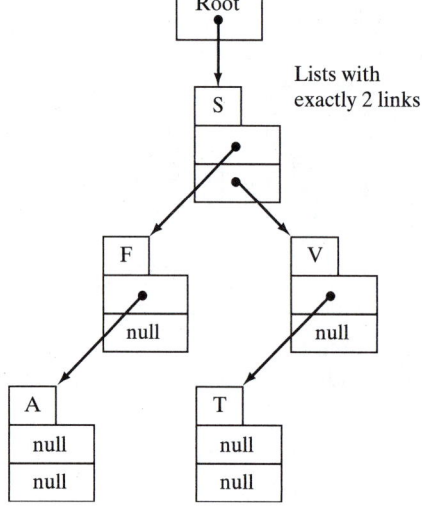

Program 10.20 shows the definition of the tree structure. If the dictionary is static, then we can define it in ROM. If it needs to be dynamic, then it must be allocated in RAM and initialized at run time. Here it is shown as a constant structure.

Program 10.20   Definition of a simple binary tree.

```
Value equ  0     name of the node
Data  equ  1     data for this node
Left  equ  2     pointer to son
Right equ  4     pointer to son
ROOT  fdb  WS    Pointer to top
NULL  equ  0     undefined address
```

```
#define NULL 0
const struct Node{
    unsigned char Value;
    unsigned char Data;
    const struct Node *Left;
    const struct Node *Right;};
```

continued

```
WS      fcb  'S',1 name,data          typedef const struct Node NodeType;
        fdb  WF    Left son            typedef NodeType * NodePtr;
        fdb  WV    Right son           #define Root WS
WV      fcb  'V',2 name,data          #define WS &Tree[0]
        fdb  WT    WT is a left son    #define WV &Tree[1]
        fdb  NULL  no right son        #define WT &Tree[2]
WT      fcb  'T',3 name,data          #define WF &Tree[3]
        fdb  NULL  no children         #define WA &Tree[4]
        fdb  NULL  no right son        NodeType Tree[5]={
WF      fcb  'F',4 name,data          { 'S',1, WF, WV},
        fdb  WA    WA is a left son    { 'V',2, WT, NULL},
        fdb  NULL  no right son        { 'T',3, NULL, NULL},
WA      fcb  'A',5 name,data          { 'F',4, WA, NULL},
        fdb  NULL  no children         { 'A',5, NULL, NULL}};
        fdb  NULL
```

Program 10.21 presents assembly and C functions that search the binary tree. To look up a word in this dictionary, one starts at the root. The following sequence is repeated until the entry is found (success) or a null point is reached (failure). If the current name matches, then quit returning the data (its definition) at that node. If the current word is not correct, then we will search left or right. If the look-up word is less than the current word, go left. If the look-up word is greater than the current word, go right. Quit with a false result if the pointer becomes null.

Program 10.21   Binary tree search functions.

```
*Inputs:   Reg A = look up letter      int Look(unsigned char letter){
*Outputs: Reg A=0 if not found,          NodePtr pt = Root;  /* top */
*              =data if found             while(pt!=NULL){  /* done when null */
Look  ldx   Root       current word         if(pt->Value == letter){
loop  cpx   #NULL                              return(pt->Data); /* good */
      beq   fail                            }
      cmpa  Value,x   Match                 if(pt->Value < letter){
      beq   found     Skip if found            pt = pt->Right;
      blo   golft                           }
      ldx   Right,x   letter>value          else{
      bra   loop                               pt = pt->Left;
golft ldx   Left,x    letter<value          }
      bra   loop                          }
fail  clra            not in tree          return NULL; /* not in tree */
      bra   exit                         }
found ldaa  Data,x    return Data
exit  rts
```

In order to add and remove nodes at run time, the tree must be defined in RAM. Program 10.22 shows how to insert a new word into the dictionary. One first searches for the word (the search should fail), then change the null pointer to point to the new list. If the search fails in the above Look subroutine, Reg X contains the address of the null pointer to be changed.

Program 10.22 Program to add a node to a binary tree.

```
*  Inputs : Reg Y points to a new word to be added to the dictionary
*     the new word is already somewhere in memory formatted e.g.,
*     fcb  'J',6
*     fdb  NULL
*     fdb  NULL
NEW   ldaa 0,Y     Reg A is the name of the new word
      bsr  LOOK
      tsta
      bne  ok      skip if already defined
      sty  0,X     Update link
OK    rts
```

Figure 10.14 shows the binary tree as the nodes J, U, G are added to the dictionary.

Figure 10.14
Nodes are added to a binary tree such that the alphabetical order is maintained.

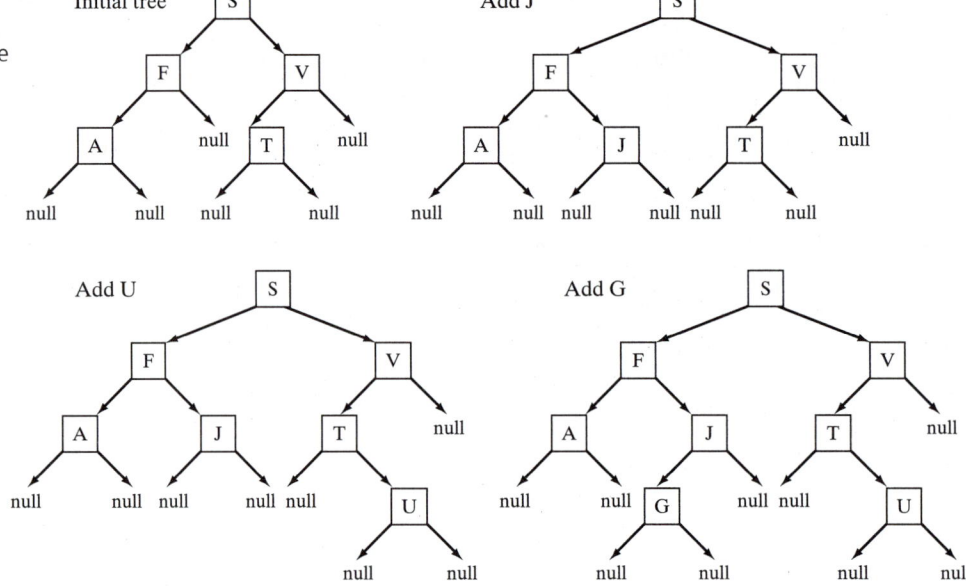

This may seem like a lot of trouble for such a simple problem. The search time for a binary tree increases as the $\log_2$ of the size of the dictionary. For a simple linear structure (e.g., table or linked list), the search time increases linearly with the dictionary size. When the dictionary is millions of words, the time savings can be extraordinary. There are similar savings in the insertion and deletion times. The dynamic efficiency (execution speed) is enhanced at the cost of static efficiency (memory storage).

Checkpoint 10.18: Consider the problem of designing a large address book where each entry as a first name, a last name, and an address field. You wish to be able to search the data base by first name and by last name. How do you organize the structure?

10.10 *Dynamically Allocated Data Structures

In order to reuse memory and provide for flexible operation, we will develop dynamic memory allocation. With a fixed allocation like the previous examples, the size of the data structures is decided in advance and specified in the source code. With a dynamic allocation the size will be determined at run time. To implement dynamic allocation we will manage a heap. The heap is a chunk of RAM that is

1. dynamically allocated by the program when it creates the data structure
2. used by the program to store information
3. dynamically released by the program when the structure is no longer needed

The heap manager provides the system two operations:

```
pt=allocate(size); /* returns a pointer to a block of size bytes */
release(pt);       /* deallocates the block at pt */
```

The implementation of this general memory manager is beyond the scope of this book. Instead, we will develop a very useful but simple heap manager with these two operations:

```
pt=Heap_Allocate(); /* returns a pointer to a block of fixed size */
Heap_Release(pt);   /* deallocates the block at pt */
```

**10.10.1
Fixed-Block
Memory Manager**

In general, the heap manager allows the program to allocate a variable block size, but in this section we will develop a simplified heap manager that handles just fixed-size blocks. In this example, the block size is fixed by `SIZE`. The initialization will create a linked list of all the free blocks (Figure 10.15).

Figure 10.15
The initial state of the heap has all of the free blocks linked in a list.

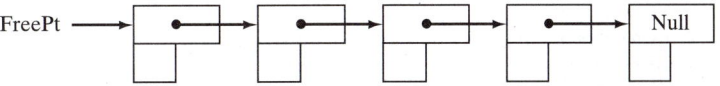

Program 10.23 shows the global structures for the heap. These entries are defined in RAM. `SIZE` is the number of 8-bit bytes in each block. All blocks allocated and released with this memory manager will be of this fixed size. `NumB` is the number of blocks to be managed. `FreePt` points to the first free block.

Program 10.23   Global structures for the fixed-block memory manager.

`SIZE     equ  4`
`NUM      equ  5`
`NULL     equ  0`
`FreePt rmb  2`
`Heap     rmb  SIZE*NUM`
	`#define SIZE 4`
	`#define NUM 5`
	`#define NULL 0  /* empty pointer */`
	`char *FreePt;`
	`char Heap[SIZE*NUM];`

Initialization must be performed before the heap can be used. Program 10.24 shows the software that partitions the heap into blocks and links them together. `FreePt` points to a linear linked list of free blocks. Initially these free blocks are contiguous and in order, but as the manager is used the positions and order of the free blocks can vary. It will be the pointers that will thread the free blocks together.

Program 10.24 Functions to initialize the heap.

```
Heap_Init ldx   #Heap
          stx   FreePt    FreePt=&Heap[0];
          ldab  #SIZE
imLoop    pshx
          puly            RegY = pt;
          aby             pt+SIZE
          sty   0,x       *pt=pt+SIZE;
          abx             pt=pt+SIZE;
          cpx   #Heap+SIZE*(NUM-1)
          bne   imLoop
          ldy   #NULL
          sty   0,x       *pt=NULL;
          rts
```

```
void Heap_Init(void){
char *pt;
  FreePt = &Heap[0];
  for(pt=&Heap[0];
      pt!=&Heap[SIZE*(NUM-1)];
      pt=pt+SIZE){
    *(short*)pt =(short)(pt+SIZE);
  }
  *(short*)pt = NULL;
}
```

To allocate a block, the manager just removes one block from the free list. Program 10.25 shows the allocate and release functions that return a pointer to a new free block. The Heap_Allocate function will fail and return a null pointer when the heap becomes empty. The Heap_Release returns a block to the free list. This system does not check to verify that a released block actually was previously allocated.

Program 10.25 Functions to allocate and release memory blocks.

```
* returns RegX points to new block
* RegX=NULL if no more available
Heap_Allocate
        ldx   FreePt    pt=FreePt;
        cpx   #NULL
        beq   aDone     if (pt!=NULL)
        ldy   0,x
        sty   FreePt    FreePt=*pt;
aDone rts
* RegX => block being released
Heap_Release
        ldy   FreePt    oldFreePt=FreePt;
        stx   FreePt    FreePt=pt;
        sty   0,x       *pt=oldFreePt;
        rts
```

```
void *Heap_Allocate(void){
char *pt;
  pt = FreePt;
  if (pt != NULL){
    FreePt = (char*) *(char**)pt;
  }
  return(pt);
}
void Heap_Release(void *pt){
char *oldFreePt;
  oldFreePt = FreePt;
  FreePt = (char*)pt;
  *(short*)pt = (short)oldFreePt;
}
```

Checkpoint 10.19: Consider a system that needs variable-size memory allocation, where the size can range from 2 to 20 bytes. How might this simple heap be used?

10.10.2
Linked List FIFO

An example application of a dynamically allocated data structure is a FIFO. In this structure `GetPt` points to the oldest node (the one to get next) and `PutPt` points to the newest node, the place to add more data. The `Fifo_Put` operation fails (full) when the heap runs out of space. The `Fifo_Get` operation fails (empty) when `GetPt` equals `NULL`. Program 10.26 shows the global variables defined in RAM. Figure 10.16 shows an example FIFO with three elements. In this example, element 1 is the oldest because it was put first.

Program 10.26   Definition of the linked list structure.

```
Next    equ  0    next
Data    equ  2    16-bit data for node

GetPt rmb  2
* GetPt  is pointer to oldest node
PutPt rmb  2
* PutPt is pointer to newest node

NULL    equ  0
```

```
struct Node{
  struct Node *Next;
  short Date;
};
typedef struct Node NodeType;
typedef NodeType *NodePtr;
NodePtr PutPt;    /* place to put */
NodePtr GetPt;    /* place to get */
#define NULL 0
```

Figure 10.16
A linked list FIFO after
putting 1,2,3.

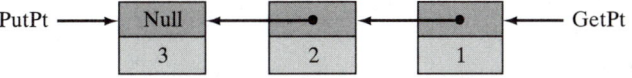

Program 10.27 shows the three functions which implement the FIFO. Figure 10.17 is a flowchart of the Put and Get functions. The FIFO is full only when the heap is full (`Heap_Allocate` returns a failure). The Put operation first allocates space for the new entry, then stores the new information into the `Data` field. Since this element will be last, its `Next` field is set to null. The last part of the Put links this new node at the end of the linked list. The Get function first checks to make sure the FIFO is not empty. Next, the `Data` field is retrieved from the node. This node is then unlinked from the linked list, and the memory block is released to the heap. A special case handles the situation where you get the one remaining node in the linked list. In this case both `PutPt` and `GetPt` point to this node. When you get this node, both `PutPt` and `GetPt` are set to null, signifying the FIFO is now empty.

Program 10.27   Implementation of the linked list FIFO.

```
Fifo_Init
    ldx  #NULL
    stx  GetPt        GetPt=NULL;
    stx  PutPt        PutPt=NULL;
    jsr  Heap_Init
    rts
* Inputs:  RegD data to put
* Outputs: V=0 if successful
*          V=1 if unsuccessful
Fifo_Put jsr  Heap_Allocate
    cpx  #NULL
    beq  Pful      skip if full
    std  Data,x    store data
    ldy  #NULL
    sty  Next,x    next=NULL
    ldy  PutPt
    cpy  #NULL     previously MT?
    beq  PMT
    stx  Next,y    link to previous
    bra  PCon
PMT stx  GetPt    Now one entry
PCon stx PutPt    points to newest
    clv            success
    bra  PDon
PFul sev           failure, full
PDon rts
* Inputs:  none
* Outputs: RegD data removed
*          V=0 if successful
*          V=1 if empty
Fifo_Get ldx GetPt
    cpx  #NULL
    beq  GMT       empty if NULL
    ldd  Data,x    read
    ldy  Next,x    pointer to next
    sty  GetPt
    cpy  #NULL
    bne  GCon
    sty  PutPt     Now empty
GCon sty  GetPt    points to oldest
    jsr  Heap_Release
    clv            success
    bra  GetDone
GMT sev            failure, empty
GDon rts
```

```c
void Fifo_Init(void){
  GetPt = NULL; /* Empty when null */
  PutPt = NULL;
  Heap_Init();
}

int Fifo_Put(short theData){
NodePtr pt;
  pt = (NodePtr)Heap_Allocate();
  if(!pt){
    return(0); /* full */
  }
  pt->Data = theData;  /* store */
  pt->Next = NULL;
  if(PutPt){
    PutPt->Next = pt; /* Link */
  }
  else{
    GetPt = pt;  /* first one */
  }
  PutPt = pt;
  return(1);   /* successful */
}




int Fifo_Get(short *datapt){
NodePtr pt;
  if(!GetPt){
    return(0); /* empty */
  }
  *datapt = GetPt->Data;
  pt = GetPt;
  GetPt = GetPt->Next;
  if(GetPt==NULL){ /* one entry */
    PutPt = NULL;
  }
  Heap_Release(pt);
  return(1);    /* success */
}
```

Figure 10.17
Flowcharts of a linked
list FIFO Put and Get
operations.

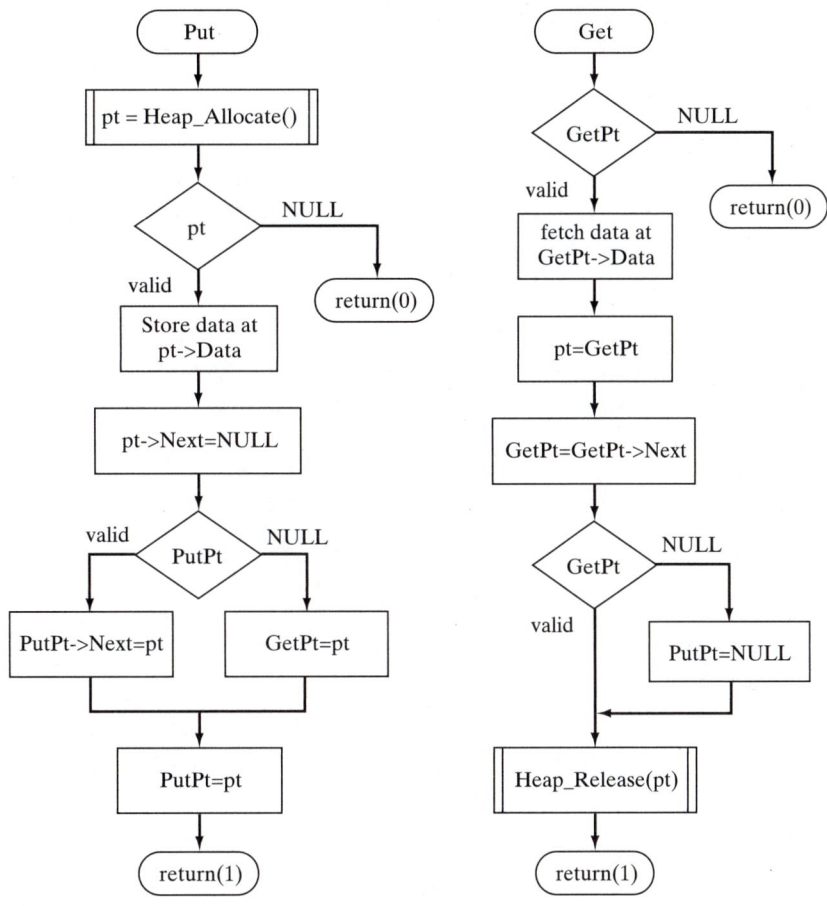

Checkpoint 10.20: Draw a picture like Figure 10.16 of a doubly linked list. How might
this more complicated structure be more efficient than the single linked list?

10.11 Tutorial 10. Command Interpreters

In this tutorial, we will study five ways to construct a command interpreter. The first mechanism does not use a data structure, and it will be called direct coding. The other four use a data
structure: array, table, linear linked list, and a binary tree. The data structure provides for a more
elegant solution. Using a data structure provides an abstraction, and should make it easier to
understand, quicker to debug, and simpler to modify. In the following examples the command
interpreter operates on a single letter basis. All examples, except the array, can be modified to
handle command strings, by replacing the simple `beq` with a string compare operation. You
will first be asked to calculate the search speed for each example. Then you will study its ease
of modification by adding additional commands to the system. The search speed of the array
structure is independent of size. Three of the systems (direct coding, table, and linear linked
list) have a linear search speed. This means the search speed is linearly related to the size of the

structure. Consequently, the average search speed will be approximately one-half the longest search speed. Assuming the binary tree is constructed with minimum depth, its search speed is related to the $\log_2$ of the size of the structure.

Action: Open the files `Chap10.us` and `Chap10.io`. ScanPoints have been added at `search`, `Accept`, `High`, `Make`, `Verify`, and `Zap`. The purpose of these ScanPoints is to measure the search time of the interpreter. Each time you type an interpreter input, there should be two numbers dumped into `TheLog.rtf` file. The difference between these numbers is the time it takes to search the structure.

The first example is **direct coding**. The interpreter part of this system is shown in Program 10.28. The complete system can be found on the CD as `Chap10a.rtf`. It is a simple and common approach to interpreter implementation. There are five commands, and each one simply prints out a message.

Program 10.28 Direct coding interpreter.

```
Exec   jsr  SCI_InChar   Input        void exec(void){char letter;
search cmpa #'A'                        while(1){
       bne  11     Check each             letter = SCI_InChar();
       jsr  Accept Execute                if(letter == 'A'){
       bra  done                            Accept();
11     cmpa #'H'                          }
       bne  12     Check next             else if(letter == 'H'){
       jsr  High   Execute                  High();
       bra  done                          }
12     cmpa #'M'                          else if(letter == 'M'){
       bne  13     Check next               Make();
       jsr  Make   Execute                }
       bra  done                          else if(letter == 'V'){
13     cmpa #'V'                            Verify();
       bne  14     Check next             }
       jsr  Verify Execute                else if(letter == 'Z'){
       bra  done                            Zap();
14     cmpa #'Z'                          }
       bne  error  not in list           else
       jsr  Zap    Execute                  error();
done   bra  Exec                         }
```

Action: Open and assemble the program `Chap10a.rtf`. Run the `Chap10a.rtf` system and execute all five commands. Try typing undefined commands.

Question 10.1. Use the ScanPoints to measure the minimum and maximum search speed (in cycles) of this system.

Question 10.2. If you were to increase the number of commands to 100, estimate the average search speed for this large interpreter.

Question 10.3. Add two new commands: B (prints Bye) and S (prints Stop).

The second example uses an **array** data structure. The 6812 version of the interpreter part of this system is shown in Program 10.29. The complete system (both 6811 and 6812) can be found on the CD as `Chap10b.rtf`. When the command names are in contiguous linear order, this is a very efficient implementation. The interpreter had to be modified to accept the letters A, B, C, D, and E in place of A, H, M, V, and Z. Otherwise this interpreter matches the other four interpreters presented in this tutorial.

Program 10.29   An array interpreter.

```
Exec    jsr  SCI_InChar  Input
search  cmpa #'E'     last one
        bhi  error
        ldx  #array   function list
        suba #'A'     now 0 to 2
        bmi  error
        lsla           calculate index
        ldx  A,X       X=> subroutine
        jsr  0,X       Execute command
        bra  Exec
array   fdb  Accept  function pt
        fdb  High
        fdb  Make
        fdb  Verify
        fdb  Zap
```

```c
const struct {void (*fnct)(void);}
    array[5]={&Accept,&Help,
    &Make,&Verify,&Zap};
void exec(void){char i;
  while(1){
    i = SCI_InChar()-'A';
    if((i<0)||(i>4))
      error();
    else
      array[i].fnct();
  }
}
```

Action: Open and assemble the program `Chap10b.rtf`. Run the `Chap10b.rtf` system and execute all five commands. Try typing undefined commands.

Question 10.4. Use the ScanPoints to measure the minimum and maximum search speed (in cycles) of this system.

Question 10.5. If you were to increase the number of commands to 100, estimate the average search speed for this large interpreter.

Question 10.6. Add two new commands: F (prints Bye) and G (prints Stop).

The third example uses a **table** data structure. The 6812 version of the interpreter part of this system is shown in Program 10.30. The complete system (both 6811 and 6812) can be found on the CD as `Chap10c.rtf`. A table provides a very good abstraction. In other words, when one looks just at the table structure, one sees very clearly what the interpreter does.

Program 10.30   A table interpreter.

```
letter equ  0    Index for command
fnct   equ  1    Index for address
Exec   jsr  SCI_InChar Input
```

```c
const struct {
  char letter;
  void (*fnct)(void);} table[6]={
```
continued

```
search  ldx   #Table-3   Commands           {'A',&Accept},
test    leax  3,x        Next entry          {'H',&Help},
        tst   letter,X   Done?               {'M',&Make},
        beq   error                          {'V',&Verify},
        cmpa  letter,X   Is it this?         {'Z',&Zap},
        bne   test                           {0,0}};
        jsr   [fnct,X]   Execute      void exec(void){char input; int i;
        bra   Exec                      while(1){
Table   fcb   'A'        letter             input = SCI_InChar()
        fdb   Accept     function            for(i=0;i<6;i++){
        fcb   'H'                               if(table[i].letter==0)
        fdb   High                                error();
        fcb   'M'                               else if(input == table[i].letter)
        fdb   Make                                table[i].fnct();
        fcb   'V'                            }
        fdb   Verify                       }
        fcb   'Z'                        }
        fdb   Zap
        fcb   0          end of table
```

Action: Open and assemble the program `Chap10c.rtf`. Run the `Chap10c.rtf` system and execute all five commands. Try typing undefined commands.

Question 10.7. Use the ScanPoints to measure the minimum and maximum search speed (in cycles) of this system.

Question 10.8. If you were to increase the number of commands to 100, estimate the average search speed for this large interpreter.

Question 10.9. Add two new commands: B (prints Bye) and S (prints Stop).

The fourth example uses a **linked list** data structure. The 6812 version of the interpreter part of this system is shown in Program 10.31. The complete system (both 6811 and 6812) can be found on the CD as `Chap10d.rtf`. The linked list, like the table, provides a very good abstraction. The advantage of a linked list over a table exists if the interpreter is dynamic, allowing commands to be added and removed.

Program 10.31   A linked list interpreter.

```
NULL    equ   0    Null pointer         #define NULL 0
letter  equ   0    Index for command    const struct Node{
fnct    equ   1    Index for address        unsigned char letter;
next    equ   3    Index for next           void (*fnct)(void);
```
continued

```
Exec    jsr  SCI_InChar Input              const struct Node *next;};
search  ldx  #LinkList  First entry    typedef const struct Node NodeType;
loop    cmpa letter,X   Is it this?    NodeType LinkedList[5]={
        beq  found                       {'A',&Accept, &LinkedList[1]},
        ldx  next,X                       {'H',&Help,   &LinkedList[2]},
        cpx  #NULL                        {'M',&Make,   &LinkedList[3]},
        bne  loop                         {'V',&Verify, &LinkedList[4]},
        bra  error                        {'Z',&Zap, 0}};
found   jsr  [fnct,X] Execute         void exec(void){char input; NodeType *pt;
        bra  Exec                        while(1){
LinkList                                   input = SCI_InChar()
CMDA    fcb  'A'       letter                pt = &LinkedList[0];
        fdb  Accept    function              while(pt!=NULL){
        fdb  CMDH       Ptr to next             if(input==pt->letter){
CMDH    fcb  'H'                                  pt->fnct();
        fdb  High                                 pt = NULL;
        fdb  CMDM                               }
CMDM    fcb  'M'                               else{
        fdb  Make                                pt = pt->next;
        fdb  CMDV                                if(pt==NULL) error();
CMDV    fcb  'V'                              }
        fdb  Verify                          }
        fdb  CMDZ                          }
CMDZ    fcb  'Z'                        }
        fdb  Zap
        fdb  NULL      end
```

Action: Open and assemble the program Chap10d.rtf. Run the Chap10d.rtf system and execute all five commands. Try typing undefined commands.

Question 10.10. Use the ScanPoints to measure the minimum and maximum search speed (in cycles) of this system.

Question 10.11. If you were to increase the number of commands to 100, estimate the average search speed for this large interpreter.

Question 10.12. Add two new commands: B (prints Bye) and S (prints Stop).

The last example uses a **binary tree** data structure, as shown in Figure 10.18. The 6812 version of the interpreter part of this system is shown in Program 10.32. The complete system (both 6811 and 6812) can be found on the CD as Chap10e.rtf. The binary tree is usual for interpreters with a large number of commands. If there are n commands, then a binary tree has a search speed proportional to $\log_2 n$. Notice the one-to-one mapping between the binary tree in Figure 10.18 and the struct in Program 10.32.

Figure 10.18
A binary tree used for an
interpreter.

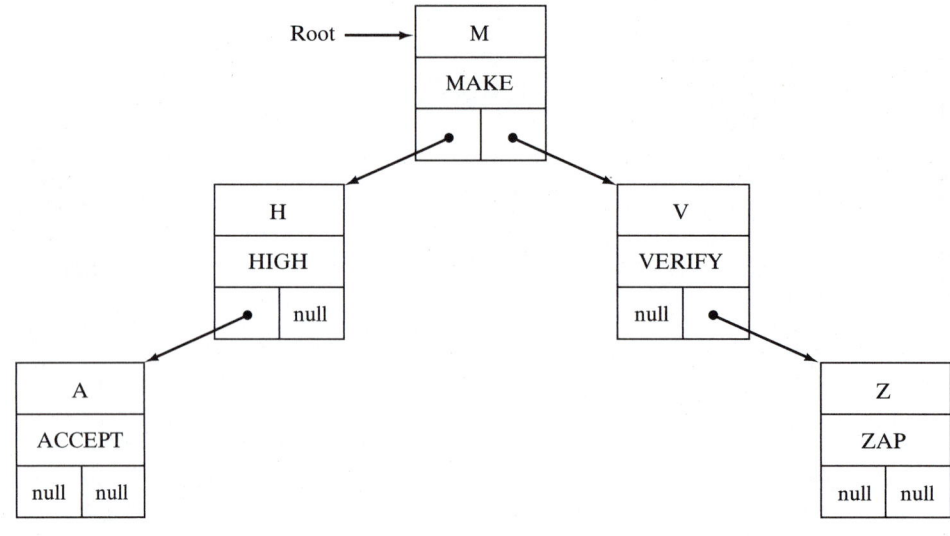

Program 10.32   A binary tree interpreter.

```
NULL    equ  0    Null pointer
letter  equ  0    command
fnct    equ  1    address
left    equ  3    left entry
right   equ  5    righ entry
Exec    jsr  SCI_InChar Input
search  ldx  Root       First entry
test    cpx  #NULL
        beq  error      Not found?
        cmpa letter,X   Is it this?
        beq  found
        bhi  goRight
goLeft  ldx  left,X     A<Letter
        bra  test
goRight ldx  right,X    A>Letter
        bra  test
found   jsr  [fnct,X]   Execute
        bra  Exec
Root fdb CMDM
CMDM fcb  'M'    Command letter
     fdb  Make   Ptr to subroutine
     fdb  CMDH   Ptr to left entry
     fdb  CMDV   Ptr to right entry
CMDH fcb  'H'    Command letter
     fdb  High   Ptr to subroutine
     fdb  CMDA   Ptr to left entry
     fdb  NULL   None to the right
CMDA fcb  'A'    Command letter
     fdb  Accept Ptr to subroutine
     fdb  NULL   None to the left
     fdb  NULL   None to the right
```

```
const struct Node{
    unsigned char letter;
    void (*fnct)(void);
    const struct Node *left;
    const struct Node *right;};
typedef const struct Node NodeType;
#define NULL 0
#define CMDM &Tree[0]
#define CMDH &Tree[1]
#define CMDA &Tree[2]
#define CMDV &Tree[3]
#define CMDZ &Tree[4]
NodeType Tree[5]={
    {'M',&Make,   CMDH, CMDV},
    {'H',&Help,   CMDA, NULL},
    {'A',&Accept, NULL, NULL},
    {'V',&Verify, NULL, CMDZ},
    {'Z',&Zap,    NULL, NULL}};
void exec(void){char input; NodeType *pt;
    while(1){
      input = SCI_InChar()
      pt = CMDM;  // start with M
      while(pt){
        if(input==pt->letter){
          pt->fnct(); // execute function
          pt = 0;
        }
        else{
          if(pt->letter < input){
            pt = pt->right;
          }
```

continued

```
CMDV fcb  'V'     Command letter
     fdb  Verify  Ptr to subroutine
     fdb  NULL    None to the left
     fdb  CMDZ    Ptr to right entry
CMDZ fcb  'Z'     Command letter
     fdb  Zap     Ptr to subroutine
     fdb  NULL    None to the left
     fdb  NULL    None to the right
```

```
            else{
              pt = pt->left;
            }
          if(pt==0) error();
        }
      }
    }
  }
}
```

Action: Open and assemble the program `Chap10e.rtf`. Run the `Chap10e.rtf` system and execute all five commands. Try typing undefined commands.

Question 10.13. Use the ScanPoints to measure the minimum and maximum search speed (in cycles) of this system.

Question 10.14. If you were to increase the number of commands to 100, estimate the average search speed for this large interpreter.

Question 10.15. Add two new commands: B (prints Bye) and S (prints Stop).

10.12 Homework Problems

Homework 10.1: Design a microcomputer-based controller using a linked-list finite state machine. The system has one input and one output.

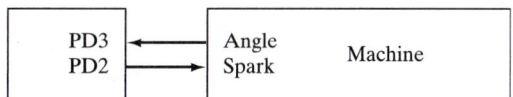

The input, Angle, is a periodic signal with a frequency of about 10 Hz (has a period of about 100ms). The output, Spark, should be a positive pulse (exactly 1 ms wide) every time Angle goes from 0 to 1. The delay between the rising edge of Angle and the start of the Spark pulse should be as short as possible. The period of Angle can vary from 1 ms to 1 sec. Since Angle is an input you can not control it, only respond to its rising edge.

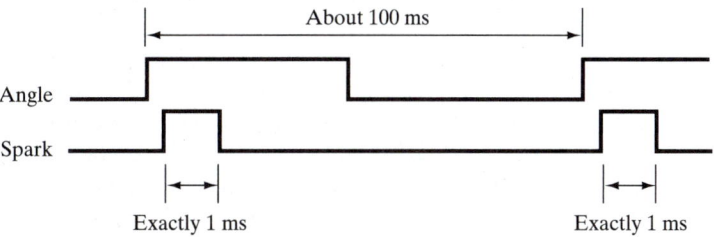

Part A.
Design the one input, one output finite state machine for this system. Draw the FSM graph. Use descriptive state names (i.e., don't call them S0,S1,S2, . . .)

Part B.

Show the assembly code to create the statically allocated linked list. Include `org` statement(s) to place it in the proper location on your microcomputer.

Part C.

Show the assembly language controller. Include ORG statement(s) to place it in the proper location on a microcomputer. Assume this is the only task that the microcomputer executes; that is, show *all* the instructions necessary. Make the program automatically start on a RESET.

Homework 10.2: Implement the following Mealy Finite State machine using linked lists. The initial state is `Stop`. Do not convert the finite state machine to an equivalent Moore; rather implement it exactly as is. There is no wait parameter for the states.

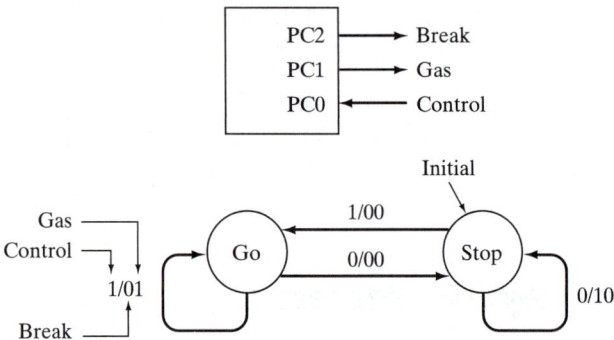

There is one input, `Control`, connected to PC0. There are two outputs, `Break` connected to PC2, and `Gas` connected to PC1. Each state has two next states and two outputs which depend on the current input. The controller continuously repeats the sequence:

> input from `Control` (PC0)
> output to `Break`, `Gas` (PC2, PC1) which depends on the input `Control`
> next state which depends on the input `Control`

For example, if the state is in `Stop`, and the `Control` is 0, then the `Break` output is 1 and the `Gas` output is 0 and the next state is `Stop`. Show *all* the assembly language software required to implement this machine on a single chip microcomputer. Use `equ` statements to clarify the data structure. Use `org` statements to implement the appropriate segmentation. COMMENTS will be graded.

Homework 10.3: Assume that we have some 6-row by 8-column matrix data structures. The precision of each entry is 16 bits. The information is stored in column-major format (the data for each column are stored contiguously) with zero indexing (i.e., the range of the row index, I, is $0 \le I \le 5$, and the range of the column index, J, is $0 \le J \le 7$. Write an assembly language subroutine that accepts a pointer to the array and the I,J indices and returns the 16-bit contents. Don't save/restore registers.

```
*Inputs   RegA row index     I=0,1,...,5
*         RegB column index J=0,1,...,7
*         RegX pointer to a 6 by 8 matrix
*Outputs RegD 16-bit contents at matrix[I,J]
```

Homework 10.4: Assume that we have some 5-row by 10-column matrix data structures. The precision of each entry is 16 bits. The information is stored in column-major format (the data for each column are stored contiguously) with zero indexing (i.e., the range of the row index, I, is 0≤I≤4, and the range of the column index, J, is 0≤J≤9. Write an assembly language subroutine that accepts a pointer to the array and the I,J indices and returns the 16-bit contents. Don't save/restore registers.

```
*Inputs  RegA row index     I=0,1,...,4
*        RegB column index J=0,1,...,9
*        RegX pointer to a 5 by 10 matrix
*Outputs RegD 16-bit contents at matrix[I,J]
```

Homework 10.5: Consider the following table structure:

```
const struct theRoom{
  unsigned char windows;    // number of windows
  unsigned char doors;      // number of doors
  unsigned short size[3];   // x,y,z dimensions
}
typedef const struct theRoom roomType;
roomType Building[4]={
  { 3,2,{16,16,8}},
  { 4,1,{20,20,10}},
  { 5,3,{32,16,12}},
  { 0,1,{18,10,8}}};
```

Part A:
Show the assembly code required to define this structure in ROM. Use `equ` to make the code easier to understand.

Part B:
Write an assembly program to return the number of windows of a room. The room number is passed by value in Register A, and the result is returned by value in Register A. For example, if the room number is 2, then the number of windows will be 5.

Part C:
Write an assembly program to return the number of doors of a room. The room number is passed by value in Register A, and the result is returned by value in Register A. For example, if the room number is 0, then the number of doors will be 2.

Part D:
Write an assembly program to return the volume of a room. The room number is passed by value in Register A, and the result is returned by value in Register D. For example, if the room number is 1, then the volume will be `20*20*10=4000`.

Homework 10.6: Consider the following table structure:

```
const struct thedesk{
  unsigned char legs;        // number of legs
  unsigned char drawers;     // number of drawers
  unsigned short size[2];    // top x,y dimensions 0.1 feet
}
```

```
typedef const struct thedesk deskType;
deskType furniture[4]={
  { 4,5,{30,50}},   // 4 legs 5 drawers
  { 4,0,{45,45}},   // square table
  { 6,7,{40,65}},
  { 4,4,{35,55}}};;
```

Part A.

Show the assembly code required to define this structure in ROM. Use equ to make the code easier to understand.

Part B.

Write an assembly program to return the number of legs of a desk. The desk number is passed by value in Register A, and the result is returned by value in Register A. For example, if the desk number is 2, then the number of legs will be 6.

Part C.

Write an assembly program to return the number of drawers of a desk. The desk number is passed by value in Register A, and the result is returned by value in Register A. For example, if the desk number is 0, then the number of drawers will be 5.

Part D.

Write an assembly program to return the area of a desk top with units in^2. The room number is passed by value in Register A, and the result is returned by value in Register D. For example, if the desk number is 3, then the desk top area will be (35*55*144)/100=1764. Worry about accuracy (divide last) and overflow (use enough bits in the multiply stage to prevent overflow.) You could factor the 144 and 100 terms to calculate (35*55*18)/25=1764. Your solution has to work for these four examples.

Homework 10.7: Write a subroutine to implement linear regression. The calling sequence is

```
    ldx  #DataSet1      ; pointer to data structure
    jsr  Regression
; Reg X = m =slope as a fixed point
; Reg Y = b =offset as a fixed point
; Reg D = e =average error  as a fixed point
```

Input x,y numbers will be in *signed* two's complement 8-bit decimal fixed point with a resolution, Δ, of *0.01*. These 8-bit numbers have effective values which range from -1.27 to 1.28. Outputs m,b,e will be in *signed* two's-complement 16-bit decimal fixed point with a resolution, Δ, of *0.0001*. Therefore, the 16-bit numbers have effective values which range from -3.2768 to 3.2766. Set all the results (m,b,e) equal to 3.2767 ($7FFF) if any data overflow or divide by zero occurs. A typical array structure looks like the following:

```
DataSet1 fdb 3          ; number of data points
         fcb 0,1        ; (x,y) = (0    , 0.01)
         fcb 10,2       ; (x,y) = (0.1  , 0.02)
         fcb 20,3       ; (x,y) = (0.2  , 0.03)
```

For this example, `b=0.01`, `m=0.1`, `e=0`, so Reg X is returned as 100, Reg Y is 1000, and Reg D is 0. Let (x_0, y_0) and (x_1, y_1) be two points; then the slope and intercept of the "y=mx+b" line through those points is given by

$$m = \frac{y_1 - y_0}{x_1 - x_0} \qquad b = \frac{y_0 \bullet x_1 - y_1 \bullet x_0}{x_1 - x_0} \qquad \text{and } e = 0$$

In general, let $x(i)$ and $y(i)$ be arrays of length $n > 2$. Each of the following sums range from $i=0$ to $n-1$:

$$m = \frac{n \bullet \Sigma(x(i) \bullet y(i)) - \Sigma x(i) \bullet \Sigma y(i)}{n \bullet \Sigma(x(i) \bullet x(i)) - \Sigma x(i) \bullet \Sigma x(i)} \text{ and } b = \frac{\Sigma y(i) \bullet \Sigma(x(i) \bullet x(i)) - \Sigma(x(i) \bullet y(i)) \bullet \Sigma x(i)}{n \bullet \Sigma(x(i) \bullet x(i)) - \Sigma x(i) \bullet \Sigma x(i)}$$

For $n > 2$, the average error is defined as

$$e = \frac{\Sigma |y(i) - m \bullet x(i) - b|}{n}$$

Homework 10.8: Write a subroutine to implement determinant. The calling sequence is

```
    ldx   #DataSet1      ; pointer to matrix
    jsr   Determinant
; Reg D = determinant of the matrix as a fixed point number
```

All matrix numbers will be in *signed* 16-bit decimal fixed point with a resolution, Δ, of *0.0001*. Therefore, the 16-bit numbers have effective values which range from -3.2768 to 3.2767. A typical matrix structure looks like the following:

```
DataSet1 fcb   4                          ; 4 by 4 matrix
         fdb      0,1000,2000,3000         ; (0.0, 0.1, 0.2, 0.3)
         fdb   4000,5000,6000,7000         ; (0.4, 0.5, 0.6, 0.7)
         fdb   8000,9000,10000,11000       ; (0.8, 0.9, 1.0, 1.1)
         fdb   12000,13000,14000,15000     ; (1.2, 1.3, 1.4, 1.5)
```

Set the result equal to 3.2767 ($7FFF) if a data overflow or divide by zero occurs. Let **A** be a square matrix of size $n+1$. Let A_{ij} be the *reduced* matrix with the ith row and jth column removed.

$$\mathbf{A} = \begin{pmatrix} a_{00} & a_{01} & \cdots & a_{0n} \\ a_{10} & a_{11} & \cdots & a_{1n} \\ \cdots & \cdots & a_{ij} & \cdots \\ a_{n0} & a_{n1} & \cdots & a_{nn} \end{pmatrix}$$

E.g., if **A** is a 4-by-4 matrix, the A_{ij} is a 3-by-3 matrix (showing just the variable parts)

If $\qquad \mathbf{A} = \begin{pmatrix} 0.0 & 0.1 & 0.2 & 0.3 \\ 0.4 & 0.5 & 0.6 & 0.7 \\ 0.8 & 0.9 & 1.0 & 1.1 \\ 1.2 & 1.3 & 1.4 & 1.5 \end{pmatrix}$ then $\mathbf{A_{21}} = \begin{pmatrix} 0.0 & 0.2 & 0.3 \\ 0.4 & 0.6 & 0.7 \\ 1.2 & 1.4 & 1.5 \end{pmatrix}$

The determinate of **A** is defined as

$$|\mathbf{A}| = \sum_{i=0}^{i=n-1} (-1)^i a_{0i} |\mathbf{A_{0i}}|$$

E.g.,

$$\begin{vmatrix} 0.0 & 0.1 & 0.2 & 0.3 \\ 0.4 & 0.5 & 0.6 & 0.7 \\ 0.8 & 0.9 & 1.0 & 1.1 \\ 1.2 & 1.3 & 1.4 & 1.5 \end{vmatrix} = 0.0 \cdot \begin{vmatrix} \begin{pmatrix} 0.5 & 0.6 & 0.7 \\ 0.9 & 1.0 & 1.1 \\ 1.3 & 1.4 & 1.5 \end{pmatrix} \end{vmatrix} - 0.1 \cdot \begin{vmatrix} \begin{pmatrix} 0.4 & 0.6 & 0.7 \\ 0.8 & 1.0 & 1.1 \\ 1.2 & 1.4 & 1.5 \end{pmatrix} \end{vmatrix}$$

$$+ 0.2 \cdot \begin{vmatrix} \begin{pmatrix} 0.4 & 0.5 & 0.7 \\ 0.8 & 0.9 & 1.1 \\ 1.2 & 1.3 & 1.5 \end{pmatrix} \end{vmatrix} - 0.3 \cdot \begin{vmatrix} \begin{pmatrix} 0.4 & 0.5 & 0.6 \\ 0.8 & 0.9 & 1.0 \\ 1.2 & 1.3 & 1.4 \end{pmatrix} \end{vmatrix}$$

The special case for n=1 is

$$|A| = \begin{vmatrix} a_{00} & a_{01} \\ a_{10} & a_{11} \end{vmatrix} = a_{00} \cdot a_{11} - a_{10} \cdot a_{10}$$

Homework 10.9. Write a subroutine that counts the frequency of occurrence of letters in a text buffer. Register X points to a null-terminated ASCII buffer. There is a 26-element array into which the frequency data will be entered. For example, the first element of Freq will contain the number of A's and a's. Count only the upper-case and lower-case letters.

```
Freq ds.w 26    16-bit counters
```

The calling sequence is

```
ldx  #buffer       ; pointer to text buffer
jsr  CalcFreq
```

10.13   Laboratory Assignments

Lab 10.1
MicroForth
Interpreter

Purpose
In this lab you will build a binary tree data structure. You will design an interpreter that performs simple arithmetic operations. Your system must handle of signed under/overflow conditions.

Description
In preparation for this assignment, review binary trees, command interpreters, and the last-in-first-out queue (stack). Perform Tutorial 10, and also see the simple binary interpreter in TREE.rtf.

The major advantage of a binary tree structure over a linear list is the speed of lookup. In the worst case, the maximum number of compares one must do to find an entry is the maximum depth of the tree. Let **size** be the number of entries and **depth** be the maximum distance from the root to any leaf. If the binary tree is full, the maximum **depth** is less than or equal to the next greatest integer of $\log_2$ **size**. For example a full tree with 1023 entries requires only 10 searches to find an entry. A linear search on the same 1023 entries would take on average 512 searches. In this assignment, we will have only 15 entries, but still will implement a linked-list binary tree. There are two basic approaches to binary searching: linked lists and indexed table. In the listed list, each entry contains a string called name, a pointer to the function to execute the called command and two pointers, left and right. If both left and right are null then the node is a leaf.

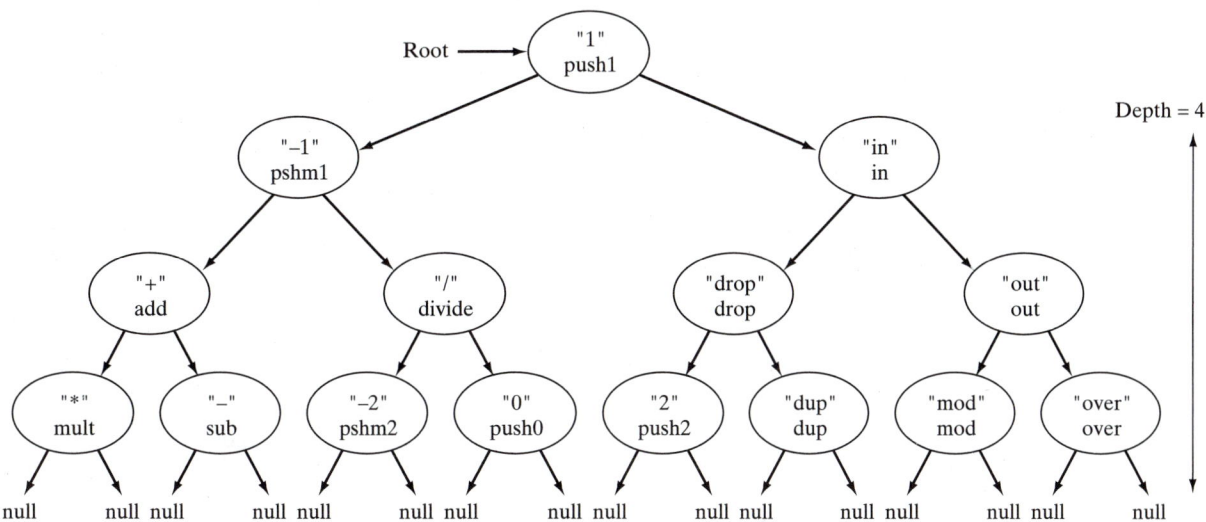

In this procedure, `input` is a string to find. We begin searching at the root.

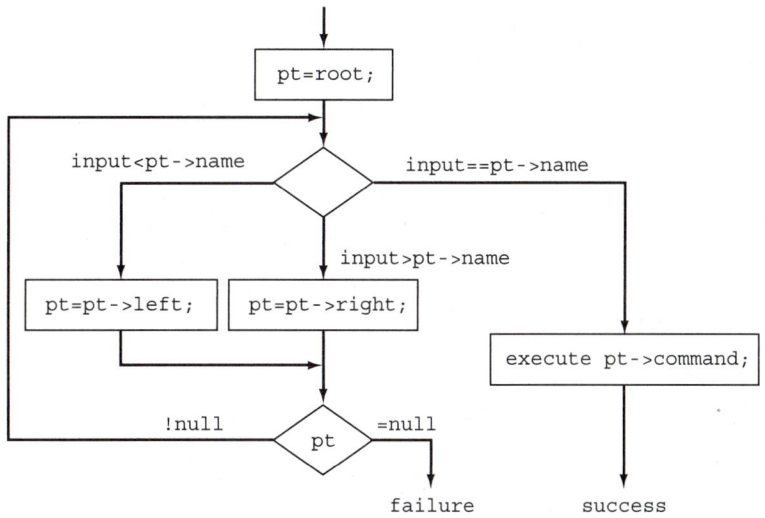

If the input is less than the name of the current node (`pt->name`) (alphabetically before) then the search will go left (`pt=pt->left`). If the input is greater than the name of the current node (`pt->name`) (alphabetically after) then the search will go right (`pt=pt->right`).

The second approach (which you will not be implementing, but is included for your consideration) is called an indexed table. In this scheme we start numbering at index 1. The Table must be sorted alphabetically. Rather than storing the pointers explicitly as we did in the previous example, notice how the index number when viewed in binary provides the same information. If the size is not exactly a power of two, we must allocate additional entries and place them alphabetically at the beginning or the end.

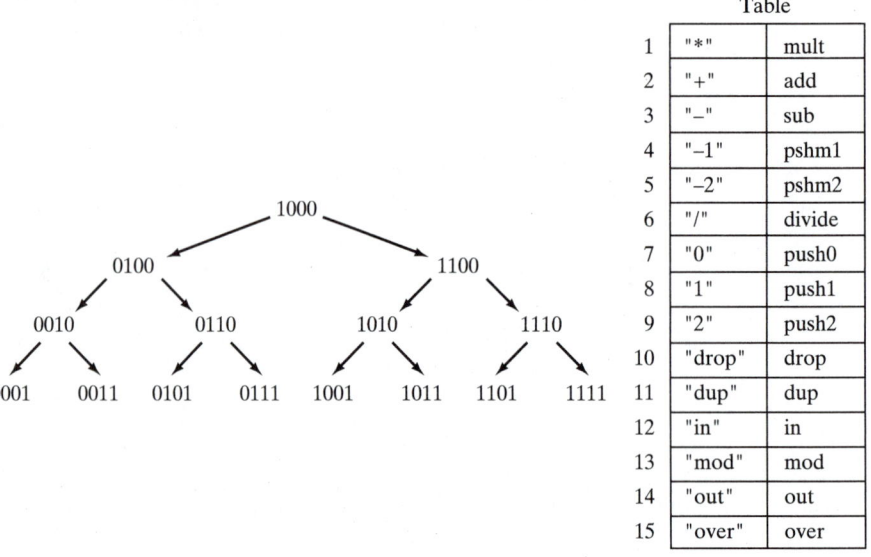

Table		
1	"*"	mult
2	"+"	add
3	"−"	sub
4	"−1"	pshm1
5	"−2"	pshm2
6	"/"	divide
7	"0"	push0
8	"1"	push1
9	"2"	push2
10	"drop"	drop
11	"dup"	dup
12	"in"	in
13	"mod"	mod
14	"out"	out
15	"over"	over

Again, input is the string that is used to match the name field of the Table. This is also a binary search because the number of tests will be less than or equal to the next greatest integer of $\log_2$ size.

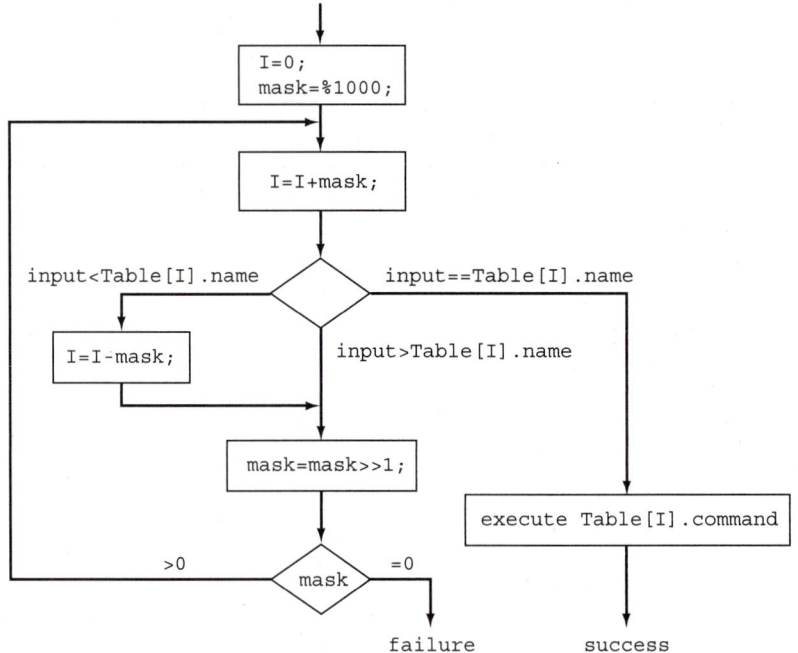

The linked list lookup will be a little faster to execute, because it is quicker to access `pt->name` than it is to access `Table[I].name`. On the other hand, it is easy to make minor changes in the indexed table. If the space is already allocated, then adding a node at run time involves shifting the entries down and adding the node so that the list remains alphabetical. Deleting a node simply involves shifting the nodes up. The only disadvantage is that the size can not increase so that it exceeds the next power of two.

The following table lists the 15 commands your FORTH interpreter will execute. Your software system will have two stacks. The return stack, pointed to by SP, will contain return addresses for the usual `jsr` and `rts` subroutine call functions. The data stack, pointed to by RegY, will contain the input/output parameters for the functions. Commands will be separated by returns (ASCII 13). The idea is to input an entire line using `InString`, then lookup the command in the tree and, if it is found, execute the function. You should display (without popping) the top data stack entry in the LCD display.

command	function
in	input 8-bit signed number from CRT keyboard, push on data stack
out	pop from data stack and output 8-bit signed number to CRT display
dup	duplicates top of data stack
over	duplicates next to top of data stack
drop	pop and discard top of data stack
+	pops two numbers from data stack, add, push result on data stack
−	pops two numbers from data stack, subtract, push result on data stack
*	pops two numbers from data stack, multiply, push result on data stack
/	pops two numbers from data stack, divide, push quotient on data stack
mod	pops two numbers from data stack, divide, push remainder on data stack
0	pushes the constant 0 onto the stack
1	pushes the constant 1 onto the stack
2	pushes the constant 2 onto the stack
−1	pushes the constant −1 onto the stack
−2	pushes the constant −2 onto the stack

We will create 10 bytes of space for the data stack (Y), separate from the hardware stack (SP). Register Y will always point into this space. You must explicitly test for data stack overflow and underflow. You must also implement *ceiling* and *floor* handling during the addition, subtraction, multiply, divide and modulo functions.

```
datastack     rmb   8
penultimate   rmb   1
ultimate      rmb   1
bottom        rmb   0
```

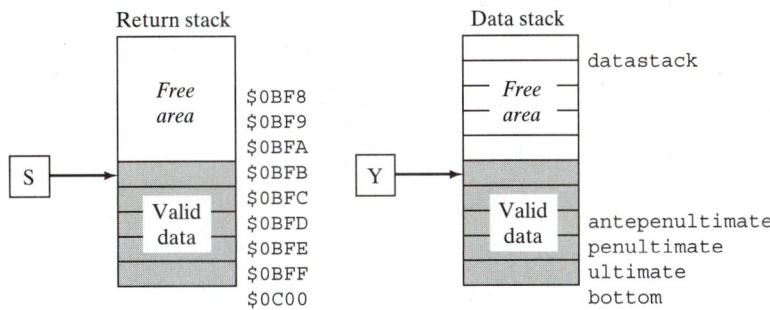

Notice that we can determine how many bytes are on the data stack by comparing Y to the fixed addresses:

if Y equals	this many bytes are on the data stack
bottom	none (empty)
ultimate	one
penultimate	two
datastack	ten (full)

Use reverse Polish format for subtraction, division. E.g., *2 1 −* is *1*, and *−2 1 /* is *−2*. The usual stack rules apply to this data stack as well.

1. Stack accesses (PUSH or PULL) should not be performed outside the allocated area;
2. Stack reads should not be performed from the *free area*;
3. Stack PUSH should first decrement Y, then store the data (not vice versa);
4. Stack PULL should first read the data, then increment Y (not vice versa.)

Here are a couple of the routines to get you started

```
* duplicate next to top
over    cpy  #ultimate  check for at least 1 element
        bhi  overend    skip if no data to duplicate
        cpy  #datastack check for full
        bls  overend    skip if stack is already full
        ldaa 1,y        copy of next to top
        staa 1,-y       push on data stack
overend rts


* push a 2 on the data stack
push2   cpy  #datastack check for full
        bls  psh2end    skip if stack is already full
        movb #2,1,-y    push 2 on data stack
psh2end rts


* multiply top two entries
mult    cpy  #penultimate  check for at least 2 elements
        bhi  overend       skip if no data to duplicate
        ldaa 1,y+       pop top of stack
        sex  a,x        X=multiplicand (-128 to +127)
        ldaa 1,y+       pop next to top
        sex  a,d        D=multiplicand (-128 to +127)
        exg  x,y        Y,D are multiplicands (X is stack pt)
        emuls           D=product (no overflow possible in 16-bit D)
        exg  x,y        Y is stack pt again, 16256 ≤ D ≤ 16384
        cpd  #127
        bgt  ceiling
        cpd  #-128
        bge  ok
floor   ldab #-128      since D>-128, set B = -128
        bra ok
ceiling ldab #127       since D>127, set B = 127
ok      stab 1,-y       push result
        rts
```

Part A.

One by one write and debug the 15 individual commands. Use stabilization to test each routine.

Part B.

Design the fixed binary tree containing the names and function addresses for all your commands. This structure will exist in EEPROM, and can not be modified unless the source code is edited and the program reassembled. Note that most FORTH interpreters place the binary tree in RAM and allow commands to be added and subtracted at run time. Use binding (equ) to make the program more readable.

Part C.

Write the main program that interprets input from the CRT keyboard and displays output back to the CRT display. Remember to display the top of the data stack on the LCD display.

11 Interrupt Synchronization

Chapter 11 objectives are to

❑ introduce the concept of interrupt synchronization,
❑ discuss the issues involved in reentrant programming,
❑ discuss the specific details of using interrupts on the 6811 and 6812,
❑ interface the SCI serial device using interrupts,
❑ design and implement background I/O for simple devices using periodic polling.

There are many reasons to consider interrupt synchronization. The first consideration is that the software in a real-time system must respond to hardware events within a prescribed time. To illustrate the need for interrupts, consider a keyboard interface where the time between new keyboard inputs might be as small as 10 ms. In this situation, the software latency is from the time the new keyboard input is ready until the time the software reads the new data. In order to prevent loss of data in this case, the software latency must be less than 10 ms. We can implement real-time software using busy-waiting polling only when the size and complexity of the system are very small. Interrupts are important for all real-time systems because they provide a mechanism to guarantee an upper bound on the software response time. Interrupts also give us a way to respond to infrequent but important events. Alarm conditions like low battery power and error conditions can be handled with interrupts. Periodic interrupts, generated by the timer at a regular rate, will be necessary to implement data acquisition and control systems. With unbuffered interfaces, the hardware and software take turns waiting for each other. Interrupts provide a way to buffer the data (using a FIFO queue) so that the hardware and software spend less time waiting. The FIFO queue placed between the interrupt routine and the main program will increase the overall bandwidth. We will begin our discussion with general issues, then present the specific details about the 6811 and 6812 microcomputers. Many interrupt examples will be presented.

11.1 Basic Concepts

An **interrupt** is the automatic transfer of software execution in response to a hardware event that is asynchronous with the current software execution. The hardware can be either an external I/O device (like the SCI input/output) or an internal event (like an op code fault, or a periodic timer). When the hardware needs service (busy- to done-state transition) it will request an interrupt. A **thread** is defined as the path of action of software as it executes. The execution of the interrupt service routine is called a background thread. This thread is created by the hardware interrupt request and is killed when the interrupt service routine executes the `rti` instruction. A new thread is created for each interrupt request. It is important to consider each individual request as a separate thread because local variables and registers used in the interrupt service routine are unique and separate from one interrupt event to the next. In a multi-threaded system we consider the threads as cooperating to perform an overall task. Consequently we will develop ways for the threads to communicate (e.g., FIFO) and synchronize with each other. Most embedded systems have a single common overall goal. On the other hand, general-purpose computers can have multiple unrelated functions to perform. A **process** is also defined as the action of software as it executes. The difference is that processes do not necessarily cooperate towards a common shared goal. Threads share access to I/O devices and global variables, while processes have separate global variables and do not share I/O devices.

The software has dynamic control over aspects of the interrupt request sequence. First, each potential interrupt source has a separate **arm** bit that the software can activate or deactivate. The software will set the arm bits for those devices it wishes to accept interrupts from, and will deactivate the arm bits within those devices from which interrupts are not to be allowed. In other words, it uses the arm bits to individually select which devices will and which devices will not request interrupts. The second aspect that the software controls is the interrupt **enable** bit, I, which is in the condition code register. The software can enable all armed interrupts by setting I=0, or it can disable all interrupts by setting I=1. The disabled interrupt state (I=1) does not dismiss the interrupt requests; rather it postpones them until a later time, when the software deems it convenient to handle the requests. We will pay special attention to these enable/disable software actions. In particular we will need to disable interrupts when executing nonreentrant code; disabling interrupts, however, will have the effect of increasing the response time of software.

The **interrupt service routine** (ISR) is the software module that is executed when the hardware requests an interrupt. There may be one large ISR that handles all requests (polled interrupts), or many small ISRs specific for each potential source of interrupt (vectored interrupts). The design of the interrupt service routine requires careful consideration of many factors. Three conditions must be true for an interrupt to be generated. A device must be armed (e.g., RIE is set), interrupts must be enabled (I=0), and an external event must occur setting a flag (e.g., new SCI input ready sets RDRF). An interrupt causes the following sequence of events. First, the current instruction is finished[1]. Second, the execution of the main program is suspended, pushing all the registers on the stack. Third, the PC is loaded with the address of the ISR (vector). Lastly, interrupts are disabled (I=1). These four steps occur automatically in hardware. Next, the software executes the ISR (background thread). When the ISR is done it executes an `rti` causing the main program execution to be resumed.

When the microcomputer accepts an interrupt request, it will automatically save the execution state of the main thread by pushing all its registers on the stack. After the ISR provides the necessary service, it will execute an `rti` instruction. This instruction pulls the registers from the stack, which returns control to the main program. Execution of the main program will then continue with the exact stack and register values that existed before the interrupt. Although interrupt handlers can create and use local variables, parameter passing between threads must be implemented using global memory variables. Global variables are also required if an interrupt thread wishes to pass information to itself, e.g., from one interrupt instance to another. The execution of the main program is called the foreground thread, and the executions of interrupt service routines are called background threads.

Many factors should be considered when deciding the most appropriate mechanism to synchronize hardware and software. One should not always use busy-waiting because one

---

[1]The 6812 instructions `rev`, `revw`, and `wav` can be interrupted in the middle of their execution.

is too lazy to implement the complexities of interrupts. On the other hand, one should not always use interrupts because they are fun and exciting. Busy-waiting synchronization is appropriate when the I/O timing is predictable, and when the I/O structure is simple and fixed. Busy-waiting should be used for dedicated single-thread systems where there is nothing else to do while the I/O is busy. Interrupt synchronization is appropriate when the I/O timing is variable, and when the I/O structure is complex. In particular, interrupts are efficient when there are I/O devices with different speeds. Interrupts allow for quick response times to important events. They can also be used for infrequent but critical events like power failure, memory faults, and machine errors. Interrupts can be used to assist program development by triggering on stack overflow, invalid op code, and breakpoints. Periodic interrupts will be useful for real-time clocks, data acquisition systems, and control systems. For extremely high bandwidth and low latency interfaces, DMA should be used. For more information on DMA see Chapter 10 of *Embedded Microcomputer Systems: Real-time Interfacing* by Jonathan W. Valvano, Brooks-Cole, 2000.

> *Checkpoint 11.1:* What three conditions must be true for an interrupt to occur?

> *Checkpoint 11.2:* How do you enable interrupts?

> *Checkpoint 11.3:* What are the steps that occur when an interrupt is processed?

11.2 First-In-First-Out Queue

The first-in-first-out circular queue (**FIFO**) is quite useful for implementing a buffered I/O interface. It can be used for both buffered input and buffered output. The order-preserving data structure temporarily saves data created by the source (producer) before it is processed by the sink (consumer). After initialization, the FIFO has two functions: *Put* (enters new data) and *Get* (removes the oldest data). You have probably already experienced the convenience of FIFOs. For example, when using an editor, you can continue to type characters while other processing is occurring. The ASCII codes are input from the keyboard as they are typed and *put* in a FIFO. When the editor is active again, it *gets* more keyboard data to process. A FIFO is also used when you ask the computer to print a file. Rather than waiting for the actual printing to occur character by character, the print command will *put* the data in a FIFO. Whenever the printer is free, it will *get* data from the FIFO. The advantage of the FIFO is that it allows you to continue to use your computer while the printing occurs in the background. To implement this magic of background printing, we will need interrupts.

There are many **producer–consumer** applications. In Table 11.1 the activities on the left are producers that create or input data, while the activities on the right are consumers that process or output data.

Table 11.1
Producer–consumer examples.

Source/Producer	Sink/Consumer
keyboard input	program that interprets
program with data	printer output
program sends message	program receives message
microphone and ADC	program that saves sound data
program that has sound data	DAC and speaker

The producer puts data into the FIFO. The `Put` operation does not discard information already in the FIFO. If the FIFO is full and the user calls `Put`, this routine will return a full error signifying that the last (newest) data was not properly saved. The consumer removes data from the FIFO. The `Get` operation will modify the FIFO. After a `Get`, the particular information returned is no longer saved on the FIFO. If the FIFO is empty and the user tries to `Get`, this routine will return an empty error signifying that no data could be retrieved. The FIFO is order preserving, such that the information returned via `Get` is in the same order as the data saved via `Put`.

Figure 11.1 shows a data-flow graph with buffered input and buffered output. FIFOs used in this book will be statically allocated global structures. Because they are global variables, it means they will exist permanently and can be carefully shared by more than one program. The advantage of using a FIFO structure for a data-flow problem is that we can decouple the producer and consumer threads. Without the FIFO we would have to produce one piece of data, then process it, produce another piece of data, then process it. With the FIFO, the producer thread can continue to produce data without having to wait for the consumer to finish processing the previous data. This decoupling can significantly improve system performance.

Figure 11.1
A data-flow graph showing two FIFOs that buffer data between producers and consumers.

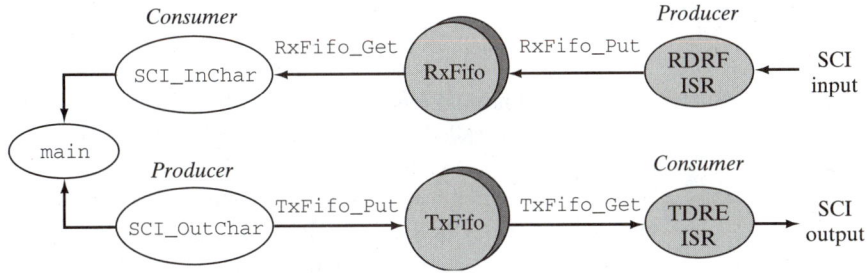

There are many ways to implement a statically allocated FIFO. We can use either a pointer or an index to access the data in the FIFO. We can use either two pointers (or two indices) or two pointers (or two indices) and a counter. The counter specifies how many entries are currently stored in the FIFO. There are even hardware implementations of FIFO queues. In this book we will show just one method. Other implementations can be found in Chapter 4 of *Embedded Microcomputer Systems: Real-time Interfacing* by Jonathan W. Valvano, Brooks-Cole, 2000.

If we were to have infinite memory, as shown in Figure 11.2, a FIFO implementation is easy. `RxGetPt` points to the data that will be removed by the next call to `RxFifo_Get`, and `RxPutPt` points to the empty space where the data will stored by the next call to `RxFifo_Put`. Program 11.1 presents the basic idea of a pointer-based FIFO implementation. To put data in the FIFO, the new data is stored at `RxPutPt`, then this pointer is incremented. To get data from the FIFO, the value at `RxGetPt` is read, then this pointer is incremented.

Figure 11.2
The FIFO implementation
with infinite memory.

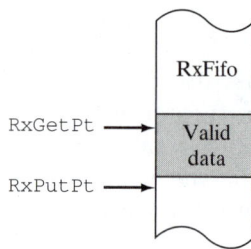

Program 11.1   Code fragments showing the basic idea of a FIFO.

```
* Reg A is data to put into the FIFO
RxFifo_Put ldx   RxPutPt
           staa ,X    store into FIFO
           inx        next place to put
           stx  RxPutPt update pointer
           rts
* Reg A returned with byte from FIFO
RxFifo_Get ldx   RxGetPt
           ldaa ,X    read from FIFO
           inx        next place to get
           stx  RxGetPt    update
           rts
```

```c
void RxFifo_Put(char data){
  *(RxPutPt++) = data;
}



  void RxFifo_Get(char *datapt){
    *datapt = *(RxGetPt++);
  }
```

Three modifications are required to these functions. If the FIFO is full when RxFifo_Put is called, then the subroutine should return a full error. Similarly, if the FIFO is empty when RxFifo_Get is called, then the subroutine should return an empty error. There is never an infinite amount of memory, so a finite number of bytes will be permanently allocated to the FIFO. Figures 11.3 and 11.4 show an example with 10 bytes allocated. The RxPutPt (abbreviated at PPt in the figures) and RxGetPt (abbreviated at GPt in the figures) must be wrapped back up to the top when they reach the bottom. The shaded blocks in these two figures represent valid data saved in the FIFO. Figure 11.3 shows how the FIFO changes as four bytes are Put into it. Figure 11.4 shows the same FIFO as Get is called four times. Observe the order-preserving nature of the FIFO.

Figure 11.3
The FIFO **Put** operation
showing the pointer
wrap.

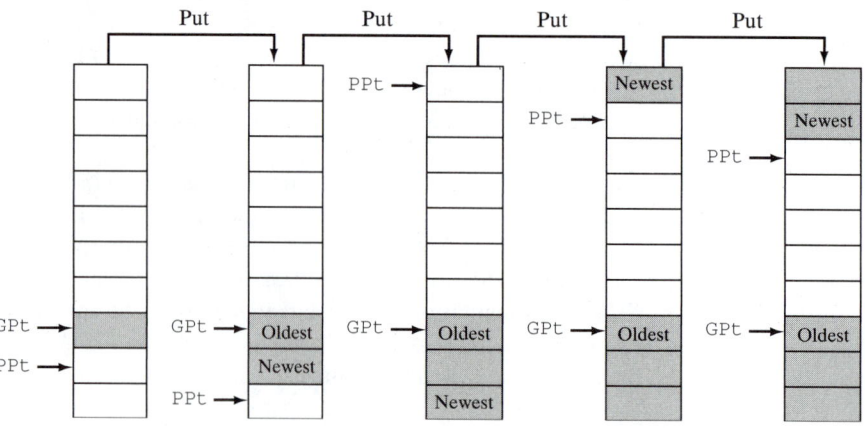

Figure 11.4
The FIFO **Get** operation showing the pointer wrap.

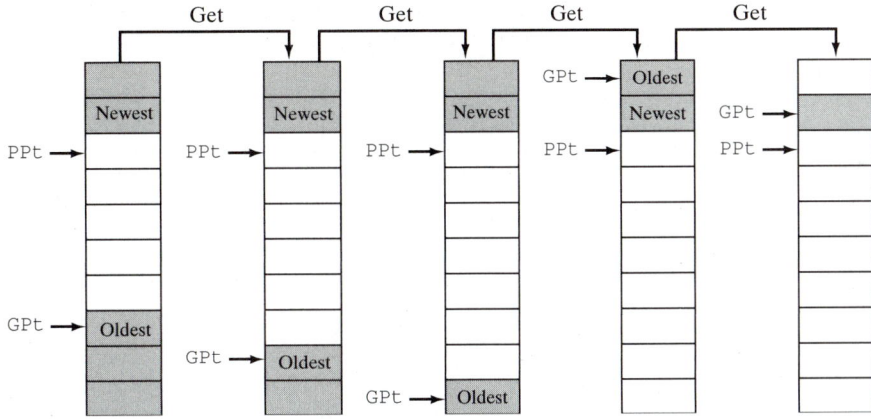

Two mechanisms determine whether the FIFO is empty or full. A simple method is to implement a counter containing the number of bytes currently stored in the FIFO. RxFifo_Get would decrement the counter and RxFifo_Put would increment the counter. The second method is to prevent the FIFO from being completely full. For example, if the FIFO had 10 bytes allocated, then the RxFifo_Put subroutine would allow a maximum of 9 bytes to be stored. If there were already 9 bytes in the FIFO and another RxFifo_Put were called, then the FIFO would not be modified and a full error would be returned. In this way if RxPutPt equals RxGetPt at the beginning of RxFifo_Get, then the FIFO is empty. Similarly, if RxPutPt+1 equals RxGetPt at the beginning of RxFifo_Put, then the FIFO is full. Be careful to wrap the RxPutPt+1 before comparing it to RxFifo_Get. This second method does not require the length to be stored or calculated. The FIFO global structures are defined in Program 11.2, and must be allocated in RAM. These definitions are private, and not accessible by programs outside the FIFO module.

Program 11.2
Global structures for a two-pointer FIFO.

```
RXFIFO_SIZE equ   10              #define RXFIFO_SIZE 10
RxPutPt     rmb   2               char *RxPutPt;
RxGetPt     rmb   2               char *RxGetPt;
RxFifo      rmb   RXFIFO_SIZE     char RxFifo[RXFIFO_SIZE];
```

The FIFO initialization is shown in Program 11.3, which is usually called once at the start of the system. The FIFO is empty if the RxPutPt equals the RxGetPt. Both pointers should always address locations within the 10-byte allocated area.

Program 11.3
Initialize both pointers to the beginning of the FIFO.

```
RxFifo_Init ldx #RxFifo          void RxFifo_Init(void){
            stx RxPutPt            RxPutPt = RxGetPt = &RxFifo[0];
            stx RxGetPt          }
            rts
```

A flowchart for the Put and Get functions is drawn in Figure 11.5. Program 11.4 shows the software that enters new data in the FIFO. To check for FIFO full, the RxFifo_Put routine attempts to put using a temporary pointer. If putting makes the FIFO look empty, then the temporary pointer is discarded and the routine is exited without saving the data. This is why a FIFO with 10 allocated bytes can hold only nine data points. If putting doesn't make the FIFO look empty, then the temporary pointer is stored into the actual `RxPutPt` saving the data as desired.

Figure 11.5
Flowcharts of the `put` and `get` operations.

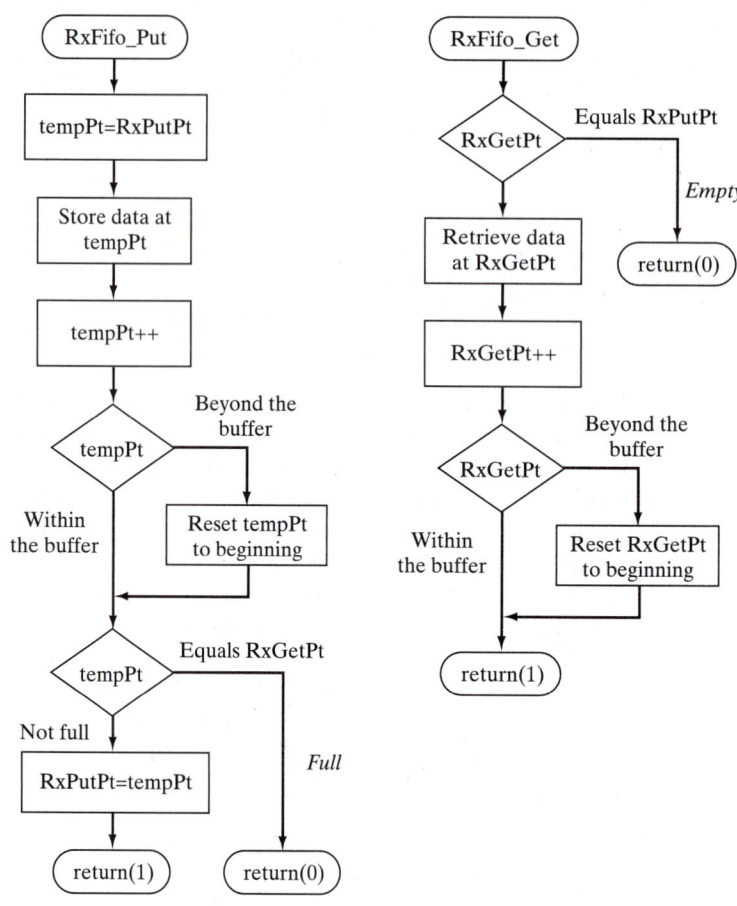

Program 11.4   Data is saved at `RxPutPt`.

```
* Input   RegA data to put         int RxFifo_Put(char data){
* Output RegB -1=OK, 0=full        char *tempPt;
RxFifo_Put                           tempPt = RxPutPt;
      pshx                           *(tempPt++) = data; /* try */
      ldx   RxPutPt    Temporary     if(tempPt==&RxFifo[RXFIFO_SIZE]){
      staa  ,x         Try to put      tempPt = &RxFifo[0];
      inx                            }
      cpx   #RxFifo+RXFIFO_SIZE      if(tempPt == RxGetPt){
      bne   skip                       return(0);        /* full! */
      ldx   #RxFifo    Wrap          }
skip  clrb                           else{
      cpx   RxGetPt    Full if same    RxPutPt = tempPt;  /* OK */
      beq   ok                         return(1);
      comb             -1 means OK   }
      stx   RxPutPt                 }
ok    pulx
      rts
```

Program 11.5 shows the software that removes the oldest data from the FIFO. To check for FIFO empty, the following `RxFifo_Get` routine simply checks to see if `RxGetPt` equals `RxPutPt`. If they match at the start of the routine, then `RxFifo_Get` returns with the "empty" condition signified. Next, the information is retreived from the FIFO. The `RxGetPt` is incremented signifying that information is no longer in the FIFO. If the add 1 to `RxGetPt` operation makes the pointer go beyond the FIFO buffer, the pointer is wrapped back to the beginning.

Program 11.5   Data is removed from `RxGetPt`.

```
* Input   none
* Output RegA data from Get
*        RegB -1=ok, 0=empty
RxFifo_Get
     pshy
     clrb
     ldy   RxGetPt
     cpy   RxPutPt    Empty?
     beq   done
     comb              -1=OK
     ldaa  ,y          Data
     iny
     cpy   #RxFifo+RXFIFO_SIZE
     bne   no          wrap?
     ldy   #RxFifo     yes
no   sty   RxGetPt
done puly
     rts
}
```

```
int RxFifo_Get(char *datapt){
  if(RxPutPt == RxGetPt){
    return(0);       /* Empty */
  }
  else {
    *datapt = *(RxGetPt++);
    if(RxGetPt==&RxFifo[RXFIFO_SIZE]){
      RxGetPt = &RxFifo[0];
    }
    return(1);
  }
}
```

Let t_p be the time (in sec) between calls to `Put`, and r_p be the arrival rate (in bytes/sec) into the system. Similarly, let t_g be the time (in sec) between calls to `Get`, and r_g be the service rate (in bytes/sec) out of the system.

$$r_g = \frac{1}{t_g} \qquad r_p = \frac{1}{t_p}$$

If the minimum time between `Put`s is greater than the maximum time between `Get`s,

$$\min t_p, \geq \max t_g$$

then a FIFO is not necessary and the data flow program could be solved with a simple global variable. On the other hand, if the time between `Put`s temporarily becomes less than the time between `Get`s because either

■ the arrival rate temporarily increases,
■ the service rate temporarily decreases,

then information will be collected in the FIFO. For example, a person might type very fast for a while followed by a long pause. The FIFO could be used to capture all the data without loss as they come in very fast. Clearly, on average the system must be able to process the data (the sink process) at least as fast as the average rate at which the data arrive. If the average input rate is larger than the average output rate,

$$\bar{r}_p > \bar{r}_g$$

then the FIFO will eventually overflow no matter how large the FIFO. If r_p is temporarily high or r_g is temporarily low, and that causes the FIFO to become full, then this problem can be solved by increasing the FIFO size.

There is fundamental difference between an empty error and a full error. Consider the application of using a FIFO between your computer and its printer. This is a good idea because the computer generates data to be printed at a very high rate followed by long pauses. The printer is like a turtle. It can print at a slow but steady rate. The computer will `Put` a byte into the FIFO that it wants printed. The printer will `Get` a byte out of the FIFO when it is ready to print another character. A full error occurs when the computer calls `Put` at too fast a rate. A full error is serious; either data will be lost or the rest of the computer pauses waiting for room in the FIFO. On the other hand, an empty error occurs when the printer is ready to print but the computer has nothing in mind. An empty error is not serious. If the FIFO is empty the printer just shuts itself off and does nothing.

> *Checkpoint 11.4:* If the FIFO becomes full, can the situation be solved by increasing the size?

11.3   Interthread Communication and Synchronization

For regular function calls we use the registers and stack to pass parameters, but interrupt threads have logically separate resisters and stack. In particular, all registers are automatically saved by the microcomputer as it switches from main program (foreground thread) to interrupt service routine (background thread). The `rti` instruction will restore the registers (including the interrupt enable bits and the PC) back to their previous values. Thus, all parameter passing must occur through global memory. One cannot pass data from the main program to the interrupt service routine using registers or the stack. The classic producer/consumer problem has two threads. One thread produces data and the other consumes data. For an input device, the background thread is the producer because it generates new data, and the foreground thread is the consumer because it uses the data up. For an output device, the data flows in the other direction so the producer/consumer roles are reversed.

One simple interthread communication scheme is the mailbox. Figure 11.6 illustrates a SCI input device interfaced using interrupt synchronization. The big arrow in this figure signifies the communication/synchronization link between the background and foreground. The mailbox structure is implemented with two global variables. `RxMail` contains data, and `RxStatus` is a flag specifying whether the mailbox is full or empty. The interrupt is requested when RDRF is set, signifying new data is ready from the input device. The ISR will read the data from the input device and store it in the global variable `RxMail`, then update its status as full. The main program will perform other calculations, while occasionally checking the status of the mailbox. When the mailbox has data, the main program will process it. This approach is adequate for situations where the input bandwidth is slow compared to the software processing speed. It is also possible to process the data within the ISR itself, and just report the results of the processing to the main program using the mailbox.

Figure 11.6
A mailbox can be used to pass data between threads.

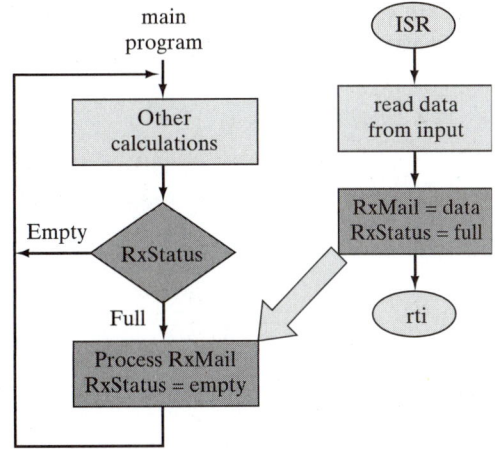

One way to visualize the interrupt synchronization is to draw a state-versus-time plot of the activities of the hardware, the mailbox, and the two software threads. Figure 11.7 shows that at time (a), the mailbox is empty, the SCI receiver is idle, and the main program is performing other tasks, because mailbox is empty. When new input data are ready, the RDRF flag will be set and an interrupt will be requested. At time (b), the ISR reads data from input device and saves these in RxMail, then it sets RxStatus to full. At time (c), the main program recognizes RxStatus is full. At time (d), the main program processes data from RxMail, sets RxStatus to empty. Notice that even though there are two threads, only one is active at a time. The interrupt hardware switches the processor from the main program to the ISR, and the rti instruction switches the processor back.

Figure 11.7
Hardware/software timing of an input interface using a mailbox.

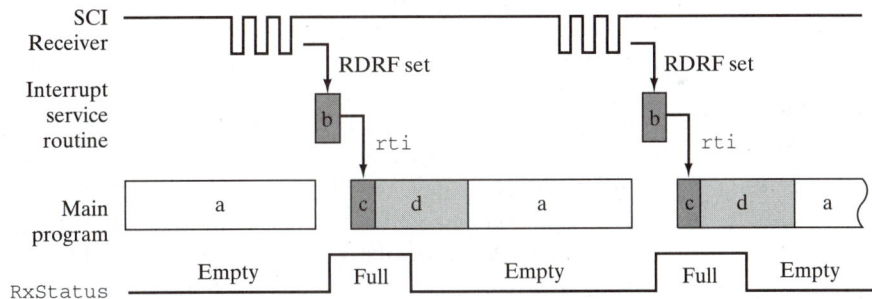

Although the input interface uses interrupts, it really isn't any better than a busy-waiting solution. If the next input data arrive before the previous data are processed, then data will be lost. When the I/O bandwidth is fast or unpredictable, it is more appropriate to pass data from the producer thread to the consumer thread using a FIFO queue. The FIFO will buffer the data between the foreground and background. The presence of the FIFO placed between the producer and consumer greatly improves performance by reducing the time each waits for the other.

Figure 11.8 shows an I/O system that uses interrupts for both input and output. When the main program wishes to output, it calls SCI_OutChar, which will put the data in the TxFifo and arm the output device. When the main program wishes to input, it calls

SCI_InChar, which will get data from the RxFifo. This example has been implemented at
tut4 as part of the **TExaS** system.

Figure 11.8 FIFO queues can be used to pass data between threads.

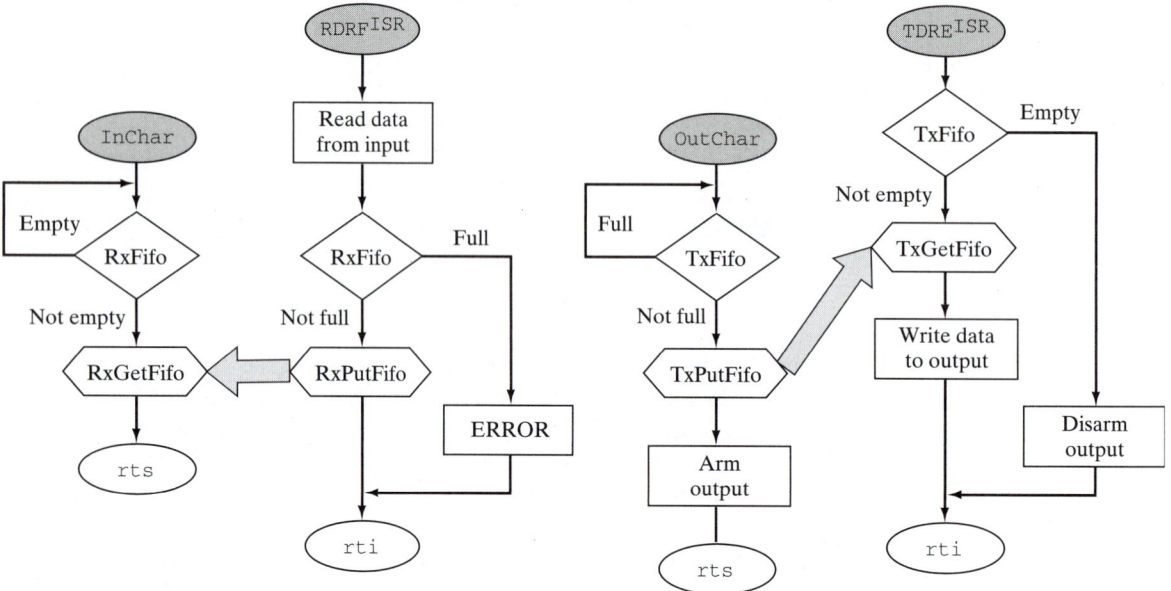

Observation: For systems with interrupt-driven I/O on multiple devices, there will be
a separate FIFO for each device.

The incoming serial data will set RDRF, requesting an interrupt. The ISR (background)
will accept the data and put it in the RxFifo. The RxFifo buffers data between the input
hardware and the main program that processes the data. If the RxFifo becomes full, then
data will be lost. FIFO full errors will always occur if the average input rate (number of
bytes arriving per second from the input hardware) exceeds the average processing rate
(number of bytes processed per second by the main program). In this situation, either the
output rate must be increased (by using a faster computer or by writing a better software
processing algorithm), or the input rate must be decreased (by slowing down the arrival rate
of data). The second way the RxFifo could become full is if there is a temporary increase
in the arrival rate or a temporary decrease in the process rate. For this situation, the full errors could be eliminated by increasing the size of the RxFifo.

TDRE signals the output device is idle, ready to output more data. If there are data
available in the TxFifo, the ISR will get it and write it to the output device. If the TxFifo is
empty, the output device is disarmed. The main program puts data in TxFifo and gets data
from the RxFifo as desired. If the TxFifo becomes full, then it is appropriate to wait for the
interrupts to make room. It is inefficient, but not catastrophic, for the main program to wait
on a full TxFifo. Efficiency can be improved for the buffered output problem by increasing
the TxFifo size. It is also inefficient, but not catastrophic, for the main program to wait on
an empty RxFifo. Efficiency can be improved for the buffered input problem by performing other tasks while waiting for data.

It is important to study the timing behavior of the I/O hardware and software processing when designing an interrupting interface. One simple way to study a problem is to

measure the number of elements in the RxFifo when new data is entered by the input ISR. If the time for the software to read and process the data is much faster than the time for the input device to create new input, then there will be very few elements in the RxFifo. For most systems, the producer and consumer rates fluctuate, but during the times when the software waits for the I/O hardware, the system is classified as **I/O bound**. For an I/O-bound input interface the RxFifo has either 0 or 1 entry, and the use of interrupts does not enhance the bandwidth over the busy-waiting implementations. Even with an I/O-bound input device, however, it may be more efficient to utilize interrupts because it provides a straightforward approach to servicing multiple devices.

If the input device generates a burst of high bandwidth activity, then there will be many elements in the RxFifo. As long as the interrupt service routine is fast enough to keep up with the input device and as long as the RxFifo does not become full, no data are lost. Recall that the ISR doesn't have to process the input data, just read it and save it in the RxFifo. In this situation, the overall bandwidth is higher than it would be with a busy-waiting implementation, because the input device does not have to wait for each data byte to be processed. This is the classic example of a "buffered" input, because data enter the system (via the interrupts), are temporarily stored in a buffer (put into the RxFifo), and are processed later (by the main program, get from the RxFifo). During the times when the I/O device is faster than the software, the system is called **CPU-bound**. A system will work only if the producer rate only temporarily exceeds the consumer rate (a short burst of high bandwidth input). If the external device sustained the high bandwidth input rate, then the RxFifo would become full and data would be lost.

For an output device, we will count the number of elements in the TxFifo when data are removed by the output ISR. If the rate for the software to generate new data is much slower than the rate for the output device to send data, then there will be very few elements in the TxFifo. During this time the system is called CPU-bound. In this situation, the TxFifo has either 0 or 1 entry, and the use of interrupts does not enhance the bandwidth over the busy-waiting implementations. Even with a CPU-bound output device, however, it may be more efficient to utilize interrupts because it provides a straightforward approach to servicing multiple devices.

If the main program generates a burst of output activity, then there will be many elements in the TxFifo. In this situation, the overall bandwidth is higher than it would be with a busy-waiting implementation, because the main program does not have to wait for each data byte to be output. This is the classic example of a "buffered" output, because data enter the system (via the main program), are temporarily stored in a buffer (put into the TxFifo) and are processed later (by the output ISR, get from the TxFifo.) During the time when the main program is faster than the output hardware, the system is called I/O-bound. Just like the input situation, a system will work only if the producer rate temporarily exceeds the consumer rate. If the main program sustained the output rate, then the TxFifo would become full and the main program would then have to wait. Again, the output situation is most efficient if the TxFifo is big enough to avoid full errors.

There are other types of interrupt that are not an input or output. For example, we will configure the computer to request an interrupt on a periodic basis. This means an interrupt handler will be executed at fixed time intervals. This periodic interrupt will be essential for the implementation of real-time data acquisition and real-time control systems. Figure 11.9 shows a data acquisition system that samples the ADC 61 times a second. The internal timer hardware requests an interrupt every 16 ms. We can use a multiple access circular queue (MACQ) to implement digital filters or digital control algorithms. The MACQ consists of a statically allocated global buffer. For example `x[0]` contains the most recent ADC sample, `x[1]` contains the previous sample, and `x[2]` contains the sample taken two time periods ago. The interrupt service routine will shift the MACQ data, sample the ADC, and calculate the digital filter. Synchronization with other threads can be performed using a FIFO or mailbox.

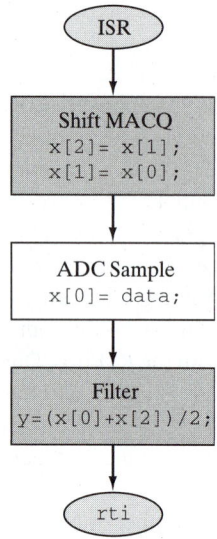

Figure 11.9
A MACQ can be used to save the most recent ADC samples.

An axiom with interrupt synchronization is that the interrupt program should execute as fast as possible. The interrupt should occur when it is time to perform a needed function, and the interrupt service routine should come in clean, perform that function, and return right away. Placing backward branches (busy-waiting loops, iterations) in the interrupt software should be avoided if possible. The percentage of time spent executing interrupt software should be minimized. For an input device, the **interface latency** is the time between when new input is available and when the software reads the input data. We can also define **device latency** as the response time of the external I/O device. For example, if we request that a certain sector be read from a disk, then the device latency is the time it takes to find the correct track and spin the disk (seek) so the proper sector is positioned under the read head. For an output device, the interface latency is the time between when the output device is idle and the time when the software writes new data. A **real-time** system is one that can guarantee a worst case interface latency.

11.4   Reentrant Programming

A program segment is **reentrant** if it can be concurrently executed by two or more threads. This issue is very important when using interrupt programming. To implement reentrant software, we place local variables on the stack and avoid storing into global memory variables. When writing in assembly, we use **registers** or the **stack** for parameter passing to create reentrant subroutines. Typically each thread will have its own set of registers and stack. A nonreentrant subroutine will have a section of code called a **vulnerable window** or **critical section**. An error occurs if

1. one thread calls the nonreentrant subroutine,
2. that thread is executing in the critical section when interrupted by a second thread,
3. the second thread calls the same subroutine.

There are a number of scenarios that can happen next. In the most common scenario, the second thread is allowed to complete the execution of the subroutine, control is then returned to the first thread, and the first thread finishes the subroutine. This first scenario is the usual case with interrupt programming. In the second scenario, the second thread executes part of it, is interrupted and then reentered by a third thread. The third thread finishes, the control is returned to the second thread, and it finishes. Lastly the control is returned to the first thread and it finishes. This second scenario can happen in interrupt programming if interrupts are reenabled during the execution of the ISR. A critical section may exist when two different subroutines access and modify the same memory-resident data structure.

Program 11.6 shows an assembly and a C function that are nonreentrant because they use a global variable, `num`. The assembly language program accepts two 16-bit signed integers in Registers D and X and returns the average in Register D. These functions could have been made reentrant by implementing `num` as a local variable, but the purpose of the example is to illustrate what can go wrong when a nonreentrant function is reentered.

Program 11.6
This function is nonreentrant because of the read-modify-write access to a global.

```
num rmb   2            short num;
AVE stx   num          short AVE(short first, short second){
    addd  num              num = first;
    asrd                   return (num+second)/2;
    rts                }
```

Checkpoint 11.5: Rewrite Program 11.6 so that it is reentrant.

A critical section exists between the `stx` and the `addd` instructions. Assume there are two concurrent threads (the main program and a background ISR) that both call this subroutine. Concurrent means that both threads are ready to run. Because there is only one computer, exactly one thread will be active (running) at a time. Typically, the operating system switches execution control back and forth using interrupts. For example, the main program might be executing when an interrupt causes the computer to switch over and execute the ISR. When the ISR is done it executes an `rti` and the control returns to the main program. An error occurs if

1. the main program calls `AVE`
2. the main program executes the `stx` instruction saving its second number in `num`
3. the OS halts the main program (using an interrupt) and starts the interrupt service routine
4. the ISR calls `AVE`
   the ISR executes the `stx` saving its number in `num`
   the ISR finishes `AVE`
5. the OS returns control to the main program
6. the main program executes the `addd` instruction but gets the wrong `num`

An **atomic operation** is one that, once started, is guaranteed to finish. In most computers, once an instruction has begun, the instruction must be finished before the computer can process an interrupt. Therefore, the following read-modify-write sequence is atomic because it can not be halted in the middle of its operation.

```
inc   counter       where counter is a global variable
```

On the other hand, this read-modify-write sequence is not atomic because it can start, then be interrupted:

```
ldaa counter        where counter is a global variable
inca
staa counter
```

In general, nonreentrant code can be grouped into three categories, all involving nonatomic writes to global variables. We will classify I/O ports as global variables for the consideration of reentrancy. We will group registers into the same category as local variables, because each thread will have its own registers and stack.

The first group is the *read-modify-write* sequence:

1. the software reads the global variable, producing a copy of the data,
2. the software modifies the copy (at this point the original variable is still unmodified),
3. the software writes the modification back into the global variable.

In the second group is the *write followed by read*, where the global variable is used for temporary storage:

1. the software writes to the global variable (this becomes the only copy of important information),
2. the software reads from the global variable expecting the original data to still be there.

In the third group, we have a *nonatomic multistep write* to a global variable:

1. the software writes part of the new value to a global variable,
2. the software writes the rest of the new value to a global variable.

Observation: When considering reentrant software and vulnerable windows, we classify accesses to I/O ports the same as accesses to global variables.

Consider the interrupting serial port interface shown in Figure 11.8, and the FIFO implementations in Programs 11.2 through 11.5. Assume also there is a `TxFifo` implementation separate, but identical to, the `RxFifo`. The details of the serial port will be presented later in the chapter, but for now let's focus on whether this usage of these two FIFOs has any critical sections. When processing input, it is possible for the main program to start to execute `RxFifo_Get`, be interrupted, and the ISR to call `RxFifo_Put`. To verify correctness of our system, we notice nothing bad would happen (e.g., crash, lost data, extra data, etc.) if the `RxFifo_Put` subroutine were to execute in between any two assembly instructions of the `RxFifo_Get` routine. A similar consideration arises when processing outputs. In this situation, the main program may start to execute `TxFifo_Put`, be interrupted, and the output interrupt routine calls `TxFifo_Get`. Again, nothing bad happens if the `TxFifo_Get` subroutine is executed in between any two assembly instructions within the `TxFifo_Put` routine. If we are processing both input and output, then two FIFO's would be used. Each FIFO routine (`RxFifo_Get`, `RxFifo_Put`, `TxFifo_Get`, or `TxFifo_Put`) is called from exactly one place in the software system. Even though the functions themselves are nonreentrant, the system has no critical sections because none of the individual functions will be reentered. They can interrupt each other, but not themselves. In conclusion, the FIFO routines as used in the system shown in Figure 11.8 (you can run this system as `tut4` on **TExaS**) have no critical sections. One the other hand, if the foreground and background threads both called the same FIFO function, then there would be a critical section.

If critical sections do exist, we can either eliminate it by rewriting the software or implement **mutual exclusion**, which simply means only one thread at a time is allowed to execute in the critical section. If we can eliminate the global variables, then the subroutine becomes reentrant. Without global variables there are no "vulnerable" windows because each thread has its own registers and stack.

Sometimes one must access global memory to implement the desired function. A simple way to implement mutual exclusion is to disable interrupts while executing the critical section. It is important to disable interrupts for as short a time as possible, so as to minimize the effect on the dynamic performance of the other threads. While we are running with interrupts disabled, time-critical events like power failure and danger warnings cannot be processed. Notice also that the interrupts are not simply disabled, then enabled. Before the critical section, the interrupt status is saved and the interrupts disabled. After the critical section, the interrupt status is restored. You cannot save the interrupt status in a global variable; rather, you should save it either on the stack or in a register. In assembly, we can use the following skeleton to implement mutual exclusion and eliminate the critical section:

```
    tpa
    psha        save CCR
    sei         disable interrupts
;execute the critical section
    pula
    tap         restore I bit to its original value
```

In C, we can use a similar approach to implement mutual exclusion and eliminate the critical section:

```
void function(void){ char saveCCR;
asm("tpa\n"                      // previous interrupt enable
    "staa %saveCCR\n"            // save previous
    "sei");                      // make atomic
// execute the critical section
asm("ldaa %saveCCR\n"  // recall previous
    "tap");                      // end critical section
}
```

Checkpoint 11.6: Consider the situation of nested critical sections. For example, a function with a critical section calls another function that also has a critical section. What would happen if you simply added a `sei` at the beginning and a `cli` at the end of each critical section?

Reentrant programming is very important in writing high-level language software too. Obviously, we minimize the use of global variables. But when global variables are necessary, we must be able to recognize potential sources of bugs due to nonreentrant code. We must study the assembly language output produced by the compiler. For example, we cannot determine whether the following read-modify-write operation is reentrant without knowing if it is atomic:

```
time++;
```

If the compiler generates the following object code, then `time++;` is atomic (therefore not critical):

```
inc  time
```

If the compiler generates the following object code, then `time++;` is not atomic (therefore critical):

```
ldd  time
addd #1
std  time
```

> **Observation:** *The ICC12 compiler generates atomic code when setting or clearing individual bits in the I/O ports (e.g.,* `PORTC&=~0x40;` *and* `PORTB|=0x20;` *will be compiled as* `bclr PORTC,#$40` *and* `bset PORTB,#$20`*).*

11.5 Interrupts on the 6811/6812

11.5.1 General Concepts

In this section we will present specific details for the 6811/6812 microcomputers. As you develop experience using interrupts, you will come to notice a few aspects that most computers share. The following paragraphs outline three essential mechanisms that are needed to utilize interrupts. Although every computer that uses interrupts includes all three mechanisms, there is a wide spectrum of implementation methods.

All interrupting systems have the **ability for the hardware to request action from the computer**. The interrupt requests can be generated using a separate connection to the microprocessor for each device, or using a negative logic wire-or request that employs open-collector logic. The Motorola microcomputers support both types.

All interrupting systems must have the **ability for the computer to determine the source**. A vectored interrupt system employs separate connections for each device so that the computer can give automatic resolution. You can recognize a vectored system because each device has a separate interrupt vector address. With a polled interrupt system, the interrupt software must poll each device, looking for the device that requested the interrupt.

The third necessary component of the interface is the **ability for the computer to acknowledge the interrupt**. Normally there is a flag in the interface that is set on the busy-to-done state transition. In essence this flag is the cause of the interrupt. Acknowledging the interrupt involves clearing this flag. It is important to shut off the request, so that the computer will not mistakenly request a second (and inappropriate) interrupt for the same condition. Some Intel systems use a hardware acknowledgment that automatically clears the request. Most Motorola microcomputers use a software acknowledgements. So when using a flag (like the RDRF and TDRE flags in the SCI interface), it will be important to know exactly what conditions will set the flag (and request an interrupt) and how the software will clear it (acknowledge) in the ISR.

There are no standard definitions for the terms *mask, enable* and *arm* in the professional, Computer Science, or Computer Engineering communities. Nevertheless, in this book we will adhere to the following specific meanings. To **arm** a device means to enable the source of interrupts. One arms a device if one is interested in interrupts. If you are not interested in interrupts for a particular device, **disarm** it. For example the STAI bit in the 6811 PIOC register is the arm bit for the 6811 STRA pin. Similarly the 6811/6812 TMSK1 has eight arm bits for the output compare and input capture interrupts. The Motorola literature calls the arm bit an "interrupt enable mask." To **enable** means to allow interrupts at this time. We **disable** interrupts if it is currently not convenient to process. In particular, to disable interrupts we set the I bit in 6811/6812 condition code register using the `sei` instruction. There are some interrupts that can't be disabled, such as `XIRQ`, `SWI`, `illegal opcode`, and `reset`. The `cli` instruction enables interupts.

> **Common Error:** *The system will crash if the interrupt service routine doesn't either acknowledge or disarm the device requesting the interrupt.*

> **Common Error:** *The interrupt service routine doesn't have to explicitly disable interrupts at the beginning (`sei`) or explicitly reenable interrupts at the end (`cli`). The disabling and enabling occur automatically.*

The sequence of events that occurs during an interrupt service is quite similar on the 6811 and 6812. The sequence begins with the **Hardware needs service (busy-to-done) transition**. This signal is connected to an input of the microcomputer that can generate an interrupt. For example, the STRA (6811 only), key wakeup, input capture, serial communication interface (SCI), and serial peripheral interface (SPI) systems support interrupt requests. Some interrupts are internally generated, like output compare, real-time interrupt (RTI), and timer overflow.

The second event is the **setting of a flag** in one of the I/O status registers of the microcomputer. This is the same flag that a busy-waiting interface would be polling on. Examples include the STRA (**STAF**), key wakeup (**KWIFJn**), serial communication interface (**RDRF** and **TDRE**), output compare (**OCnF**), real-time interrupt (**RTIF**), and timer overflow (**TOF**). In order for an interrupt to be requested, the appropriate **flag** bit must be **armed**. Examples include the STRA (**STAI**), key wakeup (**KWIEJn**), serial communication interface (**RIE** and **TIE**), output compare (**OCnI**), real-time interrupt (**RTII**), and timer overflow (**TOI**). In summary, three conditions must be met simultaneously for an interrupt service to occur:

a device is armed	e.g., STAI=1,
a microcomputer interrupt is enabled	I=0,
an interrupting event occurs that sets the flag	e.g., STAF=1.

The third event in the interrupt processing sequence is the **thread-switch**. The thread switch is performed by the microcomputer hardware automatically. The specific steps include

1. the microcomputer will finish the current instruction,[2]
2. all the registers are pushed on the stack, the **CCR** is on top, with the **I** bit still equal to 0,
3. the microcomputer will get vector address from memory and put it into the **PC**,
4. the microcomputer will set **I = 1**.

The fourth event is the software **execution of the interrupt service routine (ISR)**. For a polled interrupt configuration, the ISR must poll each possible device, and branch to specific handler for that device. The polling order establishes device priority. For a vectored interrupt configuration, you could poll anyway to check for runtime hardware/software errors. The ISR must either acknowledge or disarm the interrupt. We acknowledge an interrupt by clearing the flag that was set in the second event shown above. After we acknowledge a low-priority interrupt, we may reenable interrupts (`cli`)

---

[2]Again, the 6812 allows the instructions `rev`, `revw`, and `wav` to be interrupted.

to allow higher priority devices to go first. All ISR's must perform the necessary operations (read data, write data etc.) and pass parameters through global memory (e.g., FIFO queue).

The last event is another thread-switch in order to **return control to the thread that was running** when the interrupt was processed. In particular, the software executes an `rti` at the end of the ISR, which will pull all the registers off the stack. At the beginning of the interrupt service the CCR was pushed on the stack with I=0. Therefore, the execution of `rti` automatically reenables interrupts. After the ISR executes `rti`, the stack is restored to the state it was before the interrupt. The ISR may change global variables or I/O ports, but the registers and stack are left unchanged by the ISR.

The interrupt hardware will automatically save all registers on the stack during the thread-switch, as shown in Figure 11.10. The thread-switch is the process of stopping the foreground (main) thread and starting the background (interrupt handler). The "oldPC" value on the stack points to the place in the foreground thread to resume once the interrupt is complete. At the end of the interrupt handler, another thread-switch occurs as the `rti` instruction restores registers from the stack (including the PC).

Figure 11.10
Stack before and after an interrupt.

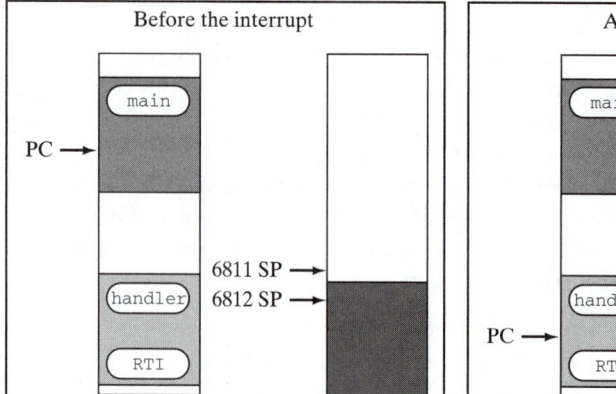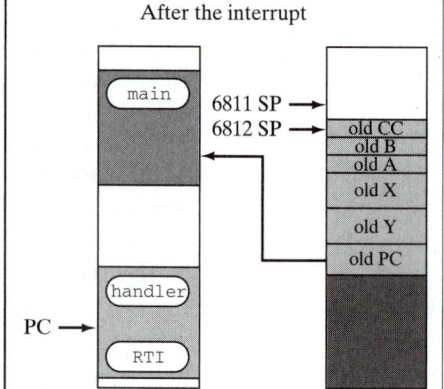

Checkpoint 11.7: What would happen if the ISR forgot to acknowledge the interrupt?

Checkpoint 11.8: If you didn't want to or couldn't acknowledge, what else might the ISR do?

11.5.2
6811 Interrupts

The 6811 has two external requests ($\overline{\text{IRQ}}$ and $\overline{\text{XIRQ}}$) that are level-zero active. Many of the internal I/O devices can generate interrupt requests based on external events (e.g., STRA, timer, serial ports). Most interrupt requests will temporarily set the I bit in the CCR during the interrupt program to prevent other interrupts (including itself). On the other hand, the XIRQ request temporarily sets both the I and X bits in the CCR during the interrupt program to postpone all other interrupt sources. The MC68HC11A8 can support

    a STRA interrupt,
    three input capture interrupts,
    five output compare interrupts,
    three timer interrupts (timer overflow, real-time interrupt, pulse accumulator),
    two serial port interrupts (SCI and SPI).

The interrupts have a fixed priority, but you can elevate one request to highest priority using the HPRIO, Hardware Priority Interrupt Register. We typically use XIRQ to interface a single highest priority device. XIRQ has a separate interrupt vector ($FFF4) and a separate enable bit (X). Once the X bit is cleared (enabled) the software cannot disable it. A XIRQ interrupt is requested when the external XIRQ pin is low and the X bit in the CCR is 0. XIRQ processing will automatically set **X = I = 1** (an IRQ cannot interrupt an XIRQ service) at the start of the XIRQ handler. Just like regular interrupts, the X and I bits will be restored to their original values by the `rti` instruction.

The priority is fixed in the order, with SCI having the lowest priority and Reset having the highest. For some interrupt sources, such as the SCI interrupts, flags are automatically cleared during the response to the interrupt requests. For example, the RDRF flag in the SCI system is cleared by the automatic clearing mechanism, consisting of a read of the SCI status register while RDRF is set, followed by a read of the SCI data register. The normal response to an RDRF interrupt request is to read the SCI status register to check for receive errors, then to read the received data from the SCI data register. These two steps satisfy the automatic clearing mechanism without requiring any additional instructions. Many of the potential interrupt requests share the same interrupt vector. For example, there are multiple possible interrupt sources (STAF and all external connections to IRQ) that all use the vector at $FFF2. Therefore, when this request is processed the software must determine which of the possible signals caused the interrupt. The entire list of potential 6811 interrupts is included in Table 11.2.

Table 11.2
Interrupt vectors for the 6811.

Vector Address	Interrupt Source, Flag	Enable	Arm
$FFFE	Reset	none	none
$FFFC	COP Clock Monitor Fail Reset	none	OPTION.CME
$FFFA	COP Failure Reset	none	CONFIG.NOCOP
$FFF8	Unimplemented Instruction Trap	none	none
$FFF6	SWI	none	none
$FFF4	XIRQ	X bit	none
$FFF2	IRQ or	I bit	External
	Parallel I/O handshake, STAF	I bit	PIOC.STAI
$FFF0	Real-time Interrupt, RTIF	I bit	TMSK2.RTII
$FFEE	Timer input capture 1, IC1F	I bit	TMSK1.IC1I
$FFEC	Timer input capture 2, IC2F	I bit	TMSK1.IC2I
$FFEA	Timer input capture 3, IC3F	I bit	TMSK1.IC3I
$FFE8	Timer output compare 1, OC1F	I bit	TMSK1.OC1I
$FFE6	Timer output compare 2, OC2F	I bit	TMSK1.OC2I
$FFE4	Timer output compare 3, OC3F	I bit	TMSK1.OC3I
$FFE2	Timer output compare 4, OC4F	I bit	TMSK1.OC4I
$FFE0	Timer output compare 5, OC5F	I bit	TMSK1.OC5I
$FFDE	Timer Overflow, TOF	I bit	TMSK2.TOI
$FFDC	Pulse Accumulator Overflow, PAOVF	I bit	TMSK2.PAOVI
$FFDA	Pulse Accumulator Input Edge, PAIF	I bit	TMSK2.PAII
$FFD8	SPI Serial Transfer Complete, SPIF	I bit	SPCR.SPI0E
$FFD6	SCI Transmit Buffer Empty, TDRE	I bit	SCCR2.TIE
	SCI Transmit Complete, TC	I bit	SCCR2.TCIE
	SCI Receiver Buffer Full, RDRF	I bit	SCCR2.RIE
	SCI Receiver Idle, IDLE	I bit	SCCR2.ILIE

How we establish the vectors depends on the memory configuration. For a stand-alone single-chip 6811 systems or multichip systems with your own PROM at $C000 to $FFFF, we can use assembly to set the vectors. For example:

```
org   $FFF0
fdb   RtiHan        Pointer to real-time interrupt handler
org   $FFF2
fdb   STRAHan       Pointer to external IRQ and STRA handler
org   $FFF4
fdb   XIRQHan       Pointer to external XIRQ handler
org   $FFFE
fdb   main          Pointer to hardware and power on reset handler
```

11.5.3
6812 Interrupts

CPU12 exceptions include resets and interrupts. Each exception has an associated 16-bit vector that points to the memory location where the routine that handles the exception is located. Vectors are stored in the upper 128 bytes of the standard 64-Kbyte address map. A hardware priority hierarchy determines which reset or interrupt is serviced first when simultaneous requests are made. Six sources are not maskable. The remaining sources have a separate bit that can be armed or disarmed. The priorities of the non-maskable sources are:

1. Power-On-Reset (POR) or regular hardware RESET pin
2. Clock monitor reset
3. Computer-Operating-Properly (COP) watchdog reset
4. Unimplemented instruction trap
5. Software interrupt instruction (`swi`)
6. XIRQ signal (if X bit in CCR = 0)

Other interrupt sources include on-chip peripheral systems and external interrupt service requests. Interrupts from these sources are recognized when the interrupt enable bit (I) in the CCR is cleared. The default state of the I bit out of reset is one, but it can be written at any time.

The 6812 interrupt hardware also automatically saves all registers on the stack during the thread-switch, see Figure 11.10. The "oldPC" value on the stack points to the place in the foreground thread to resume once the interrupt is complete. At the end of the interrupt handler, another thread switch occurs as the `rti` instruction restores registers from the stack (including the PC).

Like the 6811, the 6812 has two external requests (IRQ and XIRQ) that are level-zero active. Many of the internal I/O devices can generate interrupt requests based on external events (e.g., key wakeup, input capture, SCI, SPI, etc.) Most of the interrupt requests will temporarily set the I bit in the CCR during the interrupt program to prevent other interrupts (including itself). On the other hand, the XIRQ request temporarily sets both the I and X bits in the CCR during the interrupt program to postpone all other interrupts sources. The MC68HC812A4 can support

- 24 key wakeup interrupts (Ports D, H and J),
- 8 input capture/output compare interrupts,
  an ADC interrupt,
- 3 timer interrupts (timer overflow, real-time interrupt, pulse accumulator),
- 3 serial port interrupts (two SCI's and SPI).

The interrupts have a fixed priority, but you can elevate one request to highest priority using the HPRIO, Hardware Priority Interrupt Register ($001F). The relative priorities of the other interrupt sources remain the same.

The 6812 XIRQ interrupt system is similar to the one on the 6811. We typically use XIRQ to interface a single highest-priority device. XIRQ has a separate interrupt vector ($FFF4) and a separate enable bit (X). Once the X bit is cleared (enabled), the software cannot disable it. A XIRQ interrupt is requested when the external XIRQ pin is low and the X bit in the CCR is 0. XIRQ processing will automatically set **X = I = 1** (an IRQ cannot interrupt an XIRQ service) at the start of the XIRQ handler. Just like regular interrupts, the X and I bits will be restored to their original values by the `rti` instruction.

The priority is fixed in the order shown with *Key Wakeup H* having the lowest priority and *Reset* having the highest. Any one particular system usually uses just a few interrupts, but the entire list of potential MC68HC812A4 interrupts is included in Table 11.3.

Table 11.3
Interrupt vectors for the MC68HC812A4.

Vector Address	Interrupt Source, Flag	Enable	Arm
$FFFE	Reset	none	none
$FFFC	COP Clock Monitor Fail Reset	none	COPCTL.CME
		none	COPCTL.FCME
$FFFA	COP Failure Reset	none	COP rate selected
$FFF8	Unimplemented Instruction Trap	none	none
$FFF6	SWI	none	none
$FFF4	XIRQ	X bit	none
$FFF2	IRQ or	I bit	INTCR.IRQEN
	Key Wake Up D, KWIFD.[7:0]	I bit	KWIED.[7:0]
$FFF0	Real-time Interrupt, RTIF	I bit	RTICTL.RTIE
$FFEE	Timer Channel 0, C0F	I bit	TMSK1.C0I
$FFEC	Timer Channel 1, C1F	I bit	TMSK1.C1I
$FFEA	Timer Channel 2, C2F	I bit	TMSK1.C2I
$FFE8	Timer Channel 3, C3F	I bit	TMSK1.C3I
$FFE6	Timer Channel 4, C4F	I bit	TMSK1.C4I
$FFE4	Timer Channel 5, C5F	I bit	TMSK1.C5I
$FFE2	Timer Channel 6, C6F	I bit	TMSK1.C6I
$FFE0	Timer Channel 7, C7F	I bit	TMSK1.C7I
$FFDE	Timer Overflow, TOF	I bit	TMSK2.TOI
$FFDC	Pulse Accumulator Overflow, PAOVF	I bit	PACTL.PAOVI
$FFDA	Pulse Accumulator Input Edge, PAIF	I bit	PACTL.PAI
$FFD8	SPI Serial Transfer Complete, SPIF	I bit	SP0CR1.SPI0E
$FFD6	SCI 0 Transmit Buffer Empty, TDRE	I bit	SC0CR2.TIE
	SCI 0 Transmit Complete, TC	I bit	SC0CR2.TCIE
	SCI 0 Receiver Buffer Full, RDRF	I bit	SC0CR2.RIE
	SCI 0 Receiver Idle, IDLE	I bit	SC0CR2.ILIE
$FFD4	SCI 1 Transmit Buffer Empty, TDRE	I bit	SC1CR2.TIE
	SCI 1 Transmit Complete, TC	I bit	SC1CR2.TCIE
	SCI 1 Receiver Buffer Full, RDRF	I bit	SC1CR2.RIE
	SCI 1 Receiver Idle, IDLE	I bit	SC1CR2.ILIE
$FFD2	ATD Sequence Complete, ASCIF	I bit	ATDCTL2.ASCIE
$FFD0	Key Wakeup J, KWIFJ.[7:0]	I bit	KWIEJ.[7:0]
$FFCE	Key Wakeup H, KWIFH.[7:0]	I bit	KWIEH.[7:0]
$FF80–$FFCD	Reserved	I bit	

For some interrupt sources, such as the SCI interrupts, flags are automatically cleared during the response to the interrupt requests. For example, the RDRF flag in the SCI system is cleared by the automatic clearing mechanism, consisting of a read of the SCI status register while RDRF is set, followed by a read of the SCI data register. The normal response to an RDRF interrupt request is to read the SCI status register to check for receive errors, then to read the received data from the SCI data register. These two steps satisfy the automatic clearing mechanism without requiring any additional instructions. Many of the potential interrupt requests share the same interrupt vector (e.g., there are eight possible interrupt sources (PJ7–PJ0) that all use the vector at $FFD0. Therefore, when this request is processed the software must determine which of the eight possible signals caused the interrupt.

The priority on the MC68HC912B32 is fixed in the order shown with *BDLC* having the lowest priority and Reset having the highest. Many of the potential interrupt requests share the same interrupt vector. For example, there are four possible SCI interrupt sources (TDRE, TC, RDRF, and IDLE) that all use the vector at $FFD6. Therefore, when this request is processed, the software must determine which of the four possible signals caused the interrupt. Any one particular system usually uses just a few interrupts, but the entire list of potential MC68HC912B32 interrupts is included in Table 11.4.

Table 11.4
Interrupt vectors for the MC68HC912B32.

Vector Address	Interrupt Source Flag	Enable	Arm
$FFFE, $FFFF	Reset	None	None
$FFFC, $FFFD	COP clock monitor fail reset	None None	COPCTL.CME COPCTL.FCME
$FFFA, $FFFB	COP failure reset	None	COP rate selected
$FFF8, $FFF9	Unimplemented instruction trap	None	None
$FFF6, $FFF7	SWI	None	None
$FFF4, $FFF5	XIRQ	X bit	None
$FFF2, $FFF3	IRQ	I bit	INTCR.IRQEN
$FFF0, $FFF1	Real-time interrupt, RTIF	I bit	RTICTL.RTIE
$FFEE, $FFEF	Timer channel 0, C0F	I bit	TMSK1.C0I
$FFEC, $FFED	Timer channel 1, C1F	I bit	TMSK1.C1I
$FFEA, $FFEB	Timer channel 2, C2F	I bit	TMSK1.C2I
$FFE8, $FFE9	Timer channel 3, C3F	I bit	TMSK1.C3I
$FFE6, $FFE7	Timer channel 4, C4F	I bit	TMSK1.C4I
$FFE4, $FFE5	Timer channel 5, C5F	I bit	TMSK1.C5I
$FFE2, $FFE3	Timer channel 6, C6F	I bit	TMSK1.C6I
$FFE0, $FFE1	Timer channel 7, C7F	I bit	TMSK1.C7I
$FFDE, $FFDF	Timer overflow, TOF	I bit	TMSK2.TOI
$FFDC, $FFDD	Pulse Accumulator Overflow, PAOVF	I bit	PACTL.PAOVI
$FFDA, $FFDB	Pulse Accumulator Input Edge, PAIF	I bit	PACTL.PAI
$FFD8, $FFD9	SPI serial transfer complete	I bit	SP0CR1.SPIE
$FFD6, $FFD7	SCI 0 Transmit Buffer Empty, TDRE SCI 0 Transmit Complete, TC SCI 0 Receiver Buffer Full, RDRF SCI 0 Receiver Idle, IDLE	I bit I bit I bit I bit	SC0CR2.TIE SC0CR2.TCIE SC0CR2.RIE SC0CR2.ILIE
$FFD4, $FFD5	Reserved	I bit	–
$FFD2, $FFD3	ATD, ASCIF	I bit	ATDCTL2.ASCIE
$FFD0, $FFD1	BDLC	I bit	BCR1.IE
$FF80–$FFCF	Reserved	I bit	–

11.6 External Edge Interrupt

11.6.1
MC68HC11 STRA Interrupt

One of the simplest interrupt mechanisms on the 6811 is the STRA interrupt. This external pin can be configured to generate an interrupt on either a rising or falling edge. The hardware will latch the Port C input into the PORTCL register on the active edge of STRA. The I/O ports used in the STRA interrupt mechanism are shown in Table 11.5.

Table 11.5 6811 I/O ports used in the STRA interrupts.

address	Bit 7	6	5	4	3	2	1	Bit 0	Name
$1002	STAF	STAI	CWOM	HNDS	OIN	PLS	EGA	INVB	PIOC
$1003	PC7	PC6	PC5	PC4	PC3	PC2	PC1	PC0	PORTC
$1005	PCL7	PCL6	PCL5	PCL4	PCL3	PCL2	PCL1	PCL0	PORTCL
$1007	DDRC7	DDRC6	DDRC5	DDRC4	DDRC3	DDRC2	DDRC1	DDRC0	DDRC

Assume there is a digital signal connected to the STRA input. Program 11.7 will count the number of rising edges of this signal using interrupt synchronization. The EGA bit specifies whether a rising edge (1) or a falling edge (0) will set STAF. The arm bit is STAI. To acknowledge the interrupt (i.e., clear STAF) the software must read PIOC with STAF set followed by a read PORTCL. The Port C inputs are not used, but the data must be read in order to clear the STAF flag. A flowchart of this system, Figure 11.11, shows there are two threads: the foreground program (main) and a background thread (ISR) that runs whenever there is a rising edge of STRA. Information is passed between the threads using global variables.

Program 11.7 Edge counting using STRA interrupt.

```
            org   $0000                      unsigned char Count;
Count       rmb   1                          void STRA_Init(void){
            org   $E000                        asm(" sei");     // Make ritual atomic
STRA_Init   sei         make atomic            PIOC  = 0x42;    // Arm
            ldaa  #$42   arm STRA               DDRC  = 0x00;    // inputs
            staa  PIOC   STAI=1, EGA=1 rising   Count = 0;       // Initialize
            ldaa  #$00   PC7-PC0 inputs         asm(" cli");
            staa  DDRC                        }
            clr   Count                       #pragma interrupt_handler STRAHan()
            cli         enable IRQ            void STRAHan(void){ unsigned char dummy;
            rts                                 dummy = PIOC;      // Acknowledge
STRAHan     ldaa  PIOC                          dummy = PORTCL
            ldaa  PORTCL clear STAF             Count++;
            inc   Count  number of edges      }
            rti                               #pragma abs_address:0xfff2
            org   $FFF2                        void (*STRA_vector[])() ={
            fdb   STRAHan                        STRAHan}; // fff2 STRA
                                              #pragma end_abs_address
```

Figure 11.11
Flowchart of the STRA
interrupting system.

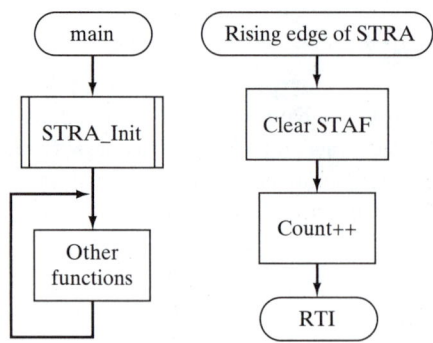

Checkpoint 11.9: How do you change Program 11.7 so it counts falling edges?

**11.6.2
MC68HC812A4
Key Wakeup
Interrupts**

One of the simplest interrupt mechanisms on the 6812 is the key wakeup interrupt. On the MC68HC812A4, key wakeups are available on Ports D, H, and J. Table 11.6 shows the I/O registers involved. Any or all of these 24 pins can be configured as a key wakeup interrupt. Each of the 24 wakeup lines has a separate I/O pin (PORTD, PORTH, PORTJ), a separate direction bit (DDRD, DDRH, DDRJ usually configured as an input), a separate flag bit (KWIFD, KWIFH, KWIFJ), and a separate arm bit (KWIED, KWIEH, KWIEJ).

Table 11.6   6812 I/O ports used in the key wakeup interrupts.

Address	Bit 7	6	5	4	3	2	1	Bit 0	Name
$0005	PD7	PD6	PD5	PD4	PD3	PD2	PD1	PD0	PORTD
$0007	Bit 7	6	5	4	3	2	1	Bit 0	DDRD
$0020	Bit 7	6	5	4	3	2	1	Bit 0	KWIED
$0021	Bit 7	6	5	4	3	2	1	Bit 0	KWIFD
$0024	PH7	PH6	PH5	PH4	PH3	PH2	PH1	PH0	PORTH
$0025	Bit 7	6	5	4	3	2	1	Bit 0	DDRH
$0026	Bit 7	6	5	4	3	2	1	Bit 0	KWIEH
$0027	Bit 7	6	5	4	3	2	1	Bit 0	KWIFH
$0028	PJ7	PJ6	PJ5	PJ4	PJ3	PJ2	PJ1	PJ0	PORTJ
$0029	Bit 7	6	5	4	3	2	1	Bit 0	DDRJ
$002A	Bit 7	6	5	4	3	2	1	Bit 0	KWIEJ
$002B	Bit 7	6	5	4	3	2	1	Bit 0	KWIFJ
$002C	Bit 7	6	5	4	3	2	1	Bit 0	KPOLJ
$002D	Bit 7	6	5	4	3	2	1	Bit 0	PUPSJ
$002E	Bit 7	6	5	4	3	2	1	Bit 0	PULEJ

For Ports D and H, a falling edge on the input pin will set the corresponding flag bit. The key wakeup interrupts on Port J can be configured on either the rising or falling edge. If the corresponding bit in the KPOLJ is 0, then a falling edge on Port J is active. Conversely, if the bit in the KPOLJ register is 1, then a rising edge on Port J will set the flag. Another convenience of Port J is the available pull-up or pull-down resistors via the PUPSJ and PULEJ configuration registers (see Table 11.7). Each of the eight pins of Port J can be configured separately.

Table 11.7
Pull-up modes of Port J.

DDRJ	PUPSJ bit	PULEJ bit	Port J mode
1	x	x	regular output
0	x	0	regular input
0	0	1	input with passive pull-down
0	1	1	input with passive pull-up

A typical application of pull-up is the interface of simple switches, see Figure 11.12. Using pull-up mode eliminates the need for an external resistor.

Figure 11.12
6812 port J supports internal passive pull-up and pull-down.

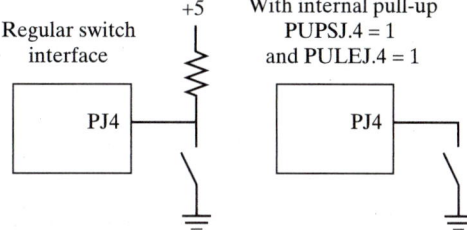

To clear a bit in the key wakeup-flag register, we write a one to it. Writing zeros to the flag registers will have no effect. The following code will clear bit 4 of the key wakeup-flag register:

```
movb #$10,KWIFJ
```

In C, we clear bit 4 of the key wakeup-flag register:

```
KWIFJ = 0x10;
```

The following code will mistakenly clear all bits in the key wakeup-flag register:

```
bset KWIFJ,#$10
```

Similarly in C, the following code would mistakenly clear all bits in the key wakeup-flag register:

```
KWIFJ |= 0x10;
```

Three conditions must be simultaneously true for a key wakeup interrupt to be requested:

- the arm bit is set, and (e.g., KWIEJ bit 4),
- the I bit in the 6812 CCR is 0,
- the flag bit is set (e.g., an edge on PJ4 sets KWIFJ bit 4).

Even though there are 24 key wakeup lines, there are only three interrupt vectors. So, if two or more wakeup interrupts are used on the same port, it will be necessary to poll. Interrupt polling is the software function to look and see which of the potential sources requested the interrupt.

Assume there is a digital signal connected to the PJ0 input. Program 11.8 will count the number of rising edges of this signal using interrupt synchronization. KPOLJ bit 0 specifies whether a rising edge (1) or a falling edge (0) will set the flag bit in the KWIFJ register. KWIEJ bit 0 is the arm bit. To acknowledge the interrupt (i.e., clear bit 0 or KWIFJ), the software must write a 1 into the bit 0 position of the KWIFJ register. The flowchart of this system is virtually identical to the STRA system shown in Figure 11.11.

Program 11.8   Edge counting using key wakeup interrupt.

```
        org   $0800                          unsigned char Count;
Count   rmb   1                              void PJ0_Init(void){
        org   $F000                            asm(" sei");    // Make ritual atomic
PJ0_Init sei              make atomic          KWIEJ = 0x01;   // Arm
        movb #$01,KWIEJ  arm PJ0               DDRJ  = 0x00;   // inputs
        movb #$00,DDRJ   PJ0 is input          KPOLJ = 0x01;   // rising edge
        movb #$01,KPOLJ  rising edge           Count = 0;      // Initialize
        clr   Count                            asm(" cli");
        cli              enable IRQ          }
        rts                                 #pragma interrupt_handler PJ0Han()
PJ0Han  movb #$01,KWIFJ  acknowledge        void PJ0Han(void){
        inc   Count      number of edges      KWIFJ = 0x01;    // Acknowledge
        rti                                    Count++;
        org   $FFD0                          }
        fdb   PJ0Han                         #pragma abs_address:0xffd0
                                            void (*PJ0_vector[])() ={
                                              PJ0Han};  // ffd0 PJ0
                                            #pragma end_abs_address
```

Checkpoint 11.10: How do you change Program 11.8 so it counts falling edges?

11.7 Serial Port Interface Using Interrupt Synchronization

**11.7.1
6811 SCI Details**   Most versions of the 6811 have one asynchronous serial port using Port D bits 1,0. Table 11.8 shows the 6811 I/O ports that implement the SCI functions. The baud rate was discussed back in Chapter 5. An interrupting SCI example can be found as tut4 within **TExaS**. The tutorial at the end of this chapter also deals with an interrupting SCI.

Table 11.8.   6811 SCI ports.

address	Bit 7	6	5	4	3	2	1	Bit 0	Name
$1028	SPIE	SPE	DWOM	MSTR	CPOL	CPHA	SPR1	SPR0	SPCR
$102B	TCLR	SCP2	SCP1	SCP0	RCKB	SCR2	SCR1	SCR0	BAUD
$102C	R8	T8	0	M	WAKE	0	0	0	SCCR1
$102D	TIE	TCIE	RIE	ILIE	TE	RE	RWU	SBK	SCCR2
$102E	TDRE	TC	RDRF	IDLE	OR	NF	FE	0	SCSR
$102F	R7/T7	R6/T6	R5/T5	R4/T4	R3/T3	R2/T2	R1/T1	R0/T0	SCDR

SCCR1 The **SCCR1** control register is used to specify the overall mode of operation. **R8** is the Receive Data Bit 8. If M bit is set, R8 stores ninth bit in receive data character. **T8** is the Transmit Data Bit 8. If M bit is set, T the software needs to set T8 as the ninth bit in the transmit data character. **M** is the Mode, where M equals 0 creates a 10-bit frame with 1 start bit, 8 data bits, 1 stop bit. If M equals 1, then there is an 11-bit frame with 1 start bit, 9 data bits, 1 stop bit (the ninth data bit is in T8/R8). **WAKE** is the Wake Up by Address Mark/Idle

bit. If WAKE equals 0, then the 6811 will wake up by IDLE line recognition. If WAKE=1, then the 6811 will wake up by address mark (most significant data bit set).

SCCR2 The **SCCR2** control register contains the bits that turn on the SCI, and contains the interrupt arm bits. **TIE** is the Transmit Interrupt Enable bit. If TIE equals 0, then the TDRE interrupts are disarmed. If TIE equals 1, then a SCI interrupt will be requested when TDRE status flag is set (armed). **TCIE** is the Transmit Complete Interrupt Enable bit. If TCIE equals 0, then TC interrupts are disarmed. If TCIE equals 1, then a SCI interrupt will be requested if TC is set to 1. **RIE** is the Receiver Interrupt Enable bit. If RIE equals 0, then RDRF and OR interrupts are disarmed. If RIE equals 1, the a SCI interrupt will be requested when the RDRF flag or the OR status flag is set (armed). **ILIE** is the Idle Line Interrupt Enable bit. If IDIE is 0, then the IDLE interrupts are disarmed. If IDIE is 1, then a SCI interrupt will be requested when IDLE status flag is set. **TE** is the Transmitter Enable bit. You turn this bit off to disable the transmitter, and set this bit to enable the transmitter. **RE** is the Receiver Enable bit. This bit should be set to 1 to activate the receiver. If RE is 0, then the receiver is disabled. **RWU** is the Receiver Wake Up Control bit. For normal serial port operation, you should clear this bit. If RWU is 1, then an incoming serial frame will wake up the 6811. **SBK** is the Send Break bit. Under normal operation, this bit too should be 0. When the transmitter is idle, a high signal is output on PD1=TxD. If you set SBK equal to 1, then a continuous low (a break) will be transmitted on PD1=TxD.

SPCR The **DWOM** bit in the SPCR register is used to activate the Port D wired-OR mode. This one bit affects all six port D pins. If DWOM is 0, then PD5-PD0 are normal CMOS signals. A normal CMOS output has two states, either driven low (0V or logic zero) or driven high (+5V or logic one). If DWOM is 1, then any port D output is open drain. *Open drain* means the output is either driven low or floats. Alternative names for the same behavior are *open collector* and *wire-or mode.*

SCSR The flags in the **SCSR** register can be read by the software, but can not be modified by writing to this register. **TDRE** is the Transmit Data Register Empty Flag. TDRE is set if transmit data can be written to SCDR. If TDRE is zero, the transmit data register contains previous data that has not yet been moved to the transmit shift register. Writing into the SCDR when TDRE is clear will result in a loss of data. TDRE is cleared by SCSR read with TDRE set followed by SCDR write. **TC** is the Transmit Complete Flag. It is set if the transmitter is idle (no data, preamble, or break transmission in progress). TC is cleared by a SCSR read with TC set followed by a SCDR write. **RDRF** is the Receive Data Register Full Flag. It is set if a received character is ready to be read from SCDR. RDRF is cleared by a SCSR read with RDRF set followed by a SCDR read. **IDLE** is the Idle Line Detected Flag. It is set if the RxD line is idle (10 or 11 consecutive logic ones). The IDLE flag is inhibited when RWU is set to one. IDLE is cleared by a SCSR read with IDLE set followed by a SCDR read. Once cleared, IDLE is not set again until the RxD line has been active and becomes idle again. **OR** is the Overrun Error Flag. It is set if a new character is received before a previously received character is read from SCDR. OR is cleared by a SCSR read with OR set followed by a SCDR read. **NF** is the Noise Error Flag. It is set if majority sample logic detects anything other than a unanimous decision. NF is cleared by a SCSR read with NF set followed by a SCDR read. **FE** is the Framing Error Flag. It is set if a zero is detected where a stop bit was expected. FE is cleared by a SCSR read with FE set followed by a SCDR read.

11.7.2
6812 SCI Details

The MC68HC812A4 has two asynchronous serial ports using Port S bits 3,2 and bits 1,0. The MC68HC912B32 has only one serial port using Port S bits 1,0. Table 11.9 shows the port names and addresses for the SCI devices on the 6812. The baud rate was discussed back in Chapter 5.

Table 11.9 6812 SCI ports.

Address	Bit 7	6	5	4	3	2	1	Bit 0	Name
$0047	LCK	PLLON	PLLS	BCSC	BCSB	BCSA	MCSB	MCSA	CLKCTL
$00C0	BTST	BSPL	BRLD	SBR12	SBR11	SBR10	SBR9	SBR8	SC0BD
$00C1	SBR7	SBR6	SBR5	SBR4	SBR3	SBR2	SBR1	SBR0	
$00C2	LOOPS	WOMS	RSRC	M	WAKE	ILT	PE	PT	SC0CR1
$00C3	TIE	TCIE	RIE	ILIE	TE	RE	RWU	SBK	SC0CR2
$00C4	TDRE	TC	RDRF	IDLE	OR	NF	FE	PF	SC0SR1
$00C5	0	0	0	0	0	0	0	RAF	SC0SR2
$00C6	R8	T8	0	0	0	0	0	0	SC0DRH
$00C7	R7T7	R6T6	R5T5	R4T4	R3T3	R2T2	R1T1	R0T0	SC0DRL
$00C8	BTST	BSPL	BRLD	SBR12	SBR11	SBR10	SBR9	SBR8	SC1BD
$00C9	SBR7	SBR6	SBR5	SBR4	SBR3	SBR2	SBR1	SBR0	
$00CA	LOOPS	WOMS	RSRC	M	WAKE	ILT	PE	PT	SC1CR1
$00CB	TIE	TCIE	RIE	ILIE	TE	RE	RWU	SBK	SC1CR2
$00CC	TDRE	TC	RDRF	IDLE	OR	NF	FE	PF	SC1SR1
$00CD	0	0	0	0	0	0	0	RAF	SC1SR2
$00CE	R8	T8	0	0	0	0	0	0	SC1DRH
$00CF	R7T7	R6T6	R5T5	R4T4	R3T3	R2T2	R1T1	R0T0	SC1DRL

SCxCR1 The **SCxCR1** control register is used to specify the overall mode of operation. **LOOPS** is the SCI LOOP Mode/Single Wire Mode Enable bit. If it is 0, then SCI transmit and receive sections operate normally. If LOOPS is 1, then the SCI receive section is disconnected from the RxD pin and the RxD pin is available as general purpose I/O. The receiver input is determined by the RSRC bit. The transmitter output is controlled by the associated DDRS bit. Both the transmitter and the receiver must be enabled to use the LOOP or the single wire mode. If the DDRS bit associated with the TxD pin is set during the LOOPS = 1, the TxD pin outputs the SCI waveform. If the DDRS bit associated with the TxD pin is clear during the LOOPS = 1, the TxD pin becomes high (IDLE line state) for RSRC = 0 and high impedance for RSRC = 1. **WOMS** is the Wired-OR Mode bit for Serial Pins for TxD and RxD. If it is 0, then the SCI output operates in a normal mode with both high and low drive capability. If WOMS is 1, then each pin operates in an open-drain fashion if that pin is declared as an output. **RSRC** is the Receiver Source bit. When LOOPS = 1, the RSRC bit determines the internal feedback path for the receiver. If RSRC is 0, then the receiver input is connected to the transmitter internally (not TxD pin). If RSRC is 1, then the receiver input is connected to the TxD pin. These configuration modes are presented in Table 11.10. **M** is the Mode bit. If M is 0, then the frame is 10 bits with one start, eight data, one stop bit. If M is 1, then the frame is 11 bits with one start, eight data, ninth data, one stop bit . The ninth bit is R8 for receiving and T8 for transmission. **WAKE** is the Wakeup by Address Mark/Idle bit. If it is 0, then the 6812 will wake up by IDLE line recognition. If WAKE is 1, then the 6812 will wake up by address mark (last data bit set). **ILT** is the Idle Line Type bit. It determines which of two types of idle line detection will be used by the SCI receiver. If ILT is 0, then Short idle line mode is enabled. If ILT is 1, then Long idle line mode is detected. **PE** is the Parity Enable bit. If it is 0, parity is disabled. If PE is 1, then the most significant data bit in the frame will be a parity bit. **PT** is the Parity Type bit.

If PT is 0, even parity is selected. An even number of ones in the data character causes the parity bit to be zero and an odd number of ones causes the parity bit to be one. If PT is 1, then odd parity is selected. An odd number of ones in the data character causes the parity bit to be zero and an even number of ones causes the parity bit to be one.

Table 11.10 Serial port modes for the 6812.

LOOPS	RSRC	DDRS1(3)	WOMS	Function of Port S Bit 1(3)
0	x	x	x	Normal Operations
1	0	0	0	LOOP mode without TxD output
1	0	0	1	LOOP mode without TxD output (TxD = HiZ)
1	0	1	0	LOOP mode with TxD output (CMOS)
1	0	1	1	LOOP mode with TxD output (open-drain)
1	1	0	x	Single wire mode without TxD output
1	1	1	0	Single wire mode with TxD output
1	1	1	1	Single wire mode for receive and transmit (open-drain)

SCxCR2 The **SCxCR2** control register contains the bits that turn on the SCI, and contains the interrupt arm bits. TIE is the Transmit Interrupt Enable bit. If it is 0, then TDRE interrupts are disarmed. We set this bit to 1 to arm SCI transmitter interrupts. If we set TIE to 1, then a SCI interrupt will be requested whenever the TDRE status flag is set (armed). **TCIE** is the Transmit Complete Interrupt Enable bit. If TC is 0, then TC interrupts are disarmed. If it is 1, then a SCI interrupt will be requested whenever the TC status flag is set. **RIE** is the Receiver Interrupt Enable bit. Setting RIE equal to 0 will prevent RDRF and OR from requesting an interrupt. We set this bit to 1 to arm SCI receiver interrupts (i.e., if RIE is 1, then a SCI interrupt will be requested whenever the RDRF status flag or the OR status flag is set). ILIE is the Idle Line Interrupt Enable bit. If it is 0, then IDLE interrupts are disarmed. If ILIE is 1, then a SCI interrupt will be requested whenever the IDLE status flag is set. **TE** is the Transmitter Enable bit, and **RE** is the Receiver Enable bit. We set both TE and RE equal to 1 in order to activate the SCI device. RWU is the Receiver Wakeup Control bit. Setting this bit to 0 will disable the wakeup function. If RWU is 1, then the wakeup function is active. The software goes to sleep by executing a `wai` instruction, and the microcomputer will wake up when a new frame is received by the SCI. SBK is the Send Break bit. Normal SCI operation occurs when SBK is 0. When the software sets this bit, a break code is generated. The SCI transmitter sends at least 10 or 11 contiguous zeros. As long as SBK is 1, break codes will be continuously sent.

SCxSR1 The flags in the **SCxSR1** register can be read by the software, but cannot be modified by writing to this register. **TDRE** is the Transmit Data Register Empty Flag. It is set by the SCI hardware if transmit data can be written to SCxDRL. If TDRE is zero, transmit data register contains previous data that has not yet been moved to the transmit shift register. Writing into the SCxDRL when TDRE is clear will result in a loss of data. On the other hand, when this bit is set, the software can begin another output transmission by writing to SCxDRL. This flag is cleared by first reading SCxSR1 with TDRE set, followed by a SCxDRL write. **TC** is the Transmit Complete Flag bit. This flag is set when the transmitter is idle (i.e., no data, preamble, or break transmission in progress). We clear by reading SCxSR1 with TC set followed by writing to SCxDRL. **RDRF** is the Receive Data Register Full bit. RDRF is set if a received character is ready to be read from SCxDR. We clear the RDRF flag by reading SCxSR1 with RDRF set followed by reading SCxDRL. **IDLE** is the Idle Line Detected Flag bit. A receiver idle line is detected, after the receipt of a minimum of 10/11 consecutive 1s. This bit will not be set by the idle line condition when the RWU bit is set. Once cleared, IDLE will not be set again until after RDRF has been set (after the line has been active and becomes idle again). **OR** is the Receiver Overrun Error Flag bit. It is set when a new byte is ready to be transferred from the receive shift register to the receive data register and the receive data register is already full (RDRF bit is set). Data transfer is inhibited until this bit is cleared. **NF** is the Receiver Noise Error Flag bit. It is set if there is noise on the start bit, any of the data bits, or on the stop bit. **FE** is the Receiver Framing Error Flag bit. This bit is set when a zero is detected where a stop bit was expected. We can clear the FE flag by reading SCxSR1 with FE set, followed by reading SCxDRL. The 6812 can generate parity and **PF** is the Receiver Parity Error Flag bit. It indicates if the parity of the received data matches parity bit. This feature is active only when parity is enabled. The type of parity is determined by the PT (parity type) bit in SCxCR1. The bit is cleared by reading SC1SR1 with PF set, then reading SCxDRL.

SCxSR2 **RAF** is the Receiver Active Flag bit. This bit is controlled by the receiver front end. It is set when a start bit is detected by the receiver. It is cleared when an idle state is detected or when the receiver circuitry detects a false start bit. A false start is generally due to noise or baud rate mismatch.

SCxDRH R8 is the Receive Bit 8; it can be read anytime. A write operation to this bit has no meaning or effect. This bit is the ninth serial data bit received when the SCI system is configured for 9-data-bit operation. T8 is the Transmit Bit 8; it can be read or written anytime. This bit is the ninth serial data bit transmitted when the SCI system is configured for 9-data-bit operation. When using 9-bit data format this bit does not have to be written for each data word. The same value will be transmitted as the ninth bit until this bit is rewritten. For example, if you enable 9-bit data mode (M=1) and leave T8 equal to 1, then the 11-bit serial protocol will have 1 start bit, 8 data bits, and 2 stop bits.

SCxDRL The **SCxDRL** register contains the data transmitted out and received in by the SCI device. Even though there are separate transmit and receive data registers, these two registers exist at the same I/O port address. Reads to SCxDRL access the 8 bits of the read-only SCI receive data register (RDR). Writes to SCxDRL access the 8 bits of the write-only SCI transmit data register (TDR). SCxDRH:SCxDRL form the 9-bit data word for the SCI. If the SCI is being used with an 8-bit data word, only SCxDRL needs to be accessed. If a 9-bit format is used, the upper register should be written first to ensure that it is transferred to the transmitter shift register with the lower register.

11.7.3
Interrupting SCI
Interface

An interrupting SCI example can be found as `tut4` within **TExaS**. A portion of that system is presented as Program 11.9. The fastest baud rate is selected to reduce real human wait time during simulation on **TExaS**. For a real microcomputer application, the baud rate must match the other module. Initially the two FIFOs are cleared, and just the receiver is armed. The transmitter will be armed when data are available. A data flow graph and a flowchart of this system were shown previously as Figures 11.1 and 11.8 respectively. An interrupt occurs when new incoming data arrive in the receiver data register (RDRF=1). An interrupt also occurs when the transmit data register is empty (TDRE=1). TDRE is 1, when the output channel is idle needing the software to supply additional data. Notice that the transmit channel is disarmed when the TxFifo is empty and rearmed when new data is put into the TxFifo. When the RxFifo becomes full, then data is lost, but when the TxFifo becomes full, the main program simply waits for space to become available.

Program 11.9   Interrupting SCI interface.

```
*6811                                       *6812
SCI_Init jsr RxFifo_Init FIFO is empty     SCI_Init jsr RxFifo_Init FIFO is empty
      jsr  TxFifo_Init  FIFO is empty           jsr  TxFifo_Init  FIFO is empty
      ldaa #$2c         arm just RDRF           movb #$2c,SC0CR2  arm just RDRF
      staa SCCR2        enable SCI              movw #1,SC0BD     baud rate=500000
      ldaa #0                                   cli
      staa BAUD      baud rate=125000           rts
      cli
      rts
* Inputs: none Outputs: RegA is ASCII       * Inputs: none Outputs: RegA is ASCII
SCI_InChar pshb                             SCI_InChar pshb
iloop jsr   RxFifo_Get B=0 if empty         iloop jsr RxFifo_Get    B=0 if empty
      tstb                                        tbeq B,iloop
      beq  iloop                                  pulb
      pulb                                        rts               A=character
      rts               A=character
* Inputs: RegA is ASCII Outputs: none       * Inputs: RegA is ASCII Outputs: none
SCI_OutChar pshb       A=character          SCI_OutCh pshb         A=character
oloop jsr  TxFifo_Put save in FIFO          oloop jsr TxFifo_Put    save in FIFO
      tstb                                        tbeq B,oloop      B=0 if full
      beq  oloop     FIFO full, so wait          movb #$AC,SC0CR2  arm TDRE,RDRF
      ldab #$AC                                   pulb
      stab SCCR2        arm TDRE,RDRF             rts
      pulb
      rts
SCIhandler ldaa SCSR1                       SCIhandler ldaa SC0SR1
      bita #$20                                   bita #$20
      beq  CkTDRE     Not RDRF set                beq  CkTDRE    Not RDRF set
      ldaa SCDR       ASCII character             ldaa SC0DRL    ASCII character
      bsr  RxFifo_Put                             bsr  RxFifo_Put
CkTDRE bita #$80                            CkTDRE bita #$80
      beq  sdone      Not TDRE set                beq  sdone     Not TDRE set
      bsr  TxFifo_Get                             bsr  TxFifo_Get
      tstb                                        tbeq B,nomore
      beq  nomore                                 staa SC0DRL     start next output
      staa SCDR       start next output           bra  sdone
      bra  sdone                           nomore movb #$2C,SC0CR2  disarm TDRE
nomore ldaa #$2C                           sdone  rti
      staa SCCR2      disarm TDRE
sdone  rti
      org  $FFD6                                  org  $FFD6
      fdb  SCIhandler                             fdb  SCIhandler
```

Observation: *Data is lost when the* RxFifo *gets full.*

Checkpoint 11.11: *Why didn't the initialization software arm* TDRE?

Checkpoint 11.12: *What bad thing would happen if the* RDRF ISR *waited for room when* RxFifo *is full as* SCI_OutChar *waits for room when* TxFifo *is full?*

Checkpoint 11.13: *Modify Program 11.9 so the baud rate is 1200 bits/sec.*

11.7.4
Distributed
Systems

Many embedded systems require the communication of command or data information to other modules at either a near or a remote location. We will limit our discussion with communication with devices within the same room. A *full-duplex* channel allows data to transfer in both directions at the same time. In a *half-duplex* system, data can transfer in both directions but only in one direction at a time. Half-duplex is popular because it is less expensive (two wires) and allows the addition of more devices on the channel without change to the existing nodes. If the distances are short, half-duplex can be implemented with simple *open-collector* or *open-drain* TTL-level logic. Open-collector logic has two output states: low and off. In the off state the output is not driven high or low, it just floats. The 10-kΩ pull-up resistor will passively make the signal high if none of the open collector outputs is low. Both the 6811 and 6812 can make their TxD serial outputs open collector. This mode allows a half-duplex network to be created without any external logic (although pull-up resistors are often used). Three factors will limit the implementation of this simple half-duplex network: (1) the number of nodes on the network, (2) the distance between nodes, and (3) the presence of corrupting noise. In these situations a half-duplex RS485 driver chip like the SP483 made by Sipex or Maxim can be used.

The master–slave system connects the master transmit output to all slave receive inputs. (See Figure 11.13.) This provides for broadcast of commands from the master. All slave transmit outputs are connected together using wire-or open-collector logic, allowing for the slaves to respond one at a time. The DWOM bit (SPCR bit 5) in the 6811 slaves should be set to activate open-collector mode on PD1. If 6812s are used, then the WOMS bit (SC0CR1 bit 6) in the slaves should be set to activate open-collector mode on PS1.

Figure 11.13
A master–slave network implemented with multiple microcomputers.

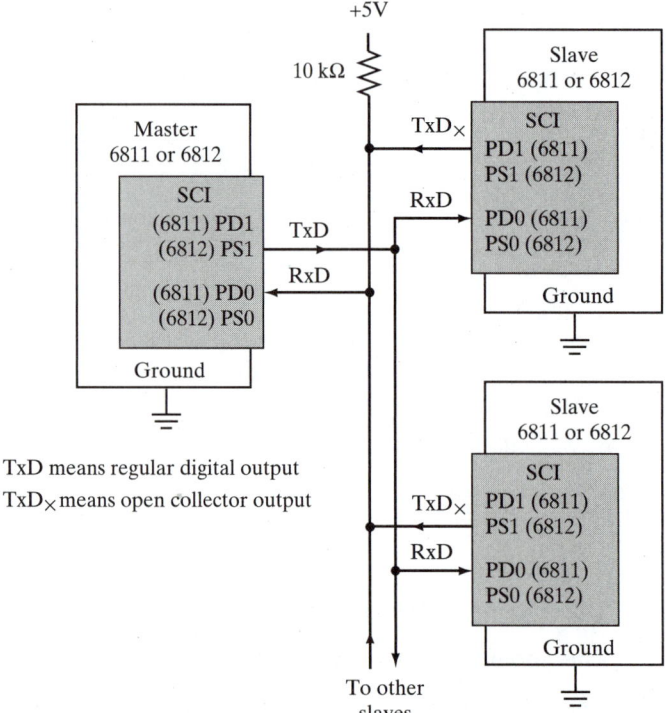

The ring network is a simple distributed approach, because it can be constructed using standard serial ports, by chaining the transmit and receive lines together in a circle, as shown in Figure 11.14. Messages will include source address, destination address, and information. If computer A wishes to send information to computer C, it sends the message to B. The software in computer B receives the message, notices it is not for itself, and re-sends the message to C. The software in computer C receives the message, notices it is for itself, and keeps the message.

Figure 11.14
A ring network implemented with 3 microcomputers.

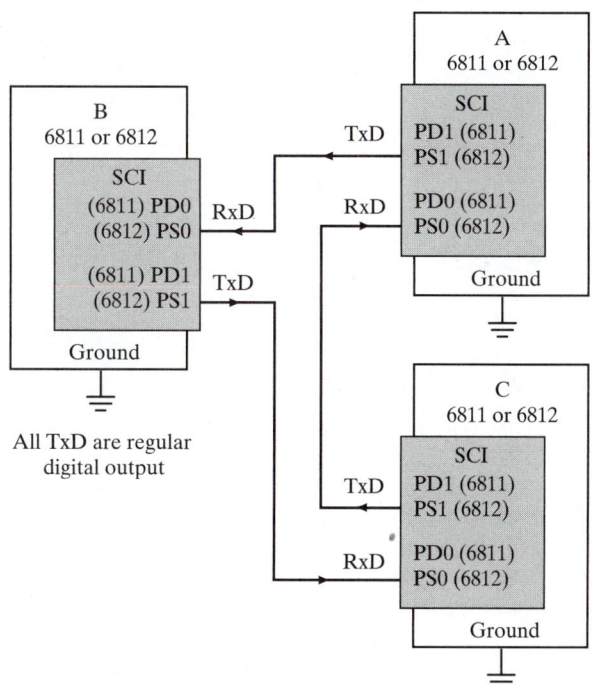

A very common approach to distributed embedded systems is called multi-drop, as shown in Figure 11.15. To transmit a byte to the other computers, the software activates the SP483 driver and outputs the frame. Since it is half-duplex, the frame is also sent to the receiver of the computer that sent it. This echo can be checked to see if a collision occurred (two devices simultaneously outputting.) If more than two computers exist on the network, we usually send address information first, so that the proper device receives the data.

Figure 11.15
Two multi-drop
networks implemented
with three
microcomputers.

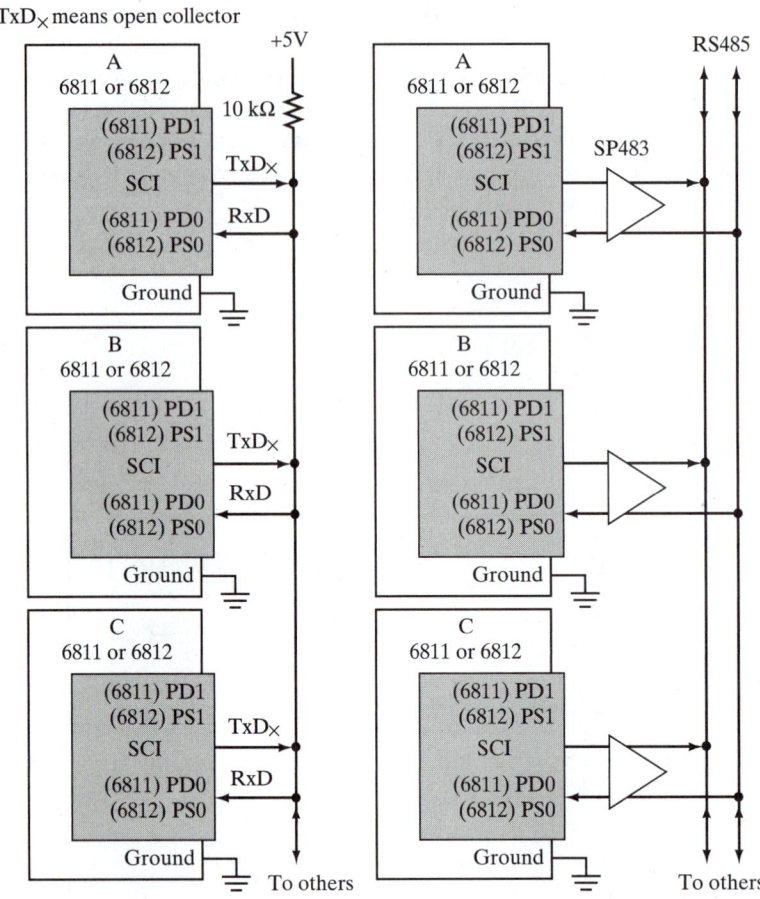

11.8   Periodic Interrupts

The purpose of this section is to present alternative methods to create periodic interrupts. A periodic interrupt is one that is requested on a fixed-time basis. This interfacing technique is required for data acquisition and control systems, because software servicing must be performed at accurate time intervals. For a data acquisition system, it is important to establish an accurate sampling rate. The time in between ADC samples must be equal (and known) in order for the digital signal processing to function properly. Similarly for microcomputer-based control systems, it is important to maintain both the sensor input and accuator output timing.

11.8.1
Periodic Polling

An application of periodic interrupts is called "intermittent polling" or "periodic polling." In regular busy-waiting, the main program polls the I/O devices continuously. With intermittent polling, the I/O devices are polled on a regular basis, established by a periodic interrupt. (See Figure 11.16.) If no device needs service, then the interrupt simply returns. This method frees the main program from the I/O tasks. Older IBM-PC computers use an 18 Hz periodic interrupt to interface its keyboard. We use periodic polling if the following two conditions apply:

1. the I/O hardware cannot generate interrupts directly;
2. we wish to perform the I/O functions in the background.

Figure 11.16
An ISR flowchart that implements periodic polling.

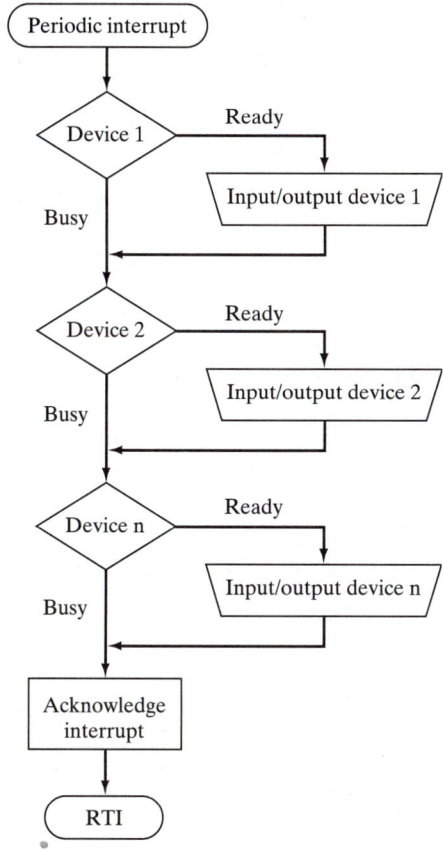

11.8.2
6811 Periodic
Interrupts

Three mechanisms on the 6811 generate periodic interrupts. Table 11.11 shows the 6811 registers used in periodic interrupts.

Table 11.11 6811 registers used to configure periodic interrupts.

address	Bit 7	6	5	4	3	2	1	Bit 0	Name
$100E	Bit 15	14	13	12	11	10	9	Bit 8	TCNT
$100F	Bit 7	6	5	4	3	2	1	Bit 0	TCNT
$1016	Bit 15	14	13	12	11	10	9	Bit 8	TOC1
$1017	Bit 7	6	5	4	3	2	1	Bit 0	TOC1
$1018	Bit 15	14	13	12	11	10	9	Bit 8	TOC2
$1019	Bit 7	6	5	4	3	2	1	Bit 0	TOC2
$101A	Bit 15	14	13	12	11	10	9	Bit 8	TOC3
$101B	Bit 7	6	5	4	3	2	1	Bit 0	TOC3
$101C	Bit 15	14	13	12	11	10	9	Bit 8	TOC4
$101D	Bit 7	6	5	4	3	2	1	Bit 0	TOC4
$101E	Bit 15	14	13	12	11	10	9	Bit 8	TI4/O5
$101F	Bit 7	6	5	4	3	2	1	Bit 0	TI4/O5
$1020	OM2	OL2	OM3	OL3	OM4	OL4	OM5	OL5	TCTL1
$1022	OC1I	OC2I	OC3I	OC4I	IC4/OC5I	IC1I	IC2I	IC3I	TMSK1
$1023	OC1F	OC2F	OC3F	OC4F	IC4/OC5F	IC1F	IC2F	IC3F	TFLG1
$1024	TOI	RTII	PAOVI	PAII	0	0	PR1	PR0	TMSK2
$1025	TOF	RTIF	PAOVF	PAIF	0	0	0	0	TFLG2
$1026	DDRA7	PAEN	PAMOD	PEDGE	DDRA3	I4/O5	RTR1	RTR0	PACTL

First, the real-time interrupt feature can be used to generate interrupts at a fixed rate. Two bits (RTR1 and RTR0) in the PACTR register determine the interrupt rate. Program 11.10 will increment a global variable, Time, every 32.767 ms. Four possible interrupt periods are possible with the 6811, as shown in Table 11.12. The flowchart depicted in Figure 11.17 shows that the background ISR thread is invoked at 30.517 Hz. To clear the RTIF flag (acknowledge the interrupt), the software writes a one to it. After acknowledging, the ISR increments a global Time variable.

Table 11.12
Real-time interrupt rates on the 6811.

RTR1	RTR0	Divide E by	Period (μs)	Frequency (Hz)
0	0	2^{13}	4096	244.14
0	1	2^{14}	8192	122.07
1	0	2^{15}	16384	61.035
1	1	2^{16}	32768	30.517

Program 11.10 Implementation of a periodic interrupt using the real-time interrupt feature.

```
         org   $0000
Time     rmb   2
         org   $E000
RTI_Init sei           make atomic
         ldaa  #3       32.768ms
         staa  PACTL
         ldaa  #$40
         staa  TMSK2    arm RTI
         ldd   #0
         std   Time
         cli            enable IRQ
         rts
```

```
unsigned short Time;
void RTI_Init(void){
  asm(" sei");     // Make ritual atomic
  PACTL = 3;       // 30.517Hz
  TMSK2 = 0x40;    // Arm
  Time = 0;        // Initialize
  asm(" cli");
}
#pragma interrupt_handler RTIHan()
void RTIHan(void){
  TFLG2 = 0x40;    // Acknowledge
  Time++;
```

```
RTIHan   ldaa #$40    acknowledge          }
         staa TFLG2   about 30.517Hz       #pragma abs_address:0xfff0
         ldd  Time                         void (*RTI_vector[])() ={
         addd #1                             RTIHan};  // fff0 RTI
         std  Time                         #pragma end_abs_address
         rti
         org  $FFF0
         fdb  RTIHan
```

Figure 11.17
Flowchart of the RTI
system showing the
foreground and
background threads.

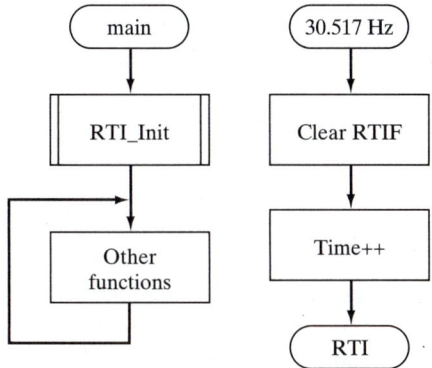

Checkpoint 11.14: How would you modify Program 11.10 to count at 61.035 Hz?

The timer overflow interrupt feature can also be used to generate interrupts at a fixed rate. (See Table 11.13.) The 16-bit TCNT register is incremented at a fixed rate. The TOF flag is set when the counter overflows and wraps back around (automatically) to zero. If armed, TOF can generate an interrupt. Two bits (PR1 and PR0) in the TMSK2 register determine the rate at which the counter will increment, hence will determine the TOF interrupt rate. To create a TOF periodic interrupt, we arm the timer overflow (TOI), and set the rate (PR1-0). (See Program 11.11.) To clear the TOF flag (acknowledge the interrupt), the software writes a one to it. This software system is very similar to the RTI flowchart shown in Figure 11.17.

Table 11.13
Timer overflow rates
assuming a 2 MHz 6811.

PR1	PR0	Divide by	TCNT clock	TOF interrupt	TOF interrupt
0	0	4	500 ns	32.768 ms	30.517 Hz
0	1	8	1 μs	65.536 ms	15.259 Hz
1	0	16	2 μs	131.072 ms	7.63 Hz
1	1	32	4 μs	262.144 ms	3.81 Hz

Program 11.11   Implementation of a periodic interrupt using timer overflow.

```
          org   $0000                        unsigned short Time;
Time      rmb   2                            void TOF_Init(void){
          org   $E000                          asm(" sei");    // Make ritual atomic
TOF_Init  sei           make atomic            TMSK2 = 0x80 ;  // Arm, 30.517Hz
          ldaa  #$80    arm, 32.768ms          Time = 0;       // Initialize
          staa  TMSK2   Set TOI=1 arm TOF      asm(" cli");
          ldd   #0                           }
          std   Time                         #pragma interrupt_handler TOFHan()
          cli           enable IRQ            void TOFHan(void){
          rts                                  TFLG2 = 0x80;      // Acknowledge
TOFHan    ldaa  #$80    acknowledge            Time++;
          staa  TFLG2   about 30.517Hz       }
          ldd   Time                         #pragma abs_address:0xffde
          addd  #1                           void (*TOF_vector[])() ={
          std   Time                           TOFHan}; // ffde TOF
          rti                                #pragma end_abs_address
          org   $FFDE
          fdb   TOFHan
```

The third mechanism to generate periodic interrupts is output compare. Program 11.12 shows a 1-kHz periodic interrupt using output compare 2. When the TCNT register matches TOC2, the output compare flag, OC2F, is set. If armed (OC2I=1), then it will request an interrupt. To clear the OC2F flag (acknowledge the interrupt), the software writes a one to it. The ISR will acknowledge the interrupt and set TOC2 to the time for the next interrupt. The interrupting period is determined by the TCNT period (set by TMSK2) multiplied by the constant RATE. Program 11.12 will increment a global variable, Time, every 1 ms. This software system is also very similar to the RTI flowchart shown in Figure 11.17.

Program 11.12   Implementation of a periodic interrupt using output compare.

```
RATE      equ   1000                         #define RATE 1000
          org   $0000                        unsigned short Time
Time      rmb   2                            void OC2_Init(void){
          org   $E000                          asm(" sei");     // Make ritual atomic
OC2_Init  sei           make atomic            TMSK1 |= 0x40;   // arm OC2F
          ldaa  #$01                           TMSK2 = 0x01;    // 1 MHz TCNT
          staa  TMSK2   1 MHz TCNT             TC2 = TCNT+RATE; // first in 1ms
          ldaa  #$40                           TFLG1 = 0x40;    // initially clear OC2F
          staa  TMSK1   arm OC2F               Time=0;          // Initialize
          ldd   #0                             asm(" cli");     // enable IRQ
          std   Time                         }
          ldd   TCNT    time now             #pragma interrupt_handler TC2handler
          addd  #RATE   first in 1ms         void TC2handler(void){
          std   TOC2                           TC2 = TC2+RATE;  // next in 1 ms
          cli           enable IRQ             TFLG1 = 0x40;    // acknowledge OC2F
          rts                                  Time++;
```

```
OC2Han   ldaa  #$40      acknowledge          }
         staa  TFLG1                           #pragma abs_address:0xffe2
         ldd   TOC2                            void (*TC2_vector[])() ={
         addd  #RATE                             TC2handler}; // ffe2 TC2
         std   TOC2     next in 1 ms           #pragma end_abs_address
         ldd   Time
         addd  #1
         std   Time
         rti
         org   $FFE2
         fdb   OC2Han
```

Checkpoint 11.15: *How would you modify Program 11.12 to count at 100 Hz?*

11.8.3
6812 Periodic Interrupts

There are three mechanisms on the 6812 that generate periodic interrupts. Table 11.14 shows the 6812 registers used in periodic interrupts.

Table 11.14 6812 registers used to configure periodic interrupts.

Address	Bit 7	6	5	4	3	2	1	Bit 0	Name
$0014	RTIE	RSWAI	RSBCK	0	RTBYP	RTR2	RTR1	RTR0	RTICTL
$0015	RTIF	0	0	0	0	0	0	0	RTIFLG
$0047	LCK	PLLON	PLLS	BCSC	BCSB	BCSA	MCSB	MCSA	CLKCTL
$0080	IOS7	IOS6	IOS5	IOS4	IOS3	IOS2	IOS1	IOS0	TIOS
$0084	Bit 15	14	13	12	11	10	9	Bit 8	TCNT
$0085	Bit 7	6	5	4	3	2	1	Bit 0	TCNT
$0086	TEN	TSWAI	TSBCK	TFFCA	PAOQE	T7QE	T1QE	T0QE	TSCR
$008C	C7I	C6I	C5I	C4I	C3I	C2I	C1I	C0I	TMSK1
$008D	TOI	0	TPU	TDRB	TCRE	PR2	PR1	PR0	TMSK2
$008E	C7F	C6F	C5F	C4F	C3F	C2F	C1F	C0F	TFLG1
$008F	TOF	0	0	0	0	0	0	0	TFLG2
$0090	Bit 15	14	13	12	11	10	9	Bit 8	TC0
$0091	Bit 7	6	5	4	3	2	1	Bit 0	TC0
$0092	Bit 15	14	13	12	11	10	9	Bit 8	TC1
$0093	Bit 7	6	5	4	3	2	1	Bit 0	TC1
$0094	Bit 15	14	13	12	11	10	9	Bit 8	TC2
$0095	Bit 7	6	5	4	3	2	1	Bit 0	TC2
$0096	Bit 15	14	13	12	11	10	9	Bit 8	TC3
$0097	Bit 7	6	5	4	3	2	1	Bit 0	TC3
$0098	Bit 15	14	13	12	11	10	9	Bit 8	TC4
$0099	Bit 7	6	5	4	3	2	1	Bit 0	TC4
$009A	Bit 15	14	13	12	11	10	9	Bit 8	TC5
$009B	Bit 7	6	5	4	3	2	1	Bit 0	TC5
$009C	Bit 15	14	13	12	11	10	9	Bit 8	TC6
$009D	Bit 7	6	5	4	3	2	1	Bit 0	TC6
$009E	Bit 15	14	13	12	11	10	9	Bit 8	TC7
$009F	Bit 7	6	5	4	3	2	1	Bit 0	TC7

First is a real-time interrupt feature that can be used to generate interrupts at a fixed rate. Three bits (RTR2, RTR1 and RTR0) in the RTICTL register ($0014) determine the interrupt rate. To clear the RTIF flag (acknowledge the interrupt), the software writes a one to it. Program 11.13 will increment a global variable, `Time`, every 32.767 ms. Seven possible interrupt periods are possible with the 6812, see Table 11.15. A flowchart for this RTI system was shown previously in Figure 11.17.

Table 11.15
6812 real-time interrupt rates.

RTR2	RTR1	RTR0	Divide E by	Period (μs)	Frequency (Hz)
0	0	0	off	off	off
0	0	1	2^{13}	1024	976.56
0	1	0	2^{14}	2048	488.28
0	1	1	2^{15}	4096	244.14
1	0	0	2^{16}	8192	122.07
1	0	1	2^{17}	16384	61.035
1	1	0	2^{18}	32768	30.517
1	1	1	2^{19}	65536	15.259

Program 11.13   Implementation of a periodic interrupt using the real-time clock feature.

```
          org   $0800
Time      rmb   2
          org   $F000
RTI_Init  sei             make atomic
          ldaa  #$86      32.768ms
          staa  RTICTL    arm RTI
          movw  #0,Time
          cli             enable IRQ
          rts
RTIHan    ldaa  #$80      acknowledge
          staa  RTIFLG    about 30.517Hz
          ldd   Time
          addd  #1
          std   Time
          rti
          org   $FFF0
          fdb   RTIHan
```

```c
unsigned short Time;
void RTI_Init(void){
  asm(" sei");      // Make ritual atomic
  RTICTL = 0x86;    // Arm,   30.517Hz
  Time = 0;         // Initialize
  asm(" cli");
}
#pragma interrupt_handler RTIHan()
void RTIHan(void){
  RTIFLG = 0x80;    // Acknowledge
  Time++;
}
#pragma abs_address:0xfff0
void (*RTI_vector[])() ={
  RTIHan};  // fff0 RTI
#pragma end_abs_address
```

Checkpoint 11.16: How would you modify Program 11.13 to count at 61.035 Hz?

The timer overflow interrupt feature can also be used to generate interrupts at a fixed rate, see Table 11.16. The 16-bit TCNT register is incremented at a fixed rate. The TOF flag is set when the counter overflows and wraps back around (automatically) to zero. If armed, TOF can generate an interrupt. Three bits (PR2, PR1 and PR0) in the TMSK2 register determine the rate at which the counter will increment, hence will determine the TOF interrupt rate. To clear the TOF flag (acknowledge the interrupt), the software writes a one to it. To create a TOF periodic interrupt, we enable the timer (TEN=1), arm the timer overflow (TOI), and set the rate (PR2-0). (See Program 11.14.) This software system is also very similar to the RTI flowchart shown in Figure 11.17.

Table 11.16
Timer overflow rates, assuming an 8 MHz 6812. The divide by 64 and the divide by 128 modes are only available on the MC68HC912D60.

PR2	PR1	PR0	Divide by	TCNT clock	TOF interrupt	TOF interrupt
0	0	0	1	125 ns	8.192 ms	122.07 Hz
0	0	1	2	250 ns	16.384 ms	61.035 Hz
0	1	0	4	500 ns	32.768 ms	30.517 Hz
0	1	1	8	1 μs	65.536 ms	15.259 Hz
1	0	0	16	2 μs	131.072 ms	7.63 Hz
1	0	1	32	4 μs	262.144 ms	3.81 Hz
1	1	0	64	8 μs	524.288 ms	1.91 Hz
1	1	1	128	16 μs	1048.576 ms	0.95 Hz

Observation: *With a PR2-0 value of 010, the TCNT will clock at the same rate as a 2MHz 6811.*

Program 11.14   Implementation of a periodic interrupt using timer overflow.

```
        org    $0800
Time    rmb    2
        org    $F000
TOF_Init sei          make atomic
        ldaa   #$B2   arm, 32.768ms
        staa   TMSK2  Set TOI=1 arm TOF
        ldaa   #$80   TEN=1
        staa   TSCR   enable TCNT
        movw   #0,Time
        cli           enable IRQ
        rts
TOFHan  ldaa   #$80   acknowledge
        staa   TFLG2  about 30.517Hz
        ldd    Time
        addd   #1
        std    Time
        rti
        org    $FFDE
        fdb    TOFHan
```

```
unsigned short Time;
void TOF_Init(void){
  asm(" sei");        // Make ritual atomic
  TMSK2 = 0xB2 ;      // Arm, 30.517Hz
  TSCR = 0x80;        // enable counter
  Time = 0;           // Initialize
  asm(" cli");
}
#pragma interrupt_handler TOFHan()
void TOFHan(void){
  TFLG2 = 0x80;       // Acknowledge
  Time++;
}
#pragma abs_address:0xffde
void (*TOF_vector[])() ={
  TOFHan};  // ffde TOF
#pragma end_abs_address
```

The third mechanism to generate periodic interrupts is output compare. Program 11.15 shows a 1 kHz periodic interrupt using output compare 6. To enable output compare, the corresponding bit in the TIOS register must be set. When the TCNT register matches TC6, the output compare flag, C6F, is set. If armed (C6I=1), then it will request an interrupt. To clear the C6F flag (acknowledge the interrupt), the software writes a one to it. The ISR will acknowledge the interrupt and set TC6 to the time for the next interrupt. The interrupting period is determined by the TCNT period (set by TMSK2) multiplied by the constant RATE. Program 11.15 will increment a global variable, Time, every 1 ms. This software system is also very similar to the RTI flow-chart shown in Figure 11.17.

Program 11.15 Implementation of a periodic interrupt using output compare.

```
RATE      equ   1000                      #define RATE 1000
          org   $0800                     unsigned short Time;
Time      rmb   2
          org   $F000
OC6_Init  sei           make atomic       void OC6_Init(void){
          movb  #$33,TMSK2   1us             asm(" sei");     // Make ritual atomic
          movb  #$80,TSCR    enable TCNT     TSCR = 0x80;
          movb  #$40,TIOS    activate OC6    TIOS |= 0x40;     // activate OC6
          movb  #$40,TMSK1   arm OC6         TMSK1 |= 0x40;    // arm OC6
          movw  #0,Time                      TMSK2 = 0x33;     // 1 MHz TCNT
          ldd   TCNT    time now             TC6 = TCNT+RATE;  // first in 1ms
          addd  #RATE   time for first       Time=0;           // Initialize
          std   TC6                          asm(" cli");      // enable IRQ
          cli           enable IRQ        }
          rts
OC6Han    movb  #$40,TFLG1   acknowledge   #pragma interrupt_handler OC6handler
          ldd   TC6                        void OC6handler(void){
          addd  #RATE                        TC6 = TC6+RATE;   // next in 1 ms
          std   TC6     next in 1 ms         TFLG1 = 0x40;     // acknowledge C6F
          ldd   Time                         Time++;
          addd  #1                         }
          std   Time                       #pragma abs_address:0xffe2
          rti                              void (*OC6_vector[])() ={
          org   $FFE2                        OC6handler};   // ffe2 TC6
          fdb   OC6Han                     #pragma end_abs_address
```

Checkpoint 11.17: How would you modify Program 11.15 to count at 100 Hz?

11.8.4
Real-Time Data
Acquisition

The goal of a data acquisition system is to sample the ADC at a regular rate. Referring to Figure 11.9, let t_i be the time the ADC creates sample number i. If the sampling rate is f_s, then we would like to have

$$(t_i - t_{i-1}) = 1/f_s$$

for all i. When using periodic interrupts to establish the sampling rate (Programs 11.10 through 11.15), there are two factors that lead to flucuations in the sample period. We define time jitter, δt, as the maximum variation in the sample-to-sample time; that is,

$$1/f_s - \delta t < (t_i - t_{i-1}) < 1/f_s + \delta t$$

The first factor is the instruction currently being executed at the time of the interrupt request. The time to execute an instruction on the 6811 varies from 2 to 41 cycles. Neglecting the `idiv` and `fdiv` instructions, the other instructions range from 2 to 10 cycles. We do not know which instruction will be executing or when during that instruction the interrupt will be requested. This uncertainty causes a maximum time jitter of about 10 cycles, or 5 μsec on the 6811. This jitter is usually acceptable. The second source of jitter can be much larger. If there are any portions of the main program that disable interrupts (e.g., because of a critical section), then the time running with interrupts disabled may cause time jitter in the sampling. In a similar fashion, if there are other interrupts, then the time to execute the other ISR may cause a time jitter.

Observation: Good software places as little processing in the ISR itself. Perform whatever functions must be done in the ISR, and shift the rest of the processing to the foreground.

Observation: Real-time systems must put an upper bound on the time the software is allowed to run with interrupts disabled.

Checkpoint 11.18: Assuming that the 6812 is running at 8 MHz, what is the time jitter caused by the uncertainty as to which instruction is being executed at the time of the interrupt?

11.9 Polled versus Vectored Interrupts

As we defined earlier, when more than one source of interrupt exists, the computer must have a reliable method to determine which interrupt request has been made. There are two common approaches, and the Motorola microcomputers apply a combination of both methods. The first approach is called vectored interrupts. With a vectored interrupt system, each potential interrupt source has a unique interrupt vector address. You simply place the correct handler address in each vector, and the hardware automatically calls the correct software when an interrupt is requested. Figure 11.18 shows an example where two serial devices are interfaced to the microcomputer.

Figure 11.18
An external I/O device is connected to the microcomputer.

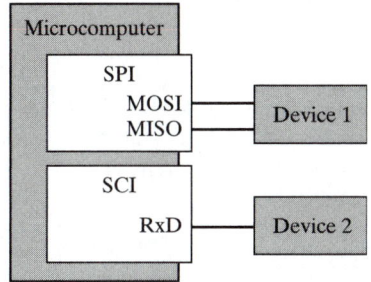

Since the two sources have separate vectors, the SPI interrupt hardware will automatically activate the SPI interrupt handler and the SCI interrupt hardware will automatically activate the SCI interrupt handler.

The second approach is called polled interrupts. With a polled interrupt system, multiple interrupt sources share the same interrupt vector address. Once the interrupt has occurred, the ISR software must poll the potential devices to determine which device needs service. Figure 11.19 shows an example where there are two external interrupt requests.

Figure 11.19
When two or more devices share an interrupt line, then the ISR must poll.

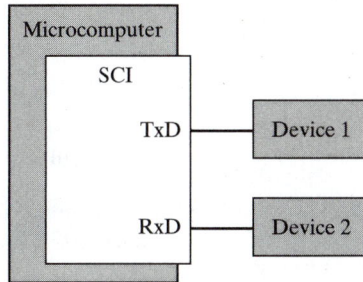

Since the two sources have the same vector, the ISR software must first determine which one caused the interrupt. The 6811/6812 systems have a separate acknowledgment, so that if both interrupts are pending, acknowledging one will not satisfy the other, so the second device will request a second interrupt and get serviced.

Common Error: *If two interrupts were requested, it would be a mistake to service just one and acknowledge them both.*

Observation: *External events are often asynchronous to program execution, so careful thought is required to consider the effect if an external interrupt request were to come in between each pair of instructions.*

Observation: *The computer automatically sets the I bit during processing, so that an interrupt handler will not interrupt itself.*

11.10   Tutorial 11. Profiling

This tutorial addresses real-time profiling of the interrupting serial port program in `Chap11.rtf`, which is a debugged version of `tut4.rtf`. Memory dumps are a minimally intrusive way to collect strategic information without affecting too much the system we are testing. Observing the FIFO will allow us to determine if the system is CPU bound or I/O bound. It will also allow us to select the optimal FIFO size. In particular, we will collect a histogram of the FIFO size at the time of a call to `SCI_OutChar`. This global variable placed in the RAM section will contain a histogram of the FIFO size during operation.

```
Dbg_Hist    rmb   TXFIFO_SIZE
```

The subroutines shown in Program 11.16 were added to `tut4.rtf`. `Debug_Init` clears the histogram, and `Debug_Histogram` is called at the beginning of `SCI_OutChar`.

Program 11.16
Debugging instruments to measure FIFO size during operation.

```
Dbg_Init                           Dbg_Histogram
      ldx    #Dbg_Hist                   pshx
      ldab   #TXFIFO_SIZE                psha
Dbg.1 clr    0,x                         pshb
      inx                                ldd    TxPutPt
      decb                               subd   TxGetPt RegB = size
      bne    Dbg.1                       bpl    Dbg.2
      rts                                addb   #TXFIFO_SIZE
                                  Dbg.2 ldx    #Dbg_Hist    histogram
                                        abx
                                        inc    0,x
                                        pulb
                                        pula
                                        pulx
                                        rts
```

Question 11.1. Open the Chap11 files, assemble, and run the new system. Describe the FIFO situation. Why doesn't the `TxFifo` ever become full?

Question 11.2. Reduce the baud rate to 9600 bits/sec. You will have to change both the `SCI_Init` and click on the IO window and execute **IO->CRT...** Run the slower baud rate system and describe the FIFO. Will it crash if the `TxFifo` becomes full?

Action: Monitors are a minimally instrustive debugging tools used to observe the execution pattern of the system. Program 11.17 shows debugging instruments that can be used to visualize when a subroutine is entered and when it is exited. The enter code is added on the first line of the profiled function and the exit code is executed as the last line.

Program 11.17
Debugging instruments to profile the system.

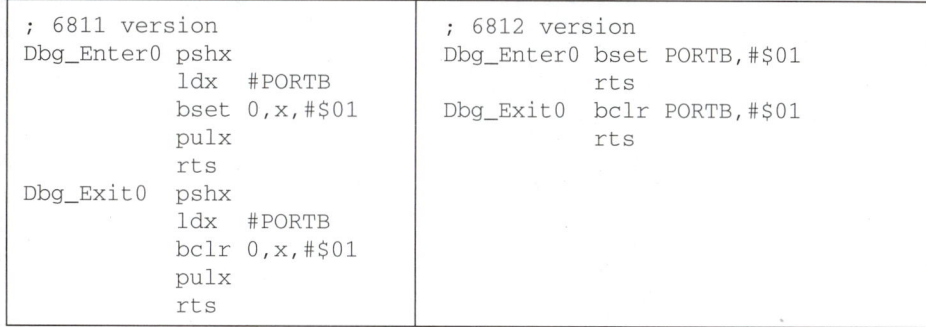

```
; 6811 version                          ; 6812 version
Dbg_Enter0 pshx                         Dbg_Enter0 bset  PORTB,#$01
           ldx   #PORTB                            rts
           bset  0,x,#$01               Dbg_Exit0  bclr  PORTB,#$01
           pulx                                    rts
           rts
Dbg_Exit0  pshx
           ldx   #PORTB
           bclr  0,x,#$01
           pulx
           rts
```

Debugging instruments like these were also added to the system. Attach the Port B output bits PB3, PB2, PB1, and PB0 to a logic analyzer. In particular, the following four functions will be profiled: `SCI_InChar`, `SCI_OutChar`, `RxFifo_Put`, and `TxFifo_Get`. You can also open the `Chap11x.scp` file if you want.

> **Observation:** *Profiling is made easier if the subroutine as a single `rts` exit point at the bottom of the function.*

Question 11.3. Figure 11.20 shows the initial activity as the system displays the welcome message. PB3 is the call to `TxFifo_Get`, and PB1 is the call to `SCI_OutChar`. Describe this initial behavior of the system. What is the sequence of execution as one character is transmitted in this CPU bound system?

Figure 11.20
Profile the system during the initial "Welcome" message.

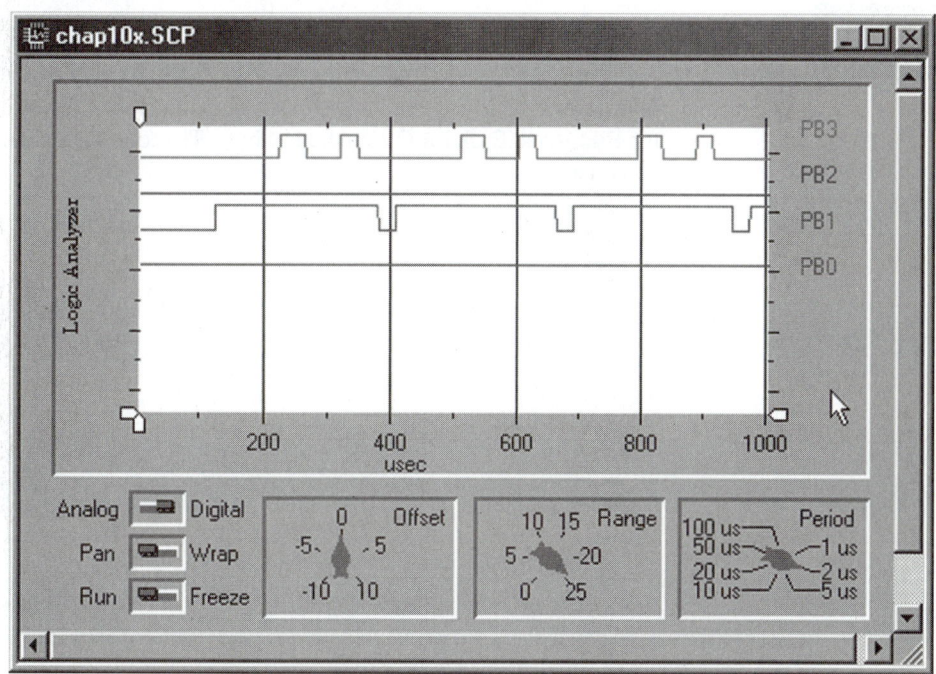

Question 11.4. Figure 11.21 shows the activity of the system after a character is typed. PB2 is the call to `RxFifo_Put`, and PB0 is the call to `SCI_InChar`. Describe this behavior.

Figure 11.21
Profile the system after a character is typed.

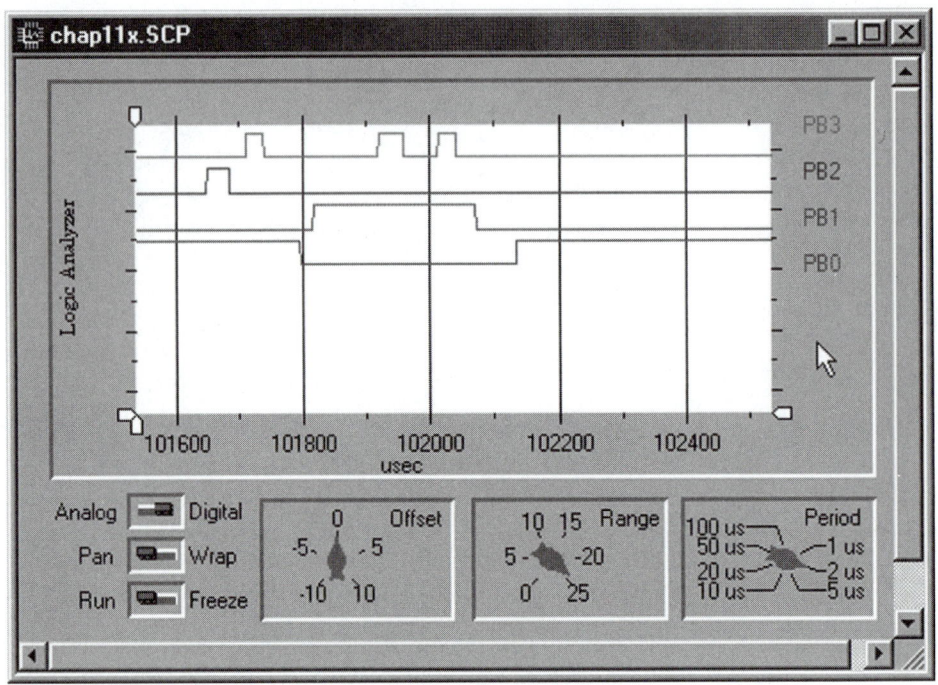

11.11 Homework Problems

Homework 11.1: Implement a FIFO system that uses indices instead of pointers to access the FIFO data. Each FIFO element is 1 byte. The following globals cannot be changed (i.e., no counter):

```
          org  RAM  place globals in RAM
MaxSize equ  10   size of FIFO (can hold up to MaxSize-1 bytes)
GetI    rmb  1    Index into FIFO where to get next
PutI    rmb  1    Index into FIFO where to put next
FIFO    rmb  MaxSize
* indices range from 0 to MaxSize-1
* FIFO[GetI] is the oldest data
* FIFO[PutI] is the free spot for next put
* If GetI==PutI, then the FIFO is empty
* If wrap(PutI+1)==GetI then the FIFO is full
```

The following subroutine initializes the FIFO

```
Initialize   clr GetI
             clr PutI
             rts
```

Part A. Write the `Put` function. The data to `Put` is passed by value on the stack. Return Reg B=1 if the FIFO was full at the time of the call, and the data could not be saved. Return Reg B=0 if the data were properly saved. COMMENTS will be graded. Save and restore Registers A,X,Y if used. Here are a couple of typical calling sequences:

```
jsr InCh    keyboard data        ldab #55
psha        by value             pshb        by value
jsr Put                          jsr  Put
ins         discard parameter    ins
tstb                             tstb
bne Error                        beq  OK
```

Part B. Write the `Get` function. The data from `Get` is returned by reference using the stack. Return Reg B=1 if the FIFO was empty at the time of the call, and the data could not be removed. Return Reg B=0 if the data were properly removed. COMMENTS will be graded. Save and restore Registers A,X,Y if used. Here are a couple of calling sequences:

```
ldx  #data reference to global   data set  0
pshx       by reference                des       allocate data
jsr Get    info goes in data           tsy       pointer to data
ins        discard parameter           pshy      by reference
ins                                    jsr  Get  info goes in data
tstb                                   ins
bne Error                              ins
                                       tstb
                                       beq  OK
```

Homework 11.2: Interrupts are a good method to create a real-time DAS. A typical approach is illustrated in the following software skeleton (let f_s be the sampling rate):

```
#pragma interrupt_handler RTIHan()
void RTIHan(void){
  RTIFLG=0x80;       // Acknowledge by clearing RTIF
  Fifo_Put(Adin());  // save new data
}
void main(void){
  Initialization();  // clear FIFO, arm RTI, enable interrupts
  while(1){
    while(Fifo_Get(&data)){};
    Process(data);
  }
}
```

The best case, average, and worst case execution times are given in the following table. Assume all other software times can be neglected. The goal of a real-time DAS is to execute `Adin()` every $1/f_s$.

Program	Best case (min)	typical (average)	Worst case (max)
`Initialization`	912 μs	912 μs	912 μs
`Adin`	25 μsec	25 μsec	25 μsec
`Process`	500 μs	1000 μs	5000 μs
`Fifo_Put`	15 μsec	15 μsec	15 μsec
`Fifo_Get`	20 μsec	20 μsec	20 μsec

What is the maximum sampling rate possible for this interrupt-based DAS? The system must be real-time continuously processing data with no delayed or lost data points. Continuous means that it runs without stopping (e.g., for years). Assume the FIFO is arbitrarily large, but not infinite size.

Homework 11.3: How do you tell if the following C code is reentrant?

```
PORTB |= 0x80;    // set PB7
```

Homework 11.4: Write interrupting software that counts a global variable at 1 Hz. Give the initialization, the ISR, and the interrupt vector.

Homework 11.5: Consider the programs of section 10.10. Are any of the following programs reentrant: `Heap_Init`, `Heap_Allocate`, `Heap_Release`, `Fifo_Init`, `Fifo_Put`, and `Fifo_Get`? How can the problem be fixed?

Homework 11.6: Consider the `Fifo_Put` and `Fifo_Get` functions from Program 10.27. Assume exactly one is called from the foreground main program and one from the background ISR. Is there a critical section? (That is, do interrupts need to be disabled to use these two functions?)

Homework 11.7: Write interrupting software that maintains the time of day. Give the initialization, the ISR, and the interrupt vector. The initial time of day is passed in when initialization is called. Register A contains the initial hour, Register B contains the initial minute. Assume the initial seconds are 0. Implement military time, where the hour goes from 0 to 23.

Homework 11.8: Consider the `RxFifo_Put` and `RxFifo_Get` functions from Programs 11.2 through 11.5. There is one foregrand thread that calls `RxFifo_Get`, and two interrupt threads that call `RxFifo_Put`. In particular, both the regular RDRF/SCI interrupt and a TOF periodic timer ISR call `RxFifo_Put` to enter data into the `RxFifo`. Is there a critical section?

11.12   Laboratory Assignments

Lab 11.1
Keyboard Device
Driver

Purpose You will design the hardware interface between a keyboard and a microcomputer, create the low-level interrupt-driven device driver, interface a single LED, and implement keyboard security system.

Description In this keyboard lab, you will design the keyboard interface using interrupt synchronization. There are two advantages of interrupts in an application like this. Placing the key input task into a background thread frees the main program to execute other tasks while the software is waiting for the operator to type something. This security system doesn't have anything else to do, but in a complex system, it is important to be able to perform multiple tasks. The second advantage of interrupts is the ability to create accurate time delays even within a complex software environment. In particular, the output compare interrupt can be used to accurately wait for the bouncing to stop. In this implementation, you will add an interrupt service routine (ISR) as a private software function inside the driver. The higher level software will not have access to this function. One way to solve the switch-bounce problem is to implement periodic polling. In this situation, an output compare is requested at a regular rate. The time in between these periodic interrupts must be longer than the bounce time of the switch, but shorter than the total time a key is touched or released. For example, if the switch has a bounce time of 500 μsec, then you could choose a 1-ms period for the output compare interrupt. During each interrupt, the ISR will scan the keyboard. If there is exactly one key typed and this key is different from the pattern observed at the time of the last interrupt, then you will put the ASCII code into the FIFO.

This experiment will illustrate how a parallel port of the microcomputer will be used to control a keyboard matrix. In each case your computer will drive the rows (output 0 or HiZ) and read the columns. The low level software (inputs, scans, debounces, and saves keys in a FIFO) runs in a background period interrupt thread. Your system must handle two-key rollover. For example, if operators were to type "1, 2, 3," they could push "1", push "2", release "1", push "3", release "2", then release "3."

Low-level *device drivers* normally exist in the BIOS ROM and have direct access to the hardware. They provide the interface between the hardware and the rest of the software. Good low-level device drivers allow

- new hardware to be installed,
- new synchronization methods to be implemented (like changing busy-waiting to interrupts),
- new algorithms to be added (error detection, data compression),
- higher level features to be built on top of the low level,

and still maintain the same software interface. In larger systems like the Workstation and IBM-PC, where the low level I/O software is compiled and burned in ROM separate from the code that will call it, it makes sense to implement the device drivers as software traps or software interrupt (`swi`) and specify the calling sequence in assembly language. In embedded systems like those we use, it is acceptable to provide a source code file that the user can assemble into their application. *Linking* is the process of resolving addresses to code and programs that have been compiled separately. In this way, the routines can be called from any program without requiring complicated linking. In other words, when the device driver is implemented with an `swi`, the linking is built into the operation of the software interrupt instruction. In our embedded system, the assembler will perform the linking. The concept of a device driver can be illustrated with a prototype device driver. You are encouraged to modify/extend this example, and define/develop/test your own format. A prototype keyboard device driver follows. The device driver software is grouped into four categories.

1. Data structures: global, private (accessed only by the device driver, not the user)

`openFlag` Boolean that is true if the keyboard port is open

initially false, set to true by `Key_Open`, set to false by `Key_Close`

static storage (global)

2. Initialization routines (called by user)

Key_Open Initialization of keyboard port

> Sets openFlag to true
> Initializes hardware
> Returns an error code in RegA if unsuccessful (already open)
> Input Parameters(none)
> Output Parameter(error code)
> Typical calling sequence

```
jsr Key_Open
tsta           ; 0 if opened correctly
bne error
```

Key_Close Release of keyboard port

> Sets openFlag to false
> Returns an error code in RegA if not previously open
> Input Parameters(none)
> Output Parameter(error code)
> Typical calling sequence

```
jsr  KeyClose
tsta           ; 0 if closed correctly
bne  error
```

3. Regular I/O calls (called by user to perform I/O)

Key_In Input an ASCII character from the keyboard port

> Waits for a key to be pressed, (there is bounce and two key rollover)
> Returns data in RegB if successful
> Returns an error code in RegA if unsuccessful device not open, hardware failure
> (probably not applicable here)
> Output Parameters: RegB is data, RegA is error code
> Typical calling sequence

```
jsr  Key_In
tsta           ; 0 if input is OK correctly
bne  error
stab data      ; save new key data
```

Key_Status Returns the status of the keyboard port

> Returns a true in RegA if a call to Key_In would return with a key
> Returns a false in RegA if a call to Key_In would not return right away, but rather
> it would wait
> Returns a true if device not open, hardware failure (probably not applicable here)

Typical calling sequence

```
loop jsr  work  ; perform work until key is typed
     jsr  Key_Status
     tsta       ; true if a key is typed
     beq  loop
     jsr  Key_In ; read and process the key
```

4. Support software (private code).

If you have any helper functions, these would be considered local to your driver and would be placed in this category. In `C++`, these helper functions would be defined as private. In C, we could define the helper functions in the `.c` file, but not place a prototype in the `.h` file. In this way, the function could only be called from functions in the .c implementation, and not by the user. In assembly language we are very careful not to call a helper function from outside the device driver. The interrupt service routine is an example of support software.

Part A. Create an I/O window and build a keyboard similar to the one shown in Figure 11.22.

Figure 11.22
0-9 keyboard, with up arrow, down arrow, 2nd, CLEAR, HELP, and ENTER.

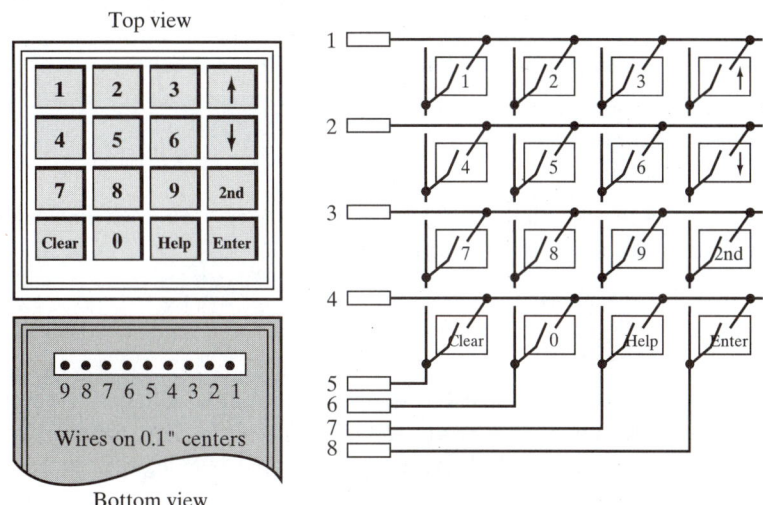

Part B. Write the low-level keyboard device driver. You will need a FIFO queue to pass the data from the background (ISR) to the foreground (main). The main program will implement an access-code-based security system. Each access code will consist of four digits between 0 and 9. The security system can recognize up to five access codes. You will specify these codes in global memory. The keyboard will be used to enter access codes. If this access code is one of the valid codes, checked by searching the access code database, the single LED is turned on. The LED will remain on until the new key is typed. The main program will need its own data structure to hold the last four keys typed. Assume "1257" and "2222" are valid codes. The following example shows the LED status (0 = off, 1 = on) after each key hit.

```
1 2 1 2 5 7 8 9 2 2 2 2 2 2 6 1 2 5 7 4
 0 0 0 0 0 1 0 0 0 0 0 0 1 1 1 0 0 0 0 1 0
```

Write a main program to test the keyboard device driver. Collect some latency data (time from key touch to `Fifo-Put`) measurements. When running the critical code test, make sure the periodic interrupt rate is fast enough to increase the probability of hitting the critical section, but not so fast that the FIFO gets full.

Part C. Build the LED display writing a simple device driver that allows you to turn the LED on and off.

Part D. Write a main program that implements the security system.

Part E. Write a new main program and periodic interrupt handler to study the concept of *critical sections.* In this system, there will be two threads implementing the producer–consumer problem. The background thread, implemented as a periodic interrupt, will produce a simple predictable data stream (e.g., 0,1,2,3, . . .) and put bytes in the FIFO. The foreground thread, implemented as the main program, will consume the data, and check for the proper sequence. Your job is the develop methodologies for *nonintrusively* detecting when a call to `Fifo_Put` (made by the producer thread) has interrupted a call to `Fifo_Get` (made by the consumer thread.) Define an 8-bit global variable called `size`, which is initialized to zero. Add code similar to the following to `Fifo_Put` each time data is actually entered (don't execute if FIFO was full):

```
ldab size
inca
stab size
```

Add code similar to the following to `Fifo_Get` each time data is actually removed (don't execute if FIFO was empty):

```
ldab size
deca
stab size
```

This creates a critical section. Show the places within `Fifo_Get` where, if the background thread were to interrupt and then call `Fifo_Put`, an error would occur (the counter becomes incorrect). In particular, develop debugging code to detect when this critical section is violated.

Lab 11.2
Real-Time
Temperature Data
Acquistion

Purpose The objective of this lab is to study analog-to-digital conversion, real-time sampling, digital filterings foreground-background communication, table lookup, and LCD display.

Description In preparation for this assignment, review fixed-point, passing/returning parameters, and ADC converters. Look up how **TExaS** simulates analog signals and the ADC using the on-help. In particular, get help for the **IO->Analog** command. Example programs that apply to this lab include `HD44780.rtf`, `LCD.rtf`, `tut3.rtf`, and `tut5.rtf`.

The ADC will be used to measure oral body temperature. The range is about 90 to 105 F, and the resolution is 0.1 F. Real-time data acquisition will use 0.5 ms periodic interrupts. The sampling rate will be 2 kHz. 500 Hz analog noise will be added. A 500-Hz digital reject filter will remove the noise. Software will convert the ADC sample into decimal fixed-point number, and the result will be displayed on a LCD. The system will implement a digital oral thermometer. The thermistor resistance is nonlinearly related to its temperature in Kelvin

$$R_T = R_0 \exp(\beta/T)$$

where R_0 is 1.03947E-07 kΩ and β is 5808.1 K for this device. An analog circuit, shown in Figure 11.23, converts the resistance to voltage, and the ADC converts voltage to digital sample. R1 is 200 kΩ, R2 is 11.1 kΩ, and the gain is 28.6.

Figure 11.23
Temperature data
acquisition circuit.

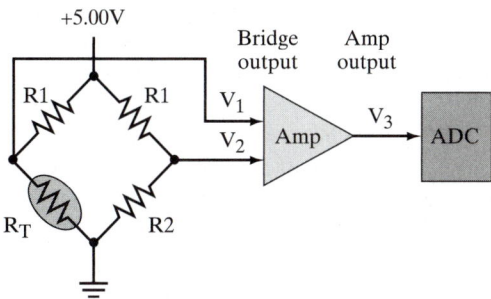

The overall response is shown in Table 11.17, and plotted in Figure 11.24. The details of these calculations can be found in the spreadsheet Lab11_02.xls. The details of the circuit design can be found in Chapters 11 and 12 of *Embedded Microcomputer Systems: Real Time Interfacing* by Jonathan W. Valvano, published by Brooks/Cole, copyright (c) 2000.

Table 11.17
Temperature calibration.

T(F)	RT (kΩ)	V1 (V)	V1-V2 (V)	V3 (V)	ADC	fixed-point
89.6	19.2	0.438	0.175	5.000	255	896
91.2	18.1	0.416	0.153	4.375	223	912
92.8	17.2	0.395	0.132	3.782	192	928
94.5	16.2	0.375	0.113	3.220	164	945
96.1	15.4	0.357	0.094	2.686	136	961
97.7	14.6	0.339	0.076	2.180	111	977
99.3	13.8	0.322	0.059	1.700	86	993
100.9	13.1	0.306	0.044	1.244	63	1009
102.6	12.4	0.291	0.028	0.813	41	1026
104.2	11.7	0.277	0.014	0.403	20	1042
105.8	11.1	0.263	0.001	0.015	0	1058

Figure 11.24
Temperature calibration.

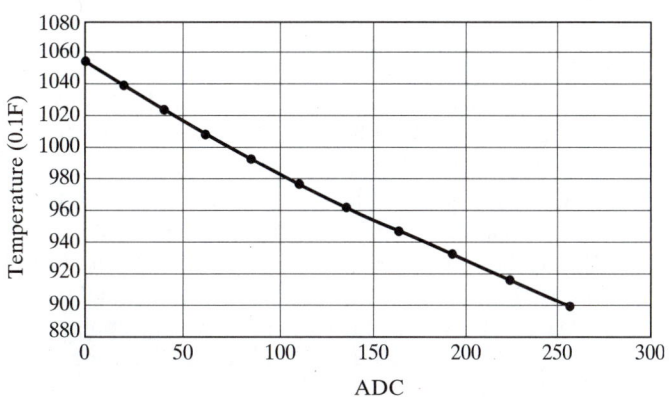

Part A. Interface an LCD display (either the simple BCD or the Hitachi HD44780), a switch, and an analog signal to the microcomputer. You should select a slow-varying sine-wave to simulate oral temperature. The peak temperature will be displayed on the LCD. The maximum voltage should be set to 5000 mV (90F) and the minimum voltage can be adjusted to change body oral temperature. For example, 1910 mV (ADC=97) represents 98.6 F. Make the sine-wave period large, e.g., 50,000 μsec. In this way, it will take a lot of ADC samples to create an entire cycle. You should add 100 mV of 500-Hz (2000-μsec period) noise.

Part B. Enter the last two columns of data from Table 11.17 into a constant data structure (allocated in ROM). Write a subroutine that converts ADC sample to decimal fixed-point temperature using table lookup and linear interpolation.

Part C. Write software that samples at 2 kHz, implements the following digital filter, and stores the digital filter output into a FIFO queue. You will need two MACQs, one for $x(n)$ and one for $y(n)$. The details of the digital filter design can be found in Chapter 15 of *Embedded Microcomputer Systems: Real Time Interfacing* by Jonathan W. Valvano, published by Brooks/Cole, copyright (c) 2000. $x(n)$ is the current ADC sample. $x(n-2)$ is the ADC sample two samples ago. $y(n-2)$ is the filter output two samples ago.

$$y(n) = (113*x(n)+113*x(n-2)-98*y(n-2))/128$$

Observation: *If the sampling rate were to be 240 Hz, this filter rejects 60 Hz.*

Part D. Write a main program that waits for the switch to be pressed. Once the switch is pressed, the interrupts are armed, data is removed from the FIFO, converted to decimal fixed point and the maximum temperature is displayed on the LCD. If the switch is released, then the interrupts are disarmed, the display is blanked, and the software waits for the switch to be pressed again.

Observation: *If you change the sampling rate, change the simulated noise period to be four times the ADC sample period.*

Lab 11.3
AC Voltmeter

Purpose In this lab you will learn how to convert, subtract, and display decimal fixed-point numbers. You will pass parameters on the stack. The analog-to-digital converter will be used to convert an analog signal into digital form.

Description In preparation for this assignment, review fixed-point, passing/returning parameters on the stack, and ADC converters. Look up how **TExaS** simulates analog signals and the ADC using the online help. In particular, get help for the **IO->Analog** command. Run and analyze the TUT3.rtf tutorial.

Five possible analog waveforms can be connected to the ADC. Your objective is to measure the analog signal 100 times and convert each ADC sample to fixed-point voltage. During the 100-sample sequence establish the minimum (**min**) and maximum (**max**). At the end of the cycle, you will measure the AC amplitude (**max-min**) and display it on the LCD display. Activate the appropriate decimal point in the LCD and add appropriate label and units. Your software should reinitialize variables and continuously repeat the 100-sample cycle.

Part A. Create an I/O file and attach an analog waveform and an LCD display. You will adjust the number of samples (e.g., 100) in your main program loop, and the period of the analog wave so that about two to five waveform periods occur in each software loop. You will adjust the minimum, maximum, and noise level during the testing steps of this problem (Part F).

Part B. Write a general-purpose ADC sampling subroutine that accepts an 8-bit call by value input parameter, channel number (0-7), and returns an 8-bit unsigned binary digital result from the ADC. Both parameters will be on the stack. Typical calling sequences are shown below. Use binding (equ) to make the subroutine more readable.

```
***** ZZ=A2D(4); ***     ******* XX=A2D(chan); ************
    ldab #4                   movb chan,1,-sp    ; push chan on the stack
    pshb
    leas -1,sp               des                 ; allocate space for result
    bsr  A2D                 jsr   A2D
    pulb                     movb 1,sp+,XX       ; get result
    stab ZZ
    leas 1,sp               ins                  ; discard input
```

Part C. Write an 8-bit unsigned binary to 16-bit unsigned decimal fixed-point conversion subroutine. The fixed-point constant is 0.01V. You may choose to pass parameters anyway you wish, but please document in the comments how parameters are passed. In the table below, the ADC converts from the first to second column. This subroutine converts the 8-bit unsigned number shown in the second column to the value shown in the third column.

Analog Voltage	ADC output	16-bit unsigned decimal fixed point	LCD display
0.0V	0	0	0.00
0.02V	1	2	0.02
1.25V	64	125	1.25
2.50V	128	250	2.50
4.98V	255	498	4.98

Part D. Write a subroutine that takes a 16-bit unsigned decimal fixed-point number (fixed constant is 0.01V), and displays it on the LCD display. You may choose to pass parameters anyway you wish, but please document in the comments how parameters are passed. In the above table, this subroutine converts the 16-bit integer shown in the third column to the LCD pattern shown in the fourth column.

Part E. Write the main program that calls the above three subroutines and performs the AC voltmeter measurements, updating the LCD at the end of each cycle.

Part F. Using the information entered in the **IO->Analog** command as truth, collect the following measurements. *At the time of checkout be prepared to discuss why the last three measurements had more errors than the first.*

waveform	noise	true max (volts)	true min (volts)	true AC (volts)	measured AC (volts)	percent error (%) = 100*(true-measured)/true
sine	none	4.000	1.000	3.000		
sine	none	1.005	1.000	0.005		
EKG	none	4.000	1.000	3.000		
sine	100mV 500μs	4.000	1.000	3.000		

Part G. In addition to the operations described in Part F, extend the main program to also measure the period of a sine wave in μsec. Display the AC amplitude on the LCD display and display the period by simply writing it to a global variable, and observing it in the **View-Box**. To implement hysteresis, we define two thresholds at 25% and 75%[3] depending on **min** and **max**:

$$\textbf{high} = (3*(\textbf{max}-\textbf{min}))/4 + \textbf{min}$$
$$\textbf{low} = (\textbf{max}-\textbf{min})/4 + \textbf{min}$$

First wait for the signal to go below **low**,
then wait for the signal to go above **high**, `first=TCNT;`
then wait for the signal to go below **low**,
then wait for the signal to go above **high**, `period=(TCNT-first)/8; first=TCNT;`
then wait for the signal to go below **low**,
then wait for the signal to go above **high**, `period=(TCNT-first)/8; first=TCNT;`

Record the TCNT each time the signal goes above **high**, and the **period** is the 16-bit unsigned difference between TCNT measurements. After subtracting, divide by 8 to get the answer in μsec. *Discuss with the TA at the time of checkout, what would happen if you divided first, then subtracted, and why two thresholds were used instead of one.* In this first example, TCNT is $1000 the first time the signal goes above **high**, and $7000 the second time. The **period** is ($7000 − $1000)/8 or 3072 μsec.

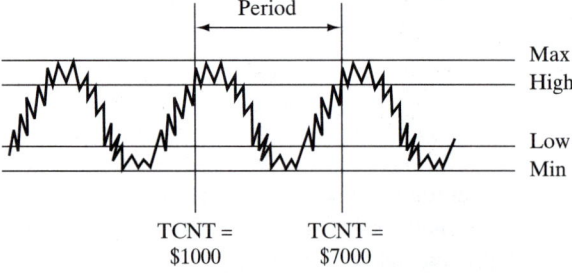

This second example illustrates that the system works even if the TCNT rolls over. In this example, the **period** is now 3584 μsec, but TCNT is $F000 the first time the signal goes above **high**, and $6000 the second time. The **period** is ($6000 − $F000)/8 or 3584 μsec. This works because both TCNT and the subtraction are unsigned 16-bit values.

---

[3]You may adjust these percentages so that you get only one trigger each period.

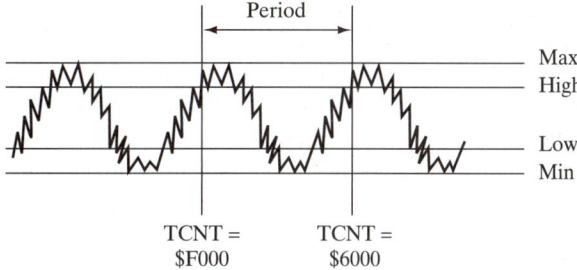

Lab 11.4
Microcomputer-
Based Motor
Controller

Purpose The objective of this lab is to study analog-to-digital conversion, digital-to-analog conversion, real-time digital control, and LCD display.

Description The objective of this problem is to design a microcomputer-based motor controller. The desired rotation speed **x*** will be selected interactively by the operator typing on keyboard, either the matrix keyboard or the SCI interface. The output information will be displayed on either an LCD or the SCI interface. You will power the motor with an analog signal from the DAC. You will estimate the motor speed (**x'**) by measuring the tachometer voltage using the ADC. Your first control software will implement an incremental control algorithm. Your second system will implement a proportional/integrator (PI) control system. The goal of the control software is to maintain the motor speed as close to **x*** as possible.

You will implement two control systems. The first one will be a simple incremental controller. Let **u** be the DAC output controlling the motor. The power to the motor is directly related to the DAC value. The value is increased by a fixed amount if it is spinning too slowly, and decreased by a fixed amount if it is spinning too fast. The incremental control algorithm executes the following at a regular rate:

$$\mathbf{u} = \min(\mathbf{255}, \mathbf{u+1}) \text{ if } \mathbf{x^*} > \mathbf{x'} \text{ (too slow)}$$

$$\text{or } \mathbf{u} = \max(\mathbf{0}, \mathbf{u-1}) \text{ if } \mathbf{x^*} < \mathbf{x'} \text{ (too fast)}$$

The min and max operations maintain the DAC output within the valid range of **0** to **255**. The disadvantage of incremental control is that it has a very slow response. The second system you will implement will be a PI controller.

We can use linear control theory to develop the digital controller. (See Figure 11.25.) We will define the tachometer voltage as the actual motor speed. This speed will be measured with the ADC. Any error in the state estimator will lead to a nonremovable controller error. Just like the data acquisition and digital filter situations, **t** is the continuous time and **n** is the discrete time. We will assume the controller is executed at a fixed interval, **Δt**.

Figure 11.25
Block diagram of a linear control system in the frequency domain.

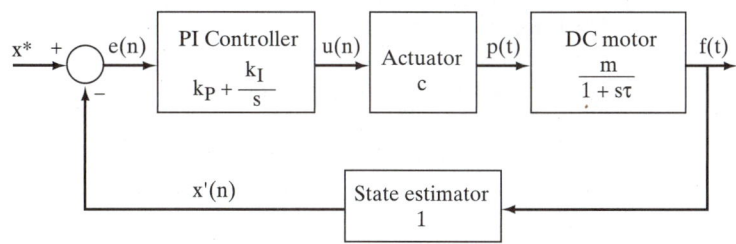

Theoretically we can choose controller constants, k_P and k_I, to create the desired controller response. Unfortunately, it can be difficult to estimate c, m and τ. If a load is applied to the motor, then m and τ will change. In addition, most motors do not follow a simple single pole relationship. The basic approach is presented in the following equations. Let $x(n)$ be the current tachometer voltage represented as a fixed-point number. Let x^* be the desired tachometer voltage, also represented as a fixed-point number. The error is

$$e(n) = (x^* - x(n))$$

The proportional term is

$$up(n) = (k_p * e(n))/100$$

The integral term is

$$ui(n) = ui(n-1) + (k_i * e(n))/100$$

$$\text{if } (ui(n) > 50) \text{ then } ui(n) = 50 \text{ (called anti-reset-windup)}$$

$$\text{if } (ui(n) < -50) \text{ then } ui(n) = -50$$

The DAC output is the combination of

$$u(n) = up(n) + ui(n)$$

$$\text{if } (u(n) > 255) \text{ then } u(n) = 255$$

$$\text{if } (u(n) < 0) \text{ then } u(n) = 0$$

All calculations are performed as 16-bit signed integers. A simple empirical method can be used to determine the controller constants. This empirical approach starts with just a proportional term (k_p). This proportional controller will generate a smooth motor speed (actuator output achieves a constant value), but the speed will not be correct. Try different k_p constants until the response times are fast enough. The response time is the delay after x^* is changed for the motor to reach a new constant speed. k_p is too big if the actuator saturates both at the maximum and minimum after x^* is changed. Steady state controller accuracy is defined as the average difference between x^* and x. The next step is to add some integral term (k_i) a little at a time to improve the steady-state controller accuracy without adversely affecting the response time. Don't change both k_p and k_i at once. Rather, you should vary them one at a time. Overshoot is defined as the maximum positive error that occurs when x^* is increased. Similarly, undershoot is defined as the maximum negative error that occurs when x^* is decreased. If the response time, overshoot, undershoot and accuracy are within acceptable limits, then a PI controller is adequate.

The foreground (main) process:
> initializes I/O ports and data structures
> explains the various interpreter commands
> maintains a display of $x(n)$

The interpreter process (using interrupting keyboard or interrupting SCI):
> can specify the desired motor speed, $0 \leq x^* \leq 5000$ mV with a resolution of 1mV

The digital controller (periodic interrupt) process:
> has a controller rate of 1 ms.
> implements a PI control system

Appendix 1

Embedded System Development Using TExaS

The goal of this appendix is to present enough information to allow a first time user to simulate a microcomputer system, using **TExaS.** As an example, we will develop a simple embedded system with four switches and four lights. Each light will activate whenever its corresponding switch is pushed. Although we will create a complicated implementation of this very simple problem, it will illustrate the process of embedded system development via the **TExaS** simulator. During the normal development process, one simulates first, then builds a real system second. Because the goal of this appendix is to introduce the simulator as a tool for developing embedded systems, we first present the completed system in Figures A1.1 and A1.2, then design, build, and test this system with the simulator. Figure A1.1 shows the actual mechanical configuration of the system, with 4 switches and 4 lights.

Figure A1.1
A simple embedded system with four switches and four lights. (Courtesy Technological Arts.)

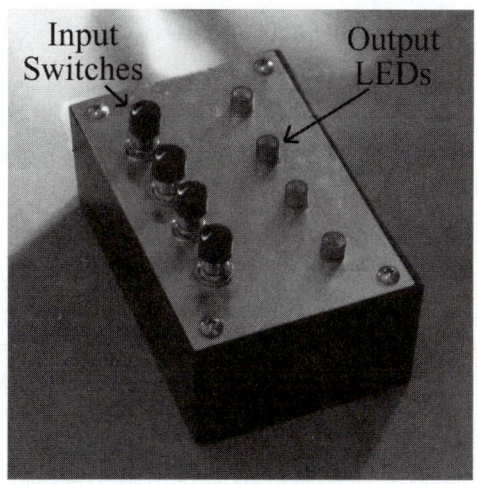

Figure A1.2 shows the electrical configuration of the 6811 and interface electronics. These electrical components either will be hidden or will be embedded into the box.

Figure A1.2
A microcomputer is used to control this embedded system. This system uses a Motorola 6811 microcomputer on the ADAPT11 board built by Technological Arts.

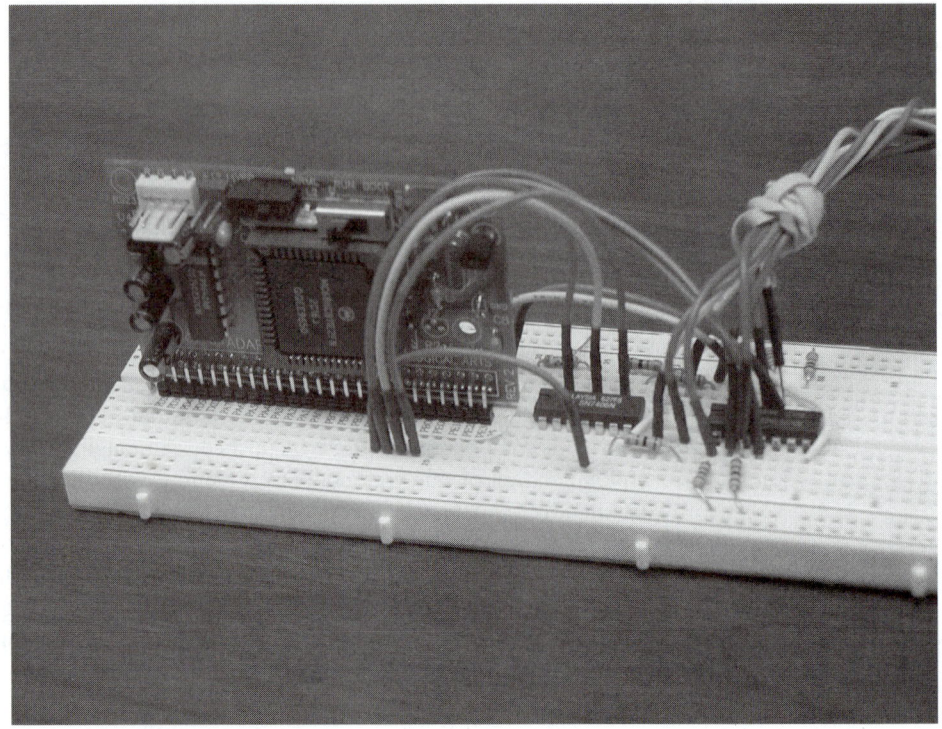

A1.1　Introduction to TExaS

TExaS is an integrated application containing an editor, an assembler, an instruction-set simulator, an I/O port simulator, and an external-device simulator. **TExaS** is an educational product. It was developed as a learning tool for real-time embedded systems. **TExaS** simulates a wide range of external devices that can be configured and attached to the microcomputer. It simulates all the machine instructions and most of the I/O ports of the 6811 and 6812 microcomputers. The interactive editor/assembler assists the new programmer. Extensive run-time checking helps to identify tricky software bugs. Students can use this application for a laboratory course on microcomputer programming or on microcomputer interfacing. **TExaS** does not allow the user to create new external devices, so its use for developing commercial products is limited. Although you have full creative control over the software you write and the I/O ports you use, you are limited to specific predefined external devices. As with many applications, the windows are used to observe and modify the state of the simulated environment. Associated with each window is a file that contains the actual information. **TExaS** supports all five phases of software development:

- defining the microcomputer type and memory configuration (Chapter 3),
- writing the program source code using an editor (Chapters 4,8,9,10),
- assembling the source code and loading the object code into memory (Chapter 4),
- interfacing external components (Chapters 5,6,11),
- debugging the program by running it on the interactive simulator (Chapter 7).

In this book, the expression **File->Open** will refer to the **Open** command in the `File menu`, as shown in Figure A1.3. There are six types of files used by the simulator.

Figure A1.3   *TExaS File->Open command.*

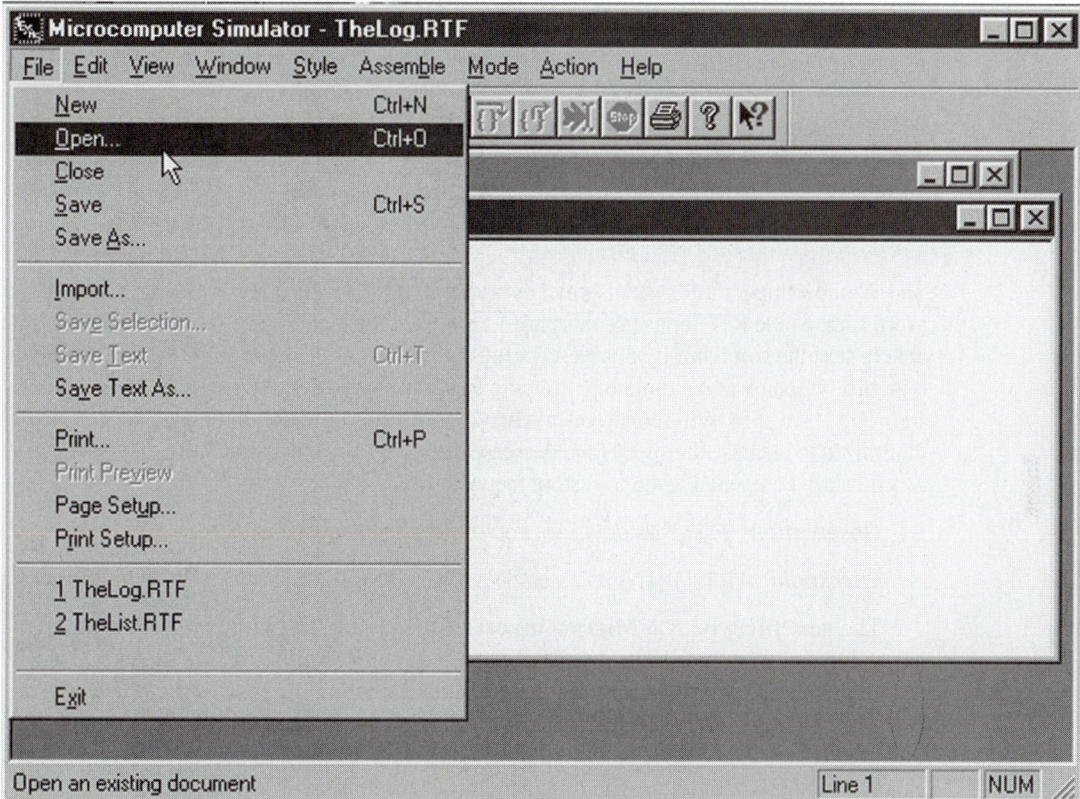

The first type is a **Program** file (`*.rtf`). Program files contain assembly source code, which includes explicit instructions for the computer in human readable format. Files of this type can be opened, edited, and saved by any application that supports standard rich-text format (RTF). Applications that support RTF include **TExaS** (of course), Microsoft Word, and WordPad. **TExaS** allows you to edit multiple RTF files simultaneously. When your program is assembled, **TExaS** puts detailed information into the assembly listing. A cathode ray tube (CRT) terminal is a common device, used by computers to input and output data. The following three RTF files are not program files, but rather have specific functions:

`TheLog.rtf`	**TExaS** logs information into this file as the program is running.
`TheList.rtf`	This file contains the assembly listing.
`TheCRT.rtf`	This file contains the input/output data of a simulated remote CRT terminal.

All other RTF files are considered assembly source files that can be edited, assembled, and executed. **TExaS** includes a cross-assembler, which runs on one computer (the Windows platform) and converts assembly source code into object code for a different computer (Motorola 6811 or 6812). To develop C programs for the 6811/6812, you will need a cross-compiler. If you are developing C language programs with a C Cross-Compiler, then you will not need any assembly source files. On the other hand, if you are developing assembly programs with this simulator, then you must have at least one assembly source file. It is into this file that you will type your assembly source code. Figure A1.4 shows typical **TExaS** icons.

Figure A1.4
TExaS icons.

tut.rtf tut.uc tut.io tut.stk tut.scp tut.plt

The Windows operating system typically assumes the RTF files are WordPad or Microsoft Word files, so the RTF icons probably won't look like those in Figure A1.4, and we won't be able to start the simulator by double-clicking the RTF icons. The file `TheCRT.rtf` will automatically open when simulating a remote terminal. Normally, the two files `TheLog.rtf` and `TheList.rtf` will remain open when running the simulator, but they can be closed or minimized to increase the speed (and decrease the visibility) of the simulation. If closed, these two files can be opened again to restore the visibility.

> **Observation:** *Hiding windows will improve the simulation speed.*

> **Checkpoint A1.1:** *What is an assembly source code?*

The next file type is a **Microcomputer** file (`*.uc`). The large chip in Figure A1.2 is a 6811 microcomputer. The details of the internal microcomputer are saved in this file. It contains the configuration inside the embedded microcomputer. In particular, the microcomputer type, the microcomputer speed, the simulation modes, the **ViewBox** entries, the *Break/ScanPoints* entries, and the formats are saved in this file. You must have exactly one Microcomputer file open to run the simulation. You will be able to start the simulator by double-clicking a microcomputer icon. Figure A1.4 shows a microcomputer icon.

The third file type is an **I/O Device** file (`*.io`). This is an optional file containing the details of the external I/O devices that are connected to the microcomputer ports. This type of file defines devices external to the microcomputer, such as switches, LEDs, LCDs, keyboard, the CRT remote terminal, motors, IR remote control, and sensors. For example, the configuration of the switches/lights shown in Figure A1.1 and the external chips shown in Figure A1.2 are stored in the I/O Device file `ChapA1.io`. You can have only one I/O Device file open at a time. You will be able to start the simulator by double-clicking an I/O Device icon, like the one shown in Figure A1.4.

The fourth file type is a **Stack** file (`*.stk`). The stack is a block of memory that the computer uses to hold temporary information. The top of the stack is defined by the register SP. In **TExaS**, the stack file (e.g., `tut.stk`) is used to create a window that you can use to observe the microcomputer stack during program execution. You can have only one Stack window open at a time. To view the stack, open an existing stack file, or create a new Stack file. The format for viewing the stack window is saved in and recalled from the Microcomputer file. There are two options for defining the bottom of the stack (the value of the SP when the stack is empty). In manual mode, you explicitly define the location of the bottom of the stack by entering its address. In automatic mode, the simulator will set the bottom of the stack to the value loaded into the SP by the `lds` assembly instruction. Figure A1.4 shows a typical Stack icon.

The fifth file type is a **Scope** file (`*.scp`). These are optional files used for debugging. You may have up to two scope files open at a time. This file is used to create a window that you can use to observe the time-dependent behavior of digital and analog signals during program execution. The scope can be configured as an *oscilloscope* or as a *digital logic analyzer*. An oscilloscope, or scope for short, is a hardware debugging tool that shows the value of an analog voltage (e.g., -12 to $+12$V) versus time. A logic analyzer is also a hardware debugging tool, but it shows digital signals (e.g., true and false) versus time. To use a scope, open an existing scope file, or create a new scope file. A typical scope icon can be seen in Figure A1.4.

> **Checkpoint A1.2:** An oscilloscope graphically displays information about an electronic circuit. Which parameter is displayed on the y-axis and which parameter is displayed on the x-axis?

> **Checkpoint A1.3:** How does a logic analyzer differ from an oscilloscope?

The last file type is a **Plot** file (`*.plt`). Files of this type are used to display graphical information about the simulation. For example, the position of a stepper motor robot is plotted in this window (run the example robot.rtf). Look for updates that include these features at: **www.ece.utexas.edu/~valvano**. You can have up to four plot files open at a time. Figure A1.4 contains a typical Plot icon.

When you open one of these six types of files, it will automatically open files of the other five types if they exist in the same directory (same name but different extension.) For example, double-clicking the `tut.uc` icon will start the application and open the existing files `tut.rtf`, `tut.uc`, `tut.io`, `tut.scp`, and `tut.stk` (no `tut.plt` file exists.) Remember, there can be no more than one Microcomputer, one I/O Device, two Scopes, and one Stack file open at a time. So, if there is already one of these types open when you open a program file (`*.rtf`), it cannot open a second Microcomputer, I/O Device, or Stack file. Notice also that, even though you can open two scope files, only one can be opened automatically. The other scope file will have to be opened manually.

In summary, it is necessary to have at least one program file (`*.rtf`) and one Microcomputer file (`*.uc`) open in the application to develop assembly-language software by using the simulator. For the developing of software with a C compiler, only the Microcomputer file (`*.uc`) is needed. If you need devices external to the microcomputer (such as switches, LED, LCD, keyboard, or CRT terminal), then an I/O file will be required. Also, having the two files `TheLog.rtf` and `TheList.rtf` open will greatly enhance the interactivity of the application. In the virtual world of simulation, all of these files together constitute the embedded system.

A1.2 Major Components of TExaS

Next, we will review the five major components of this application, and then we will outline the steps required to develop software by using the simulator. The components allow you to write software, convert the software into machine code, simulate the execution by the processor, simulate the I/O ports of the microcomputer, and simulate external devices connected to the 6811 or 6812 microcomputer.

Editor. The **TExaS** editor is used to create assembly source files. If you are developing C language programs with a C Cross-Compiler (such as ICC11/ICC12), then you will not need this editor, but, rather, you will use the editor associated with the compiler. On the other hand, if you are developing assembly-language programs, you will create your assembly-language source code by using this editor. In some ways, this editor is very similar to WordPad. This editor supports fonts, styles, multiple files, and the importing and exporting of text files. In addition to the usual editor functions, there are two special features that assist assembly-language development. The first feature is the ability to embed figures into the assembly source files. These figures can be flowcharts or circuit diagrams, and they can be quite useful in documenting your system. The `tut2.rtf` file is a good example of how to use embedded figures to document software. The second special feature is the automatic recolor feature. Enabling this automatic behavior provides real-time feedback while you are typing in the source window. Because a complete 2-pass assembly is performed to recolor the file after each change, it is recommended to enable this option only for short programs. For medium-sized programs, you can activate the "single-line recolor" mode, which performs a partial syntax check on a few lines above and below the editor focus. For long programs, the assembler will color the source code only when the program is explicitly assembled. The source colors are very helpful in identifying syntax errors.

Assembler. The **TExaS** assembler converts source code (human-readable instructions, such as `ldaa $10`) into object code (microcomputer-readable instructions, such as $96 $10). There is a simple and direct translation between assembly instructions given in the source code and machine instructions given in the object code. During the first pass, the *symbol table* is created. The symbol table is a mapping from a symbolic name to its corresponding 16-bit address (e.g., PORTB on the 6812 is $0001). During the second pass, the object code and assembly listing are created. The machine code is automatically loaded into the simulated memory. You may also create Motorola-standard S19 object code files, if desired. The simulated execution does not require S19 object files. On the other hand, S19 object-code files would be needed if you were to use the editor and assembler to create software to run on an actual microcomputer. The listing file, created in the `TheList.rtf` file, contains both the source and object codes, combined in a human-readable format. Once again, there are a couple of unique features of this assembler. If enabled, the assembler will add color to the source code, showing individual items like labels, opcodes, pseudo-op codes, numbers, strings, and comments (see the **Assemble->Options. . .** command). You can configure the assembler to identify illegal items by a special color or format. The specific colors for each item can be specified with the **Assemble->TextFormat. . .** command. The second unique feature is the extensive error reporting that is included when a syntax error is found. Rather than reporting the usual terse "syntax error" message, this assembler attempts to suggest ways to correct the error. The third unique feature is its expression evaluator, which applies the standard rules of precedence and parentheses. For example,

```
five    equ '0'|(3+4*5/10)
```

The application will automatically create object code for the microcomputer you have selected (6811 or 6812). There are many assemblers on the market for the 6811 and 6812. This assembler attempts to support many of their syntax and commands. This means you can import software originally written for these other assemblers into **TExaS**.

Checkpoint A1.4: What is object code?

Instruction-set simulator. This part of the **TExaS** application implements the basic central processing unit (CPU) of the 6811 or 6812. The *program counter,* or PC, is a special register located in the CPU used to control program execution. Software is a sequence of specific instructions to be executed. These instructions are fetched/executed per the current value of the program counter, PC. The simulated memory supports the appropriate amount of RAM, EEPROM, or ROM. The *bus* is a set of digital signals that connect the CPU, memory, and I/O devices. The unique aspects of the instruction-set simulator are the bus-cycle activity and the extensive error checking. The exact read/write cycles of the 6811 can be viewed. The three-element 16-bit instruction queue of the 6812 will be explained in Chapter 3. **TExaS** simulates the 6812 memory-bus activity, using simplified 8-bit memory read/write cycles similar to those of the 6811. Although the software/hardware timing is accurate for simulating a 6812, the memory-bus activity is shown in a simplified format. Information collected during simulation is recorded in the `TheLog.rtf` file. The `Mode menu` commands are used to configure the instruction-set simulator. There is a clear trade-off between the simulation speed (your program runs faster) and the amount of information you can observe (your program runs slower.) Some `Mode menu` commands are illustrated in Figure A1.5.

Figure A1.5
TExaS Mode menu commands.

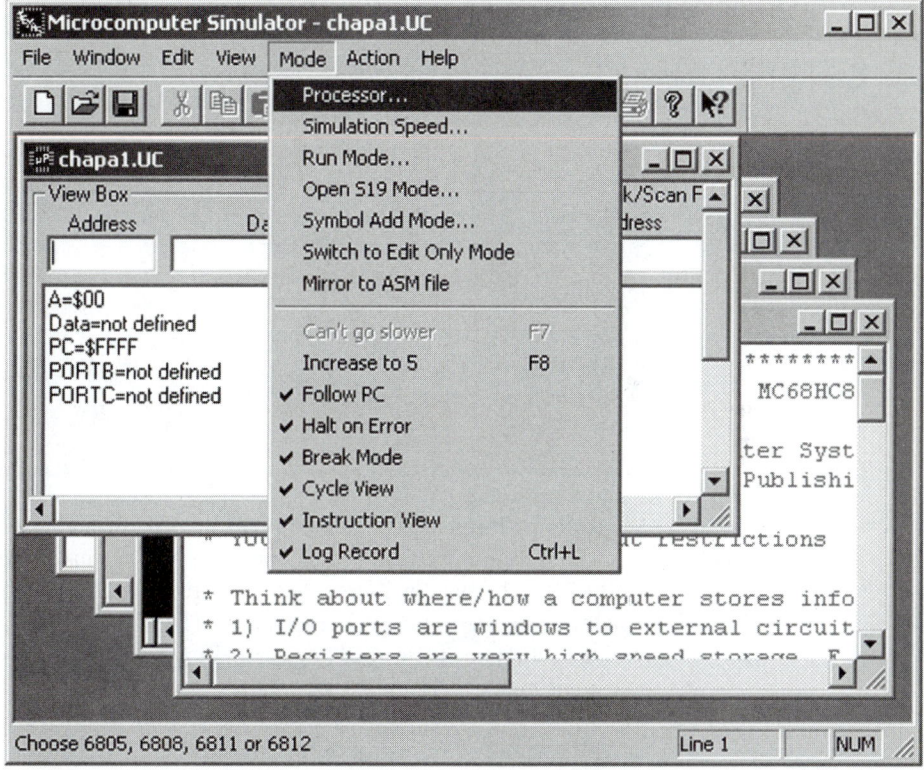

The **Mode->Processor** command allows you select the processor and memory configuration of the microcomputer. The **Mode->SimulationSpeed** command allows you to choose the number of instructions executed between screen updates. All of the `Mode menu` commands can be set in one dialog box by use of the **Mode->RunMode** command. You use the **Mode->OpenS19Mode** command to specify how **TExaS** imports object code created with a cross-compiler. Toggle the **Mode->FollowPC** command to enable and disable highlighting of the current instruction being executed in the `TheList.rtf` window. Toggle the **Mode->CycleView** command to enable and disable showing of memory-bus cycles in `TheLog.rtf` during execution. Toggle the **Mode->InstructionView** command to enable and disable displaying of instructions in the `TheLog.rtf` as they are executed. Toggle the **Mode->LogRecord** command to enable and disable recording of strategic information in The `Log.rtf` during execution. The `Mode menu` is used to configure the simulation; the `Action menu` initiates activity. Some `Action menu` commands are shown in Figure A1.6.

Figure A1.6
TExaS `Action menu` *commands.*

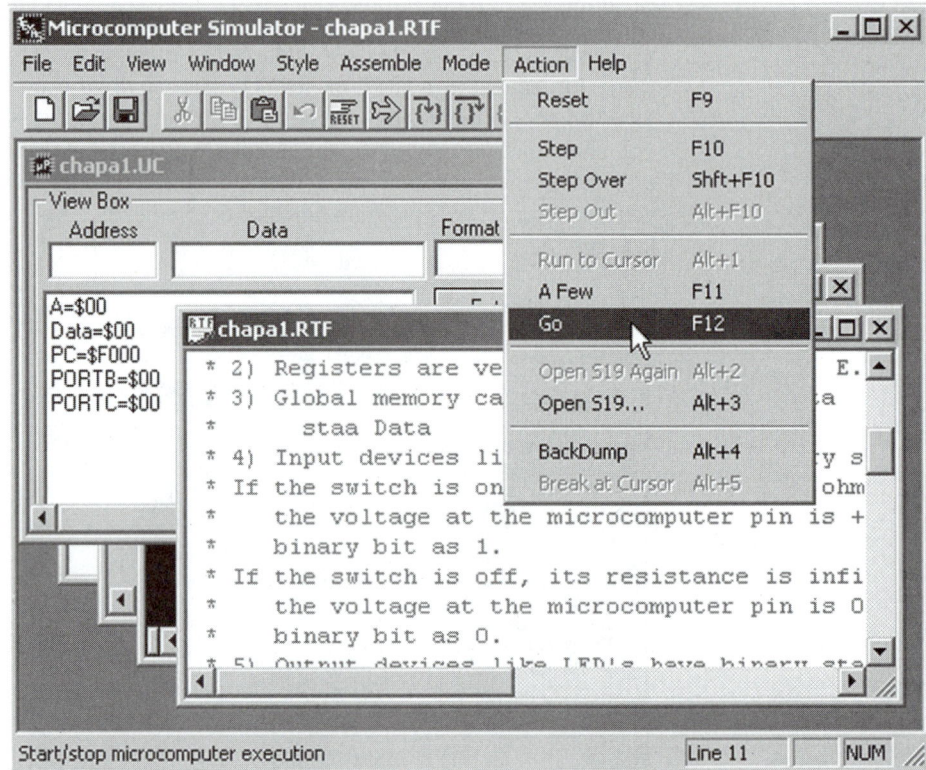

The **Action->Reset** (F9) command performs a hardware reset on the microcomputer system. The **Action->Step** (F10) command executes one instruction. The **Action->StepOver** (Shft+F10) command will execute one instruction. If that instruction is a subroutine call, then the entire subroutine will be executed. The **Action->StepOut** (Alt+F10) will execute until the subroutine is finished, then stop. Toggle the **Action->Go** (F12) command to start and stop simulation. You use the **Action->OpenS19** command to import object code created with a cross-compiler. Executing the **Action->BackDump** (Alt+4) will display the activity generated by the instructions most recently executed.

The second unique aspect of this simulator is the error checking. Examples of illegal activity include the following:

- execution of an illegal instruction,
- read/write to an undefined address,
- stack underflow (causing a read/write from unimplemented memory),

- write to ROM, EPROM, EEPROM,
- read from unprogrammed ROM, EPROM, EEPROM,
- read from RAM that has not yet been written to,
- read from an unimplemented I/O port.

These error-checking operations will catch many run-time programming errors. Real 6811 and 6812 microcomputers will execute an unimplemented-instruction trap interrupt (like a software interrupt) when the processor attempts to execute an illegal instruction. The **TExaS** application gives you the option of halting simulation or executing the trap interrupt (like the real microcomputer.) You select this option by using the **Mode->RunMode...** command. Software bugs often results in one of the illegal activities shown above. Whereas a real computer gives garbage data, then continues on, when executing an illegal read or write, this simulator will report the error and stop. Debugging, the process of identifying and removing software errors, is an important aspect of embedded system developing, and this simulator has many powerful debugging tools.

I/O port simulator. This part of the application simulates many of the I/O peripherals on the microcomputer. Simple peripherals include the parallel I/O ports, with direction registers as appropriate. Other functions like the timer, timer overflow, input capture, output compare, key wakeup, serial communications interface, and ADC are available, as supported by the actual microcomputer device. Most, but not all, peripherals on the 6811 and 6812 are supported. For a complete list of implemented features, see the `Port11.rtf` and `Port12.rtf` files created when **TExaS** is installed. Interrupts (flags, masks, priority, and vectors) are accurately simulated. The application will automatically simulate the I/O ports for the specific microcomputer you have selected (e.g., MC68HC11A8 or MC68HC812A4). For the latest implementation details, see the `readme.txt` file. If your software accesses an unimplemented I/O port, a run-time error is generated. As we saw earlier, the command **Mode->HaltOnError** will enable/disable the reporting of run-time errors. Figure A1.7 shows a simple microcomputer with Port C configured as an input and Port B configured as an output.

External device simulator. This is one of the most complex and important parts of the **TExaS** application. What makes embedded system programming interesting is its interaction with physical devices external to the microcomputer. These devices are configured by using the commands in the `IO menu`. An IO file must be open to create external I/O devices. The user interacts with these devices (e.g., toggling a switch) by using the IO window. Logic probes and voltmeters are automatically attached, as appropriate. A logic analyzer and oscilloscope can be added to provide visual information about signals outside the microcomputer chip. Figure A1.7 shows four switches and four light-emitting diodes (LEDs) interfaced to a microcomputer.

A1.3 Developing Assembly Software

Next, we will discuss the process of software development. There are two approaches when first starting to use this application. The first approach involves modifying an existing example. The second approach is to start from scratch. As we saw earlier, 6811

Figure A1.7
Simple microcomputer
system with four inputs
and four outputs.

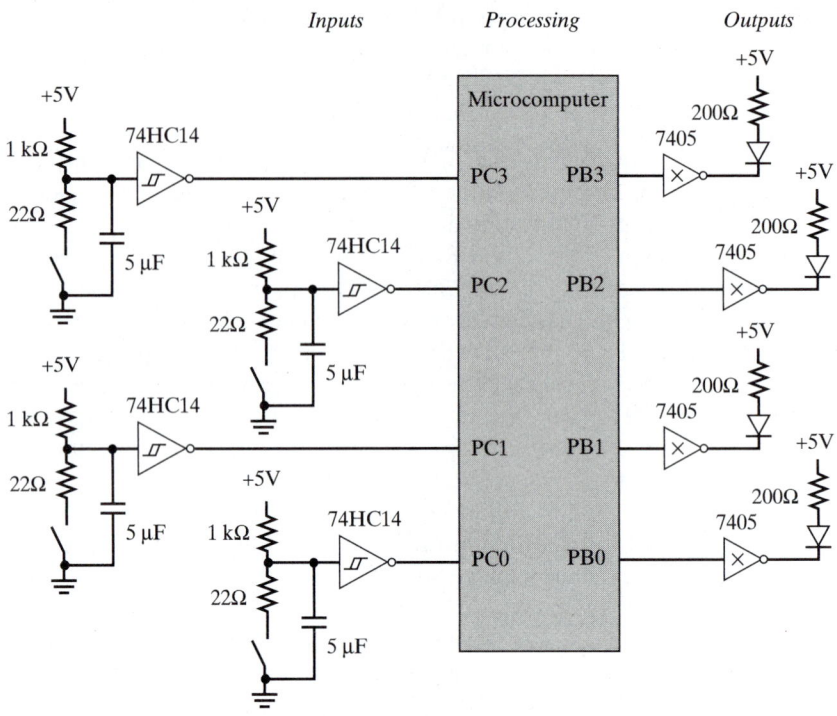

and 6812 microcomputers have separate assemblers, instruction simulators, and I/O port simulators. Consequently, the first decision we must settle is the microcomputer type. If you change the microcomputer type in the middle of a software development project, then you must rebuild the I/O devices in the IO file.

One way to develop assembly software is to first write the software in a high-level language like C, then convert the software by hand into assembly. Program A1.1 shows the C code for the simple system shown previously in Figures A1.1, A1.2, and A1.7. The line numbers are not part of the program, but were added for this example in order to help with the explanation. Characters between /* and */ are comments and are added as documentation. Line 2 defines symbols for all the I/O ports on the 6811. These symbols make the software easier to read. For example, line 8 could have been written as

```
*(unsigned char volatile *)(0x1004)=Data;
```

Line 3 defines a global variable. Line 5, when executed, defines Port C as an input port. The `while(1)` code causes lines 7 and 8 to be executed over and over. Executing the code `Data=PORTC;` will bring a copy of the 8 input pins of Port C into the global variable. The code `PORTB=Data;` stores the value from the global out to the output Port B, changing the pattern on the LED lights. Lines 11 through 14 define the reset vector, which specifies where the software will begin when power is applied or when the reset button is pushed.

Program A1.1   C language program for the MC68HC811E2 written in Imagecraft ICC11.

```
1    /* ********ChapA1.c********************/
2    #include "HC11.h"
3    unsigned char Data;
4    void main(void){
5      DDRC=0x00;        /* Port C in an input */
6      while(1){
7        Data=PORTC;     /* read switch value into Data */
8        PORTB=Data;     /* write value to LEDs */
9      }
10   }
11   extern void_start();     /* entry point */
12   #pragma abs_address:0xfffe
13   void (*reset_vector[])() = { _start };
14   #pragma end_abs_address
```

Program A1.2 shows the assembly code for the simple system shown previously in Figures A1.1, A1.2, and A1.7. The line numbers are not part of the program, but were added for this example in order to help with the explanation. Lines that begin with * are comments and are added as documentation. A line of assembly code has four fields. The label field is optional; it starts in the leftmost column. The next field contains the opcode (such as `ldaa`) or a pseudo-op code (such as `equ`). Opcodes contain actual instructions to be executed by the computer. For example, `ldaa` brings a value from memory or I/O port into Register A, `staa` sends a value from Register A out to memory or I/O port, and `bra` causes the program to branch. Pseudo-op codes give instructions to the assembler and are not executed by the computer. The third field is the operand field, which contains information needed by the instruction. The instruction `ldaa #n` will load the number `n` into Register A. For example, the `#$00`[1] operand field in line 8 specifies that the data to be used in the instruction is the value $00. On the other hand, the instruction `ldaa N` will load the contents of memory location `N` into Register A. For example, the `PORTC` operand field in line 10 specifies that the place to read the data from will be PORTC. The instruction `staa N` will store the contents of Register A out to memory location `N`. The last field, which is the rightmost field, contains comments. Comments are ignored by the computer, but are used by the programmer to clarify the program operation. Tabs and/or spaces delimit the fields, which are nicely lined up in this example. Lines 2, 3, and 4 use the pseudo-op `equ` to define symbols for three of the I/O ports on the 6811. Just like symbols in the C code, these symbols make the software easier to read. For example, line 12 could have been written as `staa $1004`, but notice how much easier it is to understand `staa PORTB`. Lines 5 and 7 use the pseudo-op `org` to place the variables in RAM and the program in ROM. Line 6 defines a global variable. Lines 8 through 13 give the actual instructions the 6811 will execute. Lines 8 and 9, when executed, define Port C as an input port. The instruction `ldaa #$00` places the value

---

[1]The # means immediate mode (take this as data) and the $ means hexadecimal.

$00 into Register A. The instruction `staa DDRC` stores the value from Register A into the I/O register DDRC, making Port C an input. Executing the instruction `ldaa PORTC` will bring a copy of the 8 input pins of Port C into Register A. The instruction `staa PORTB` stores the value from Register A out to the output Port B, changing the pattern on the LED lights. The `bra loop` instruction causes lines 10–13 to be executed over and over. Lines 14 and 15 define the reset vector, which specifies where the software will begin when power is applied or when the reset button is pushed.

Program A1.2   Assembly language program for the MC68HC811E2.

```
 1   * Appendix 1 tutorial program for the MC68HC811E2
 2   PORTB equ  $1004    PortB outputs 3,2,1,0 connected to LED lights
 3   PORTC equ  $1003    PortC inputs 3,2,1,0 connected to switches
 4   DDRC  equ  $1007    direction register for PortC
 5         org  $0000    globals go in RAM
 6   Data  ds   1        copy of Input from Port C switches
 7         org  $F800    object code goes in ROM
 8   main  ldaa #$00
 9         staa DDRC     make all pins of Port C input
10   loop  ldaa PORTC    read switch values
11         staa Data     save a copy in global variable
12         staa PORTB    output to lights
13         bra  loop     repeat
14         org  $FFFE
15         fdb  main     starting address after a RESET
```

There are many examples, so you might find one similar to the problem you are attempting to solve. The `readme.txt` file contains a list and brief description of the examples. You will find it much easier to adapt an example from the same microcomputer you are interested in. In other words, if you wish to develop a 6811 program, you will find it harder to adapt a 6812 example than to adapt a 6811 example. The example files are grouped into directories according to microcomputer type. In addition, there are three additional directories that can be used as starting points for developing software in C.

A1.3.1 Running an Existing Assembly-Language Program

The simplest place to begin using the **TExaS** simulator is to run an existing configuration. In fact, the tutorial, located at the end of this chapter, will step you through this process in detail. In this section, however, the general approach to running an existing system is presented. First, you choose the general topic of interest. For example, if you were interested in microcomputer software that used the serial port, you could choose the `sci` example. Next you choose the microcomputer—either the 6811 or the 6812. If you choose the 6811, open the `sci.rtf` file located in the MC6811 subdirectory. The example assembly-language files for the 6812 can be found in the MC6812 subdirectory. When any of the `sci.*` files is opened, **TExaS** will automatically open the related files. For example, `sci.rtf` is the source code, `sci.uc` is the microcomputer file, and `sci.io` is the external I/O devices. On most computers, double-clicking `sci.rtf` will incorrectly start WordPad or Microsoft Word. So, to start **TExaS** with these files, you could double-click the `sci.uc` icon.

The program must be assembled before it can be executed. The assembly process involves converting the human-readable assembly source code (`sci.rtf`) into the machine-readable object code. Click on the source code (the `sci.rtf` window) and execute the **Assemble->Assemble** command (ctl–B). The **TExaS** assembler automatically loads the object code into simulated memory.

Next, you run the simulation. You start the simulation by executing the **Action->Go** command (F12). There are many windows you can observe during execution. The **View-Box** in the microcomputer window shows strategic information during execution. If the *FollowPC* mode is active (execute the **Mode->FollowPC** command to toggle this mode on and off), then the `TheList.rtf` window will show the current position of the executing software. The `TheLog.rtf` window can be configured to show a wide range of results during simulation. If the *CycleView* mode is active (execute the **Mode->CycleView** command to toggle this mode on and off), then the address/data-bus activity will be logged. If the *InstructionView* mode is active (execute the **Mode->InstructionView** command to toggle this mode on and off), then the executed instruction will be logged. If the *LogRecord* mode is active (execute the **Mode->LogRecord** command to toggle this mode on and off), then the parameters of the **ViewBox** will be dumped into the `TheLog.rtf` during execution. The status of external devices is shown in the IO device window. If a serial port is active, then the input/output of this external device is shown in the `TheCRT.rtf` window.

A1.3.2 Modifying an Existing Assembly-Language Program

The first step is to find and open an existing system. Again, a list of examples can be found in the `readme.txt` file, and the assembly-language examples themselves can be found in the MC6811 and MC6812 subdirectories. Because you plan to modify these files, it makes sense to give them new names. If you didn't change the names, and were to upgrade to a newer version of **TExaS**, then the install process might overwrite your programs. Use the **File->SaveAs** command to change the names of all the files you will be using. You should maintain the appropriate extension (*.rtf, *.uc, *.io, *.stk, *.plt, *.scp), place these new files all in the same directory (but the directory may be different from the original directory that contained the existing example), and give them all the same first part of the filename name. For example, you could save the source file as `My.rtf`, save the microcomputer file as `My.uc`, and save the external I/O file as `My.io`. The scope and stack view files, if needed, are saved as `My.scp` and `My.stk`, respectively. Leave the `TheList.rtf`, `TheLog.rtf`, and `TheCRT.rtf` files alone, because these are special files that must maintain these exact file names.

The second step is to **reconfigure the processor and external I/O ports** as needed. For example, you may wish to switch from one type of 6811 to another. Use the **Mode->Processor...** command to select the new microcomputer. To reconfigure the external devices, click on the I/O window, and execute the appropriate command from the IO menu. Chapter 5 describes the process in detail. If you are converting an example from one microcomputer to another, you will have to rebuild all the external I/O and scope connections to be compatible with the I/O port names of the new microcomputer. Because the I/O devices are at different memory-mapped locations, switching between the 6811/MC68HC711D3/6812 will require you to rebuild the I/O and scope connections.

The third step is to **write assembly code** by editing the `My.rtf` file. For small programs, you may wish to enable automatic recolor; for large programs, you may wish to disable it. The program must be assembled (ctl–B) before it can be executed. For an overview of the assembly-language development process, see Chapter 4.

The fourth step is **configuring the simulation modes**. A basic tradeoff exists between simulation speed and the ability to observe the system behavior. The commands that affect simulation are grouped in the `Mode menu`. These configuration settings are saved and restored with the microcomputer file (e.g., in the file `My.uc`).

The last step is **running and debugging** your software. The **ViewBox**, Stack Window, IO window, oscilloscope, and logic analyzer provide visualization of your running program. The action commands are appropriately grouped into the `Action menu`. The usual debugging commands, **Action->Reset**, **Action->Go**, **Action->Step**, **Action->StepOver**, and **Action->StepOut**, are available. Breakpoints can be set on any address (even I/O ports). When a read or write access occurs to a breakpoint address, the simulation can be configured to stop (halt mode) or to simply copy the **ViewBox** parameters into the `TheLog.rtf` window (scan mode). Some special debugging features include **Action->Few** (executes a few instructions, then stops) and **Action->BackDump** (displays the simulator state for the previous instructions). You can perform right-click commands in the listing window: **Action->RuntoCursor**, and **Action->BreakatCursor**.

A1.3.3 Developing an Assembly Program from Scratch

Sometimes, it is more efficient to build a system from scratch. To develop assembly programs, you need at least one source-code file and one microcomputer file. You will almost always wish to have an I/O file, too. The first step is to create new files as needed. After creating a new microcomputer file, use the **Mode->Processor. . .** command to select the desired microcomputer and specify the clock period.

The second step is to perform **File->SaveAs** operations on the files you will be developing. You should maintain the appropriate extension (`*.rtf`, `*.uc`, `*.io`, `*.stk`, `*.scp`), place these new files all in the same directory, and give them all the same first part of the filename name. Next, follow the development cycle described in steps 3, 4, and 5 of the previous section.

A1.4   *Developing C Language Software

In this section, we will discuss the process of C language software development. Just like with assembly language, you can adapt or start from scratch. Again, the first decision you must settle is the microcomputer type. If you change the microcomputer type in the middle of a software development project, then you must rebuild the I/O devices in the IO file and reconnect signals to the oscilloscope. For C language development, no assembly source program (`*.rtf`) is needed, but you will need a microcomputer (`*.uc`). You will almost always wish to have an I/O file (`*.io`), too. The basic approach when developing C language software is to use the compiler to create the object code. When using ICC11 or ICC12, the compiler will create a listing file (`*.lst`) and an object code file (`*.s19`).

A1.4.1 Running an Existing C Program

The existing examples are already compiled, as you can run them directly from the **TExaS** application. You start by opening one of the microcomputer files that you can find in the ICC11, ICC11A, or ICC12. When one of these files is opened, **TExaS** will automatically open the related files. The next step is to load the object code created by the compiler. Execute the **Action->OpenS19. . .** command, and select the corresponding `S19` object code file. The command will load the object code into the simulated microcomputer memory. You start the simulation by executing the **Action->Go** command (F12).

A1.4.2 Adapting an Existing C Program

The first step is to select the best example. For example, if you are interested in the analog-to-digital converter, then you could start with the `tut3.*`. Use **TExaS** to open the `tut3.uc` microcomputer file. **TExaS** will automatically open the other related files. Next, remove any unwanted files. For example close the stack window and/or scope window if either of these views is not needed. Next, use the compiler to open the C source-code files. With the freeware ICC11, you can use **NotePad** to edit C source programs. The freeware ICC11 compiler is installed in the ICC11 folder as part of the **TExaS** installation. Back in **TExaS,** you use the **Mode->Processor. . .** command to select the processor. Use the **File->SaveAs. . .** command in both the compiler and **TExaS** to rename the file names.

The second step is to **configure the external I/O ports** as needed. Click on the I/O window, and execute the appropriate command from the `IO menu`. For more information, see Chapter 5 and the help system for the each of the I/O devices.

Next, **write C code** by editing the C source files. The editing, compiling and assembling are performed outside the scope of the **TExaS** simulator. For more information, see the documentation supplied with ICC11 or ICC12. There is an extensive html document describing the development of C programs for embedded systems via ImageCraft's ICC11 or ICC12. This html document can be found on the **TExaS** CD under the **embed** directory. The latest version of this document can also be found on

`http://www.ece.utexas.edu/~valvano/embed/toc1.htm`

The compiler must be configured to generate an S19-format object file and an assembly-listing file.

The fourth step is **configuring the simulation modes**. If you are developing C programs with ICC11 or ICC12, then observe the listing file to identify the column number of the absolute address and label fields. We define the first column as number 0. For example, the left side of Figure A1.8 shows a fragment of a listing generated by ICC12 version 5. The spaces have been changed to the ~ character so that you can more readily count column numbers. In ICC12 version 5, the address field begins in column 1, and the label field begins in column 23. Another example, shown on the right side of Figure A1.8, is a fragment of a listing generated by the ICC11 freeware compiler. In this example, the address field begins in column 5, and the label field begins in column 29. These two parameters must be entered in the **Mode->OpenS19. . .** dialog box. These configuration settings are saved and restored with the microcomputer file (e.g., in the file `My.uc`).

Figure A1.8

Two examples illustrating how to determine column numbers when importing object code generated by a cross-compiler.

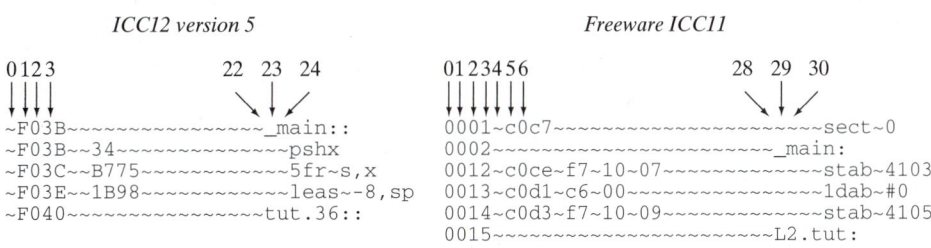

The fifth step is **running and debugging** your software. The **ViewBox**, Stack Window, IO window, oscilloscope, and logic analyzer provide visualization of your running program. The usual debugging commands described earlier are available. When using the `ICC11` or `ICC12` compiler, you can perform right-click commands in the listing window: **RuntoCursor**, and **BreakatCursor**.

For additional information on using this Simulator to run C programs compiled by ICC11/ICC12, use the help system to look up the **Mode->OpenS19Again** command. There are three subdirectories containing examples of C programs, which can be run on the **TExaS** simulator, that will be created when you install **TExaS**. The names of these subdirectories are `ICC11`, `ICC11A`, and `ICC12`. For more information on using one of the cross compilers with **TExaS**, see the `ReadMe.txt` file in the corresponding subdirectory.

A1.4.3
Developing a C
Program from
Scratch

To develop C programs, you need a microcomputer file. You will almost always wish to have an I/O file, too. The first step is to create new files as needed. After creating a new microcomputer file, use the **Mode->Processor. . .** command to select the desired microcomputer and specify the clock period.

The second step is to perform **SaveAs** operations on the files you will be developing. You should maintain the appropriate extension (`*.uc`, `*.io`, `*.stk`, `*.scp`), place these new files all in the same directory, and give them all the same first part of the filename name (e.g., `My.uc` is a microcomputer file, `My.io` is an external I/O file, `My.stk` is a stack view file, `My.scp` is an oscilloscope file). Next, follow the development cycle described in steps 3, 4, and 5 of the previous section.

A1.5 Tutorial A1. Getting Started

The purpose of this tutorial is to introduce the first-time user to **TExaS**. The specific objectives of this tutorial include the following:

1. how to launch the simulator,
2. how to modify input switches,
3. how to edit, assemble, and run a 6811 or 6812 program,
4. how to modify display format in the **ViewBox**,
5. how to move and resize windows,
6. how to get on-line help.

A second objective is to visualize four places information can be stored on a computer. You will learn later in Chapter 8 how to create and use local variables. But, in this example, data can exist in

1. external switches, input devices that hold information in the switch positions;
2. registers, high-speed temporary storage inside the processor;
3. global variables, to hold information that is easy to access by all programs;
4. external LEDs, output devices that hold information in the light patterns.

Action: Install the **TExaS** application.
The **TExaS** application must be installed before it can be used. To install **TExaS**, click on the `Setup.exe` file in the `Texas` folder on the CD that accompanies this book. You could check the web site at **www.ece.utexas.edu/~valvano** to see whether there is an upgrade available.

Action: Copy the tutorial programs onto a hard drive.
The tutorial directory on the CD contains files needed to perform the tutorials in this book. If you are studying the 6811, then copy the `Mc6811T` subdirectory from the CD onto

a hard drive. Likewise, if you are studying the 6812, then copy the `Mc6812T` subdirectory from the CD onto a hard drive. **TExaS** needs to read and write files during execution, so you cannot simulate microcomputer software directly from the CD. The movies do not need to be copied, because they can be viewed simply by double-clicking the movie file on the CD itself.

Action: How to launch the **TExaS** application.

There are three ways to start the simulator. `TExaS.exe` is the simulator application. The first way to launch **TExaS** is to double-click its icon, as shown in Figure A1.9.

You can also start the application by selecting **TExaS** from the start menu. The third way to start **TExaS** is to double-click on a microcomputer file. Microcomputer files have a `*.uc` extension. If you start **TExaS** by double-clicking the `TExas.exe` icon or by using the start menu, you should get a dialog screen similar to the one shown in Figure A1.10. If not, then reread the installation instructions, and try installing again. If you start **TExaS** by double-clicking the file `ChapA1.uc`, then you'll skip the new dialog and go directly to Question A1.1.

TExaS.exe

Figure A1.9
TExaS *application icon.*

Figure A1.10
TExaS *New file dialog.*

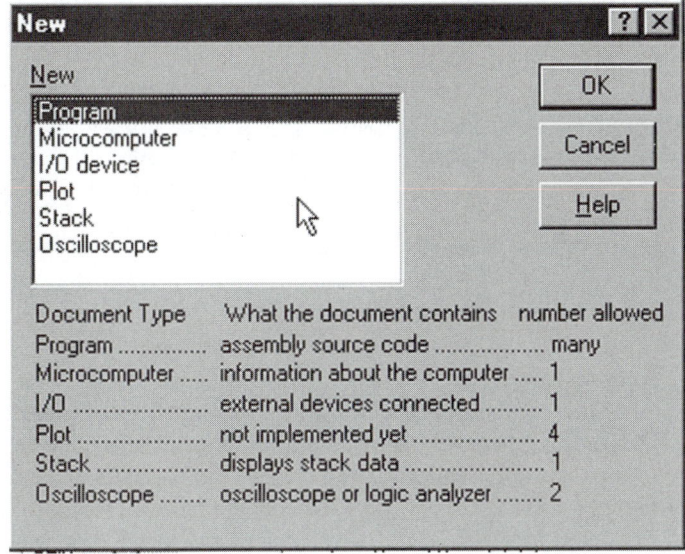

The *New file dialog* is asking whether you wish to create a new Program, Microcomputer, I/O device, Stack, Oscilloscope, or Plot file. For this tutorial, choose **Cancel**, which creates no new file. There should now be two windows open, `TheLog.rtf` and `TheList.rtf`. Next, open the `ChapA1.rtf` file, using the command **File->Open**. For the 6811, look in the MC6811T subdirectory; for the 6812, look in the MC6812T subdirectory.

Question A1.1. List the names of the six files that are now open.

Action: The third way to start the simulator.

Quit the application by executing **File->Exit**. For this example, it doesn't matter whether you save or don't save the files. A very convenient way to start the simulator is to double-click on one of its icons (e.g., `ChapA1.uc`). Although you could start **TExaS** by double-clicking any of its icons, it is best to open a microcomputer file (`*.uc`) first. Try double-clicking on the `ChapA1.uc` icon. If you were to double-click `ChapA1.rtf`, it would probably start **WordPad** or **Microsoft Word**. Double-clicking the `ChapA1.uc` icon should open the same six windows you answered in **Question A1.1**.

Program A1.3 shows the 6812 version of the chapter 1 tutorial program. It is quite similar to the 6811 version shown as Program A1.2. There are two minor differences. First, the I/O ports, the RAM, and the ROM on the MC68HC812A4 are at different locations. Second, the 6812 has a direction register for Port B, which must be initialized to $FF, defining this port as an output.

Program A1.3
Assembly-language program for the MC68HC812A4.

```
* Appendix A1 tutorial program for the MC68HC812A4
PORTB equ  $0001   PortB outputs 3,2,1,0 connected to LED lights
DDRB  equ  $0003   direction register for PortB
PORTC equ  $0004   PortC inputs 3,2,1,0 connected to switches
DDRC  equ  $0006   direction register for PortC
      org  $0800   globals go in RAM
Data  ds   1       copy of Input from Port C switches
      org  $F000   object code goes in ROM
main  ldaa #$00
      staa DDRC    make all pins of Port C input
      ldaa #$FF
      staa DDRB    make all pins of Port B output
loop  ldaa PORTC   read switch values
      staa Data    save a copy in global variable
      staa PORTB   output to lights
      bra  loop    repeat
      org  $FFFE
      fdb  main    starting address after a RESET
```

Action: How to assemble a program.

There are four phases of software development on an embedded microcomputer: source-code creation, hardware interfacing, assembly, and debugging. In this tutorial, the source code (`ChapA1.rtf`) has already been created and the external devices have already been connected (`ChapA1.io`), so you will begin in phase 3, assembly. To assemble a program, you first select the source code, by clicking on the source code window (in this case the source code is `ChapA1.rtf`). If you are having trouble finding the `ChapA1.rtf` window, try the command **Window->ChapA1.RTF**, to bring that window to the top. Notice the colors of the **Start**, **ExecuteOne**, and **StepOver** tools in the toolbar, as shown in Figure A1.11. The gray color signifies that function can not be run at this point. Run the mouse over the top of the tools, observing the popup. Also, notice the explanations in the footer bar, at the bottom.

Figure A1.11 *TExaS toolbar.*

Execute the command **Assemble->Assemble**, which also has a shortcut **Ctrl+B**. If there were assembly errors they would be shown in the listing window (`TheList.rtf`). When there are no assembly errors, the object code has been loaded into memory and the program is now ready to run.

Question A1.2. What happened to the colors of the **Start, ExecuteOne**, and **StepOver** tools in the tool bar after the program was assembled? Why?

Action: How to modify input switches.

The program simulates a Motorola single-chip computer connected to various input and output devices. There are many different types of input and output devices that the system can simulate. Input ports are pins (wires) that allow information to enter the computer. Output ports are pins (wires) that allow information to exit from the computer. In this tutorial, Port C is an input port and Port B is an output port. In particular, there are four toggle switches connected to Port C pins PC3, PC2, PC1, and PC0. There are four LED lights connected to Port B pins PB3, PB2, PB1, and PB0. To modify the position of one of the toggle switches, you simply click on the switch in the `ChapA1.io` window with the mouse button. If you are having trouble finding the `ChapA1.io` window, execute the command **Window->ChapA1.IO**, to bring that window to the top. Use the mouse to close the switches PC3 and PC1 (PC3 means Port C bit 3) and open the switches PC2 and PC0. The I/O window can be drawn with either a black or a white background. If you execute **IO->WhiteBackground**, then you will see something like the window shown in Figure A1.12. This figure shows PC3/PC1 closed and PC2/PC0 open. The rectangular voltage probes can measure high (+5V), low (0V) or floating (z). Floating is a logic state that is not driven.

Figure A1.12
TExaS IO document window, showing 4 switches and 4 lights.

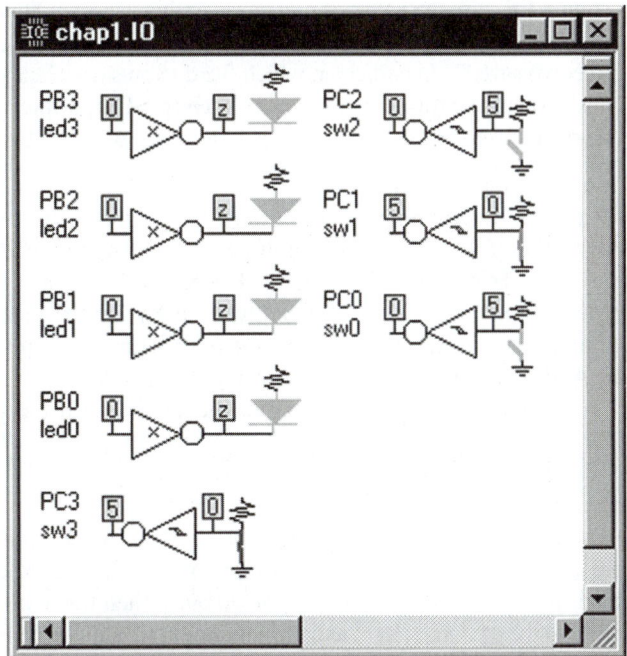

Question A1.3. Describe the relationship between the switch position (open or closed) and

    1. the voltage across the switch (probe near the switch),
    2. the voltage at the output of the 74HC14 not gate (the "circle-end" of the not gate),
    3. the digital value of the input port (look in the `ChapA1.uc` window).

Action: Observing the details of execution.

    During the early debugging stages of a project, you need access to a lot of information. Hide (not close) all the windows except `TheList.rtf` and `TheLog.rtf`. `TheList.rtf` will highlight the current instruction, and details of the simulation are dumped into `TheLog.rtf`. Move and resize windows so both are fully visible. In general, unwanted windows can be minimized (not closed) to speed up the simulation. Start execution, using the **Start** button in the tool bar or the **Action->Run** command (shortcut is F12). After a while, stop execution, using the **Stop** button in the tool bar or the shortcut F12. Observe the information recorded in `TheLog.rtf`.

Question A1.4. Explain the behavior of each of the following modes.

    Mode->FollowPC
    Mode->CycleView
    Mode->InstructionView
    Mode->LogRecord

Each can be toggled on and off by using the corresponding command in the `Mode menu`. In particular, activate these settings one at a time, and run the program. Look up the *on-line help* for these functions. Single-click the `Mode menu` bar to stick the menu, roll the mouse over the command, then type F1.

    Observation: The simulation speed (real human time to run your program) is greatly improved by turning these four modes off.

Question A1.5. Use the help system to answer this question: "Into which file are the parameters specified by the `Mode menu` commands (e.g., **FollowPC, CycleView, InstructionView, LogRecord**) saved?" In particular, which file do you save when you want to remember these settings? To answer this question, execute **Help->HelpTopics**, double-click **Menus**, then double-click `Mode menu`.

Action: Backdump.

    Reset the microcomputer (tool bar or F9), and clear the `TheLog.rtf` window (click on the `TheLog.rtf` window, type **ctrl+A,** then **Delete**). Turn off the modes **CycleView, InstructionView,** and **LogRecord,** which cause information to be recorded in the `TheLog.rtf` window. The **FollowPC** mode can be left on. Run the program for a short time, then stop execution with the **Stop** tool bar button or F12. Execute **Action->BackDump** and observe the `TheLog.rtf` window. Backdump is a very good mechanism for observing the behavior just prior to a crash or a breakpoint.

    Debugging Tip: Execute the Backdump command after your software crashes.

Action: Setting breakpoints.

    There are three ways to set breakpoints. Make sure the **Mode->BreakMode** is checked before continuing. When checked, the program will stop when a breakpoint address is reached. If it were not checked (i.e., *ScanPoint*), then the **ViewBox** data would simply be copied into the `TheLog.rtf`, and the simulation would not stop. The first way is to type the absolute address into the **Break/ScanPoints** box, then push the *Add* button. This box is in the

`ChapA1.uc` window. Try adding a breakpoint at $F805 for the 6811 or $F008 for the 6812. Reset the microcomputer (tool bar or F9) and start execution (tool bar or F12). After the breakpoint is reached, the program should stop. Notice that it does finish the complete instruction during which the breakpoint address was encountered. Some debuggers allow breakpoints only on the first address of an instruction and will break *before* executing the instruction. In contrast, **TExaS** breaks *after* executing the instruction. When execution stops, the program counter is usually pointing to the instruction on the line after the instruction that caused the breakpoint. Restart execution one more time; it should stop again at that same point. Delete that breakpoint by clicking on the breakpoint and hitting the *Remove* button.

The second way is to type the symbolic name (case sensitive) into the **Break/ ScanPoints** box, then push the *Add* button. Any symbol that your software defines (labels in column 1 of the program) can be used. Try adding a breakpoint at `loop` and running the program. After the breakpoint is reached, the program should again stop. Setting the breakpoint at $F805 for the 6811 ($F008 for the 6812) or `loop` should have exactly the same effect. Delete that breakpoint, using the *RemoveAll* button.

The third way to set a breakpoint is the easiest. First, with the left mouse button, click on the line in the `TheList.rtf` listing file where you want a breakpoint; then, with the right mouse button, click and hold. A popup menu allows you to set a breakpoint at this address. Try this procedure to set 2 or 3 breakpoints within the program. Notice that, if the line has a label, it is used to define the break; otherwise, the absolute address is used. Delete these breakpoints, using the *RemoveAll* button.

You can break on any address, not just addresses that are machine instructions. Add a breakpoint on PORTC, and run the program. The program will stop on a read or write access to PORTC. Delete this breakpoint, using the *RemoveAll* button.

Action: Observing the microcomputer.

The current state of the microcomputer can be observed in the `ChapA1.uc` microcomputer window. Bring this window to the top, then hit the reset button, and it should look something like the window shown in Figure A1.13. The figures in this tutorial were created

Figure A1.13
TExaS ViewBox window, showing register and memory values.

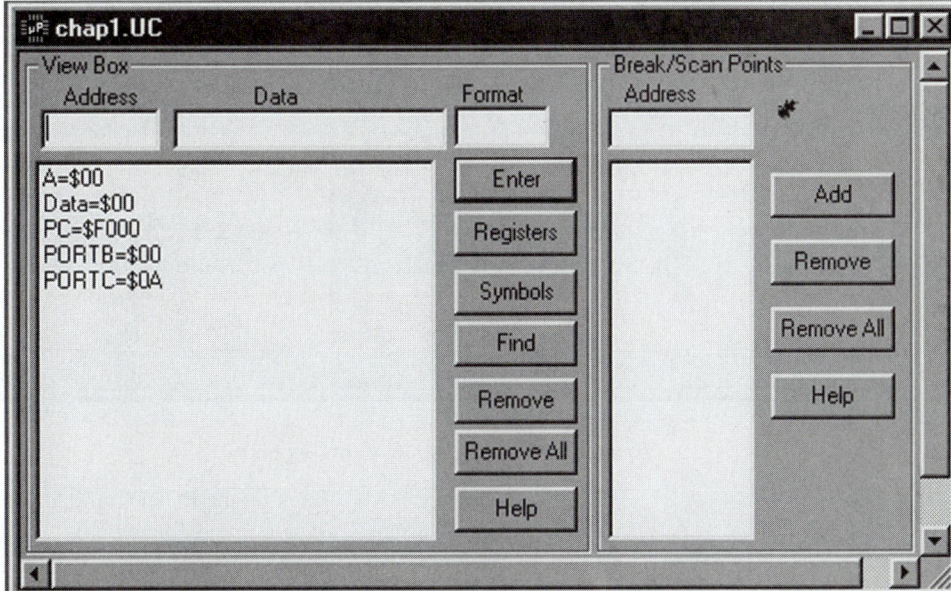

with the MC6812 simulation. Except for the PC value, the 6811 simulation should be identical. The expression `PORTC=$0A` means the value of `PORTC` is $0A. Using the `V` format specification, you can observe the address of a parameter, but usually the **ViewBox** is used to display the contents of parameters.

Action: Rearranging the windows.

During the debugging phase of a project, you need access to where the program is executing (`TheList.rtf`), the values of strategic parameters (`ChapA1.uc`), and the state of the external devices (`ChapA1.io`). Hide (not close) all the windows except `TheList.rtf ChapA1.uc` and `ChapA1.io`. Move and resize windows so most of these three windows are visible.

Question A1.6. What is the program counter, PC? How is its initial value (right after reset) established?

Question A1.7. What does the value PORTC=$0A in the **ViewBox** window mean?

Action: Changing the format of the **ViewBox** data.

Single-step (tool bar or F10) the program until the value in PORTB just changes to $0A. On the 6811, the PC will be $F80D just after PORTB changes. On the 6812, the PC will be $F00F just after PORTB changes. If you step too many times, just hit reset and try it again. The current state of the microcomputer can be observed in the `ChapA1.uc` microcomputer window. It should now look something like the window shown in Figure A1.14.

Figure A1.14   *TExaS ViewBox window, showing changed register and memory values.*

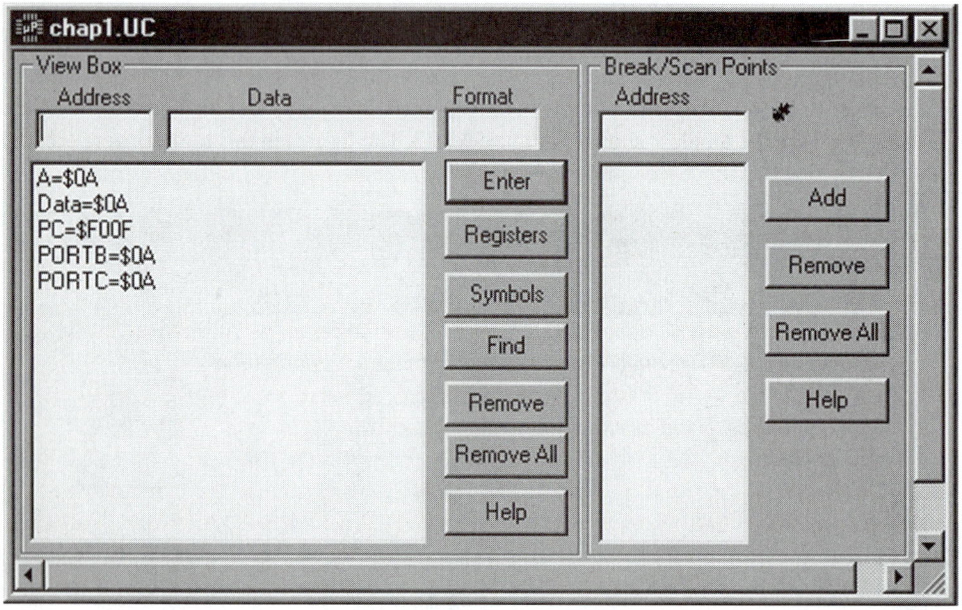

To change the format of a **ViewBox** entry:

1. select the entry   click on the PORTC=$0A line in the **ViewBox**
2. give the new format   type d in the *Format* box
3. enter the change   type \<Enter\>, or click on the *Enter* button

Question A1.8. Using this procedure, click on the PC, and give the values for the following formats:

H	16-bit unsigned hexadecimal (should be $F80D on the 6811 and $F00F on the 6812)
+H	16-bit signed hexadecimal (+H and −H are the same format)
D	16-bit unsigned decimal
+D	16-bit signed decimal (+D and −D are the same format)
B	16-bit unsigned binary
+B	16-bit signed binary (+B and −B are the same format)

Using this procedure, click on `Data=$0A`, and give the values for the following formats:

h	8-bit unsigned hexadecimal (should be $0A)
+h	8-bit signed hexadecimal (+h and −h are the same format)
d	8-bit unsigned decimal
+d	8-bit signed decimal (+d and −d are the same format)
b	8-bit unsigned binary
+b	8-bit signed binary (+b and −b are the same format)
2d	two unsigned decimal numbers

Action: Changing the value in the **ViewBox** data.

The current state of the microcomputer can be changed via this window. To change the value of a **ViewBox** entry:

1. select the entry	click on the A=$0A line in the **ViewBox**
2. give the new value	type 100 in the *Data* box
3. enter the change	type <Enter>, or click on the *Enter* button

To add new entries in the **ViewBox** (step 2 is optional):

1. give the Address	for the 6811 type the address $0080; for the 6812 type the address $0900
2. give the new value	type "**TExaS**" in the *Data* box (including the " ")
3. enter the new format	type 6d in the *Format* box (try also s and 6c)

Debugging Tip: Choose the formats in the ViewBox to simplify debugging.

Debugging Tip: While debugging software, adjust the parameters in the ViewBox to include important information and exclude unimportant information.

Action: Simulating the microcomputer hardware/software.

During the final testing phase of a project, you need access to the state of the external devices (`ChapA1.io` and `ChapA1.scp`). Hide (not close) all the windows except `ChapA1.io` and `ChapA1.scp`. Move and resize windows so both are visible. Start execution, using the **Start** button in the tool bar or the **Action->Run** command (shortcut is F12). Toggle the switches, and observe that the LED values match the corresponding switches.

Observation: The simulation speed (real human time to run your program) is greatly improved by closing (not merely hiding) scope windows.

Appendix 2

Glossary of Terms

1/f noise A fundamental noise in resistive devices arising from fluctuating conductivity. Same as pink noise.

2's complement See *two's complement.*

60-Hz noise An added noise from electromagnetic fields caused by either magnetic-field induction or capacitive coupling.

accumulator High-speed memory located in the processor used to perform arithmetic or logical functions. The accumulators on the 6811/6812 are A and B.

accuracy A measure of how close our instrument measures the desired parameter, referred to the NIST.

acknowledge Clearing the interrupt flag bit that requested the interrupt.

actuator Electro-mechanical or electro-chemical device that allows computer commands to affect the external world.

ADC Analog-to-digital converter, an electronic device that converts analog signals (e.g., voltage) into digital form (i.e., integers).

address bus A set of digital signals that connect the CPU, memory, and I/O devices, specifying the location to read or write for each bus cycle. See also *control bus* and *data bus.*

aliasing When digital values sampled at f_s contain frequency components above 0.5 f_s, then the apparent frequency of the data is shifted into the 0 to 0.5 f_s range. See *Nyquist theorem.*

alternatives The total number of possibilities; for example, an 8-bit number scheme can represent 256 different numbers. An 8-bit digital-to-analog converter (DAC) can generate 256 different analog outputs.

arithmetic logic unit (ALU) Component of the processor that performs arithmetic and logic operations.

arm Activate a particular device so that interrupts are requested.

ASCII American Standard Code for Information Interchange, a code for representing characters, symbols, and synchronization messages as 7-bit, 8-bit or 16-bit binary values.

assembler System software that converts an assembly-language program (human-readable format) into object code (machine-readable format).

assembly directive Operations included in the program that are not executed by the computer at run time, but rather are interpreted by the assembler during the assembly process. Same as pseudo-op.

assembly listing Information generated by the assembler in human-readable format, typically showing the object code, the original source code, any assembly errors, and the symbol table.

asynchronous communications interface adapter (ACIA) Device to transmit data with asynchronous serial communication protocol (same as UART and SCI).

asynchronous protocol A protocol where-in the two devices have separate and distinct clocks.

atomic Software execution that cannot be divided or interrupted. Once started, an atomic operation will run to its completion without interruption. On most computers, the assembly-language instructions are atomic.

background mode A 6812 mode with the background debug module (BDM) active.

bandwidth The information transfer rate, the amount of data transferred per second. Same as throughput.

bang-bang A control system where the actuator has only two states, and the system "bangs" all the way in one direction or "bangs" all the way in the other; same as binary controller.

basis Subset from which linear combinations can be used to reconstruct the entire set.

baud rate In general, the baud rate is the total number of bits (information, overhead, and idle) per time that are transmitted; in a modem application, it is the total number of sounds per time that are transmitted.

bidirectional Digital signals that can be either input or output.

biendian The ability to process numbers in both big and little endian formats.

big endian Mechanism for storing multiple-byte numbers such that the most significant byte exists first (in the smallest memory address). See also *little endian*.

binary A system that has two states, on and off.

binary operation A function that produces its result from two input parameters. For example, addition, subtraction, and multiplication are binary operations.

bipolar stepper motor A stepper motor where the current flows in both directions (in/out) along the interface wires; a stepper with 4 interface wires.

bit Basic unit of digital information, taking on the value of either 0 or 1.

bit time The basic unit of time used in serial communication.

blind cycle A software/hardware synchronization method where the software waits a specified amount of time for the hardware operation to complete. The software has no direct information (blind) about the status of the hardware.

borrow During subtraction, if the difference is too small, then we use a borrow to pass the excess information into the next higher place. For example, in decimal subtraction, $36 - 27$ requires a borrow by the ones from the tens place, because $6 - 7$ is too small to fit into the 0-to-9 range of decimal numbers.

break or **trap** A break or a trap is an instrument that halts the processor. The **TExaS** application will halt both software and hardware simulation when a specific address is encountered. With a resident debugger, the break is created by replacing a specific opcode with a software interrupt instruction. When encountered, it will stop your program and jump into the debugger. Therefore, a break halts the software. The condition of being in this state is also referred to as a break.

breakpoint The place where a break is inserted, the time when a break is encountered, or the time period when a break is active.

BUFFALO A 6811-based resident debugger developed in the mid-1980s. BUFFALO stands for Bit User's Fast Friendly Aid to Logical Operation.

buffered I/O A FIFO queue is placed in between the hardware and software in an attempt to increase bandwidth by allowing hardware and software to run in parallel.

burn The process of programming a ROM, PROM, or EEPROM.

bus A set of digital signals that connect the CPU, memory, and I/O devices, consisting of address signals, data signals, and control signals. See also address bus, control bus, and data bus.

bus interface unit (BIU) Component of the processor that reads and writes data from the bus.

busy-waiting A software/hardware synchronization method where the software continuously reads the hardware status, waiting for the hardware operation to complete. The software usually performs no work while waiting for the hardware. Same as *gadfly*.

byte Digital information containing 8 bits.

carry During addition, if the sum is too large, then we use a carry to pass the excess information into the next higher place. For example, in decimal addition, $36 + 27$ requires a carry from the ones to tens place, because $6 + 7$ is too big to fit into the 0 to 9 range of decimal numbers.

cathode ray tube (CRT) terminal An I/O device used to input data from a keyboard and output character data to a screen. The electrical interface is usually asynchronous serial.

ceiling Establishing an upper bound on the result of an operation. See also *floor.*

closed-loop control system A control system that includes sensors to measure the current state variables. These inputs are used to drive the system to the desired state.

CMOS A digital logic system called complementary metal-oxide semiconductor. It has properties of low power and small size. Its power is a function of the number of transitions per second.

compiler System software that converts a high-level-language program (human-readable format) into object code (machine-readable format).

condition code register (CCR) Register in the processor that contains the status of the previous ALU operation, plus some operating mode flags, such as the interrupt enable bit.

control bus A set of digital signals that connect the CPU, memory, and I/O devices, specifying when to read or write for each bus cycle. See also *address bus* and *data bus.*

control unit (CU) Component of the processor that determines the sequence of operations.

CPU-bound A situation where the input or output device is faster than the software. In other words it takes less time for the I/O device to process data than for the software to process data. The bandwidth is limited by CPU speed.

crisp input An input parameter to the fuzzy-logic system, usually with units such as cm, cm/sec, or °C.

crisp output An output parameter from the fuzzy-logic system, usually with units such as dynes or watts.

critical section Locations within a software module; if an interrupt were to occur at one of these locations, then an error could occur (e.g., data lost, corrupted data or program crash). Same as vulnerable window.

cross-assembler An assembler that runs on one computer but creates object code for a different computer.

cross-compiler A compiler that runs on one computer but creates object code for a different computer.

DAC Digital-to-analog converter, an electronic device that converts digital signals (i.e., integers) to analog form (e.g., voltage).

data acquisition system A system that collects information; same as *instrument.*

data bus A set of digital signals that connect the CPU, memory, and I/O devices, specifying the value that is being read or written for each bus cycle. See also *address bus* and *control bus.*

defuzzification Conversion from fuzzy-logic output variables to crisp outputs.

denormalized A denormalized number is an unnormalized floating-point number having an exponent of the smallest possible value. An unnormalized number has a mantissa value less than 1. The mantissa of a normalized floating point number is greater than or equal to 1, but strictly less than 2.

desk checking or **dry run** We perform a desk check (or dry run) by determining in advance, either by analytical algorithm or explicit calculations, the expected outputs of strategic intermediate stages and final results for a typical inputs. We then run our program to compare the actual outputs with this template of expected results.

device driver A collection of software routines that perform I/O functions.

digital signal processing Processing of data with digital hardware or software after the signal has been sampled by the ADC—e.g., filters, detection, and compression/decompression.

direct An addressing mode where the data or address value for the instruction is located in memory at address $0000 to $00FF.

direction register A bidirectional-port-configuration register that determines whether the port will be an input or an output.

disarm Deactivate a particular device so that interrupts are not requested.

DMA Direct Memory Access is a software/hardware synchronization method where the hardware itself causes a data transfer between the I/O device and memory at the appropriate time when data needs to be transferred. The software usually can perform other work while waiting for the hardware. No software action is required for each individual byte.

double byte Two bytes containing 16 bits. Same as *word*.

double-pole switch Two separate and complete switches that are activated together; same as *two-pole*. Contrast with *single-pole*.

double-throw switch A switch with three contact connections. The center contact will be connected to exactly one of the other two contacts. Contrast with *single-throw*.

download The process of transferring object code from the host (e.g., the PC) to the target microcomputer (e.g., the 6812.)

drop out An error that occurs after a right shift or a divide; the consequence is that an intermediate result loses its ability to represent all of the values—e.g., I = 100*(N/51) can result in only the values 0, 100, or 200, whereas I = (100*N)/51 properly calculates the desired result.

dummy PC A computer bus cycle that fetches data pointed to by the PC, but the data is not used.

dummy SP A computer bus cycle that fetches data pointed to by the SP, but the data is not used.

duty cycle For a periodic digital wave, it is the percentage of time the signal is high.

dynamic RAM Volatile read/write storage built from a capacitor and a single transistor, having a low cost, but requiring refresh. Contrast with static RAM.

EEPROM Electrically erasable programmable read-only memory that is nonvolatile and easy to reprogram. Typically, EEPROM can be erased and reprogrammed 10,000 times.

effective address register (EAR) A register that contains the address for the current memory cycle.

embedded computer system A system that performs a specific dedicated operation where the computer is hidden or embedded inside the machine.

emulator An in-circuit emulator is an expensive debugging hardware tool that mimics the processor pin outputs. To debug with a 6811 emulator, you would remove the 6811 processor chip and attach the emulator cable into the 6811 processor socket. The emulator would sense the processor input signals and recreate the processor output signals on the socket as if a 6811 chip were actually there running at 2 MHz. Inside the emulator, you have internal read/write access to the registers and processor state. Most emulators allow you to visualize/record strategic information in real time without halting the program execution. You can also remove ROM chips and insert the connector of a ROM-emulator. This type of emulator is less expensive, and it allows you to debug ROM-based software systems.

EPROM programmer System hardware/software that burns the object code into the microcomputer's EPROM.

EPROM Same as PROM. Electrically programmable read-only memory that is nonvolatile and requires external devices to erase and reprogram. It is usually erased by UV light.

erase The process of clearing the information in a PROM or EEPROM. The information bits are usually all set to logic 1.

EVB Evaluation Board, a Motorola product used to develop microcomputer software.

even parity A communication protocol where the number of ones in the data plus a parity bit is an even number. Contrast with *odd parity*.

expanded mode The mode where some of the I/O ports are used to create an external data bus (control, address, data) allowing external memory to be connected.

extended An addressing mode where the data or address value for the instruction is located anywhere in memory.

fan out The number of inputs that a single output can drive if the devices are all in the same logic family.

fast clear A 6812 timer mode where the associated flag is automatically cleared when the timer register is accessed.

filter In the debugging context, a filter is a boolean function or conditional test used to make run-time decisions. For example, if we print information only if two variables (x,y) are equal, then the conditional (x==y) is a filter. Filters can involve hardware status as well. For example, if we halt when the serial port has an overrun error, then (SCSR&0x08) is the filter, and `if (SCSR&0x08)asm(" swi");` would be the entire instrument.

finite impulse response filter (FIR) A digital filter where the output is a function of a finite number of current and past data samples, but not a function of previous filter outputs.

fixed-point A technique where calculations involving nonintegers are performed via a sequence of integer operations: for example, 0.123*x is performed in decimal fixed-point as (123*x)/1000 or in binary fixed-point as (123*x)>>10.

flash EEPROM Electrically erasable programmable read-only memory that is nonvolatile and easy to reprogram. Flash EEPROMs are typically larger than regular EEPROM and have fewer erase–reprogram cycles.

floating A logic state where the output device does not drive high or pull low. The outputs of open collector and tristate devices can be in the floating state. Same as HiZ.

floor Establishing a lower bound on the result of an operation. See also *ceiling*.

frame A complete and distinct packet of bits occuring in a serial communication channel.

framing error An error when the receiver expects a stop bit (1) and the input is 0.

friendly Friendly software modifies just the bits that need to be modified, leaving the other bits unchanged.

full-duplex channel Hardware that allows bits (information, error checking, synchronization, or overhead) to transfer simultaneously in both directions. Contrast with *simplex* and *half-duplex* channels.

full-duplex communication A system that allows information (data, characters) to transfer simultaneously in both directions.

functional debugging The process of detecting, locating, or correcting functional and logical errors in a program and the process of instrumenting a program for such purposes is called functional debugging or often simply debugging. Contrast with *performance debugging*.

fuzzification Conversion from the crisp inputs to the fuzzy-logic input variables.

fuzzy logic Boolean logic (true/false) that can take on a range of values from true (255) to false (0). Fuzzy-logic **and** is calculated as the minimum. Fuzzy-logic **or** is the maximum.

gadfly A software/hardware synchronization method where the software continuously reads the hardware status, waiting for the hardware operation to complete. The software usually performs no work while waiting for the hardware. Same as *busy-waiting*.

general-purpose computer system A system like the IBM-PC or Macintosh, with a keyboard, a disk, and a display, that can be programmed for a wide variety of purposes.

half-duplex channel Hardware that allows bits (information, error-checking, synchronization, or overhead) to transfer in both directions, but in only one direction at a time. Contrast with *simplex* and *full-duplex* channels.

half-duplex communication A system that allows information to transfer in both directions, but in only one direction at a time.

handshake A software/hardware synchronization method where control and status signals go in both directions between the transmitter and receiver. The communication is interlocked, meaning each device will wait for the other.

hard real-time system One that can guarantee that a process will complete a critical task within a certain specified time period. In data-acquisition systems, "hard real-time" means there is an upper bound on the latency between when a sample is supposed to be taken (every 1/fs) and when the ADC converter is actually started. "Hard real-time" also implies that no ADC sample is missed.

hexadecimal A number system that uses the base 16.

HiZ A logic state where the output device does not drive high or pull low. The outputs of open-collector and tristate devices can be in the floating state. Same as *floating*.

hold time When latching data into a device with a rising or falling edge of a clock, the hold time is the time after the active edge of the clock that the data must continue to be valid. See *setup time*.

hysteresis A condition in which the output of a system depends not only on the input, but also on the previous outputs—e.g., a transducer that follows a different response curve when the input is increasing from that when the input is decreasing.

I/O bound A situation where the input or output device is slower than the software. In other words, it takes longer for the I/O device to process data than it does for the software to process data. The bandwidth is limited by I/O speed.

I/O device A computer component capable of bringing information from the external environment into the computer (input device) or of sending data out from the computer to the external environment (output device).

I/O port A hardware device that connects the computer with external components.

IEEE488 A medium-speed, handshaking, parallel-I/O standard, used for desktop instruments.

immediate An addressing mode where the operand is a fixed data or address value.

incremental control system A control system where the actuator has many possible states, and the system increments or decrements the actuator value according to whether the error is positive or negative.

indexed An addressing mode where the data or address value for the instruction is located in memory pointed to by an index register.

infinite impulse response filter (IIR) is a digital filter where the output is a function of an infinite number of past data samples, usually as a function of previous filter outputs.

inherent An addressing mode where there is no operand or where the operand is implied (not explicitly stated).

input capture A mechanism to set a flag and capture the current time (TCNT value) on the rising, falling, or rising&falling edge of an external signal. The input capture event can also request an interrupt.

instruction register (IR) Register in the control unit that contains the opcode for the current instruction.

instrument An instrument is the code injected into a program for debugging or profiling. This code is usually extraneous to the normal function of a program and can be temporary or permanent. Instruments injected during interactive sessions are considered to be temporary, because these instruments can be removed simply by terminating a session. Instruments injected in source code are considered to be permanent, because removal requires editing and recompiling the source. An example of a temporary instrument occurs when the debugger replaces a regular op code with the `swi` instruction. This temporary instrument can be removed dynamically by restoring the original opcode. A print statement added to your source code is an example of a permanent instrument, because removal requires editing and recompiling.

instrument A system that collects information; same as data acquisition system.

instrumentation The process of injecting or inserting an instrument.

interrupt A software/hardware synchronization method where the hardware causes a special software program (interrupt handler) to execute when its hardware operation is complete. The software usually can perform other work while waiting for the hardware.

interrupt flag A status bit that is set by the hardware to signify that an external event has occurred.

interrupt mask An individual control bit that, if programmed to 1, will cause an interrupt request when the associated flag is set. Same as *arm*.

interrupt service routine (ISR) Program that runs as a result of an interrupt.

interrupt vector 16-bit values at the end of memory, specifying where the software should execute after an interrupt request. There is a unique interrupt vector for each type of interrupt.

IRQ An interrupt mechanism on the 6805, 6808, 6811, and 6812.

latch As a noun, it means a register. As a verb, it means to store data into the register.

latched input port An input port where the signals are latched (saved) on an edge of an associated strobe signal. E.g., PORTCL on the 6811 is a latched input port.

latency In this book, latency usually refers to the response time of the computer to external events—for example, the time between new inputs becoming available and the time the input is read by the computer; or the time between an output device's becoming idle and the time the computer writes new data to it. There can also be a latency for an I/O device, which is the response time of the external I/O device hardware to a software command. For a data-acquisition system, the time between the time when it should be sampled and the time the ADC is started.

LCD Liquid Crystal Display, where the computer controls the reflectance or transmittance of the liquid crystal; characterized by flexible display patterns, low power, low cost, and slow speed.

LED Light Emitting Diode, where the computer controls the electrical power to the diode; characterized by simple display patterns, medium power, and high speed.

linear filter Means the output is a linear combination of its inputs.

little endian Mechanism for storing multiple-byte numbers such that the least significant byte exists first (in the smallest memory address). Contrast with *big endian*.

loader System software that places the object code into the microcomputer's memory. If the object code is stored in EPROM, the loader is also called a EPROM programmer.

logic analyzer A hardware debugging tool that allows you to visualize many digital logic signals versus time. Real logic analyzers have at least 32 channels and can have up to 200 channels, with sophisticated techniques for triggering, saving, and analyzing the real-time data. In **TExaS,** logic analyzers have only 8 channels and simply plot digital signals versus time.

LSB The least significant bit in a number system is the bit with the smallest significance, usually the rightmost bit. With signed or unsigned integers, the significance of the LSB is 1.

maintenance Process of verifying, changing, correcting, enhancing, and extending a system.

mark The digital value *true* or logic 1. Contrast with *space*.

mask As a verb, mask is the operation that selects certain bits out of many bits, using the logical *and* operation. The bits that are not being selected will be cleared to zero. When used as a noun, mask refers to the specific bits that are being selected.

measurand A signal measured by a data-acquisition system.

membership sets Fuzzy-logic variables that can take on a range of values from *true* (255) to *false* (0).

memory A computer component capable of storing and recalling information.

memory-mapped I/O A configuration where the I/O devices are interfaced to the computer in a manner identical to the way memories are connected. From an interfacing perspective, I/O devices and memory modules share the same bus signals; from a programmer's point of view, the I/O devices exist as locations in the memory map, and I/O device access can be performed via any of the memory-access instructions.

microcomputer An electronic device capable of performing input/output functions and containing a microprocessor, memory, and I/O devices.

microcontroller A single-chip microcomputer like the Motorola 6811, Motorola 6816, Intel 8051, Intel 8096, PIC16, or the Texas Instruments TMS370.

mnemonic The symbolic name of an operation code, such as `ldaa`, `psha`, or `stx`.

monitor or **debugger window** A monitor is a debugger feature that allows us to (passively) view strategic software parameters during the real-time execution of our program. An effective monitor is one that has minimal effect on the performance of the system. When debugging software on a windows-based machine, we can often set up a debugger window that displays the current value of certain software variables.

MSB The most significant bit in a number system is the bit with the greatest significance, usually the leftmost bit. If the number system is signed, then the MSB signifies positive (0) or negative (1).

multiple-access circular queue (MACQ) A data structure used in data-acquisition systems to hold the current sample and a finite number of previous samples.

multi-threaded A system with multiple threads (e.g., main program and one or more interrupt service routines) that cooperate towards a common overall goal.

negative logic A signal where the *true* value has a lower voltage than the *false* value. In digital logic, *true* is 0, and *false* is 1; in TTL logic, *true* is less than 0.7 volts, and *false* is greater than 2 volts; in RS232 protocol, *true* is -12 volts, and *false* is $+12$ volts. Contrast with *positive logic*.

nibble 4 binary bits or 1 hexadecimal digit.

nonatomic Software execution that can be divided or interrupted. Most lines of C code require multiple assembly-language instructions to execute; therefore, an interrupt can occur in the middle of a line of C code.

noninstrusive A collection of information where the collection process itself does not affect the parameters being measured. Nonintrusivenes is the characteristic or quality of a debugger that allows the software/hardware system to operate normally, as if the debugger did not exist. Intrusiveness is used as a measure of the degree of perturbation caused in program performance by an instrument. For example, a print statement added to your source code, or single-stepping, is very intrusive, because each significantly affects the real-time interaction of the hardware and software: When a program interacts with real-time events, the performance is significantly altered. On the other hand, an instrument that outputs a strategic variable on the LED's (that requires just 50μs to execute) is much less intrusive. A logic analyzer that passively monitors the address and data bus is completely nonintrusive. An in-circuit emulator is also nonintrusive, because the software input/output relationships will be the same with and without the debugging tool.

noninvasive/invasive Noninvasiveness is the characteristic or quality of a debugger that makes the order of invocation immaterial. The debugger and the user program coexist in the same global environment. On the other hand, an invasive debugger requires the user program to execute within an environment defined by the debugger. The debugger is invoked first, and the program is then loaded either by the debugger or by the user from within the debugger. Invasiveness is also a measure of the degree of source-code modification to debug or monitor a program. A resident debugger like BUFFALO is invasive, because it exists first and then your program is loaded on top of it. This program-development environment is very invasive, because the 6811 in expanded mode with BUFFALO is very different from the eventual single-chip embedded application. An in-circuit emulator is noninvasive because it can coexist (be added or deleted) from our system without changing the way the system runs.

nonlinear filter A filter whose output is not a linear combination of its inputs. For example, median, minimum, and maximum are examples of nonlinear filters.

nonreentrant A software module that, once started by one thread, cannot be interrupted and executed by a second thread. A nonreentrant module usually involves nonatomic accesses to global variables or I/O ports: read–modify–write, write followed by read, or a multistep write.

nonvolatile A condition where information is not lost when power is removed. When power is restored, then the information is in the state that existed when the power was removed.

nonvolatile RAM Read/write storage that achieves its long-term storage ability because it includes a battery.

normalized The mantissa of a normalized floating-point number is greater than or equal to 1, but strictly less than 2.

null cycle A computer bus cycle that fetches data at address $FFFF, but the data is not used.

Nyquist Theorem If a input signal is captured by an ADC at the regular rate of fs samples/sec, then the digital sequence can accurately represent the 0-to-0.5fs frequency components of the original signal.

object code Programs in machine-readable format created by the compiler or assembler. The S19 records are examples of object code.

odd parity A communication protocol where the number of ones in the data plus a parity bit is an odd number. Contrast with *even parity*.

op code, **opcode**, or **operation code** A specific instruction executed by the computer. The combination of opcode and operand completely specifies the function to be performed. In assembly-language programming, the opcode is represented by its mnemonic, such as `ldaa`. During execution, the opcode is stored as a machine code loaded in memory. The `ldaa` instruction with immediate addressing has the machine code $86.

open collector A digital-logic output that has two states low and HiZ.

open-loop control system A control system that does not include sensors to measure the current state variables.

operand The second part of an instruction; it specifies either the data or the address for that instruction. An assembly instruction typically has an opcode (e.g., `ldaa`) and an operand (e.g., #55). Instructions that use inherent addressing mode have no operand field.

operating system System software for managing computer resources and facilitating common functions, such as input/output, memory management, and file management.

oscilloscope A hardware debugging tool that allows you to visualize one or two analog signals versus time. In **TExaS,** oscilloscopes can plot up to 8 channels.

output compare A mechanism to cause a flag to be set and an output pin to change when the TCNT matches a preset value. The output-compare event can also request an interrupt.

overflow An error that occurs when the result of a calculation exceeds the range of the number system. For example, with 8-bit unsigned integers, $200+57$ will yield the incorrect result 1.

overflow When TCNT increments from $FFFF back to $0000, setting the TOF flag. This overflow event can also request an interrupt.

overrun error An error that occurs when the receiver gets a new frame but the data register and shift register already have information. One frame is lost.

parallel port A port where all signals are available simultaneously. In this book, the parallel ports are 8 bits wide.

PC relative An addressing mode where the effective address is calculated by its position relative to the current value of the program counter.

performance debugging or **profiling** The process of acquiring or modifying timing characteristics and execution patterns of a program and the process of instrumenting a program for such purposes is called performance debugging or profiling. Contrast with *functional debugging.*

periodic polling A software/hardware synchronization method that is a combination of interrupts and busy-waiting. An interrupt occurs at a regular rate (periodic) independent of the hardware status. The interrupt handler checks (polls) the hardware device to determine whether its operation is complete. The software usually can perform other work while waiting for the hardware.

personal computer system A small general-purpose computer system having a price low enough for individual people to afford and used for personal tasks.

PID Controller A control system where the actuator output depends on a linear combination of the current error (P), the integral of the error (I), and the derivative of the error (D).

poll for zeros and ones An interrupt handler that checks both for the interrupt flag and for the presence of other ones and zeros in the status register.

polling A software function to discover which of the potential sources requested the interrupt.

port External pins through which the microcomputer can perform input/output. Same as *I/O port.*

positive logic A signal where the *true* value has a higher voltage than the *false* value. In digital logic, *true* is 1, and *false* is 0; in TTL logic, *true* is greater than 2 volts, and *false* is less than 0.7 volts; in RS232 protocol, *true* is $+12$ volts, and *false* is -12 volts. Contrast with *negative logic.*

precision For an input signal, it is the number of distinguishable input signals that can be reliably detected by the measurement. For an output signal, it is the number of different output parameters that can be produced by the system. For a number system, precision is the number of distinct or different values of a number system in units of "alternatives". The precision of a number system is also the number of binary digits required to represent all its numbers in units of "bits".

priority When two requests for service are made simultaneously, priority determines the order in which to process them.

private Can be accessed only by software modules in its local group.

private variable A global variable that is used by a single module and is not shared with other modules.

process The execution of software that does not necessarily cooperate with other processes.

producer–consumer A multithreaded system where the producers generate new data and the consumers process or output the data.

program counter (PC) A register in the processor that points to the memory containing the instruction to execute next.

PROM Same as EPROM. Programmable read-only memory that is nonvolatile and requires external devices to erase and reprogram. It is usually erased by UV light.

promotion Increasing the precision of a number for convenience or to avoid overflow errors during calculations.

pseudo-code A shorthand for describing a software algorithm. The exact format is not defined, but many programmers use their favorite high-level language syntax (like C) without paying rigorous attention to the punctuation.

pseudo op Operations included in the program that are not executed by the computer at run time, but rather are interpreted by the assembler during the assembly process. Same as *assembly directive.*

public Can be accessed by any software module.

public variable A global variable that is shared by multiple modules.

pulse-width modulation A technique to deliver a variable signal (voltage, power, energy) by using an on/off signal with a variable percentage of time the signal is on (duty cycle). Same as *variable duty cycle.*

qualitative DAS A DAS that collects information not in the form of numerical values, but rather in the form of the qualitative senses—e.g., sight, hearing, smell, taste and touch. A qualitative DAS may also detect the presence or absence of conditions.

quantitative DAS A DAS that collects information in the form of numerical values.

RAM Random Access Memory, a type of memory where information can be stored and retrieved easily and quickly. It is volatile, so the information is lost when power is removed.

range Includes both the smallest possible and the largest possible signal (input or output). The difference between the largest and smallest input that can be measured by the instrument. The units are in the units of the measurand. When precision is in alternatives, range=precision•resolution.

real-time A system that can guarantee an upper bound (worst case) on latency. A system where time-critical operations occur when needed.

recursion A programming technique where a function calls itself.

reentrant A software module that can be started by one thread, then interrupted and executed by a second thread. A reentrant module allow both threads to execute the desired function properly.

registers High-speed memory located in the processor. The registers on the 6811/6812 are CCR, A, B, X, Y, SP, and PC.

reproducibility (or **repeatability**) A parameter specifying how consistent over time the measurement is when the input remains fixed.

reset vector The 16-bit value at memory locations $FFFE and $FFFF, specifying where the software should start after power is turned on or after a hardware reset.

resolution For an input signal, it is the smallest change in the input parameter that can be detected reliably by the measurement. For an output signal, it is the smallest change in the output parameter that can be produced by the system; range equals precision times resolution.

ritual Software, usually executed once at the beginning of the program, that defines the operational modes of the I/O ports.

ROM Read-Only Memory, a type of memory where information is programmed into the device once, but can be accessed quickly. It is of low cost, must be purchased in high volume, and can be programmed only once. See also *EPROM, EEPROM,* and *flash EEPROM.*

roundoff The error that occurs in a fixed-point or floating-point calculation when the least significant bits of an intermediate calculation are discarded so the result can fit into the finite precision.

sampling rate The rate at which data is collected in a data-acquisition system.

Scan or **ScanPoint** Any instrument used to produce a side effect without causing a break (halt) is a scan. Therefore, a scan may be used to gather data passively or to modify functions of a program. Examples include software added to your source code that simply outputs or modifies a global variable without halting. A ScanPoint is triggered in a manner similar to a breakpoint, but a ScanPoint simply records data at that time without halting execution.

scope A logic analyzer or an oscilloscope—hardware debugging tools that allows you to visualize multiple digital or analog signals versus time.

SCSI Small Computer Systems Interface, a high-speed, handshaking, parallel-I/O standard.

sensitivity The sensitivity of a transducer is the slope of the output versus input response. The sensitivity of a qualitative DAS that detects events is the percentage of actual events that are properly recognized by the system.

serial communication A process where information is transmitted one bit at a time.

serial communications interface (SCI) A device to transmit data with asynchronous serial communication protocol (same as UART and ACIA).

serial peripheral interface (SPI) Device to transmit data with synchronous serial communication protocol.

serial port An I/O port where the bits are input or output one at a time.

setup time When latching data into a device with a rising or falling edge of a clock, the setup time is the time before the active edge of the clock that the data must be valid. See *hold time.*

signed two's-complement binary A mechanism to represent signed integers where 1 followed by all 0's is the most negative number, all 1's represents the value -1, all 0's represents the value 0, and 0 followed by all 1's is the largest positive number.

sign-magnitude binary A mechanism to represent signed integers where the most significant bit is set if the number is negative, and the remaining bits represent the magnitude as an unsigned binary.

simple poll An interrupt handler that simply checks the interrupt flag.

simplex channel Hardware that allows bits (information, error-checking, synchronization, or overhead) to transfer in only one direction, used to synchronize the receiver shift register with the transmitter clock. Contrast with *half-duplex* and *full-duplex* channels.

simplex communication A system that allows information to transfer in only one direction.

simulator A simulator is a software application, like **TExaS,** that simulates or mimics the operation of a processor or computer system. Most simulators recreate only simple I/O ports and often do not effectively duplicate the real-time interactions of the software/hardware interface. On the other hand, they do provide a simple and interactive mechanism to test software. Simulators are especially useful when learning a new language, because they provide more control and access to the simulated machine than one normally has with real hardware.

single-pole switch One switch that acts independently from other switches in the system. Contrast with *double-pole.*

single-throw switch A switch with two contact connections. The two contacts may be connected or disconnected. Contrast with *double-throw.*

software interrupt vector The 16-bit value at memory locations $FFF6 and $FFF7, specifying where the software should go after executing a software interrupt instruction, `swi`.

software maintenance Process of verifying, changing, correcting, enhancing, and extending software.

source code Programs in human-readable format, created with an editor.

space A digital value of *false* or logic 0. Contrast with *mark.*

specificity The specificity of a transducer is the relative sensitivity of the device to the signal of interest versus the sensitivity of the device to other unwanted signals. The specificity of a qualitative DAS that detects events is the percentage of events detected by the system that are actually true events.

stabilize The process of stabilizing a software system involves specifying all its inputs. When a system is stabilized, the output results are consistently repeatable. Stabilizing a system with multiple real-time events, such as input devices and time-dependent conditions, can be difficult to accomplish. It often involves replacing input hardware with sequential reads from an array or disk file.

stack Last-in-first-out data structure located in RAM and used to save information temporarily.

stack pointer (SP) A register in the processor that points to the RAM location of the stack.

start bit Overhead bit(s) specifying the beginning of the frame, used in serial communication to synchronize the receiver shift register with the transmitter clock. See also *stop bit, even parity* and *odd parity.*

static RAM Volatile read/write storage built from three transistors having fast speed, and not requiring refresh. Contrast with *dynamic RAM.*

stepper motor A motor that moves in discrete steps.

stop bit Overhead bit(s) specifying the end of the frame, used in serial communication to separate one frame from the next. See also *start bit, even parity* and *odd parity.*

string A sequence of ASCII characters, usually terminated by a zero.

symbol table A mapping from symbolic names to their corresponding 16-bit addresses, generated by the assembler in pass one and displayed in the listing file.

synchronous protocol A system where the two devices share the same clock.

tachometer A sensor that measures the revolutions per second of a rotating shaft.

thread The execution of software that cooperates with other threads. A thread embodies the action of the software. One concept describes a thread as the sequence of operations including the input and output data.

throughput The information transfer rate; the amount of data transferred per second. Same as *bandwidth.*

time constant The time to reach 63.2% of the final output after the input is instantaneously increased.

time profile and **execution profile** Time profile refers to the timing characteristic of a program; execution profile refers to the execution pattern of a program.

transducer A device that converts one type of signal into another type.

tristate The state of a tristate logic output when HiZ or not driven.

tristate logic A digital logic device that has three output states: low, high, and HiZ.

truncation The act of discarding bits as a number is converted from one format to another.

two-pole switch Two separate and complete switches, which are activated together; same as *double-pole*.

two's complement A number system used to define signed integers. The MSB defines whether the number is negative (1) or positive (0). To negate a two's complement number, one first complements (flips from 0 to 1 or from 1 to 0) each bit, then adds 1 to the number.

unary operation A function that produces its result from a single input parameter; for example, negate, increment, and decrement are unary operations.

unbuffered I/O The hardware and software are tightly coupled, so that each waits for each other during the transmission of data.

unipolar stepper motor A stepper motor where the current flows in only one direction (on/off) along the interface wires; a stepper with 5 or 6 interface wires. Compare to *bipolar stepper motor*.

universal asynchronous receiver/transmitter (UART) A device to transmit data with asynchronous serial communication protocol; same as SCI and ACIA.

unnormalized An unnormalized floating-point number has a mantissa value less than one. The mantissa of a normalized floating point number is greater than or equal to 1, but strictly less than 2.

unsigned binary A mechanism to represent unsigned integers where all 0's represents the value 0 and all 1's represents the largest positive number.

vector An address at the end of memory containing the location of the interrupt service routines. See also *reset vector* and *interrupt vector*.

volatile A condition where information is lost when power is removed.

vulnerable window Locations within a software module; if an interrupt were to occur at one of these locations, then an error could occur (e.g., data lost, corrupted data, or program crash). Same as *critical section*.

white noise A fundamental noise in resistive devices arising from the uncertainty about the position and velocity of individual molecules. Same as Johnson noise and thermal noise.

word Two bytes containing 16 bits. Same as *double byte*.

workstation A powerful general-purpose computer system having a price in the $10K to $50K range and used for handling large amounts of data and performing many calculations.

XIRQ A high-priority interrupt mechanism available on the 6811 and 6812.

XON/XOFF A protocol used by printers to feed the printer status back to the computer. XOFF is sent from the printer to the computer in order to stop data transfer; XON is sent from the printer to the computer in order to resume data transfer.

Appendix 3

Solutions Manual

A3.1 Checkpoint Solutions

Checkpoint 1.1: The software in a digital watch must maintain time by using a real-time clock, output the current time on the LCD display, respond to button pushes (updating parameters as required), and check on whether the current time matches the alarm time.

Checkpoint 1.2: It failed because employees were rewarded for poor behavior. It is much better to punish poor behavior and reward good behavior.

Checkpoint 1.3: An embedded system is a microcomputer with mechanical, chemical, and electrical devices attached to it, programmed for a specific dedicated purpose, and packaged as a complete system.

Checkpoint 1.4: A microcomputer is a small computer that includes a processor, memory, and I/O devices.

Checkpoint 1.5: Typical input devices include the keys on the keyboard, the mouse and its buttons, joystick, CD reader, and microphone. The floppy disk can be used for input and output.

Checkpoint 1.6: Typical output devices include the LEDs on the keyboard, monitor, speaker, printer, CD burner, and speaker. The floppy disk can be used for input and output.

Checkpoint 1.7:

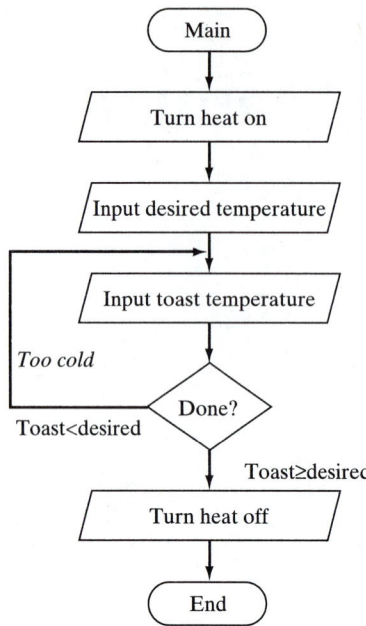

Checkpoint 1.8: Both terms refer to parameters of a system, but the differences lie in the level of detail used to describe the parameter. A requirement is usually defined in general terms, whereas a specification entails detailed engineering rigor. A requirement often refers to an objective of the system, while a specification describes how well the actual device works.

Checkpoint 2.1: 3½ decimal digits is about 2000 alternatives, which is about 11 bits.

Checkpoint 2.2: The rule of thumb says 2^{60} is about 10^{18}, which is 18 decimal digits. 2^4 is 16, which is about 1½ decimal digits. Together, we have 19½ decimal digits.

Checkpoint 2.3: The computer engineering answer is 2^{41}; the scientific answer is $2*10^{12}$.

Checkpoint 2.4: $45

Checkpoint 2.5: $CAB

Checkpoint 2.6: %01000000

Checkpoint 2.7: %011000111111

Checkpoint 2.8: Four bits for each hex digit, so 24 bits.

Checkpoint 2.9: Rotary switches, like the wiper speed switches on cars, have multiple positions.

Checkpoint 2.10: 64+32+8+2 is 106.

Checkpoint 2.11: 4*16+5 is 69.

Checkpoint 2.12: 45=32+8+4+1, so 45 = %00101101 = $2D.

Checkpoint 2.13: 200=128+64+8, so 200 = %11001000 = $C8.

Checkpoint 2.14: −128+62+32+8+2 = −22.

Checkpoint 2.15: They are the same, because bit 7 is zero.

Checkpoint 2.16: −45 = −128+64+16+2+1 = %11010011 = $D3.

Checkpoint 2.17: Because the range of 8-bit signed numbers is −128 to +127.

Checkpoint 2.18: $30.

Checkpoint 2.19: `$48656C6C6F20576F726C6400`.

Checkpoint 2.20: Each four bits represent a single decimal digit, `%00100101`.

Checkpoint 2.21: 8192+64+32+8+2=8298.

Checkpoint 2.22: 1*4096+2*256+3*16+4=4660.

Checkpoint 2.23: 1234 = 4*256+13*16+2 = $04D2.

Checkpoint 2.24: 10000 = 8192+1024+512+256+16 = %0010011100010000.

Checkpoint 2.25: 1*4096+2*256+3*16+4 = 4660.

Checkpoint 2.26: −32768 + 2*4096+11*256+12*16+13 = −21555.

Checkpoint 2.27: 1234 = 4*256+13*16+2 = $04D2.

Checkpoint 2.28: −10000 = −32768 +16384+4096+2048+128+64+32+16 = %1101100011110000.

Checkpoint 2.29: `$3456`.

Checkpoint 2.30:

```
ldaa  10
staa  20
```

Checkpoint 2.31:

```
ldaa   #%11000111
staa   PORTB
```

Checkpoint 2.32: Out = not(eor(A,B))

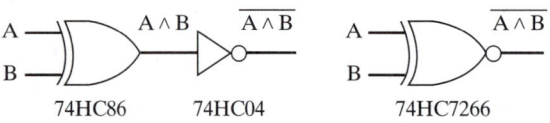

```
    A ∧ B      ‾‾‾‾‾               ‾‾‾‾‾
A                A ∧ B    A         A ∧ B
B                         B
   74HC86     74HC04         74HC7266
```

Checkpoint 2.33:

```
ldaa   PORTB
anda   #$FD clear bit 1
staa   PORTB
```

Checkpoint 2.34:

```
ldaa   PORTB
oraa   #$80 set bit 7
staa   PORTB
```

Checkpoint 2.35:

```
anda   #$0F result is 0 to 9
```

Checkpoint 2.36:

Checkpoint 2.36:

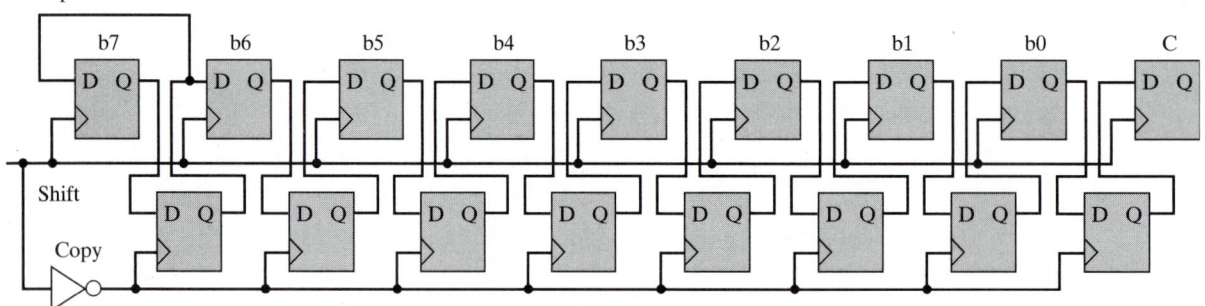

Checkpoint 2.37: 9 bits. If $0 \leq x \leq 255$ and $0 \leq y \leq 255$ then $0 \leq (x+y) \leq 510$.

Checkpoint 2.38: 9 bits. If $-128 \leq x \leq 127$ and $-128 \leq y \leq 127$ then $-256 \leq (x+y) \leq 254$.

Checkpoint 2.39: 16 bits. If $0 \leq x \leq 255$ and $0 \leq y \leq 255$ then $0 \leq (x*y) \leq 65025$.

Checkpoint 2.40: 16 bits. If $-128 \leq x \leq 127$ and $-128 \leq y \leq 127$ then $-16256 \leq (x*y) \leq 16384$.

Checkpoint 2.41: $-100+64 = -36$, so V=0. $156+64 = 220$, so C=0. N=1 (negative) and Z=0 (not zero).

Checkpoint 2.42: $-100-64 = -164$, so V=1 (overflow). $156-64 = 92$, so C=0. N=0 (positive) and Z=0 (not zero).

Checkpoint 2.43: $-56-64 = -120$, so V=0. $200-64 = 136$, so C=0. N=1 (negative) and Z=0 (not zero).

Checkpoint 2.44: $-56+64 = 8$, so V=0. $200+64 = 264$, so C=1 (overflow). N=0 (positive) and Z=0 (not zero).

Checkpoint 2.45: To avoid ambiguity, because there may be two solutions to the equation $N = M*Q+R$.

Checkpoint 2.46: $\pi*1000$ is about 3141.59, so the variable integer part is 3142.

Checkpoint 2.47: $\pi*256$ is about 804.2477, so the variable integer part is 804.

Checkpoint 2.48: $F = (461 \bullet C)/256+32$.

Checkpoint 2.49: $y = (1000 \bullet x - 53 \bullet x_1 + 1000 \bullet x_2 + 51 \bullet y_1 - 903 \bullet y_2)/1000$.

Checkpoint 2.50: Simply, $R_3 = (R_1*R_2)/(R_1+R_2)$, because the fixed constants factor out.

Checkpoint 2.51: You need more than 80 bits to convert, and most calculators do have this precision.

Checkpoint 3.1: 2^{16}, which is 65536 locations.

Checkpoint 3.2: The 6812 can access individual bytes. Each 8-bit byte has a unique address.

Checkpoint 3.3: CU means control unit, DMA means direct memory access, BIU stands for bus interface unit, and ALU means arithmetic logic unit.

Checkpoint 3.4: I=1 will disable interrupts.

Checkpoint 3.5: The addressing mode defines the format for the effective address for that instruction.

Checkpoint 3.6: They look different, but they are the same instruction, and therefore perform exactly the same operation when executed. The only difference is programming style. We use hexadecimal for addresses and numbers that have binary meanings.

Checkpoint 3.7: `ldaa #36` loads the value 36. `ldaa 36` loads the 8-bit memory contents at address 36.

Checkpoint 3.8: `ldx #$0801` loads the value $0801. `ldx $0801` loads the 16-bit memory contents at address $0801.

Checkpoint 3.9: Both perform the same operation, which is to load the memory contents of $0012–$0013 into Register X. The extended-mode instruction does require more program memory space and will execute a little less quickly.

Checkpoint 3.10: For both the 6811 and 6812, the `bra` instruction is two bytes long. The destination address is equal to the location of instruction, so the **rr** field is −(size of the instruction), which is −2 (or $FE). The opcode is $20, so the object code is $20,$FE.

Checkpoint 3.11: Because of the pipeline, the real 6812 first executes its operation, then fetches the opcodes for the next instruction.

Pipeline before	cycle	address	R/W	LSTRB	data	executing
$B608,$0006	r	$0800	Read	1	$55	ldaa $0800
$B608,$0006	O	$FFFE	Read	0	****	
$B608,$0006	P	$F004	Read	0	$F000	
$0006,$F000	P	$F006	Read	0	****	jmp loop
	P	$F000	Read	0	$B608	
$B608	P	$F002	Read	0	$0006	

Checkpoint 3.12: For either the 6811 or 6812:

```
xgdx
addd    #100
xgdx
```

For just the 6812:

```
leax    100,x
```

Checkpoint 3.13: The values in Registers A and B range from 0 to 255, so their product must range from 0 to 65025. Therefore, all potential results will fit in Register D.

Checkpoint 3.14: dividend=quotient*divisor+remainder.

Checkpoint 3.15:

```
ldaa    N
ldab    #7
mul         D=7*N
ldx     #31
idiv        X=(5*N)/31
xgdx
stab    M
```

Checkpoint 3.16:

```
ldaa    N
asla
asla
staa    M
```

Checkpoint 3.17:

```
anda    #$0F
andb    #$3C
```

Checkpoint 3.18:

```
xgdx
oraa  #$12
orab  #$34
xgdx
```

Checkpoint 3.19:

```
ldaa  N
anda  #$EF clear bit 4
staa  N
```

Checkpoint 3.20:

```
ldaa  M
psha
ldaa  N
staa  M
pula
staa  N
```

Checkpoint 3.21:

```
      ldaa N
      cmpa #25
      bne  next      ; skip if not equal
      jsr  isEqual   ; N==25
next
```

Checkpoint 3.22: It determines whether to use the `bls` or `ble` instruction:

```
;unsigned version            ;signed version
   ldaa N                        ldaa N
   cmpa #25                      cmpa #25
   bls  next      ; skip if N<=25   ble  next      ; skip if N<=25
   jsr  isGreater ; N>25            jsr  isGreater ; N>25
next                          next
```

Checkpoint 3.23:

```
loop ldd  N
     cpd  #25
     beq  next      ; stop when N==25
     jsr  body      ; execute body of the while loop
     bra  loop
next
```

Checkpoint 3.24:

```
     leay 10,x
```

Checkpoint 3.25:

```
; MC68HC711E9                          ; MC68HC912B32
      org  $0000 ;RAM                        org  $0800 ;RAM
cnt   rmb  1      ;global             cnt   rmb  1      ;global

      org  $B600 ;EEPROM                     org  $8000 ;EEPROM
const fcb  5       ;amount to add     const fcb  5       ;amount to add

      org  $D000 ;ROM                 init  movb #$FF,DDRC ;outputs
init  ldaa #$FF                             clr  cnt
      staa DDRC   ;outputs                   rts
      clr  cnt                        main  lds  #$0C00 ;sp=>RAM
      rts                                   bsr  init
main  lds  #$01FF ;sp=>RAM            loop  ldaa cnt
      bsr  init                             staa PORTC      ;output
loop  ldaa cnt                              adda const
      staa PORTC ;output                     staa cnt
      adda const                            bra  loop
      staa cnt
      bra  loop                             org  $FFFE      ;EEPROM
                                            fdb  main       ;reset vector
      org  $FFFE    ;ROM
      fdb  main     ;reset vector
```

Checkpoint 4.1: The assembler builds the symbol table.

Checkpoint 4.2: The assembler creates the object code and listing file.

Checkpoint 4.3: RAM can be altered, so the information in RAM can be changed. The contents of ROM are fixed (programmed once), so the information cannot be changed.

Checkpoint 4.4: The *opcodes* are shown in blue; the *pseudo-op codes* are shown in gray.

Checkpoint 4.5: $2+4*6/5+1=2+24/5+1=2+4+1=6+1=7$

Checkpoint 4.6: Using parentheses makes the expression easier for humans to understand, so, from a style perspective, the second one is better.

Checkpoint 4.7: With `ldaa #5+6`, the addition occurs in the IBM-PC at assemble time; with the second case, the addition occurs in the 6811/6812 microcomputer at run time.

Checkpoint 4.8: The assembler must create the symbol table during pass 1, so it must know the size of each assembly line during pass 1. A forward reference would prevent the assembler from knowing how many bytes to allocate during pass 1. It would probably create a phasing error.

Checkpoint 4.9: The loader stores the data $F0, $80, $20, $FE at addresses $F026–$F029.

Checkpoint 4.10: No, the checksum should be 66, making it

```
S10CF0DD80CCF0E616F4BC20B866
```

Checkpoint 4.11: `S10556781234E6`

Checkpoint 5.1:

```
#define TCNT *(unsigned short volatile *)(0x100E)
```

Checkpoint 5.2:

```
#define TCNT *(unsigned short volatile *)(0x0084)
```

Checkpoint 5.3: MC68HC711E9

Input only	Output only	Bidirectional
PA2-PA0	PA6-PA4	PA7, PA3
PE7-PE0	PB7-PB0	PC7-PC0
		PD5-PD0

Checkpoint 5.4: MC68HC812A4 8-bit bidirectional ports are A, B, C, D, H, J, S, and T. MC68HC912B32 8-bit bidirectional ports are A, B, P, S, and T.

Checkpoint 5.5:

```
*6811                          *6812
* Make PC1 an output pin       * Make PC1 an output pin
InitPC1 ldaa  DDRC             InitPC1 bset DDRC, #$02
        oraa  #$02                     rts
        staa  DDRC             * Make PC2 an output pin
        rts                    InitPC2 bset DDRC, #$04
* Make PC2 an output pin               rts
InitPC2 ldaa  DDRC
        oraa  #$04
        staa  DDRC
        rts
```

Checkpoint 5.6: P=4 and BR=8, so the baud rate is 2,000,000/(16*4*8) = 3906 bits/sec.

Checkpoint 5.7: P*BR is 2,000,000/16/2604 = 48, so P=3, BR=16. BAUD =$14.

Checkpoint 5.8: The baud rate is 8,000,000/(16*35) = 14286 bits/sec.

Checkpoint 5.9: BR is 8,000,000/16/2604 = 192, so SC0BD =192.

Checkpoint 5.10: $V_{out} = 5*D_{in}/256$ or $V_{out} = 5*D_{in}/255$, which is the same as $V_{out} = D_{in}/51$.

Checkpoint 5.11: $D_{out} = 256*V_{in}/5$ or $D_{out} = 51*V_{in}$.

Checkpoint 5.12:

```
* 6811                          * 6812
Init  ldaa #3 PR1=PR0=1         Init movb #$80,TCSR
      staa TMSK2                     movb #5,TMSK2 PR210=101
      rts                            rts
```

Checkpoint 5.13: 4 μs*65536 is about 256 ms.

Checkpoint 5.14: The frequency resolution is 1 Hz. It means that, if the input frequency changes by 1 Hz, then the system will be able to detect the change.

Checkpoint 5.15: Change the call to `Wait1sec` so that it waits only 1 ms.

Checkpoint 5.16: The 6811 pulse-width resolution is 32 μsec; the 6812 pulse-width resolution is 8 μsec. It means that if the input pulse width changes by this amount, then the system will be able to detect the change.

Checkpoint 5.17: The 6811 function will return the value 1234.5/32, or 38; the 6812 function will give the result 1234.5/8, or 154.

Checkpoint 6.1: The resistor is used to create the +5V (logic high) when the switch is open.

Checkpoint 6.2: The capacitor is used to filter out fluctuations caused by switch bounce. The 22Ω resistor prevents sparks from occurring when the capacitor is discharged when the switch is pressed.

Checkpoint 6.3: The resistor is (5−1.5−0.5V)/5mA, which is 600Ω.

Checkpoint 6.4: Yes, connect 6811 TxD to 6812 RxD, and 6811 RxD to 6812 TxD. The grounds also must be connected.

Checkpoint 6.5: Yes, if the 6811 and 6812 are far apart or if there is noise in the wires connecting them together. No, if they are on the same circuit board or in the same box.

Checkpoint 6.6: $9906

Checkpoint 6.7: The resolution is 5V/256, which is about 20 mV.

Checkpoint 6.8: It appears that key 'f' is pressed, regardless of whether 'f' is actually pressed.

Checkpoint 6.9: Because a piano must be able to distinguish multiple keys pressed simultaneously.

Checkpoint 6.10: The actuator value would wrap around. For example, if the power were already 255 and the controller wished to add more power, then the power would be incremented, giving the result 0. Similarly, if the power were already 0 and the controller wished to subtract more power, then the power would be decremented, giving the result 255. In both cases, the exact opposite result occurred.

Checkpoint 6.11: The controller would respond twice as fast, but the error would double. It is faster because it achieved the correct value twice as fast. The error is worse because the system loses fine control over the applied power.

Checkpoint 6.12: The physical motion of the human operator is not of constant velocity.

Checkpoint 6.13:

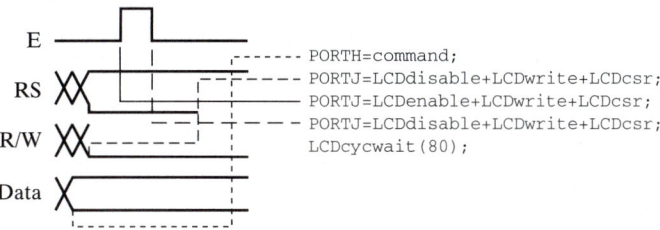

Checkpoint 7.1: In general, the presence of a minimally intrusive debugging instrument itself only has minimal effect on the parameter being measured. One criterion is that the total execution time required to perform the instrumentation be small compared to the execution times of the original target operation.

Checkpoint 7.2: Run-time debugging can be activated in final production systems. Run-time debugging is quicker to activate/deactivate because an edit/assemble/download cycle is not needed. Assembly-time debugging produces a final production system that runs faster and requires less memory.

Checkpoint 7.3: It makes it easy to activate/deactivate debugging in final production systems. The debugging statements can be used to evaluate the proper operation of the system.

Checkpoint 7.4: In Section 3.2.4, we translated the sequence of executed 6811 instructions into the memory cycles required to execute the instructions. In Section 3.2.6, we translated the sequence of executed 6812 instructions into the memory cycles required to execute the instructions. The translation is 1–1, so, for this situation, we perform the reverse translation and convert the memory cycles into the sequence of instructions.

Checkpoint 7.5:

```
* 6811 or 6812
SetB3 psha
      ldaa PORTB
      oraa #$08
      staa PORTB
      pula rts
ClrB3 psha
      ldaa PORTB
      anda #$F7
      staa PORTB
      pula
      rts
```

```
* 6812
SetB3 bset  PORTB,#$08
      rts
ClrB3 bclr  PORTB,#$08
      rts
```

Checkpoint 8.1: Define the variable within the scope of the function—for example,

```
void MyFunction(void){ short myLocalVariable;
}
```

Checkpoint 8.2: Define the variable outside the scope of the function—for example,

```
short myGlobalVariable; // accessible by all programs
void MyFunction(void){
}
```

Checkpoint 8.3: A `static` local has permanent allocation, which means it maintains its value from one call to the next. It is still local, meaning it is accessible only from within the function. In the following example, *count* contains the number of times `MyFunction` is called:

```
void MyFunction(void) { static short count;
   count++;
}
```

A `static` global has reduced scope. Regular globals can be accessed from any function in the system; a `static` global can be accessed only by functions within the same file. `Static` globals are private. Functions can be static also, meaning they can be called only from other functions in the file—for example,

```
static short myPrivateGlobalVariable; // accessible by this file only
void static MyPrivateFunction(void){
}
```

Checkpoint 8.4: A `const` global is read only. It is allocated in the ROM portion of memory.

Checkpoint 8.5: You don't explicitly define the size of the stack. It can be implicitly calculated if the global variables are contiguously allocated starting at the beginning of RAM and the stack pointer is initialized to the end of RAM; then the stack size is defined as the total RAM bytes minus the size of global variables. Some systems have a heap, used by `malloc` and `free`, that also is defined in RAM.

Checkpoint 8.6: The local variable name can be used in other subroutines (just as in C).

Checkpoint 8.7:

```; 6811 or 6812```
``` pshx```
``` psha   allocate 3 locals```
```; access locals```
``` ins```
``` ins```
``` ins   deallocate```
	```; 6812 only```
	``` leas -3,sp allocate 3 locals```
	
	```; access locals```
	``` leas 3,sp deallocate locals```

Checkpoint 8.8: ICC11 and ICC12 use Register D for the first parameter passed, but it is more efficient to use Register X to pass the reference pointer.

```
;Reg X is pointer
next ldd  0,x  access
     addd #1
     std  0,x  (*pt)++
     cpd  #200
     bne  skip
     ldd  #0
     std  0,x  (*pt)=0
skip rts
main lds  #$0c00 $00FF for 6811
     ldx  #0
     pshx           allocate angle
     jsr  Stepper_Init
loop jsr  Stepper_Step
     tsx            pointer to angle
     jsr  next
     bra  loop
```

Checkpoint 8.9: ICC11 and ICC12 use Register D for the first parameter and the stack for the second parameter, but it is more efficient to use Registers X and Y to pass the reference pointers.

```
      org  $0800    (0 for 6811)
xx    rmb  2
yy    rmb  2
myX   rmb  2
myY   rmb  2
      org  $F000    ($D000 for MC68HC711E9)
;Reg X is Xpt pointer
;Reg Y is Ypt pointer
where ldd  xx
      std  0,x  write access, (*Xpt) = xx
      ldd  yy
      std  0,y  write access, (*Ypt) = yy
      rts
main  lds  #$0c00  $00FF for 6811
      ldx  #myX
      ldy  #myY
      jsr  where
      stop
```

Checkpoint 8.10: ICC11 and ICC12 also uses Register D to pass/return a single parameter.

```
;Reg D is both input and output
next  addd #1
      cpd  #200
      bne  skip
      ldd  #0
skip  rts
```

```
main    lds    #$0c00    $00FF for 6811
        ldx    #0
        pshx             allocate angle
        jsr    Stepper_Init
loop    jsr    Stepper_Step
        tsx              pointer to angle
        ldd    angle,x copy of the variable
        jsr    next
        std    angle,x
        bra    loop
```

Checkpoint 8.11: Each time requires 3 bytes. It is called five times, so 15 bytes are required.

Checkpoint 8.12: Whichever is easiest to understand. The more natural one.

Checkpoint 8.13: Convert Register B to Register D.

Checkpoint 8.14: The Z-bit is set if and only if the most significant byte of the result is zero.

Checkpoint 8.15: Add (long) before one of the terms. For example P=(N+M)/10; becomes P=(long)(N+M)/10;

Checkpoint 8.16: The one on the left is a simple 32-bit divide by 2, which is 100001/2=50000. The one on the right rounds up, because the number is odd; 100001/2+1=50001.

Checkpoint 8.17:

```
;6811                                    ;6812
Sub    ldaa   0,y    ;subtract top two   Sub    ldaa   1,y+   ;add top two,pop
       iny           ;pop                       suba   0,y
       suba   0,y                               staa   0,y
       staa   0,y                               rts
       rts                               Div    ldab   1,y+   ;top is divisor
Div    ldab   0,y    ;top is divisor            tfr    b,x    ; X is divisor
       clra                                     ldab   0,y    ;dividend
       iny           ;pop                       clra
       xgdx          ; X is divisor             idiv
       ldab   0,y    ;dividend                  tfr    x,b    ;RegB is quotient
       clra                                     stab   0,y
       idiv                                     rts
       xgdx          ;RegD is quotient
       stab   0,y
       rts
```

Checkpoint 8.18: Write an interrupt handler for it similar to `swi`. Define the vector at $FFF8 to point to this ISR. To execute an illegal instruction, you could use `fcb`. On the 6811:

```
fdb $1A40 ; illegal instruction
```

On the 6812:

```
fdb $1840 ; illegal instruction
```

Via **TExaS,** illegal instructions can be created with a `trap` opcode; see `trap.rtf`.

```
trap #$40 ; illegal instruction
```

Checkpoint 8.19: To make it easier to understand.

Checkpoint 9.1: Coupling can be parameters passed, globals shared, functions called, and I/O devices shared.

Checkpoint 9.2: They are permanently allocated and can be accessed by any function.

Checkpoint 9.3: Use a "does-call" graph, so only one module actually does access the I/O device.

Checkpoint 9.4: Change to `#define DEBUG 0` to remove the printf statements.

```
#define DEBUG 1
unsigned int Median(unsigned int u1,unsigned int u2,unsigned int u3) {
unsigned int result;
#if DEBUG
  printf("The inputs are %d, %d, %d.\n",u1,u2,u3);
#endif
  if(u1>u2)
    if(u2>u3)    result=u2;    // u1>u2,u2>u3            u1>u2>u3
      else
        if(u1>u3) result=u3;   // u1>u2,u3>u2,u1>u3      u1>u3>u2
        else      result=u1;   // u1>u2,u3>u2,u3>u1      u3>u1>u2
  else
    if(u3>u2)    result=u2;    // u2>u1,u3>u2            u3>u2>u1
      else
        if(u1>u3) result=u3;   // u2>u1,u2>u3,u1>u3      u2>u1>u3
        else      result=u1;   // u2>u1,u2>u3,u3>u1      u2>u3>u1
#if DEBUG
  printf("The median is %d.\n",result);
#endif
  return(result);;}
```

Checkpoint 9.5: Public functions have an underline (e.g., `SCI_OutString`).

Checkpoint 9.6: Local variables begin with a lower-case letter—e.g., `myKey`. Global variables begin with an upper-case letter—e.g., `TheKey`.

Checkpoint 9.7: There are long conditional branch instructions (`lbeq`), long branch to subroutine (`lbsr`), and PC relative addressing (`ldaa data,PCR`).

Checkpoint 10.1: A global variable is permanently defined and can be accessed by multiple programs. A local variable is temporarily defined and can be accessed only by a single program.

Checkpoint 10.2: Yes, you can power the RAM with a battery.

Checkpoint 10.3: Because the compiler must make two copies of the structure. The initial copy is stored in ROM, and the working copy is stored in RAM. At start-up, the initial copy is transferred into the working copy.

Checkpoint 10.4: The precision is 32 bits or 4 bytes, the length is 5, and the total size is 20 bytes.

Checkpoint 10.5: The termination code might exist in the data itself.

Checkpoint 10.6:

```
J     set  0     Column index
READ pshb        Allocate and initialize
     ldab #3     number of columns
     mul         3*I
     addb j,s    3*I+J
     ldaa B,X    read value at M[I,J]
     ins
     rts
```

Checkpoint 10.7:

```
READ aslb        Reg B = 2*J
     abx         Reg X = base + 2*J
     ldaa A,X    Read a byte from array(I,J)
     rts
```

Checkpoint 10.8:

```
Access pshb          allocate and initialize J
       pshx          allocate and initialize Base
       ldab #n       Reg B is the number of columns
       mul           n*I
       addb J,s      n*I+J
       adca #0
       lsld          2*(n*I+J)
       addd Base,s   Reg D = Base + 2*(n*I+J)
       xgdx          Reg X = Base + 2*(n*I+J)
       leas 3,s
       rts
```

Checkpoint 10.9:

```
Access psha          allocate and initialize I
       pshx          allocate and initialize Base
       ldaa #m       Reg A is the number of rows
       mul           m*J
```

```
addb I,s    m*J+I
adca #0
addd Base,s  Reg D = Base + m*J+I
xgdx         Reg X = Base + m*J+I
leas 3,s
rts
```

Checkpoint 10.10: The compiler will skip bytes to align the 16-bit elements.

$F950	$15
$F952	$0001
$F954	$82
$F956	"PORTB",0,0,0,0,0

Checkpoint 10.11: It is allocated contiguously, and a simple equation can be used to calculate the address of each entry.

Checkpoint 10.12: $a = (v[0] - 2 \ast v[1] + v[2])/\delta t^2$

Checkpoint 10.13: $a = (d[0] + 3 \ast d[1] - 3 \ast d[2] - d[3])/(6 \ast \delta t^2)$

Checkpoint 10.14: There are two ways a FIFO can get full. If the average rate at which data is put in the FIFO is larger than the average rate data is got from the FIFO, then the FIFO will always fill up. If the temporary rate at which data is put in the FIFO is larger than the temporary rate data is got from the FIFO, and the FIFO size is small, then the FIFO might fill up.

Checkpoint 10.15: Tree.

Checkpoint 10.16: Yes, there can be no cycles.

Checkpoint 10.17: General graph.

Checkpoint 10.18: A double binary tree: one for first name, one for last name.

Checkpoint 10.19: Create it using a 20-byte size, and just waste the space when less than 20 bytes are requested.

Checkpoint 10.20:

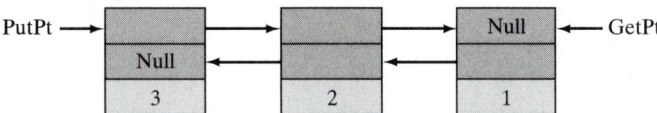

Checkpoint 11.1: 1) external event occurs, 2) the condition is armed, 3) the microcomputer is enabled.

Checkpoint 11.2: Clear the I bit in the CCR with a `cli` instruction.

Checkpoint 11.3: 1) Finish the instructions, 2) push registers onto the stack, 3) get interrupt vector, 4) execute the ISR, 5) execute `rti` instruction, returning to the problem executing at the time of the interrupt.

Checkpoint 11.4: Yes, if the average service rate exceeds the average request rate. No, if the average service rate is less than the average request rate.

Checkpoint 11.5: Convert global to local

```
AVE pshx              short AVE(short first, short second){ short num;
    tsx                 num = first;
    addd 0,x            return (num+second)/2;
    asrd              }
    pulx
    rts
```

Checkpoint 11.6: It would reenable interrupts too early.

Checkpoint 11.7: The software would crash, because the ISR would interrupt over and over again.

Checkpoint 11.8: The software could disarm.

Checkpoint 11.9: Set PIOC to $40.

Checkpoint 11.10: Set KPOLJ to $00.

Checkpoint 11.11: The TxFifo is empty, and there is nothing to print.

Checkpoint 11.12: The software would crash because the ISR is spinning with interrupts disabled, and no software has the chance to empty the RxFifo.

Checkpoint 11.13: For the 6811, set Baud equal to $33. For the 6812, set Baud equal to 417.

Checkpoint 11.14: Set PACTL to 2.

Checkpoint 11.15: Set RATE to 10000.

Checkpoint 11.16: Set RTICTL to $85.

Checkpoint 11.17: Set RATE to 10000.

Checkpoint 11.18: The time required to execute a 6812 instruction varies from 1 to 13 (emacs) cycles. At 8 MHz, this is 1.625 μsec. The `rev`, `revw`, and `wav` instructions are not considered because they can be interrupted.

Checkpoint A1.1: An assembly source code is software in human-readable format created with an editor; it gives explicit instructions to the computer. It includes both specific functions to perform and the order in which to perform them.

Checkpoint A1.2: An oscilloscope plots voltage level on the y-axis versus time on the x-axis.

Checkpoint A1.3: A logic analyzer also plots voltage level versus time, but the difference is that the voltage level takes on only the two digital logic states, high and low.

Checkpoint A1.4: Object code is software in machine-readable format loaded into memory; it gives explicit instructions to the computer.

A3.2 Tutorial Solutions

Answer 1.1: Read PORTA, then read PORTB (or the two data values together), and write the result to PORTC.

Answer 1.2: Four switches are connected to the PORTA input, four switches are connected to the PORTB input, and four LED lights are connected to the PORTC output.

Answer 1.3: The red arrow signifies the location of the executing program.

Answer 1.4: The different parts of the program are automatically colored. For example, labels, comments, opcodes, pseudo-op codes, and operands have unique colors. Also, misspelled opcodes and other syntax errors are flagged in red.

Answer 1.5: MC68HC812A4.

Answer 1.6: To look back in time to see the previously executed instructions.

Answer 1.7: **TExaS** simulates a small number of hardware devices as well as the software.

Answer 2.1: The value is outside the range of 8-bit numbers. It will be truncated, and only the lower 8 bits will be stored; RegA will be improperly set to 0.

Answer 2.2: As an 8-bit binary number, -1 is 11111111; thus, RegA will be properly set to 11111111_2. Because the format is set to unsigned decimal (d), RegA will be shown as 255 in the ViewBox. If the format were set to signed decimal ($+$d), RegA would have been shown as -1 in the ViewBox. This points out the reality that, within the computer all data is binary. The data can be interpreted in a multitude of ways.

Answer 2.3: The value is outside the range of 16-bit numbers. It will be truncated, and only the lower 16 bits will be stored; RegX will be improperly set to 0.

Answer 2.4: As a 16-bit binary number, -1 is 1111111111111111; thus, RegX will be properly set to 1111111111111111_2. Because the format is set to unsigned decimal (D), RegX will be shown as 65535 in the ViewBox. If the format were set to signed decimal ($+$D), RegX would have been shown as -1 in the ViewBox.

Answer 2.5: As an 8-bit binary number, 128 is 10000000; thus, RegA will be properly set to 10000000_2. Since the format is set to signed decimal (+d), RegA will be shown as -128 in the ViewBox. If the format were set to unsigned decimal (d), RegA would have been shown as 128 in the ViewBox.

Answer 2.6: The value is outside the range of 8-bit numbers. It will be truncated, and only the lower 8 bits will be stored; RegA will be improperly set to 127.

Answer 2.7: As a 16-bit binary number, 32768 is 1000000000000000; thus, RegX will be properly set to 1000000000000000_2. Because the format is set to signed decimal (+D), RegX will be shown as -32768 in the ViewBox. If the format were set to unsigned decimal (D), RegX would have been shown as 32768 in the ViewBox.

Answer 2.8: The value is outside the range of 16-bit numbers. It will be truncated, and only the lower 16 bits will be stored; RegX will be improperly set to 32767.

Answer 2.9: R is the result loaded into Register A. N is set if the result is negative, N=R7. Z is set if the result is zero. Z=not(R7)•not(R6)•not(R5)•not(R4)•not(R3)•not(R2)•not(R1)•not(R0). V is cleared to 0. There can be no overflow. The carry bit is not altered.

Answer 2.10:

```
$0F&$85 = $05   N=0,   Z=0
$0F|$85 = $8F   N=1,   Z=0
0F $85  = $8A   N=1,   Z=0
~$0F    = $F0   N=1,   Z=0
```

Answer 2.11:

```
155>>1  = 77    N=0,   Z=0,   C=1
50<<1   = 100   N=0,   Z=0,   C=0
96+64   = 160   N=1,   Z=0,   C=0
224+64  = 32    N=0,   Z=0,   C=1
160-64  = 96    N=0,   Z=0,   C=0
32-64   = 224   N=1,   Z=0,   C=1
```

Answer 2.12:

```
-101>>1 = -51   N=1,   Z=0,   V=0
-50<<1  = -100  N=1,   Z=0,   V=0
-32+64  = +32   N=0,   Z=0,   V=0
96+64   = -96   N=1,   Z=0,   V=1
32-64   = -32   N=1,   Z=0,   V=0
-96-64  = +96   N=0,   Z=0,   V=1
```

Answer 3.1: Solutions for the MC68HC711E9 and MC68HC812A4 are already shown in Program 3.5.

Answer 3.2: Here is the simulator output.

```
MC68HC711E9                                   MC68HC812A4
Opcode fetch    R 0xD000 0x8E from ROM        Opcode fetch    R 0xF000 0xCF from EEPROM
Operand fetch   R 0xD001 0x01 from ROM        Operand fetch   R 0xF001 0x0C from EEPROM
Operand fetch   R 0xD002 0xFF from ROM        Operand fetch   R 0xF002 0x00 from EEPROM
0xD000 lds #$01FF                             0xF000  lds #$0C00
A=$00 PC=$D003 PORTC=$03 SP=$01FF             A=$00 PC=$F003 PORTC=$03 SP=$0C00
Opcode fetch    R 0xD003 0x8D from ROM        Opcode fetch    R 0xF003 0x07 from EEPROM
Operand fetch   R 0xD004 0x0D from ROM        Operand fetch   R 0xF004 0x0B from EEPROM
Null Cycle      R 0xFFFF 0x00 from ROM        Stack store msbW 0x0BFE 0xF0 to RAM
Dummy PC fetch R 0xD012 0x86 from ROM         Stack store lsbW 0x0BFF 0x05 to RAM
Stack store msbW 0x01FE 0xD0 to RAM          0xF003  bsr $F010
Stack store lsbW 0x01FF 0x05 to RAM          A=$00 PC=$F010 PORTC=$03 SP=$0BFE
0xD003 bsr $D012                              Opcode fetch    R 0xF010 0x86 from EEPROM
A=$00 PC=$D012 PORTC=$03 SP=$01FD             Operand fetch   R 0xF011 0xF0 from EEPROM
Opcode fetch    R 0xD012 0x86 from ROM        0xF010  ldaa #$F0
Operand fetch   R 0xD013 0xF0 from ROM        A=$F0 PC=$F012 PORTC=$03 SP=$0BFE
0xD012  ldaa #$F0                             Opcode fetch    R 0xF012 0x5A from EEPROM
A=$F0 PC=$D014 PORTC=$03 SP=$01FD             Operand fetch   R 0xF013 0x06 from EEPROM
Opcode fetch    R 0xD014 0xB7 from ROM        Store using EARW 0x0006 0xF0 to I/O port
Operand fetch   R 0xD015 0x10 from ROM        0xF012  staa $06
Operand fetch   R 0xD016 0x07 from ROM        A=$F0 PC=$F014 PORTC=$03 SP=$0BFE
Store using EARW 0x1007 0xF0 to I/O port      Opcode fetch    R 0xF014 0x3D from EEPROM
0xD014 staa $1007                             Stack read msb R 0x0BFE 0xF0 from RAM
A=$F0 PC=$D017 PORTC=$03 SP=$01FD             Stack read lsb R 0x0BFF 0x05 from RAM
Opcode fetch    R 0xD017 0x39 from ROM        0xF014  rts
Dummy PC fetch R 0xD018 0xFF from ROM         A=$F0 PC=$F005 PORTC=$03 SP=$0C00
Dummy SP read  R 0x01FD 0x00 from RAM         Opcode fetch    R 0xF005 0x96 from EEPROM
Stack read msb R 0x01FE 0xD0 from RAM         Operand fetch   R 0xF006 0x04 from EEPROM
Stack read lsb R 0x01FF 0x05 from RAM         Fetch using EARR 0x0004 0x03 from I/O
0xD017 rts                                    0xF005  ldaa $04
A=$F0 PC=$D005 PORTC=$03 SP=$01FF             A=$03 PC=$F007 PORTC=$03 SP=$0C00
Opcode fetch    R 0xD005 0xB6 from ROM        Opcode fetch    R 0xF007 0x41 from EEPROM
Operand fetch   R 0xD006 0x10 from ROM        0xF007  coma
Operand fetch   R 0xD007 0x03 from ROM        A=$FC PC=$F008 PORTC=$03 SP=$0C00
Fetch using EARR 0x1003 0x03 from I/O         Opcode fetch    R 0xF008 0x48 from EEPROM
0xD005 ldaa $1003                             0xF008  lsla
A=$03 PC=$D008 PORTC=$03 SP=$01FF             A=$F8 PC=$F009 PORTC=$03 SP=$0C00
Opcode fetch    R 0xD008 0x43 from ROM        Opcode fetch    R 0xF009 0x48 from EEPROM
Dummy PC fetch R 0xD009 0x48 from ROM         0xF009  lsla
0xD008 coma                                   A=$F0 PC=$F00A PORTC=$03 SP=$0C00
A=$FC PC=$D009 PORTC=$03 SP=$01FF             Opcode fetch    R 0xF00A 0x48 from EEPROM
Opcode fetch    R 0xD009 0x48 from ROM        0xF00A  lsla
Dummy PC fetch R 0xD00A 0x48 from ROM         A=$E0 PC=$F00B PORTC=$03 SP=$0C00
0xD009 lsla                                   Opcode fetch    R 0xF00B 0x48 from EEPROM
A=$F8 PC=$D00A PORTC=$03 SP=$01FF             0xF00B  lsla
Opcode fetch    R 0xD00A 0x48 from ROM        A=$C0 PC=$F00C PORTC=$03 SP=$0C00
Dummy PC fetch R 0xD00B 0x48 from ROM         Opcode fetch    R 0xF00C 0x5A from EEPROM
0xD00A lsla                                   Operand fetch   R 0xF00D 0x04 from EEPROM
A=$F0 PC=$D00B PORTC=$03 SP=$01FF             Store using EARW 0x0004 0xC0 to I/O port
Opcode fetch    R 0xD00B 0x48 from ROM        0xF00C  staa $04
```

continued

```
Dummy PC fetch R 0xD00C 0x48 from ROM        A=$C0 PC=$F00E PORTC=$C3 SP=$0C00
0xD00B lsla                                  Opcode fetch   R 0xF00E 0x20 from EEPROM
A=$E0 PC=$D00C PORTC=$03 SP=$01FF             Operand fetch  R 0xF00F 0xF5 from EEPROM
Opcode fetch   R 0xD00C 0x48 from ROM        0xF00E  bra $F005
Dummy PC fetch R 0xD00D 0xB7 from ROM         A=$C0 PC=$F005 PORTC=$C3 SP=$0C00
0xD00C lsla
A=$C0 PC=$D00D PORTC=$03 SP=$01FF
Opcode fetch   R 0xD00D 0xB7 from ROM
Operand fetch  R 0xD00E 0x10 from ROM
Operand fetch  R 0xD00F 0x03 from ROM
Store using EARW 0x1003 0xC0 to I/O port
0xD00D staa $1003
A=$C0 PC=$D010 PORTC=$C3 SP=$01FF
Opcode fetch   R 0xD010 0x20 from ROM
Operand fetch  R 0xD011 0xF3 from ROM
Null Cycle     R 0xFFFF 0x00 from ROM
0xD010 bra $D005
A=$C0 PC=$D005 PORTC=$C3 SP=$01FF
```

Answer 4.1:

```
              ; MC68HC11A8                            ; MC68HC812A4
              PORTC equ  $1003                        PORTC equ  $0004
              DDRC  equ  $1007                         DDRC  equ  $0006
              org   $E000                              org   $F000
immediate     main lds  #$00FF       immediate     main lds  #$0C00
PC relative        bsr  init         PC relative        bsr  init
extended      loop staa PORTC        direct        loop staa PORTC
inherent           inca              inherent           inca
PC relative        bra  loop         PC relative        bra  loop
immediate     init ldaa #$FF         immediate     init ldaa #$FF
extended           staa DDRC         direct             staa DDRC
inherent           clra              inherent           clra
inherent           rts               inherent           rts
                   org  $FFFE                           org  $FFFE
                   fdb  main                            fdb  main
```

Answer 4.2: Determine the size of each instruction in Program 4.6.

```
          ; MC68HC11A8                        ; MC68HC812A4
          PORTC equ  $1003                    PORTC equ  $0004
          DDRC  equ  $1007                     DDRC  equ  $0006
          org   $E000                          org   $F000
3    main lds  #$00FF        3    main lds  #$0C00
2         bsr  init          2         bsr  init
3    loop staa PORTC         2    loop staa PORTC
1         inca               1         inca
2         bra  loop          2         bra  loop
2    init ldaa #$FF          2    init ldaa #$FF
3         staa DDRC          2         staa DDRC
1         clra               1         clra
1         rts                1         rts
          org  $FFFE                   org  $FFFE
2         fdb  main          2         fdb  main
```

Answer 4.3: Use the sizes and the `org` to find the address of each line.

```
              ; MC68HC11A8                        ; MC68HC812A4
  $1003       PORTC equ  $1003      $0004        PORTC equ  $0004
  $1007       DDRC  equ  $1007      $0006        DDRC  equ  $0006
              org   $E000                        org   $F000
  $E000       main  lds  #$00FF     $F000        main  lds  #$0C00
  $E003       bsr   init            $F003        bsr   init
  $E005       loop  staa PORTC      $F005        loop  staa PORTC
  $E008       inca                  $F007        inca
  $E009       bra   loop            $F008        bra   loop
  $E00B       init  ldaa #$FF       $F00A        init  ldaa #$FF
  $E00D       staa  DDRC            $F00C        staa  DDRC
  $E010       clra                  $F00E        clra
  $E011       rts                   $F00F        rts
              org   $FFFE                        org   $FFFE
  $FFFE       fdb   main            $FFFE        fdb   main
```

The symbol table is

```
  DDRC   $1007              DDRC   $0006
  PORTC  $1003              PORTC  $0004
  init   $E00B              init   $F00A
  loop   $E005              loop   $F005
  main   $E000              main   $F000
```

Answer 4.4: The object code for each instruction in Program 4.6 is

```
                 ; MC68HC11A8                           ; MC68HC812A4
  $1003          PORTC equ  $1003      $0004            PORTC equ  $0004
  $1007          DDRC  equ  $1007      $0006            DDRC  equ  $0006
                 org   $E000                            org   $F000
  $E000 8E00FF   main  lds  #$00FF     $F000 CF0C00     main  lds  #$0C00
  $E003 8D06     bsr   init            $F003 0705       bsr   init
  $E005 B71003   loop  staa PORTC      $F005 5A04       loop  staa PORTC
  $E008 4C       inca                  $F007 42         inca
  $E009 20FA     bra   loop            $F008 20FB       bra   loop
  $E00B 86FF     init  ldaa #$FF       $F00A 86FF       init  ldaa #$FF
  $E00D B71007   staa  DDRC            $F00C 5A06       staa  DDRC
  $E010 4F       clra                  $F00E 87         clra
  $E011 39       rts                   $F00F 3D         rts
                 org   $FFFE                            org   $FFFE
  $FFFE $E000    fdb   main            $FFFE $F000      fdb   main
```

Answer 4.5: During pass 1, we would not be able to tell whether the `staa PORTC` and `staa DDRC` instructions should use direct or extended addresses. Without knowing the size of these instructions, we couldn't build the symbol table during pass 1. This uncertainty is the cause of the "Phasing Error."

Answer 5.1: Flowcharts are a convenient way to describe computer algorithms.

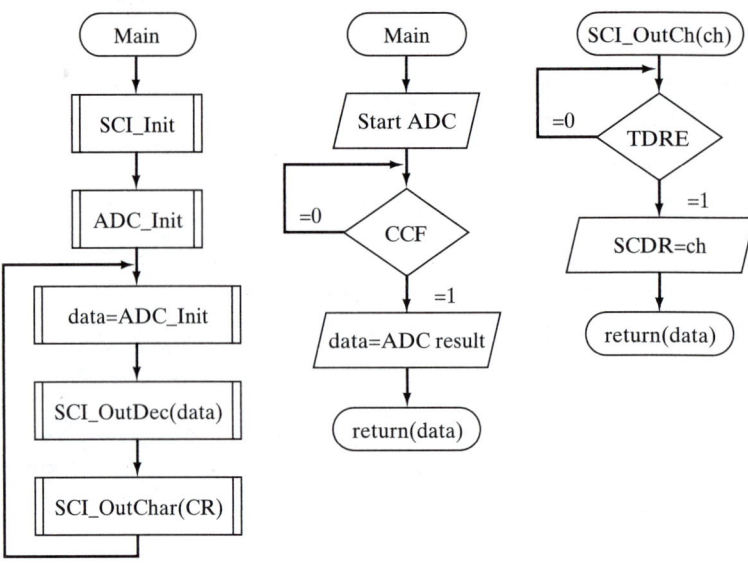

Answer 5.2: Call graphs are used to visualize software hierarchy.

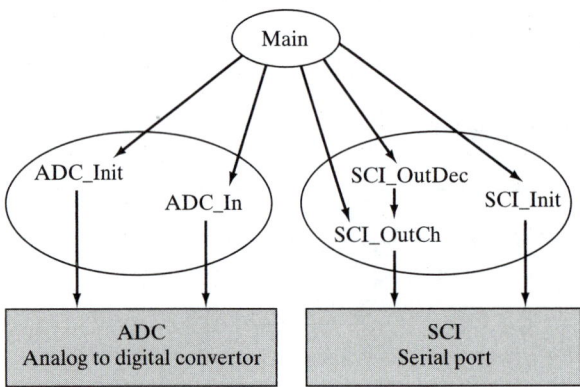

Answer 5.3: A data-flow graph illustrates the data as it flows from input to output.

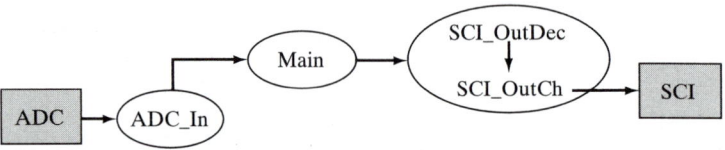

Answer 5.4: For the 6811, the TCNT default period is 500 ns. For the 6812, the TCNT default period is 125 ns.

Answer 5.5: For the 6811, 2000 cycles occur in 1 ms. For the 6812, 8000 cycles occur in 1 ms.

Answer 5.6: For the 6811, $3411 - 2661$ is 750 cycles, which is 375 μsec. For the 6812, $3110 - 2309$ is 801 cycles, which is about 100 μsec.

Answer 5.7: For the 6811, you need to wait $2000 - 750 = 1250$ cycles; in particular, call the `Wait` function shown in Program 5.16 with Register D=1250. For the 6812, you need to wait $8000 - 800 = 7200$ cycles; in particular, call the `Wait` function shown in Program 5.16 with Register D=7200.

```
*6811
main  lds  #$00FF
      jsr  SCI_Init
      jsr  ADC_Init
loop  ldaa #0
      jsr  ADC_In
      clra
      jsr  SCI_OutDec
      ldaa #CR
      jsr  SCI_OutCh
      ldd  #1250
      bsr  Wait
      bra  loop
* Reg D is the time to wait in cycles
Wait  addd TCNT end of wait time
wloop cpd  TCNT EndT-TCNT<0 when EndT<TCNT
      bpl  wloop
      rts
```

```
*6812
main  lds  #$0C00
      jsr  SCI_Init ; initialize SCI
      jsr  ADC_Init
loop  ldaa #0       ; sample channel 0
      jsr  ADC_In
      jsr  SCI_OutDec
      ldaa #CR
      jsr  SCI_OutCh
      ldd  #7200
      bsr  Wait
      bra  loop
* Reg D is the time to wait in cycles
Wait  addd TCNT end of wait time
wloop cpd  TCNT EndT-TCNT<0 when EndT<TCNT
      bpl  wloop
      rts
```

Answer 6.1: There are other possibilities, but these are the ones used most commonly in this book.

	MC68HC11A8	MC68HC711E9	MC68HC812A4	MC68HC912B32
RAM	$0000-$00FF	$0000-$01FF	$0800-$0BFF	$0800-$0BFF
ROM	$E000-$FFFF	$D000-$FFFF	$F000-$FFFF	$8000-$FFFF

Answer 6.2:

	6811	6812
PORTC	$1003	$0004
DDRC	$1007	$0006

Answer 6.3:
Sw6=off, Sw5=on, Sw4=off, Sw3=off, Sw2=off, Sw1=on, Sw0=on.

Answer 6.4:
```
key   fcb   %00001111   ; key code
```

Answer 7.1: There are two stack pushes, instead of a push at the top and a pull at the bottom. Change the second `pshx` to `pulx`.

Answer 7.2: A breakpoint dumps the **ViewBox** contents into the `TheLog.rtf` file and halts simulation. A ScanPoint dumps the **ViewBox** contents into the `TheLog.rtf` file, but continues simulation.

Answer 7.3: **Input** is the variable part of a binary fixed-point number. **Meaning** is the effective value of the **Input. Meaning**=**Input**/16. **Sqrt** is the correct result: **Sqrt** =sqrt(**Meaning**). **Truth** is the variable part of the binary fixed-point result: **Truth** is **Sqrt***16, rounded to the closest integer. **Output** is the result of the software. As you can see, there are no errors in the program for these input values.

Input	Meaning	Sqrt	Truth	Output	Error
0	0.0000	0.0000	0	0	0
1	0.0625	0.2500	4	4	0
2	0.1250	0.3536	6	6	0
3	0.1875	0.4330	7	7	0
4	0.2500	0.5000	8	8	0
5	0.3125	0.5590	9	9	0
6	0.3750	0.6124	10	10	0
7	0.4375	0.6614	11	11	0
8	0.5000	0.7071	11	11	0
9	0.5625	0.7500	12	12	0

Answer 7.4: The `psha` instruction caused the stack overflow.

Answer 7.5: The overflowing stack pushed data into the global RAM area.

Answer 7.6: For the 6811 system, it is $3B00. For the 6812 system, it is $3000.

Answer 7.7: The instructions are as follows:

```
clrb
tsta
beq
pula
rts
```

Answer 7.8: Purple means that instruction has been executed only once. It means the error occurred the first time this subroutine was called.

Answer 8.1: The 6811 static efficiency is 55 bytes. The 6812 static efficiency is 49 bytes.

Answer 8.2: The 6811 execution time is 31 to 46 cycles. The 6812 execution time is 19 to 31 cycles.

Answer 8.3: See file `Chap8as.rtf`.

Answer 8.4: The 6811 static efficiency is 99 bytes. The 6811 execution time is 31 to 66 cycles. The 6812 static efficiency is 89 bytes. The 6812 execution time is 19 to 47 cycles.

Answer 8.5: The 6811 static efficiency is 22 bytes. The 6812 static efficiency is 19 bytes.

Answer 8.6: The 6811 execution time is always 26 cycles. The 6812 execution time is always 15 cycles.

Answer 8.7: See file `Chap8bs.rtf`.

Answer 8.8: The 6811 static efficiency is 33 bytes. The 6811 execution time is still always 26 cycles. The 6812 static efficiency is 32 bytes. The 6812 execution time is still always 15 cycles.

Answer 9.1:

> $F000-$F3FF Module 1, plus `swi` vector
> $F400-$F7FF Module 2, plus `trap` vector
> $F800-$FF00 Module 3, plus reset vector

Answer 9.2: Yes, module 2 can be moved, the object code for modules 1 and 3 would not change.

Answer 9.3: Divide up RAM, just as ROM was divided.

Answer 9.4: Using registers.

Answer 9.5: Using registers, but the value must be stored on the stack, so that when `rti` is executed the register is loaded.

Answer 9.6: See `Chap9s.rtf`.

Answer 10.1: The 6811 search speed is 12 cycles minimum, 32 cycles maximum. The 6812 search speed is 7 cycles minimum, 23 cycles maximum.

Answer 10.2: Let n be the entry number, where the first entry is n=0. The 6811 search speed is approximately $12 + 5*n$. When there are 100 commands, the average number would be 50, so the estimated average search speed is 262 cycles. The 6812 search speed is approximately $7 + 4*n$. When there are 100 commands, the average number would be 50, so the estimated average search speed is 207 cycles.

Answer 10.3: See the file `Chap10as.rtf`.

Answer 10.4: The 6811 search speed is 32 cycles minimum, 32 cycles maximum. The 6812 search speed is 15 cycles minimum, 15 cycles maximum. Notice that the lookup speed is not data-dependent.

Answer 10.5: The 6811 search speed is fixed at 32 cycles. The 6812 search speed is fixed at 15 cycles.

Answer 10.6: See the file `Chap10bs.rtf`.

Answer 10.7: The 6811 search speed is 35 cycles minimum, 119 cycles maximum. The 6812 search speed is 19 cycles minimum, 67 cycles maximum.

Answer 10.8: Let n be the entry number, where the first entry is n=0. The 6811 search speed is approximately $35 + 21*n$. When there are 100 commands, the average number would be 50, so the estimated average search speed is 1085 cycles. The 6812 search speed is approximately $19 + 12*n$. When there are 100 commands, the average number would be 50, so the estimated average search speed is 619 cycles.

Answer 10.9: See the file `Chap10cs.rtf`.

Answer 10.10: The 6811 search speed is 17 cycles minimum, 97 cycles maximum. The 6812 search speed is 15 cycles minimum, 63 cycles maximum.

Answer 10.11: Let n be the entry number, where the first entry is n=0. The 6811 search speed is approximately $17 + 20*n$. When there are 100 commands, the average number would be 50, so the estimated average search speed is 1017 cycles. The 6812 search speed is approximately $15 + 12*n$. When there are 100 commands, the average number would be 50, so the estimated average search speed is 615 cycles.

Answer 10.12: See the file `Chap10ds.rtf`.

Answer 10.13: The 6811 search speed is 28 cycles minimum, 78 cycles maximum. The 6812 search speed is 18 cycles minimum, 50 cycles maximum.

Answer 10.14: Let n be the tree depth, where the root is n=0. The 6811 search speed is approximately $28 + 50*n$. When there are 100 commands, the average depth would be 6, so the estimated average search speed is 328 cycles. The 6812 search speed is approximately $18 + 32*n$. When there are 100 commands, the average depth would be 6, so the estimated average search speed is 210 cycles.

Answer 10.15: See the file `Chap10es.rtf`.

Answer 11.1: Because the system is CPU bound—i.e., the SCI output rate (consumer) is much faster than the software generation rate (producer).

Answer 11.2: The TxFifo becomes full, because the system is now I/O bound. It does not crash, just runs more slowly as the main program waits inside `SCI_OutChar` for there to be space in the FIFO.

Answer 11.3: Because the system is CPU bound, each call to `SCI_OutChar` (PB1) is followed by an interrupt (call to `TxFifo_Get` (PB3)). There are two calls to `TxFifo_Get`, because there are two TDRE interrupts for each character. The first gets the data; the second occurs because the transmission hardware is idle (TDRE=1) but the TxFifo is empty.

Answer 11.4: The main program is waiting in `SCI_InChar` (PB0), then a RDRF interrupt occurs (call to `RxFifo_Put` (PB2)), then a call to `SCI_OutChar` (because the character is echoed by the main program) is followed by a TDRE interrupt (two calls to `TxFifo_Get` (PB3)). At the end, you can also see the next call to `SCI_InChar`. The first call to `TxFifo_Get` (PB3) occurs during the RDRF interrupt as the ISR checks the TxFifo and finds it empty.

Answer A1.1: There are six windows open.

> `TheLog.rtf`   will contain information about the simulation.
> `TheList.rtf` is the assembly listing.
> `ChapA1.io`   shows the four switches and four LEDs interfaced to the microcomputer.
> `ChapA1.uc`   shows the current state of the microcomputer.
> `ChapA1.scp`   is the digital logic analyzer.
> `ChapA1.rtf`   is the assembly source code.

Answer A1.2: The color of the **Start, ExecuteOne,** and **StepOver** commands changed, signifying that these commands can now be executed. In particular, it means the program is ready to run.

Answer A1.3: When the switch is open, the resistance across it is infinite. The pull-up resistor makes the voltage across the switch +5 V. Because the 74HC14 is a *not* gate, its output will be 0 V, making the digital value of the input port 0.

   When the switch is closed, the resistance across it is almost zero. Because the resistance of the closed switch (about 0.1Ω) is much smaller than the resistance of the pull-up resistor (about 5 kΩ), the voltage across the switch will be very close to 0 V. The 74HC14 is a *not* gate, so its output will be +5 V, making the digital value of the input port 1.

Answer A1.4: Each of the following commands will toggle the corresponding mode. If the mode was enabled, the command will disable it. If the mode was disabled, then command will enable it.

Mode->FollowPC	The current instruction is highlighted in the `TheList.rtf` window
Mode->CycleView	Memory access cycles are displayed in the `TheLog.rtf` window
Mode->InstructonView	Executed instructions are displayed in the `TheLog.rtf` window
Mode->LogRecord	The **ViewBox** data is dumped into the `TheLog.rtf` window

Answer A1.5: These settings are saved in the microcomputer document (e.g., `ChapA1.uc`). When you quit **TExaS,** some of these settings are also saved as default settings in the `Texas.ini` file, so that, when you launch **TExaS** and create a new microcomputer, **TExaS** can use them as default settings.

Answer A1.6: The PC is a 16-bit address register that points into memory at where the program is currently executing. The initial program counter for 6811 simulation is $F800 and for the 6812 simulation is $F000. In each case, this value is the starting point of our program. With the 6811/6812, the initial value of the PC is determined by the EEPROM locations $FFFE and $FFFF. This value was specified by the assembly-language lines at the end of `ChapA1.rtf`:

```
org $FFFE
fdb main
```

Answer A1.7: The **$** means hexadecimal. According to the discussion presented in answer 3, the value PORTA=$0A in the ViewBox window means the following:

sw3 (PA3) is closed;
sw2 (PA2) is open;
sw1 (PA1) is closed;
sw0 (PA0) is open.

Note that we haven't connected anything yet to Port C bits 7,6,5,4. In a real CMOS embedded system, we usually connect unused inputs to ground, so that they won't oscillate at 60 Hz and consume battery power. In the simulator, unconnected inputs will remain at zero (as if they were grounded.) With many real digital circuits, unconnected inputs will float high. It would be poor design to assume that the unconnected input is either high or low.

Answer A1.8: The PC value is

Format	6811	6812
H	$F80D	$F00F
+H	−$07F3	−$0FF1
D	63501	61455
+D	−2035	−4081
B	%1111100000001101	%1111000000001111
+B	−%0000011111110011	−%0000111111110001

The `Data` values are

h	$0A
+h	+$0A
d	10
+d	+10
b	%00001010
+b	+%00001010
2d	10,0

Index

A

T

U

		Bits 4 to 6							
		0	1	2	3	4	5	6	7
	0	NUL	DLE	SP	0	@	P	`	p
B	1	SOH	DC1	:	1	A	Q	a	q
i	2	STX	DC2	!	2	B	R	b	r
t	3	ETX	DC3	#	3	C	S	c	s
s	4	EOT	DC4	$	4	D	T	d	t
	5	ENQ	NAK	%	5	E	U	e	u
0	6	ACK	SYN	&	6	F	V	f	v
	7	BEL	ETB	'	7	G	W	g	w
t	8	BS	CAN	(	8	H	X	h	x
o	9	HT	EM	)	9	I	Y	i	y
	A	LF	SUB	*	:	J	Z	j	z
3	B	VT	ESC	+	;	K	[	k	{
	C	FF	FS	,	<	L	\	l	;
	D	CR	GS	-	=	M	]	m	}
	E	SO	RS	.	>	N	^	n	~

Standard 7-bit ASCII

Menu	Key	Commands
File	Ctrl+N	create a new document
File	Ctrl+O	open an existing document
File	Ctrl+S	save the document
File	Ctrl+T	save the document as a text file
File	Ctrl+P	print the document
Edit	Ctrl+Z	undo change
Edit	Ctrl+X	cut to clipboard
Edit	Ctrl+C	copy to clipboard
Edit	Ctrl+V	paste from clipboard
Edit	Ctrl+A	select all
Edit	Ctrl+F	find
Edit	F3	Find next
Edit	Ctrl+H	find and replace
Edit	Ctrl+D	insert date
Edit	Alt+Enter	Object properties
Style	Ctrl+Shft+B	**Bold**
Style	Ctrl+Shft+I	*Italics*
Style	Ctrl+Shft+U	<u>Underline</u>
Style	Ctrl+Shft+S	~~Strikeout~~
Assemble	Ctrl+B	Assemble
Mode	F7	fewer instructions/screen update
Mode	F8	more instructions/screen update
Mode	Ctrl+L	deactivate/activate log record
Action	F9	Reset
Action	F10	Execute one instruction
Action	Shft+F10	Step over
Action	Alt+F10	Step out
Action	Alt+1	Run to cursor
Action	F11	Execute a few instructions
Action	F12	Go/Stop execution
Action	Alt+2	Open S19 again
Action	Alt+3	Open S19
Action	Alt+4	Backdump
Action	Alt+5	Add breakpoint at cursor
Help	F1	Help

Shortcuts for the TExaS application